D0845898

MAY 1 8 2000

THE BIOMECHANICS

OF INSECT FLIGHT

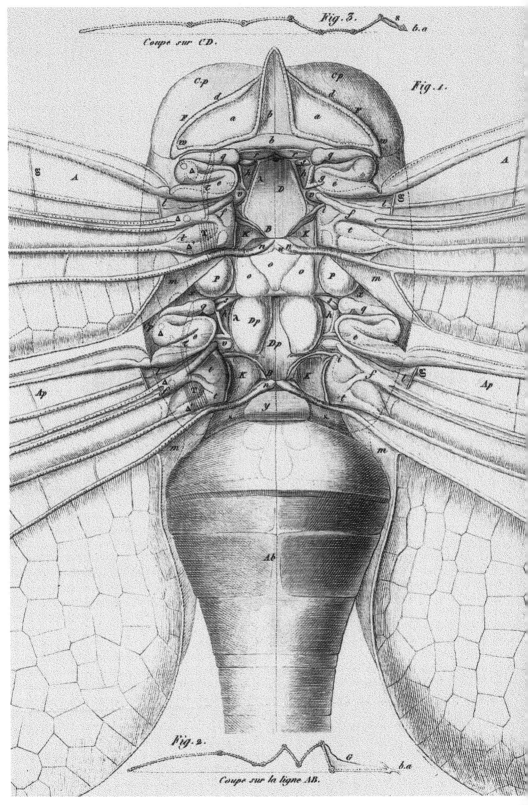

DRAGONFLY THORACIC ANATOMY AND WING CROSS SECTIONS, FROM THE FIRST BOOK (1822) ON INSECT FLIGHT (J. CHABRIER: *ESSAI SUR LE VOL*

INSECTES ET OBSERVATIONS SUR QUELQUES PARTIES DE LA MÉCANIQUE DES MOUVEMENS PROGRESSIFS DE L'HOMME ET DES ANIMAUX VERTÉB

THE BIOMECHANICS

OF INSECT FLIGHT

FORM, FUNCTION, EVOLUTION

Robert Dudley

PRINCETON UNIVERSITY PRESS

PRINCETON, NEW JERSEY

Copyright © 2000 by Princeton University Press
Published by Princeton University Press, 41 William Street,
Princeton, New Jersey 08540
In the United Kingdom: Princeton University Press,
Chichester, West Sussex
All Rights Reserved

Library of Congress Cataloging-in-Publication Data

Dudley, Robert
The biomechanics of insect flight : form, function,
evolution / Robert Dudley.
p.   cm.
Includes bibliographical references (p.   ) and index.
ISBN 0-691-04430-9 (cloth : alk. paper)
1. Insects—Flight.   I. Title.
QL496.7.D83   1999
573.7'98157—dc21   99-29653

This book has been composed in Palatino

The paper used in this publication meets the
minimum requirements of ANSI/NISO Z39.48-1992
(R1997) (*Permanence of Paper*)

http://pup.princeton.edu

Printed in the United States of America

10 9 8 7 6 5 4 3 2 1

For my wife, Lu Min

CONTENTS

ACKNOWLEDGMENTS

THE PRACTICE of science, like other human endeavors, relies upon cultural transmission of information. I am particularly indebted to my parents, Bettina Dudley and the late Ted Dudley (*botaniste extraordinaire*), for regular exposure to diverse features of natural history and geographical exploration. I thank Geerat Vermeij for an early introduction to concepts of biogeography and evolutionary biology, and Pierre Sprey for his inspirational roles as aerodynamicist and connoisseur. Steve Vogel provided wonderfully riotous instruction and incitement during my undergraduate tenure at Duke University. Peter Klopfer, Horst Meyer, and Steve Wainwright at the same institution were also generous with their advice and expertise. I thank Charlie Ellington at the University of Cambridge for his tutelage during the course of my graduate studies. At the Smithsonian Tropical Research Institute in Panama, I have benefited substantially from collaborations with Greg Adler and Stan Rand; numerous discussions with Egbert Leigh and the late Alan Smith have also been informative. I am deeply indebted to Carl Gans for teaching me electromyography and for providing scholarly if not rabbinical perspectives on diverse issues; our continuing interactions invariably convince me of the primacy of the organism in evolutionary biology. Ongoing scientific collaborations with Peng Chai, Phil DeVries, Evandro Oliveira, and Bob Srygley encourage me to consider animal flight performance in natural environments, the ultimate testing ground for biomechanical design.

Numerous colleagues have commented either on particular chapters (chapter 3: Charlie Ellington, Hao Liu, Sandy Willmott; chapter 4: Tim Casey, Bob Full, Bob Josephson, Raul Suarez; chapter 5: Jim Marden; chapter 6: Conrad Labandeira, Jim Marden, Riley Nelson, Stan Rand, Geerat Vermeij, Mary Jane West-Eberhard; chapter 7: Phil DeVries, Bill Eberhard, Larry Gilbert, Evandro Oliveira, Bob Srygley, Geerat Vermeij, Mary Jane West-Eberhard), or on the entire book (Greg Adler, Doug Altshuler, Claire Balint, Peng Chai, Wai Pang Chan, Mark Denny, Michael Dickinson, Erica Feuerbacher, Carl Gans, Dmitry Grodnitsky, Jon Harrison, Rebecca Johnston, Kate Loudon, Don Pick, Steve Roberts, Sanjay Sane, Jocelyn Staunton, Swifty Stevenson, Steve Vogel, Robin Wootton, Diana Wu, and Lijiang Zeng). I am grateful for their constructive advice and criticism. Gwen Gage and Kristina Schlegel helped extensively with illustrations; Dmitry Grodnitsky and Vadim Rossman have kindly assisted with the procure-

S	reference area of wing or body; total wing area
v	virtual wing mass
V	relative fluid velocity; insect airspeed
V_{down}	mean wing velocity during downstroke
V_i	induced velocity
$V_R(r)$	wing relative velocity at radial distance r
V_{up}	mean wing velocity during upstroke
V_{vert}	vertical component of the body velocity
V_x, V_y, V_z	velocity component in the (x), (y) or (z) dimension, respectively
α	angle of attack
β	stroke plane angle
Γ	circulation
ε	muscle strain
θ	wing elevational angle; glide angle
$\theta_{\text{horizontal}}$	horizontal error angle
θ_{vertical}	vertical error angle
μ	dynamic viscosity
ν	kinematic viscosity
ρ	air density; density of muscle
σ	myofibrillar stress
φ	positional angle of the wing
φ_{max}	maximum positional angle
φ_{min}	minimum positional angle
$\bar{\varphi}$	mean positional angle
$(d\varphi/dt)_{\text{max}}$	maximum angular velocity of wing during a half-stroke
Φ	stroke amplitude
χ	body angle

THE BIOMECHANICS
OF INSECT FLIGHT

taxa of flying insects were substantially larger than under present conditions (see chapter 6).

Such restrictions on insect body size are more than offset, however, by a remarkable capacity for mobility and maneuverability in air. Rapid energy-efficient flight has permitted access to and exploitation of distant ecological resources. This option has been unavailable to other numerically abundant invertebrate taxa such as crustaceans, mites, and the soil-inhabiting collembolans and nematodes. Aerial agility permits rapid escape from predators, and also has become incorporated into components of many insect mating systems. As arthropods (animals with exoskeletons that undergo molts), insects possess a high degree of architectural plasticity in expression of their external morphology. The physical advantages of flight as a mode of locomotion have also become coupled in insects to the general arthropodan characteristic of high reproductive rates. Flying insects have accordingly diversified to an extent not witnessed in other metazoan taxa. Among the arthropods, perhaps only mite diversity rivals that of the insects, and much of mite speciation may derive from obligate associations with individual insect species. Furthermore, terrestrial dominance by winged insects has been a feature of the earth's biota probably since the mid- to late Carboniferous (325 MYBP [million years before present]), the presence of tetrapods notwithstanding.

If major evolutionary events can be thought of as transitions between distinct states, switches between aquatic, terrestrial, and aerial environments are clearly of major significance for subsequent patterns of animal diversification. Terrestrialization by plants, the literal and figurative emergence of air-breathing arthropods, and finally the origin of vertebrate amphibians represent important transitions between life in water and life in air, the latter albeit always in physical contact with the surface of the earth. Note also that the initial stages of these events involved amphibiotic adaptations—the realm of water could not be fully abandoned, and remains obligatory for reproductive purposes in most extant amphibians as well as in mosses. The origin of flight in all animal taxa is similar in that no flier can remain continuously aloft, and the physical substrate of the earth remains essential for most features of life history. Flight in many contexts is best viewed as merely a means of moving between two points on earth; this is in fact the dominant human perspective on aerial locomotion. Indeed, "flight" and "air travel" are often used synonymously, explicitly placing flight within the context of terrestrial endpoints.

Insects are no exception in utilizing flight to move about, but they also use air for three-dimensional behaviors that serve purposes in both natural and sexual selection. A major thesis of this book is that much of winged insect diversity arises not merely as a consequence of enhanced displacement, but also through synergistic interaction between flight and the need to evade predators, to effect intrasexual competition for mates (aggressive male-male interactions), and to effect intersexual selection (male-female localization, courtship, and mating). Interaction among these different selective factors, when combined with aerial dispersal and exploitation of distant ecological resources, has historically produced the remarkable functional and morphological diversity of the winged insects. Evolution of flight and subsequent exploitation of the aerial environment by insects must rank comparably in significance to the colonization of land by both plants and animals. Explaining the functional morphology, biomechanics, and physiology of insect flight can correspondingly yield insight into a predominant theme of biotic diversity.

1.1 INSECT DIVERSITY

To place the numerical abundance and morphological radiations of winged insects in biological context, the taxonomic definition of insects, their evolutionary history (phylogeny), and pattern of body segmentation (tagmosis) relative to other arthropods must be considered. The phylogeny of insects specifies, within a specific geological time frame, the relatedness of insects to other arthropods and the relationships of different insect lineages to one another. The evolutionary history of the insects must be established because, in spite of the presence of numerous and diverse arthropod lineages, the flapping flight of pterygote insects appears to have a single (monophyletic) origin best evaluated with respect to contemporaneous taxa. Similarly, contemporary insect diversity is most pronounced when expressed relative to that of other extant invertebrate and vertebrate taxa. Body organization and morphological features common to all arthropods have been successfully co-opted and combined by insects with the flight machinery to yield a unique animal design.

1.1.1 Historical Origins

To understand the biological underpinnings of winged insect diversity, phylogenetic relationships of insect taxa must be evaluated (fig. 1.1). True insects are six-legged (hexapodan) arthropods with protruding mouthparts (Kristensen, 1991). Reconstructions of arthropod phylogeny have historically been characterized by a profusion of alternative data sources, informed perspectives, and opinionated

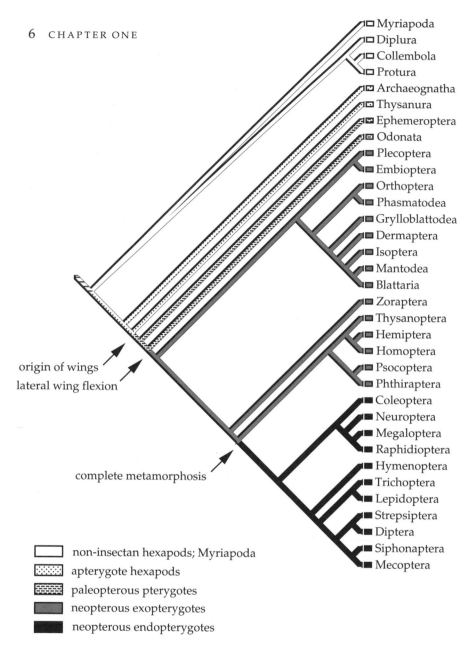

FIG. 1.1. Phylogeny of the Superclass Hexapoda (primarily from Kristensen, 1991; Pashley et al., 1993; Whiting et al., 1997). Myriapoda may be the the sister taxon of all hexapods, although this placement is equivocal (section 1.1.1). The insect order Strepsiptera has classically been recognized as the sister taxon of Coleoptera, although recent analyses suggest affinity with Diptera (Whiting and Wheeler, 1994; Carmean and Crespi, 1995; Chalwatzis et al., 1996; Whiting et al., 1997). Note also that the classical taxon Homoptera is now recognized as a paraphyletic assemblage (see section 7.1.1).

interpretations that to this day have yielded no consistent or convergent result; a diversity of references is accordingly provided for in-
terested readers. The precise geological origin of hexapods remains
unclear, but possibly represents a split from myriapod (millipede) relatives by the late Silurian (420–400 MYBP; see Tiegs, 1949; Wille, 1960;
Birket-Smith, 1984; Labandeira et al., 1988; Kukalová-Peck, 1991;
Wheeler et al., 1993; Carroll, 1995; Wheeler, 1997). From the perspective of locomotion, the basic feature linking myriapods and insects is
the unbranched (uniramous) nature of the appendages. Alternative
views of hexapod origins are variously provided by Manton (1966),
Dohle, 1988, 1997, Schram and Emerson (1991), McNamara and Trewin (1993), Averof and Akam (1995), Boore et al. (1995), Friedrich and
Tautz (1995), and Fryer (1996). These latter studies generally identify
the crustaceans and the hexapods as sister taxa, although crustaceans
and related taxa, in contrast to insects, are characterized by branched
(usually biramous) appendages. Although the subject of insect origins
is currently the topic of intense research, convergent and universally
accepted cladistic analyses remain to be carried out. A combined analysis of all available morphological and molecular data, however, does
suggest a hexapod/myriapod sister taxon relationship (Wheeler,
1997). Monophyly of both hexapods and of true insects is generally
accepted by entomologists (Kristensen, 1981, 1991; see fig. 1.1). Fortunately, no controversy characterizes the identification of wingless
apterygote insects and the winged pterygotes as sister taxa within the
hexapods.

All noninsect hexapods (e.g., the Collembola or springtails) together
with the apterygote insects orders Archaeognatha and Thysanura (see
fig. 1.1) are primarily if not exclusively terrestrial, and the precursors
of winged insects were in all likelihood similarly confined to land (see
chapter 6). Winged lineages subsequently diverged from apterygotes
possibly in the early Carboniferous, diversifying rapidly into about fifteen described orders by the end of the period (285 MYBP). (Note that
the Devonian structures described as insect body parts by Rohdendorf
[1961, 1970a] are in fact crustacean in origin [Rohdendorf, 1972].) The
evolutionary genesis of wings is still poorly understood (chapter 6),
but in absence of contrary evidence a monophyletic origin seems likely
(Hennig, 1981; Kristensen, 1991; see, however, the alternative views of
Lemche, 1940; Matsuda, 1970; La Greca, 1980, 1981).

Structurally, hexapods can be recognized as arthropods with three
tagmata or distinct body regions: the head, thorax, and abdomen. Each
such region represents functional integration among what were once
discrete segments in an ancestral arthropod (insect head: equivocally
six segments; thorax: three segments; abdomen: eleven segments).
Cephalization in arthropods, or the emergence of a distinct head and

centralization of the nervous system, is indicated in part by fusion of nerve ganglia among anterior body segments. In hexapods, the head contains the brain (the superesophageal ganglion, ancestrally comprising multiple segmental ganglia), a subesophageal ganglion, and some smaller ganglia. Externally, the head expresses the particularly important sensory structures of the eyes and the antennae. Posterior to the head, each thoracic and abdominal segment contains, at least ancestrally, a single ganglion; ganglia of adjacent segments are often fused. The thorax is the tagma principally concerned with locomotion (legs and wings), whereas the abdomen contains primarily digestive, excretory, and reproductive structures.

Similarities in tagmosis among the noninsect hexapods and the basal insect orders are striking. As the locomotor tagma, the thorax of noninsect hexapods (Diplura, Collembola, Protura) is tripartite and accommodates the leg musculature (one leg pair per segment) required for gressorial (walking) and cursorial (running) locomotion. This function is still a necessary feature of the thorax in all winged insects, but hypertrophy of the thorax to accommodate the flight musculature is particularly evident among the basal pterygote lineages relative to their apterygote counterparts. In present-day pterygotes, wings are located on the second (meso-) and third (meta-) thoracic segments; the first, or prothoracic, segment bears only a pair of legs and no wings. Given the number of ancestral segments within each tagma and the degree of cephalization among insects, the mass and anteroposterior length of the head and thorax are typically less than the abdominal mass and length, respectively. One biomechanical feature suggested by this pattern of pterygote tagmosis is that the center of body mass is typically located behind the wing base and within the abdomen, an observation originally detailed by Plateau (1872) and Demoll (1918). Gravity acts downward through the center of body mass, whereas aerodynamic forces generated by the flapping wings will typically act anterior to this point. Thus, the wings must produce not only vertical force to offset the body mass, but also a nose-downward pitching moment to balance gravitational torque. Successful flight therefore depends not only on the magnitude of aerodynamic forces generated by the wings, but also on the orientation and subtle regulation of these forces (section 5.1).

1.1.2 Morphological Diversification

As impressive as insect diversity is today, even more remarkable is the fact that most major morphological innovations and indeed insect orders were present before the Mesozoic (245–65 MYBP). Upper Carboniferous and Permian radiations of winged insects resulted in about twenty-eight orders immediately prior to the end-Permian extinctions around 245 MYBP (Rohdendorf and Rasnitsyn, 1980; Carpenter, 1992; Labandeira and Sepkoski, 1993). The present-day fauna is obviously different from early insect faunas in many ways, but representatives of many extant orders would not be out of place in a late Paleozoic eco-system. Ordinal-level classification of insects is not universally codi-fied, but approximately thirty pterygote orders exist today (table 1.1). Insect diversity is thus both ancient and persistent.

The ability to move in the three-dimensional aerial environment is regularly cited in entomological textbooks as a causal factor promoting taxonomic richness of the winged insects. Pterygotes have diversified and exploited novel habitats and niches far more effectively than have wingless apterygotes, the latter evidently possessing only limited op-portunities for mobility or ecological diversification. Monophyly of in-sect flight renders any hypothesis linking wings and taxonomic diver-sity impossible to test statistically, but pterygote diversification would obviously be consistent with any evolutionary scenario that derives the emergence of new species from dispersal to novel environments. Although secondary wing loss and flightlessness are taxonomically widespread among the pterygote insects, such cases are often confined to ecologically restrictive environments (section 6.2). Obvious counter-examples to this trend are the generally nonflying ants and termites, taxa that represent the majority of animal biomass in tropical rain for-ests (e.g., Lévieux, 1982; Fittkau and Klinge, 1973). Even such social insects, however, obligately retain wings in order to disperse and en-gage in reproductive behavior during nuptial flights (Wilson, 1971; Hölldobler and Wilson, 1990). Aerial vagility and the capacity for dis-persal have probably been general factors contributing both to insect species diversity and to overall individual abundance.

In addition to presence of wings, several other features have been classically associated with insect diversity. Martynov (1923, 1925; see also Crampton, 1924; Schwanwitsch, 1943) noted a fundamental taxo-nomic division of winged insects into two major groups, the Paleop-tera and the Neoptera. The latter taxon is characterized by the ability to flex the wings laterally against the abdomen when the insect is not flying. Only two orders, the Ephemeroptera (mayflies) and the Odo-nata (dragonflies and damselflies) compose the extant Paleoptera. Also, at least one extinct order of paleopterous insects (the Diaphanop-terodea) was apparently capable of lateral wing flexion; the particular mechanism of wing articulation was, however, distinct from that of the Neoptera (Carpenter, 1963; Kukalová-Peck, 1974; Kukolavá-Peck and Brauckmann, 1990). The ability to flex wings laterally may increase

TABLE 1.1
Taxonomy of the Class Insecta (Superclass Hexapoda)

Subclass	Order	Common Name	Approximate No. of species
Apterygota (<0.1%)	Archaeognatha (<0.03%)	Bristletails	250
	Thysanura (<0.05%)	Silverfish/firebrats	400
Pterygota (>99.9%)			
Infraclass Paleoptera	Ephemeroptera (0.28%)	Mayflies	2400
(~0.9%)	Odonata (0.63%)	Dragonflies/ damselflies	5400
Infraclass Neoptera (~99%)			
Division	Plecoptera (0.22%)	Stoneflies	1900
Exopterygota	Embioptera (0.02%)	Webspinners	200
(~14%)	Orthoptera (2.04%)	Grasshoppers/crickets	17,500
	Phasmatodea (0.24%)	Walking sticks	2100
	Grylloblattodea (<0.01%)	Rock crawlers	20
	Dermaptera (0.13%)	Earwigs	1100
	Isoptera (0.26%)	Termites	2200
	Mantodea (0.23%)	Mantids	2000
	Blattaria (0.50%)	Cockroaches	4300
	Zoraptera (<0.01%)	Zorapterans	24
	Thysanoptera (0.64%)	Thrips	5500
	Hemiptera (4.07%)	True bugs	35,000
	Homoptera (5.24%)	Cicadas/hoppers/ aphids	45,000
	Psocoptera (0.27%)	Booklice	2300
	Phthiraptera (0.41%)	Lice	3500
Division	Coleoptera (40.79%)	Beetles	350,000
Endopterygota	Neuroptera (0.50%)	Lacewings and antlions	4300
(~85%)	Megaloptera (0.03%)	Alderflies/dobsonflies	250
	Raphidioptera (0.03%)	Snakeflies	300
	Hymenoptera (12.92%)	Ants/bees/wasps/ sawflies	111,000
	Trichoptera (1.45%)	Caddisflies	12,500
	Lepidoptera (15.71%)	Butterflies/moths	135,000
	Strepsiptera (0.03%)	Twisted-wing parasites	300
	Diptera (13.03%)	Flies	112,000
	Siphonaptera (0.26%)	Fleas	2200
	Mecoptera (0.06%)	Scorpionflies	500

Notes: Approximate numbers for described extant species are primarily from Arnett (1985), CSIRO (1991), and Gaston (1991); percentages in parentheses indicate relative fraction of the total extant fauna. The two paleopteran orders and all exopterygote orders except Thysanoptera exhibit incomplete metamorphosis; metamorphosis is complete in the endopterygote orders. Four endopterygote orders (Coleoptera, Diptera, Lepidoptera, and Hymenoptera) are numerically predominant. The extant Lepidoptera, given their relatively larger body size, are probably more completely described than are other major orders. Although the order Homoptera has recently been shown to be paraphyletic (see fig. 4.3A), ordinal-level ranking is used here and through much of the book in accordance with classical usage.

terrestrial mobility and reduce apparency to predators (Carpenter, 1976), but direct consequences of wing flexion for species diversification remain unclear, as does any statistical proof of this particular hypothesis. Contemporary Paleoptera are, however, few in number relative to their neopterous counterparts (table 1.1). Note also that modern insect phylogenies place the Ephemeroptera as a sister taxon to all other winged insects (fig. 1.1), indicating that the Paleoptera is probably a paraphyletic assemblage that does not include all taxa with common ancestors (Kristensen, 1975, 1989, 1991). Alternative phylogenetic perspectives of the Insecta are provided by Lemche (1940), Schwanwitsch (1946), Hennig (1981), and Matsuda (1981); these phylogenies incorporate fewer relevant morphological data than are represented in figure 1.1. Present-day insect phylogenies consistently place lateral wing flexion as a basal pterygote condition, but this fact alone cannot unequivocally associate the trait with insect diversity.

Metamorphosis, or the possession of a complex lifestyle (*sensu* Moran, 1994) is another trait traditionally associated with selective advantage and causally linked with insect diversification. The holometabolous condition, with its distinct larval, pupal, and adult life stages, has regularly been cited as an important factor promoting adaptive radiation in the Insecta (e.g., Hinton, 1948, 1963; Carpenter, 1953; Hennig, 1981). The four largest insect orders (Coleoptera, Diptera, Hymenoptera, and Lepidoptera) all exhibit complete metamorphosis and a pupal stage (holometabolous development), whereas those species with incomplete metamorphosis (hemimetabolous development) represent only about 15% of the extant fauna (table 1.1). An inactive stage similar to the holometabolous pupa has also evolved in the Thysanoptera and in some homopterans. Selective advantages of holometaboly and related strategies such as diapause are indisputable in fluctuating environments (Tauber et al., 1986), but interactions between this physiological mechanism and speciation are unclear. As with wings and lateral wing flexion, holometaboly probably arose only once (fig. 1.1), so that any hypothesis causally linking these factors with diversity is impossible to test statistically within a phylogenetic context. Radical morphological changes in ontogeny clearly do not require complete metamorphosis, however; the change of paleopterous mayfly and dragonfly larvae from aquatic larvae to free-flying adults is an impressive transformation by any standard.

Miniaturization (i.e., the evolution of small body size) has been a dominant and repeated theme in the historical dynamics that have resulted in the contemporary insect fauna. Anthropomorphic bias causes humans to focus primarily on insects greater than 1 cm in body length (e.g., butterflies), yet such insects are unrepresentative of the fauna as a whole. Average adult insect body length is on the order of 4–5 mm, and body size distributions more generally are skewed toward smaller sizes (May, 1978; Blackburn and Gaston, 1994; see section 7.1.1). By contrast, most biomechanical studies of insect flight have primarily been confined to relatively large insects, such as locusts, sphingid moths, bumblebees, and the like. Relative to the massive literature on the genetics and molecular biology of *Drosophila* (typically 2.5–4.5 mm in body length), studies of free-flight kinematics and aerodynamics in these small insects are limited in number. Moreover, ecophysiology and flight performance of *Drosophila* in natural environments (as distinct from vineyards and orchards) is essentially unstudied. Wings and bodies of such small insects operate in physical regimes of fluid flow substantially different from those of more readily studied insects, and aerodynamic mechanisms during flight are correspondingly distinct (Horridge, 1956; see section 7.1.3). The paucity of flight studies on miniaturized insects is particularly unfortunate given their disproportionate ecological significance for the contemporary fauna. Heightened dispersal by ambient winds is also an important consequence of small body size in insects (section 7.4).

Parallel to comparative treatments of insect anatomy (most notably the work of Snodgrass, 1935, and Matsuda, 1970), the magnitude of insect species richness has historically been a topic of intense interest, description, and speculation. One recent approach to estimating the number of undescribed insect species has been to fog tree canopies with insecticide and then to count the number of morphologically distinct and presumably new species that are taken. Preexisting data on typical host-specific associations between insect and angiosperm (flowering plant) species can then be used to estimate, based on the number of extant angiosperm species worldwide, the number of undescribed insect morphospecies. Such numerical extrapolation from site-specific sampling in tropical tree canopies has produced suggestions of 30 million and more undescribed species worldwide (Erwin, 1982, 1988). More conservative estimates for the total number of insect species range from 5 to 10 million (Gaston, 1991; see also May, 1988; Stork, 1988; Hodkinson and Casson, 1991). Relative to an extant vertebrate fauna of about 42,000 species, and primates such as ourselves numbering a mere 230 species or so, insect species richness truly appears overwhelming. The largest beetle family (Curculionidae, or the true weevils) alone contains more described species than exist vertebrate species. Also, most extant insect species are as yet undescribed (Gaston, 1991) whereas vertebrate faunas worldwide are fairly well known, even in lesser studied tropical regions.

Any evaluation of insect abundance must, in addition to considering species diversity, also consider numbers of individuals. By this criterion, insects are superabundant relative to other animals of comparable body mass. Large locust swarms, for example, have been estimated to contain from 10^8 to 10^{10} individuals (Williams, 1958; Rainey, 1989). Diurnal netting from airplanes in temperate ecosystems suggests that from 1 to 2 million insects can be found aloft above each square kilometer of land in summertime (see Hardy and Milne, 1938; Glick, 1939; Johnson, 1957). Such numerical abundance suggests that insects play a predominant role in the trophic structuring of contemporary ecosystems, acting as primary consumers of plant productivity and additionally serving as a nutritional resource for diverse arthropod and vertebrate taxa. A rough indication of this trophic influence is given by an estimate of a global insect biomass of 10^{12} kg; approximately 10^{18} insects may exist worldwide (Williams, 1960), and mean insect body mass can be conservatively estimated at 1 mg. Remarkably, this estimate for insect biomass substantially exceeds a comparable estimate for the contemporary human population (2×10^{11} kg, assuming $\sim 5 \times 10^9$ individuals with a mean body mass of 40 kg averaged across ontogenetic stages). Flying insects to this day thus challenge human populations at primary trophic levels within terrestrial ecosystems.

1.1.3 Insect Diversity and Flight

Biomechanical prerequisites for flapping flight in insects include a means of force generation (the thoracic musculature), a mechanism to transmit this force to the wings (the axillary apparatus, or articulation between the wing and thorax), and the wings themselves that generate useful aerodynamic forces. Plasticity of insect form is evident at all levels of biomechanical design in the flight apparatus. A relatively simple tripartite body, for example, belies an extraordinary range of shapes and aerodynamic configurations (plate 1). Such structural diversity is matched by functional divergence in wingbeat kinematics and the aerodynamics of flight (see chapter 3). Much of this variation is allometric in character and is dependent on insect size and shape. Wingbeat frequencies, for example, typically show an inverse relationship with body size. Wings of large butterflies oscillate up and down at frequencies as low as 5 Hz (cycles/second), whereas minute ceratopogonid flies oscillate their wings at frequencies as high as 1000 Hz. Perhaps with the exception of tongue motions in insectivorous amphibians and reptiles (e.g., chameleons), such quick motions are essentially unknown from the world of vertebrates. On longer temporal scales, insect flight trajectories and body kinematics are highly diverse. Some insects exhibit straight flight paths, whereas others fly only irregularly and erratically. Various insect taxa are capable of precise hovering as well as of seemingly arbitrary three-dimensional translations in fractions of a second (e.g., dragonflies); other insects seem incapable even of making a controlled landing. Unpalatable danaine butterflies high in the Indo-Malayan rain forest can glide for minutes on end without a single wingbeat. By contrast, an emergent male termite flaps its wings continuously before mating and dying in a brief but glorious nuptial period that lasts only several hours.

Such differences in flight performance are likely the target of both natural and sexual selection (chapter 7). Flying insects are regularly the subject of attack by insectivorous birds and bats, and evasive maneuverability is often required for survival. Many insects are aerial predators themselves, most notably odonates, bittacid mecopterans, and diverse lineages of flies (section 7.3). Sexual selection for enhanced flight performance also acts on both males and females of many insect species. Males typically hover or patrol, and then attempt to capture mates in high-speed chases; subsequent evasion by females can be an important feature of mate choice (Thornhill and Alcock, 1983). Maneuverability and performance traits that are under sexual selection may be particularly labile and subject to rapid evolution (West-Eberhard, 1983, 1984; Andersson, 1994). The functional diversity of insect mating systems and morphology is well known; less appreciated is the intimate association between reproductive behavior and flight performance over both short- and long-term timescales.

For winged insects, intense selection on aerodynamic performance and underlying biomechanics is thus imposed in both intra- and interspecific contexts. Such diverse selective forces can, in general, interact synergistically and result in reciprocal escalation between selective agent and the target of selection (Vermeij, 1987). Among all animal taxa, insects may best exemplify the morphological diversification that potentially arises as a consequence of such evolutionary escalation. On ecological timescales, dispersal behavior and flight energetics also exert a major influence on patterns of habitat selection, niche partitioning, and resource utilization. Such considerations indicate that the various selective factors molding insect flight performance are both diverse in character and potent in magnitude. To understand fully the associated evolutionary responses by insects, flight as a physical form of locomotion and those features of aerodynamics that influence wing and body design must be considered in detail.

1.2 Basic Aerodynamics

Fluid mechanics is a topic little studied by entomologists in spite of its importance in such diverse areas of insect biology as tracheal respiration, circulation of hemolymph (insect blood), feeding and locomotion by aquatic insects, and of course flight. Moving fluids often behave nonintuitively, and the combination of wing flapping with body movement through the air engenders complex patterns of airflow that vary dynamically. Understanding of aerodynamic forces and of the physical medium within which insects fly is correspondingly necessary to understand the basic physical requirements of flight. Denny (1993) and Vogel (1994) provide detailed introductions to the various interactions between the mechanics of fluids and the design of biological structures.

1.2.1 Physical Characteristics of Fluids

Flight by animals as well as by airplanes necessitates transfer of momentum from flapping wings to the surrounding air. The physical characteristics of air, in turn, determine the subsequent reaction forces on insect wings and bodies that effect weight support and propulsion. Most fundamental to the aerodynamic analysis of flight is the fluid density (the mass per unit volume) and the fluid viscosity (a measure of the internal resistance to flow). Air density (ρ) is a principal determinant of force production by airfoils, biological and otherwise; aerodynamic forces on structures tend to increase linearly with air density. Conveniently, the density of air can effectively be considered to be constant in many biological circumstances. Temperature-dependent variation in air density is small, as is that with relative humidity (see Denny, 1993), and for entomological purposes only the variation in density with altitude is of practical significance. Sea-level density of air at 20°C is about 1.21 kg/m^3, decreasing to 0.95 kg/m^3 at 2000 m elevation and 0.74 kg/m^3 at 4000 m, the latter being a 40% reduction relative to the sea-level value. Both density reduction across elevational gradients and the parallel decline in oxygen partial pressure at higher altitudes may influence insect physiology and flight-related morphology (Mani, 1962, 1968; see section 8.2.1). Flying insects are regularly found at altitudes above 2000 m, either flying freely in the air column or participating as members of high-elevation communities on mountains (see section 7.4). Insect responses to altitudinal variation in air density may be expected over behavioral, ecological, and evolutionary timescales.

Air, like other fluids, responds to applied forces by resisting not deformation per se but rather the rate of deformation. Resistance to this latter quantity is determined by the dynamic viscosity (μ) of a fluid. Fluids with high viscosity exhibit high intermolecular forces that act to resist externally imposed force; glycerol, for example, is of the same density as water but is much more viscous. Although it is small relative to that of water, the dynamic viscosity of air is nonnegligible and contributes substantially to the characteristics of aerodynamic flow around insect wings. Dynamic viscosity and fluid density are commonly conflated as the kinematic viscosity ν (= μ/ρ), a parameter that indicates the magnitude of internal cohesive forces relative to the inertial characteristics of a fluid. The kinematic viscosity of air exhibits an approximately 28% increase over a temperature range of 0–40°C, driven mostly by the thermal reduction in air density over this range (see Vogel, 1994). Flight in variable air temperatures thus potentially involves both biomechanical compensation for changes in the physical characteristics of air as well as physiological responses to a variable thermal regime.

In order for an insect to fly, forces must be generated by the flapping wings that create a momentum flux and thus a useful reaction force in the surrounding air. Three general categories of forces act on objects such as insect wings and bodies moving within air–drag, lift, and the acceleration reaction. Flapping wings can theoretically use each type of force to generate a momentum flux that effects body displacement; drag and the acceleration reaction also potentially act to retard body motion. Although they are not well studied, acceleration reactions appear to be of little importance in insect flight mechanics (section 3.2.2.1).

1.2.2 Drag Forces

The net physical force acting on an object moving within a fluid is a vector quantity that can be, in a two-dimensional perspective, resolved into two orthogonal components. The component of force parallel to flow is termed drag and, in absence of an acceleration reaction, represents the combined effects of pressure drag and viscous drag. Pressure is highest at the most anterior point of contact between object and moving fluid, and is lower behind the object. Presence of the object removes momentum and thus kinetic energy from the moving fluid system, and acts to slow the object relative to flow. Pressure drag, also known as inertial drag, therefore derives from disruption of the inertial characteristics of the moving flow field and varies with fluid density, relative fluid velocity, and the object's shape.

Viscous drag, by contrast, emerges from boundary interactions between the object and the surrounding fluid. Because fluid particles directly at the surface of a moving object cannot penetrate the surface and do not move relative to the object (the no-slip condition; see Vogel, 1994), a velocity gradient termed the boundary layer defines the spatial transition from object surface to free stream flow. Viscous forces ensure resistance to flow between adjacent layers of fluid, and any shear within the boundary layer engenders a reaction force throughout the fluid that terminates at the surface of the moving object. Viscous drag (also known as skin friction) acts to slow the object, and varies not with density but primarily with fluid viscosity, the relative fluid velocity, and the total wetted surface area of the object. Total drag forces acting on insect wings and bodies thus derive from the independent and summed effects of the pressure and viscous drags. The relative magnitude of these two components is obviously of importance if structural design is to be appropriately analyzed. For example, skin friction will be much greater on a long flat plate than on a sphere of equivalent volume.

Given the combined influence of density and viscosity on total drag, the relative magnitude of pressure drag and viscous drag can be quantitatively estimated for any nonaccelerating fluid system. Pressure and viscous drag have in common a dependence on relative fluid velocity V and on object dimensions, but differ fundamentally in their dependence on fluid density and viscosity, respectively. This observation suggests that the ratio of inertial to viscous forces will vary with ratio of density and viscosity, and must additionally incorporate object dimensions and relative velocity. The simplest (but not the only) dimensionless formulation of these four parameters is the Reynolds number (Re), which is proportional to the ratio of inertial to viscous forces acting on objects moving within fluids:

$$Re = \frac{\rho l V}{\mu} = \frac{l V}{\nu}, \tag{1.1}$$

where l is a characteristic dimension of the object in question. Because values of Re are proportional to the ratio of inertial to viscous forces, use of the Re permits quantitative characterization of a broad continuum of dynamically variable flow regimes. Flow is usually laminar below Re of 10^3–10^4, whereas turbulence becomes more pronounced at higher Re. For large objects moving at high speeds, inertial effects dominate and boundary layer interactions are of minimal importance. Very high Re regimes (e.g., airplanes at $Re > 10^6$) exhibit highly turbulent and inertially driven flows; viscous effects are much reduced in importance. The product of characteristic dimension and velocity

TABLE 1.2
Representative Values of the Reynolds Number for the Body (Re_{body}) and for the Wing Chord (Re_{wing}) for Species of Body Mass m during Flight at a Forward Velocity V, or While Hovering

Order	Genus and Species	m (mg)	V (m/s)	Re_{body}	Re_{wing}
Odonata	Anax junius[a]	900	7.5	33,000	10,500
Orthoptera	Locusta migratoria[b]	1500	4.6	12,300	2500
Hymenoptera	Bombus terrestris[c]	175	(hovering)	n/a	1210
			1	1240	1360
			2.5	3100	1940
			4.5	5580	2990
Hymenoptera	Encarsia formosa[d]	0.025	(hovering)	n/a	15–18
	Encarsia formosa[e]		(hovering)	n/a	0.07
Lepidoptera	Urania fulgens[f]	350	3.4	4480	5270
Diptera	Drosophila virilis[g]	2	2	300	350

[a] Maximum speed in nature (Rüppell, 1989; additional data from May, 1991). Value for Re_{wing} is based on the forewing chord and assumes that the stroke amplitude and the stroke plane angle equal 90°.

[b] Flight in migratory swarm (Baker et al., 1981; additional data from Faure, 1932, and Baker and Cooter, 1979a). Value for Re_{wing} is based on the forewing chord and assumes a stroke amplitude of 100° and a stroke plane angle of 70°.

[c] Bumblebee worker flying freely in jet of wind tunnel (Dudley and Ellington, 1990a).

[d] Hovering flight in laboratory (Weis-Fogh, 1973).

[e] Re_{wing} based on wing setal diameter (Ellington, 1975).

[f] Migratory flight in nature (Dudley and DeVries, 1990).

[g] Tethered in jet of wind tunnel (Vogel, 1966).

determines Re, however, so that even large structures moving at very low speeds (e.g., an airplane moving at 10^{-9} m/s) experience almost exclusively viscous forces in the boundary layer surrounding their external surface area. Most importantly, different situations of fluid flow are physically equivalent if the corresponding Re's are approximately the same. That is, patterns of fluid flow at the same Re are dynamically equivalent and are independent of the physical context of motion. This important conclusion enables scaled physical models (attained by varying l), physically variable fluids (varying ρ and μ), and different flow velocities to be employed in dynamic reconstructions of otherwise experimentally intractable flow situations (Vogel, 1994).

Operationally, flying insects can be characterized by two distinct but correlated estimates of Re, one for the wings and one for the body (table 1.2). Selection of the characteristic dimension used in calculation of the Re is somewhat arbitrary, but this choice is more readily understood in light of the comparative and not absolute nature of the Re. Body length serves as the characteristic dimension for the Re of the body in forward flight, whereas the mean wing chord (the mean distance between the leading and trailing edge of the wing, calculated as the wing area divided by the wing length) serves as the relevant quantity for the Re of wings. In general, both wings and bodies of larger insects operate at higher Re and are positively correlated, but this relationship is typically nonlinear and varies with relative wing size, wingbeat kinematics, and flight speed. Many studies of insect flight have worked with fairly large taxa operating in an Re range of 10^2–10^4 (table 1.2). This range is generally understood to be characterized mostly by inertially driven flows (Vogel, 1994). Because most insect species are only 4–5 mm in body length (section 7.1), however, the Re for flying insects most typically falls within the less-studied range of 10^0–10^2 (section 7.1.3). At such low Re, flow is essentially viscous in character, and locomotor mechanisms used at higher Re (e.g., circulation-based lift; see below) are much less effective. Finally, flight in variable air temperatures may, in addition to potential thermal effects on muscle performance (section 4.3.1), also influence flight aerodynamics through correlated variation not only in air density and viscosity but also, by implication, in the Re of the wings and body.

The total fluid drag D acting on an object represents the summed effects of pressure and viscous drag, and can be measured empirically for different shapes and sizes using force transducers mounted in either flow tanks or wind tunnels. Comparison of the drag of different objects (i.e., the relative extent of streamlining) is facilitated through definition of a nondimensional drag coefficient C_D such that:

$$C_D = \frac{2D}{\rho S V^2},\qquad(1.2)$$

where ρ is the fluid density, S a reference area of the object (most typically the cross-sectional area), and V the velocity of fluid relative to object (Vogel, 1994). Drag coefficients are not, in general, constant for any given object, but vary substantially both with Re and with object orientation relative to flow (section 3.2.1). This variation arises in part from the interaction between turbulent and viscous flows around the object as flow speed changes, and in part from the differential dependence of inertial and viscous drag on object dimensions. Accordingly, extrapolation of drag coefficient data beyond the Re range of empirical validation is not warranted for most biological structures. Drag coefficients of both insect wings and bodies do, however, tend to decline at higher Re (section 3.2.1). Also, equation (1.2) refers only to motion at constant velocity. For objects accelerating within fluids, a force supplemental to body drag is also required to accelerate the added or virtual mass of surrounding air entrained by the object's motion (Denny,

1993). Because of the low density of air and the small size of most insects, however, forces associated with acceleration of this added mass are likely to be insignificant in comparison with viscous and pressure drag.

Drag forces on the bodies of flying animals have been classically termed parasite drag, signifying the drag supplemental to that associated with force production by the flapping wings. For bilaterally symmetric insect bodies oriented directly into flow, the dextral (right) and sinistral (left) components of parasite drag offset one another and the total drag imposes a net backward force. Insects flying with the body not oriented in the direction of motion, however, experience a supplemental component of drag force that is not parallel to the insect's body and that will impose a lateral deceleration in absence of compensation by the wings. Similarly, the drag force on a wing in two-dimensional flow (i.e., with no spanwise movement of air toward the wing tip) is termed the profile drag and acts parallel to the relative wind velocity, tending to decelerate the wing. Profile drag can be further represented as the sum of pressure drag (associated with flow separation and momentum losses in the wake of the wing) and of skin friction (derived from shear forces within the boundary layer on both dorsal and ventral wing surfaces). The mechanical power required to overcome both profile drag on the wings and parasite drag on the body is a major component of the total power expenditure in flapping flight. Drag coefficients for both insect wings and bodies have correspondingly been the subject of empirical investigation (section 3.2.1).

1.2.3 Lift Forces

Just as drag is the force parallel to flow, lift is defined as the component of force orthogonal to flow and thus perpendicular to drag in a two-dimensional perspective. All objects positioned asymmetrically in moving fluids experience lift as well as drag, but the magnitude of lift forces thus generated varies considerably with object geometry. For most objects in biology, lift forces are small relative to drag. Structures that, on the contrary, produce high lift relative to drag are termed airfoils if they are technological in character; wings are simply biotic airfoils. High-lift structures such as wings represent one extreme of the lift-drag continuum, whereas the other extreme is exemplified by high-drag and low-lift objects such as insect bodies. Note, however, that the distinction between lift and drag is purely conceptual, and that aerodynamic forces can be equivalently viewed as a single net force vector of variable magnitude and direction acting on insect wings and bodies (see Dickinson, 1996). Force components perpendicular to airflow are

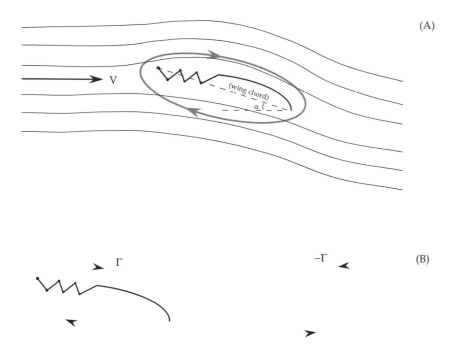

F<small>IG</small>. 1.2. (A) Airflow over a wing moving at relative velocity V. The angle of attack α indicates the orientation of the wing chord (line segment connecting the leading and trailing edges) with respect to the relative air velocity. (B) Two-dimensional perspective of circulation Γ around a translating wing. Circulation of magnitude equal to that of the bound vortex but of opposite sign is shed at the start of wing translation.

somewhat counterintuitive, however, and a brief description of the physical mechanisms underlying lift production is necessary to appreciate the nature of the aerodynamic forces generated by flapping wings.

A simplified representation of a flapping insect wing is to consider a cambered wing momentarily moving at a fixed orientation and constant speed relative to the surrounding air (fig. 1.2A). This condition can be termed steady-state flow, as opposed to unsteady conditions characterized by changes over short time intervals in wing speed, geometry, and orientation. At low angles of attack under steady-state conditions, air flows smoothly over both dorsal and ventral wing surfaces. Because pressure drag of well-designed airfoils is low, momentum extraction from the moving fluid is minimized and the dorsal airstream merges with that over the ventral surface near the trailing edge of the airfoil. Shear stress within the boundary layer at the trailing edge prevents the dorsal airstream from flowing directly to the trailing edge (with a necessarily ventral component of velocity) and from instantaneously rejoining the mainstream flow that is orthogonal to the dorso-ventral axis of the wing. Instead, the dorsal airflow separates from the wing slightly anterior to the trailing edge, whereas the ventral airstream diverges from the wing precisely at the trailing edge. A small turbulent wake reflecting the slightly premature departure of the dorsal airstream is thus present even at very low angles of attack relative to oncoming flow.

The positive camber of the airfoil and the merging of dorsal and ventral airstreams near the trailing wing edge yield slightly different translational velocities for the two airstreams—airflow is slightly faster above and slower beneath the airfoil. Bernoulli's Principle indicates that this difference in dorsal and ventral airstream velocities results in a pressure gradient and net lift directed across the surface area of the wing. The difference in translational velocities of the dorsal and ventral airstreams is physically equivalent to net movement of air from the ventral wing surface around the leading edge and then over the dorsal surface of the wing. In a two-dimensional perspective, then, air appears to move anteriorly from the ventral to dorsal wing surface and to yield a net rotational movement of air around the wing (fig. 1.3B). This motion of air is equivalent to a rotating flow field (or vortex) that circulates around the wing and that is centered about the wing itself (usually at a point 25% along the chord length from the leading edge). Any such bound vortex is an irrotational vortex with the quantitative characteristic that, at any point within the vortex, the product of the local tangential velocity and the distance between the point and the vortex core (i.e., the radius) is a constant (see Vogel, 1994). The rotational intensity of the vortex is simply all tangential components of velocity summed around the circumference of the vortex, a quantity termed the circulation (Γ). The physical origin of lift thus lies within creation of net air circulation about a moving wing.

Circulation, as a form of angular momentum, requires kinetic energy in order to be initiated. For a wing section that is initially at rest and then begins to translate, circulation increases continuously in accordance with that required to satisfy the equilibrium value associated with flow separation from the dorsal wing surface. Conservation of energy for a rotating system (i.e., the conservation law of circulation known as Kelvin's theorem) dictates that circulation of comparable magnitude but of opposite sense must be generated in the surrounding fluid. Thus, a bound vortex on a wing must always be associated with a so-called starting vortex of circulation $-\Gamma$ that remains in the

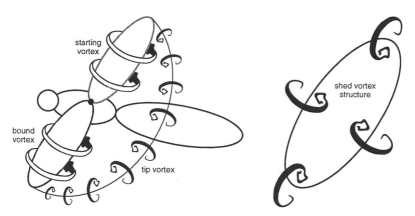

FIG. 1.3. Three-dimensional perspective of the vortex structure generated during the downstroke. The starting, tip, and stopping vortices are linked to form a complete vortex structure; vortex structures of each wing pair likely combine to produce a single vortex ring (section 3.2.2). Vorticity shed consecutively at the end of down- and upstrokes generate the vortex wake of a flying insect, although vortical interactions between down- and upstroke are not well resolved.

region of initial acceleration of the airfoil (fig. 1.2B). The presence of this vortex in the airfoil's wake results in a mathematical field of vorticity that, at any point in space, is proportional to the angular velocity of the local air particles. The angular momentum represented by this vorticity field must be supplied by the wing and represents a transient energetic input associated with the initiation of circulation about the wing.

Once a bound vortex is in place, additional work is necessarily associated with sustained lift production. Circulation must be maintained in the face of viscous dissipation that characterizes on all vortices. More importantly, the ventral momentum flux associated with a bound vortex represents a continuous transfer of energy from the wing to the surrounding air. For three-dimensional airfoils, this momentum flux derives from yet another vortex in addition to the bound and starting vortices. As a wing translates in space, the pressure gradient underlying lift production yields airflow not only around spanwise wing sections (the bound circulation), but also around the wing tip. This pattern of airflow creates an additional vortex, the tip vortex (fig. 1.3), that is unavoidably associated with lift production by three-dimensional wings. The tip vortex produces a net dorsoventral air flow (the induced velocity) that yields a useful flux of air but also that imposes an additional force on the airfoil, termed the induced drag.

For lift generation to persist, the wing must actively supply energy to the surrounding air in order to overcome this induced drag. The total energetic costs of lift production are thus positively correlated with the circulation of the wing tip vortex and with the magnitude of the induced velocity. Wings that are long relative to the mean wing chord (i.e., of high aspect ratio) generate tip vortices that are small in magnitude relative to the bound circulation. The induced velocity is relatively low in such cases, and the associated momentum flux of air derives primarily from a greater mass of air moving ventrally at lower velocities. Relative to wings of lower aspect ratio, high aspect ratio wings are associated with a reduced power expenditure to create comparable lift forces.

Given that a three-dimensional vortex is generated by a translating wing, the associated reactive lift force on the wing can be estimated. The tip vortex is linked to the starting and to the bound wing vortices (fig. 1.3) and creates a closed vortex loop that exerts a momentum flux on the surrounding air. Intuitively, the momentum induced by the presence of the vortex should be proportional to the mass of the air moved ventrally (i.e., to the air density) and to the velocity at which the air moves (i.e., the circulation). The Kutta-Joukowski law relates the magnitude of lift produced per unit wing span (L_{span}) to the air density, the translational velocity of the wing, and the bound circulation:

$$L_{span} = \rho V \Gamma. \tag{1.3}$$

The dependence of lift production on wing velocity can best be understood by realizing that the area of the closed vortex loop increases linearly with the translational velocity, in essence sweeping the bound vortex through space over a greater area at higher velocities (fig. 1.4). Circulatory lift is thus inextricably associated with the momentum flux produced by any vortex structure (see Dickinson, 1996). Furthermore, the instantaneous force that any vortex can produce is directly proportional to the rate at which it can be moved through space. More rapid flapping motions by wings, for example, will yield greater lift.

As with drag, the total lift L on a structure can be nondimensionalized through definition of a lift coefficient C_L:

$$C_L = \frac{2L}{\rho S V^2}, \tag{1.4}$$

where ρ, S, and V are as in equation (1.2). For comparative purposes, the reference area used for airfoils is typically the plan area of the wing (i.e., the horizontal projection of wing area). Inertial characteristics of

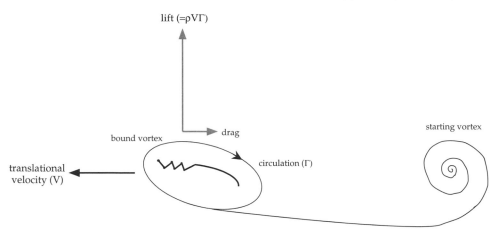

Fig. 1.4. Two-dimensional cross section of the vortex wake generated by a moving wing. The instantaneous translational velocity (V) of the wing interacts with the bound circulation Γ to produce a downward momentum flux in the wake and an instantaneous force on the airfoil.

flow dorsally and ventrally over wings are directly influenced by air density, whereas airflow separation from the trailing edge of the wing is constrained by viscous effects within the wing's boundary layer. The *Re* at which a wing is operating thus exerts a strong influence on lift production. Under highly viscous circumstances, vortex generation becomes more difficult as intermolecular stickiness progressively impedes rotational motions of airflow (section 3.2.1). Circulatory lift becomes increasingly more costly to maintain at low *Re*, and very small insects may rely on noncirculatory mechanisms of force production to stay aloft (section 7.1.3).

Bound vortices are, however, probably characteristic of most insects several millimeters and larger in wing length. At the higher *Re* characteristic of such wings, cambered profiles characterize effective airfoils as such shapes facilitate smooth merging of dorsal and ventral airflows downstream of the airfoil at low angles of attack. The bound circulation is of a magnitude comparable to the local airspeed, and lift forces increase in approximate proportion to the square of the translational velocity of the wing (see eq. 1.3). As the angle of attack increases, lift forces increase linearly but then peak at angles typically between 20° and 30°. Separation of the dorsal airstream from the wing then becomes more pronounced and occurs more anteriorly at higher angles of attack, ultimately leading to disruption of circulatory flow and finally to stall, a sharp decline in lift production (section 3.2.1).

Lift is a tremendously useful force because the lift : drag ratios of well-designed airfoils, including insect wings, are well in excess of unity. Lift forces over much of the *Re* range relevant to insect flight also increase in approximate proportion to the square of wing velocity. In addition to the induced drag and power expenditure associated with maintenance of the tip vortex, however, profile drag is always present on wings and incurs energetic cost. In return, however, the resultant force is substantially higher than would otherwise be available through use of drag-based propulsion alone. Large aerodynamic forces can thus be generated with minimal energetic expenditure. Wing lift in fact predominates the overall force balance of flapping insect wings; only in the takeoff flight of pierid butterflies (Ellington, 1980a, 1984a; Sunada et al., 1993b) has pressure drag been found to substantially exceed wing lift. Furthermore, the directionality of lift is easily altered on a flapping airfoil. Simply by altering wing orientation relative to flow and/or the wing's relative airspeed through variation in wingbeat kinematics, rapid change in the magnitude and direction of resultant forces is possible. Wing lift is not necessarily directed vertically; lateral forces as well as horizontal thrust can also be generated by a moving wing. As with drag, net lateral component of lift on bilaterally symmetric bodies and on bilaterally paired wings or wing pairs is minimal. During maneuvers, however, asymmetric wing kinematics can result in substantial imbalance between opposite (contralateral) wings, generating sideways displacement and body rotation (see section 5.2.2). Insect bodies can themselves, somewhat surprisingly, generate useful lift forces at positive angles of attack (section 3.2.1), but the magnitude of these forces is small relative to lift on wings.

1.2.4 Force Production through Wing Flapping

The preceding discussion has referred to aerodynamic forces acting on an airfoil operating at constant velocity and angle of attack (i.e., under steady-state nonaccelerating conditions). Insect wings, by contrast, are flapped continuously about the wing base. This reciprocating motion generates a spanwise velocity gradient along the wings—flapping velocity is greatest at the wing tip and declines monotonically to zero at the wing base. Because wings are reciprocated at continuously varying angular velocities (in contrast, for example, to a continuously rotating helicopter blade), the translational velocity, acceleration and higher positional derivatives of any point along the wingspan vary continuously. Moreover, periodic rotation of wings about their longitudinal axis results in time-dependent angles of orientation relative to the stroke plane. Wing rotation occurs mostly at the beginning and at the end of a wing's half-stroke, and these rotations are thus approximately 90° out of phase with respect to wing oscillation about the wing base. The angle of attack may also vary through the wingbeat via smaller-scale rotational adjustments. Both timing and the angular extent of wing rotation at ends of half-strokes can be actively controlled and may exhibit substantial variation according to aerodynamic demand (section 3.1.2).

These kinematic deviations from the above steady-state portrayal of lift generation render impossible the accurate prediction of the aerodynamic forces on flapping wings. At the very least, wing orientation and relative velocity must be assumed to vary continuously and to yield correlated variation in both the magnitude and direction of wing lift and drag. The unsteady aerodynamic effects associated with wing rotation and wing acceleration also yield instantaneous forces that differ substantially from those predicted by equations (1.2) and (1.4). In sum, aerodynamic forces on flapping wings are not well understood, although unsteady lift appears to be a fundamental motive force underlying insect flight and maneuverability (see section 3.2.2).

Further complexities arise from consideration of the vortex flows associated with multiple reciprocating wingbeats. As a wing accelerates and then decelerates in a half-cycle of a full wingbeat, vorticity is initially shed spanwise across the full length of the wing's trailing edge (the starting vortex), a tip vortex is shed continuously as the wing translates, and finally termination of motion results in release of the bound circulation from the wing. A complete vortex loop is formed that then translates in space away from the point of formation (fig. 1.3). Depending on the extent of lift generation, a similar vortex structure may be produced during the upstroke (section 3.2.2.5). Vortex loops of the down- and upstroke may link together or move in close proximity, each influencing the other's motion. More importantly, preexisting vorticity in the fluid surrounding a wing may enhance or detract from the generation of new bound circulation. Consecutive wingbeats generate series of such vortex structures that translate and interact with one another and with the bound circulation of the wings in as yet undescribed ways; the three-dimensional geometry of the vortex wake is in general unknown for arbitrary configurations of continuously oscillating and rotating wings. For insects with two pairs of wings (the apparent ancestral condition), this situation is further complicated by interaction of the shed vortex sheets between ipsilateral (same-side) and potentially between contralateral wings. Although little is known

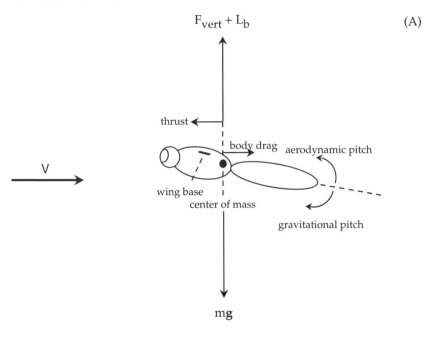

$F_{vert} + L_b$ (A)

thrust

body drag   aerodynamic pitch

V

wing base
center of mass

gravitational pitch

mg

(B)

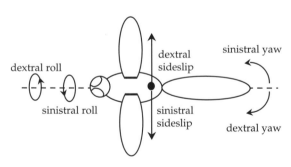

dextral roll

sinistral yaw

dextral
sideslip

sinistral roll

sinistral
sideslip

dextral yaw

FIG. 1.5. Force and moment balance for a free-flying insect: (A) lateral perspective, and (B) dorsal perspective. Forces and torques through and about the center of body mass are balanced in nonaccelerated horizontal flight. F_{vert}: net vertical forces produced by wing flapping; L_b: body lift. Body weight mg is usually much larger in magnitude than the body drag.

about the vortex wake of free-flying insects, extensive experimental work has been carried out on vortex production by tethered insects (see section 3.2.2.5).

Independent of the mechanisms of force generation, an overall balance of forces and torques characterizes the steady nonaccelerating forward flight of insects (fig. 1.5). Because there are three rotational and three translational axes of movement for any object in space, a total of six degrees of freedom characterizes body motion of a free-flying insect. Most experimental work on insect flight mechanics has evaluated only a subset of these kinematic quantities (section 5.2). For stable forward flight, net vertical force production must balance the body weight less any body lift forces, whereas thrust (net horizontal force) must offset parasite drag forces on the body. Equilibrium of rotational moments also applies in steady flight: the net pitching moment about the center of mass must equal zero, whereas left and right wing motions must be bilaterally symmetric such that yaw and roll moments are balanced (see fig. 1.5). Stability in flight therefore involves a wingbeat-to-wingbeat maintenance of this force and moment balance. Such dynamic precision clearly requires continuous sensory transduction and kinematic compensation in the face of unexpected turbulence and intrinsic instability of underlying aerodynamic mechanisms. Transient disruption of the force and moment balance, by contrast, provides for the rapid body rotations and accelerations that are the essence of insect flight maneuverability (section 5.2).

1.3 WHY STUDY FLIGHT BIOMECHANICS?

A central theme of this book is that diverse features of insect biology are united by the commonality of biomechanical performance during flight. This thesis can be briefly illustrated by examining morphological diversity of the extant fauna, flight performance in relation to particular features of insect ecology, and the largely unexplored realm of unsteady aerodynamic force production associated with the flapping and rotation of flexible wings.

1.3.1 Evolution of Morphological Diversity

The extraordinary morphological diversity of the winged insects is best evaluated in the context of functional utility. Genetic and abiotic constructional constraints notwithstanding, wing and body morphologies often appear to correlate with various taxon-specific aspects of flight performance and locomotor capacity. Many scenarios of physiological adaptation have been criticized for assumptions of optimized matching between structure and function (see Garland and Huey, 1987; Dudley and Gans, 1991; Weibel et al., 1998). The biomechanical evaluation of extant structures does not, however, imply either particular evolutionary trajectories or current selective advantage. Rather, discussions of aerodynamic performance and biomechanical feasibility can be used to frame hypotheses of adaptation that can then be tested rigorously using a phylogenetic framework. This approach is specifically used in chapters 6 and 7 to test various hypotheses relating flight and insect diversification. Throughout the book, correlational statements relating functional utility to performance within particular behavioral and ecological contexts are intended primarily as hypotheses for evaluation within the context of modern comparative biology.

Pterygote insects are defined taxonomically by the presence of wings, and wing morphology in particular exhibits myriad functional modifications (section 2.2). The phylogenetically ancestral wing condition is generally presumed to be that of homonomy, namely of iteratively homologous wing pairs that share equivalent ontogenetic origin, shape, and function. Homonomous wings are best illustrated by damselflies (Odonata: Zygoptera; see plate 2). In this taxon, fore- and hindwings are morphologically and kinematically highly similar, operating at similar beat frequencies and serving comparable aerodynamic functions. Homonomous wing pairs can be found in a number of insect orders (section 2.2.3), but in general fore- and hindwings have diverged historically both in terms of aerodynamic contributions to flight and in their nonaerodynamic biological roles. One major trend in insect evolution has been relative reduction in area of one wing pair and functional or actual coupling of the two ipsilateral wings to yield one continuous aerodynamic surface. Relative wing reduction is typically mirrored by a reduction in the size of the corresponding thoracic segment. Major orders exhibiting this trend of coupling and transformation to one effective wing pair include the Hymenoptera and the Lepidoptera, together almost 30% of the extant insect fauna (table 1.1). In addition to functioning aerodynamically, the overlapping and in most cases physically linked ipsilateral wings of the Lepidoptera perform numerous nonaerodynamic roles in such diverse contexts as crypsis, sexual selection, thermoregulation, and mimicry.

A second major theme of wing transformation at the ordinal level is conversion of forewings to protective devices with reduced and usually minimal aerodynamic function. The forewings of beetles, for example, are termed elytra and are heavily thickened and sclerotized, serving primarily to protect the wings and abdomen when the insect is not in flight. Forewings in the order Hemiptera are termed hemelytra and are also partially thickened to serve, at least in part, a protective function. Forewings of various other orders are slightly sclerotized and leathery, and are known as tegmina (section 2.2.3). A general tendency toward tegminization and elytrization of the forewings is thus widespread among the winged insects. Dramatic reduction in the size of one wing pair has also occurred in various taxa. Most notably, the hindwings of flies (the halteres) are miniaturized and function as sensory rather than aerodynamic structures; analogous miniaturization of one wing pair can be found in at least two other orders (see section 2.2.3). Many insect taxa are rendered secondarily flightless through substantial reduction and even through complete loss of both wing pairs (section 6.3.1). Although functional correlates of wing differentiation have not been systematically investigated among insect orders, major differences in flight performance and ecology are clearly associated with interordinal patterns of wing modification and reduction.

1.3.2 Ecological Contexts of Flight: Specific Examples

Numerous features of insect ecology depend either directly or indirectly on the ability of insects to fly. Selection on flight performance has apparently facilitated adaptive radiation of insects into such varied ecological roles as herbivores, pollinators, blood feeders, and long-range dispersers. Brief consideration of each of these ecological habits is therefore instructive from the unifying perspective of aerial locomotion.

All nonaquatic insects live within a structural context defined by the presence primarily of angiosperms and secondarily of gymnosperms. Roots, stems, seeds, fruits, shoots, flowers, and leaves provide a grand diversity of habitats for insect eggs, larvae, and adults. Herbivory and pollination are the two dominant forms of ecological interaction between insects and plants. Most insects are phytophagous (plant eating) for at least some portion of their life cycle, and many insects are involved in pollen transfer between conspecific angiosperm flowers. Both herbivory and pollination are at least in part influenced by pterygote flight capacity. Many plant viruses are also vectored by insects. Spatial mobility is essential for location of widely dispersed plant resources, either for purposes of immediate consumption or for oviposition if herbivory is confined to the larval stages (as is characteristic of most Lepidoptera). Host specificity of many herbivore-plant interactions indicates that aerial search for the appropriate species of host plant is commonplace. Similarly, pollination of angiosperms has often involved reciprocal coevolutionary interactions between insect locomotor capacity and floral structure and location. Specific coevolutionary interactions between pollination and flight performance are explored in greater detail in section 7.2.

Hematophagy, or the ingestion of blood, is a highly specialized feeding mode that superficially appears to be of little relevance in a discussion of insect flight biomechanics. After all, hematophagy has been a major factor associated with wing loss in two insect orders that are obligate ectoparasites on vertebrates (Phthiraptera and Siphonaptera, or the lice and fleas, respectively; see Waage, 1979). Also found among hematophagous insect species, however, are numerous dipteran vectors involved in transmission of bacterial, viral, and protozoan diseases. Malaria, dengue, and leishmaniasis are some of the major diseases that are transmitted to humans by flies. Such detrimental behavior is not confined to the Diptera: various disease-transmitting hemipteran Reduviidae have acted as vectors of Chagas disease (a degenerative cardiac condition of bacterial origin) to millions of humans throughout Central and South America. The advent of global warming may serve to hasten the spread of many of these pathogens and of their vectors; malaria in particular is increasing rapidly in frequency and is thought conservatively to infect hundreds of millions of people worldwide.

Selection on flight performance plays a major role in disease transmission by pathogen-bearing insects in at least two distinct ways. Initial host location often involves extended dispersal to locate hosts, as well as implementation of aerial search strategies typically based on olfactory cues (Lehane, 1991). Maneuverability, once the potential host is found, is essential to avoid aversive efforts that interfere with parasitism. Humans worldwide spend considerable amounts of their time both outdoors and indoors swatting at flies. Hematophagous insects are well equipped aerodynamically to evade their vertebrate hosts, and the rapidity of such flight is in part the product of millions of years of unintentional selection by vertebrates for increased flight performance. After feeding, temporary ectoparasites must fly away from the host with a greatly increased body mass. Tsetse flies, for example, increase their body mass by a factor of 2–3 during feeding (Langley, 1970), and the flight apparatus must be capable of offsetting such heavy blood loads (section 5.4.1). Flight metabolism of these hematophagous insects is correspondingly specialized: water from the blood meal is rapidly excreted, whereas amino acids derived from ingested blood are used as metabolic fuels (Bursell, 1981). This latter feature distinguishes hematophagous insects from other winged insects, which generally power flight using carbohydrates and lipids (section 4.1.2). Hematophagy has thus imposed specific performance demands on the insect flight apparatus that are of biomechanical, physiological, and medically anthropocentric interest.

A final flight-related theme of insect ecology is long-distance dispersal, a behavior characteristic of many insects and most typically associated with resource location. On timescales ranging from seconds to months, insects implement spatial movements typically to relocate within or among appropriate habitats. Wings are not strictly essential for such behavior (most arthropods disperse to varying degrees, and some insects migrate via walking or hopping), but flight increases the geographical range available for relocation by orders of magnitude. Many insects can utilize ambient air motions to aid long-distance dispersal, the most extreme form of which is intercontinental transport. Flying insects have, for example, been routinely trapped over all major oceans (Bowden and Johnson, 1976). As a corollary of widespread insect dispersal, a continuous and substantial fallout of the aerial insect fauna occurs over most terrestrial habitats, facilitating colonization of such remote regions as mountaintops and oceanic islands. The aerodynamics of insect dispersal and associated flight behaviors are treated in chapter 7.

1.3.3 Insect Mastery of Unsteady Aerodynamics

As discussed above in section 1.2.4, generation of lift forces on flapping wings involves continuous shedding of vorticity into the surrounding air and generation of a complex flow field termed the vortex wake. Production and manipulation of these unsteady vortex flows forms the basis for extraordinary maneuverability. In contrast to the staid flight paths typical of airplanes, most insects are capable of exceedingly rapid changes in flight speed and direction. Transient generation of forces and torques results in near-instantaneous turns, accelerations, and directional changes, the dynamic characteristics of which are in most cases unmatched by technology. Small body size facilitates rapid accelerations for allometric reasons (see section 5.4.1), but the maneuvers attained by some insects are nonetheless remarkable. For example, a 180° reversal of flight direction can be carried out by various flies in 40–100 ms, whereas backwards and even sideways flight is at least transiently implemented by many insects (see chapter 5). Tabanid flies can rapidly reverse flight direction using an Immelmann turn (a vertical half-loop followed by a half-roll; see Wilkerson and Butler, 1984). Many of these amazing maneuvers occur during events of aerial predation or mate selection, contexts for which successful evasion or capture has direct consequences for either mortality or reproductive fitness, respectively.

The kinematic correlates of insect maneuverability have not been fully elucidated, although controlled instability is likely a general feature not only of such maneuvers but also of apparently steady forward flight (chapter 5). Unsteady forces on flapping wings can readily be manipulated through subtle but significant changes in wing contour and orientation, and asymmetric wing motions typically form the basis of maneuverability in insects. Leg and abdominal motions can also contribute to the generation of advantageous torques during maneuvering flight; additional flexibility is provided by aerodynamic interactions between opposite wing pairs. Neither the complete range of maneuverability nor the diverse kinematic means of force regulation has been fully described for any insect. Moreover, precise regulation and control of unsteady forces may be of more than zoological interest. Transient application of aerodynamic forces supplemental to weight support and thrust generation is otherwise known as vectored maneuverability, currently the subject of intense development for technological applications (e.g., Gal-Or, 1990). Although the *Re* range relevant to insect flight is orders of magnitude below that of conventional aeronautics, ongoing interest in the construction of microair vehicles may soon render the study of insect flight of more than biological interest.

1.4 SUMMARY

The ability to fly has been central to the evolutionary diversification of insects. Initial wing evolution in the late Paleozoic led to extensive adaptive radiations throughout the Carboniferous and Permian, establishing most of the ordinal-level diversity in flight-related morphology that exists today. Flight has been a key enabling character underlying major adaptive trends in contemporary insect diversity, including miniaturization, aerial predation, dispersal, and mating systems. Winged insects are both quantitatively and qualitatively important components of terrestrial ecosystems. Pollinating and herbivorous insects impose intense selective regimes on plants; flying insects are also preyed upon by a variety of animals and the occasional insectivorous plant. Forces of both natural and sexual selection have contributed synergistically to the evolution of insect flight performance and maneuverability. In spite of the apparent advantages of flight, secondary flightlessness is also widespread in the pterygote Insecta.

Insects fly in regimes of fluid flow that are neither purely inertial nor purely viscous in character, but rather that represent the combined effects of these two forces. Use of the Reynolds number provides a nondimensional parametrization to evaluate the relative influence of fluid inertia and viscosity; the Re for flying insects ranges from 10^0 to 10^4. Vortices and circulation-based lift about airfoils can be effectively generated by many flying insects, but equally the viscous features of air impose substantial resistive forces on wings and bodies. The tremendous diversity in insect shape, size, and wingbeat kinematics indicates comparable variation in Re and in the aerodynamic mechanisms used to effect flight. Unsteady aerodynamic flows invariably result, however, from the flapping and rotating wings characteristic of most volant animals.

Chapter Two

MORPHOLOGY OF THE FLIGHT APPARATUS

INSECTS CAN FLY only if the necessary aerodynamic machinery is expressed morphologically. Anatomical structures required for flight include wings to generate aerodynamic force, thoracic musculature, and an articulated juncture between wings and thorax that permits the transmission of muscular force and active control of wing orientation. This chapter treats first the structure of the thorax, then the functional morphology of the wings, and finally the ancillary morphological structures used during flight. Modern reviews of the functional morphology of wings are provided by Brodsky and Ivanov (1983a), Wootton (1979, 1981b, 1992), and Grodnitsky and Morozov (1994). The flight apparatus is evaluated here in a context of biomechanical design rather than phylogenetic diversity. Because variation in flight-related morphology among the orders is so extreme, I here use primary descriptive sources for individual taxa rather than secondary ones. Comprehensive ordinal-level sources on wing and thoracic morphology include works by Comstock (1918), Maki (1938), Rohdendorf (1949), Séguy (1959, 1973), Matsuda (1970, 1979), Brodsky (1994), and Grodnitsky (1999). In general, the partial or complete modification of one set of wings relative to the other has dramatically defined new patterns of morphological evolution in the Insecta.

2.1 Thoracic Design

The structural components required for flight derive from the morphological underpinnings of insect body design. Intrinsic to biomechanical performance of the insect flight apparatus are the flexible cuticular properties of the arthropod exoskeleton. The material characteristics of arthropod cuticle derive from the presence of polysaccharide chitin microfibers embedded in a protein matrix. Such material design is termed a fibrous composite, a structure that combines the intrinsic strength of the embedded fibers with a high toughness (the ability to absorb energy) derived from the binding forces between fibers and matrix (see Wainwright et al., 1976; Neville, 1993). Fibrous composites are therefore both strong and elastically flexible, and arthropod cuticle is the finest zoological example of this mechanical design. Flexibility is essential for the insect thorax, as internal muscular forces cause the cuticle to deform and either directly or indirectly transmit forces to the wings. Elastic return of stored energy in the thorax is also essential to minimize the total energetic cost during wing flapping. The wings themselves are very thin cuticular structures that, particularly in larger insects, act as semiflexible airfoils that often substantially bend and change shape during flight. For biomechanical purposes, wings can be considered to be inanimate structures that are activated by the insect only when it applies muscular forces at the axillary apparatus.

2.1.1 Segmentation and Cuticular Anatomy

Much of arthropod biology derives from the functional specialization of sets of adjacent body segments. Insects by definition possess three thoracic segments, each of which in both larval and adult stages bears a pair of legs. In adult insects, the meso- and metathoracic segments are collectively termed the pterothorax, upon which the wings are joined and within which the flight muscle is found (fig. 2.1A). Segments are composed externally of sclerites, well-defined cuticular plates with distinct boundaries that can either be membranous or sutural in character. In arthropods generally, the dorsal sclerite of any given segment is termed the tergum. For thoracic segments of insects, the tergum (or tergal sclerite) is specifically termed the notum (or notal sclerite). The notum, furthermore, is subdivided in many insect taxa into an anterior scutum and a posterior scutellum. Ventrally, each thoracic segment is bounded by the sternum; two pleural sclerites or pleura (singular: pleuron) define the lateral limits of the thoracic segment (fig. 2.1B). A dorsoventral pleural suture divides each pleuron into two distinct regions, an anterior episternum and a posterior epimeron. An internal pleural ridge runs parallel to the external pleural suture and strengthens the pleural sclerite dorsoventrally. Dorsally, the pleural suture ends in the pleural wing process that acts as a fulcrum directly beneath the wing base (fig. 2.1B). Two small sclerites, the basalar and the subalar, are located directly anterior and posterior to the pleural wing process, respectively. These sclerites serve as insertion points for muscles that act to flap the wing dorsoventrally as well as to rotate the wing about its longitudinal axis. Legs originate at the ventral base of the pleuron. The leg is an integral component of the flight apparatus in that some dorsoventral flight muscles terminate ventrally within the first and second leg segments (the coxa and trochanter, respectively).

In contrast to the ventral location of insect legs, wings of insects connect to the thorax in the dorsal region of the pleuron. Two processes (small projections) extend laterally from the notum, one anterior to and one posterior to the pleural wing process (fig. 2.2A). The base of

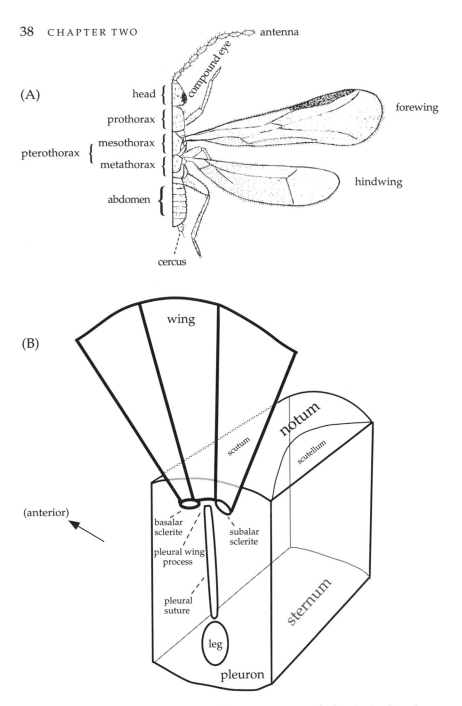

FIG. 2.1. (A) Pterygote tagmosis and thoracic segmental identity in dorsal perspective (Zoraptera: *Zorotypus brasilensis*; modified from Choe, 1992). The abdominal cercus is a small appendage that bears mechanosensory hairs. (B) Generalized anatomy of a pterothoracic segment in lateral oblique perspective.

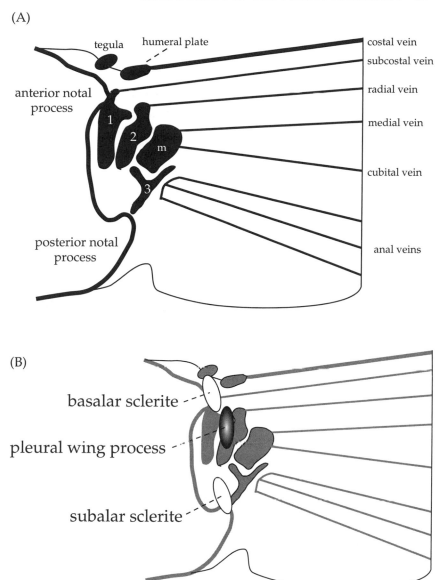

FIG. 2.2. Generalized anatomy of a pterothoracic segment and wing articulation in dorsal (A) and ventral (B) perspective (modified from Snodgrass, 1935). The first, second, and third axillary sclerites are indicated by their corresponding numerals; a median sclerite is indicated with an *m*.

the wing inserts on the thorax among these three processes. A median wing process is present in some orders and projects between the anterior and posterior notal processes. (Alternative nomenclature for notal processes was introduced by La Greca 1947 and was used by Matsuda 1970 in his comprehensive anatomical monograph.) The wing articulation is comprised of a set of small axillary sclerites lying between the wing and the notum. Additional preaxillary sclerites termed the humeral plate and the tegula may be contained within the wing articulation anterior to the anterior notal process (see fig. 2.2A); these two sclerites and the axillary sclerites are collectively termed the pteralia or the axillary apparatus. Wing articulation and the number of axillary sclerites exhibit substantial variation even at the basal levels of pterygote phylogeny. This observation is consistent with ancestral diversity in the anatomical structures that underlie flight in winged insects.

Specific evolutionary transformation of the ancestral insect wing base is still the subject of considerable speculation. For example, the precise identity of axillary sclerites in the Paleoptera is not well established. The Odonata have only a humeral plate (lying at the base of and supporting the costal wing vein; see below) and one axillary sclerite; the homology (or common phylogenetic origin) of this latter sclerite with that in other orders is uncertain (Tannert, 1958; Matsuda, 1970; Pfau, 1986, 1991). The three axillary sclerites of the Ephemeroptera, by contrast, appear to be homologous with those of the Neoptera (Matsuda, 1970; Kukolavá-Peck, 1985). In the Neoptera, the first axillary sclerite lies directly lateral to the anterior notal process, whereas the second axillary sclerite is joined laterally with the first axillary sclerite and lies directly above the pleural wing process (fig. 2.2B). The second axillary sclerite and the subalar sclerite are mechanically coupled by internal cuticular connections, whereas the basalar sclerite attaches directly to the axillary apparatus at the base of the most anterior vein, the costa. Median sclerites may lie laterally to the second axillary sclerite and the base of wing veins. The third axillary sclerite is supported by the posterior notal process (see fig. 2.2A) and forms the attachment point for muscles that connect to the pleuron. These pleuroalar muscles flex the wing against the abdomen when the insect is not flying, but may also influence wing orientation during flight (e.g., Heide, 1971a, b). The first, second, and third axillary sclerites are joined distally with the bases of the subcostal, radial, and anal wing veins, respectively (see below). A fourth axillary sclerite, which arises from the posterior notal process, is found in the Hymenoptera and Orthoptera (Matsuda, 1970). The secondary musculature associated with this sclerite may subtly influence the kinematic characteristics of the wingbeat (see Pringle, 1968).

2.1.2 Muscle Configuration and Function

Contraction of intrinsic flight musculature generates forces that are transmitted to the surrounding cuticle of the exoskeleton and to the the base of insect wings. Two different sets of thoracic muscles are primarily responsible for wing motions. Wing depressors and wing elevators effect the downstroke and the upstroke, respectively. Muscles may insert either directly on the wing base and sclerites of the axillary apparatus, or may act indirectly to move the wings via indirect thoracic deformation. In either case, muscles connect to the cuticle directly via tonofibrillae (cuticular microfibrils) without intervening tendons (see Korschelt, 1932; Boettiger, 1960; Auber, 1963; Lai-Fook, 1967). Elastic tendon-like structures are, however, widespread in flight muscles of odonates of the suborder Anisoptera (Clark, 1940). The rotational axis for wing depression and elevation is predominantly defined by dorsoventral movement of the first axillary sclerite against the notum, and particularly against the anterior notal process (fig. 2.2). Thoracic muscles that effect dorsoventral wing motion act either directly on sclerites at the wing base, or activate the wing indirectly via displacement of the notum and transmission of force to the wings through the axillary apparatus. In some insect taxa, both modes of force transmission may be present depending on the particular muscle under consideration.

Direct action of flight muscles on the wing base is phylogenetically ancestral. Among extant taxa, the Odonata (Pfau, 1986, 1991) and the Blattaria (Tiegs, 1955) use primarily direct muscles to generate the downstroke. In these taxa, the dorsolongitudinal muscles are small relative to the direct basalar and subalar muscles as well as to the indirect dorsoventral muscles that effect the upstroke. Action of direct depressor muscles lowers the wings through direct transmission of applied force along axillary sclerites to the basal wing venation. The direct wing depressors of Odonata (i.e., the basalar and subalar muscles) insert via tendons lateral to the pleural ridge and rotate the wing downwards about the pleural wing base (fig. 2.3A). These wing depressors originate on a cuticular brace (the furca) located at the base of each thoracic segment and above the leg musculature (Sargent, 1917; fig. 2.3A). Direct flight muscles also are prominent in the Ephemeroptera (Matsuda, 1970), Orthoptera (Tiegs, 1955; Pfau, 1978) and some Coleoptera (Pringle, 1957; Schneider, 1987). Such flight muscles generally insert on the basalar and subalar sclerites and terminate within the coxa, although some trochanteral muscles can also act directly on the wing base. Many direct dorsoventral muscles in the Orthoptera (Wilson, 1962) and Blattaria (Fourtner and Randall, 1982) have been suggested to contribute additionally to leg motions during cursorial

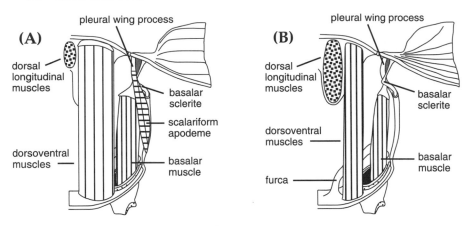

FIG. 2.3. Direct and indirect flight muscles in cross-sectional perspective. (A) Generalized odonate pterothorax (modified from Sargent, 1917; Pfau, 1991). (B) Generalized pterothoracic segment with both direct (basalar) muscles and indirect flight muscles (modified from Snodgrass, 1935). In both (A) and (B), the subalar muscle and corresponding subalar sclerite are directly behind the basalar muscle and basalar sclerite, respectively.

locomotion, blurring the functional distinction between muscle origin and insertion. Leg position during flight may also be influenced by tonic (constant force) activity in direct flight muscles (see Kutsch and Usherwood, 1970). Such bifunctionality (use in flight and in running) may be confined to only one of such muscles (Burrows, 1996), however, and different patterns of central neural control characterize activation of these bifunctional muscles (Ramirez and Pearson, 1988). In addition to originating in the leg, muscles that act to displace the basalar and subalar sclerites may be pleural in character, originating on the episternum and the epimeron, respectively.

In addition to their role in the downstroke, direct flight muscles also influence rotation of the wing about its longitudinal axis. Because the basalar and subalar sclerites are not coincident with the pleural wing process (see fig. 2.2B), contraction of the corresponding basalar and subalar muscles generates not only downwards displacement of the wing but also torque and concomitant wing rotation. Pronation refers to a nose-downward rotation of the leading edge of the wing, whereas supination conversely indicates wing rotation of the opposite sense. In the locust, near-simultaneous contraction of the basalar and subalar muscles produces the downstroke, whereas the extent of wing pronation is correlated with variation in contraction of the basalar muscle (Pfau, 1977, 1978; Wolf, 1990). Differential activity of the basalar and subalar muscles also effects the fore-aft (anteroposterior) wing motions known as promotion and remotion. In beetles, for example, subalar muscles serve to regulate wing angle of attack as well as more subtle kinematic features involved in flight control (Kammer, 1971; Govind, 1972; Pfau and Honomichl, 1979). Fore-aft motions as well as the extent of wing rotation effected by the direct flight muscles are largely defined by rotational mobility of the second axillary sclerite about the pleural wing process (fig. 2.2B).

Activity of additional direct flight muscles can also influence wing orientation. In locusts, pleuroalar muscles inserting on the third axillary sclerite act antagonistically to the basalar and subalar muscles during the upstroke. Particularly during the upstroke, the pleuroalar muscles regulate wing supination and angle of attack relative to the dorsoventral plane of motion (Nachtigall, 1981a; Pfau and Nachtigall, 1981; Heukamp, 1984). Similar functions characterize action of the third axillary muscle in the lepidopteran *Manduca sexta*, which functions not only in wing folding but also contracts periodically at the wingbeat frequency (i.e., phasically) to regulate wing promotion and remotion (Rheuben and Kammer, 1987). Bilateral asymmetry in activation of this muscle also influences axial and rotational maneuverability (Kammer, 1971; Kammer and Nachtigall, 1973; Wendler et al., 1993; see section 5.2.2). Thus, the third axillary muscle provides not only for wing folding but also for subtle yet important features of flight control.

In contrast to the phylogenetically ancestral case of direct muscle action at the wing base, indirect flight musculature effects wing motions via notal displacement. Some phylogenetically basal insect orders (e.g., Ephemeroptera, Plecoptera) nonetheless possess a highly derived flight apparatus in that the ventral wing motions appear to be generated indirectly via thoracic deformation (see Matsuda, 1970; Brodsky, 1994). In many groups with indirect flight muscles, contraction of the dorsolongitudinal muscles produces the downstroke through indirect arching and vertical displacement of the notum between opposite wing bases of a segment (fig. 2.3B). This motion causes the lateral edge of the notum and the adjacent axillary apparatus to rise above and rotate about the pleural wing process, thereby depressing the wings. The Odonata, although generally considered to utilize only direct muscles in the downstroke, do possess limited dorsolongitudinal musculature that may have ancestrally arched the notum dorsally to assist with wing depression (Pfau, 1986; see fig. 2.3A). In present-day odonates, these muscles now function in promotion or remotion and exhibit different roles in the meso- and metathoracic segments (Pfau, 1991). Muscle function during both the down- and the upstroke is thus highly variable among extant winged insects. Rather than a strict dichotomy between direct and indirect musculature as has been classically assumed, a more complete description would entail evaluation of the relative contribution of different muscles to wing depression, elevation, rotation, and fore-aft movements. Most evaluations of muscle action have been made strictly on the anatomical bases of insertion point and relative muscle volume. Patterns of neuromuscular activation are much less studied, and the experimental manipulation of muscles (e.g., denervation) to investigate subsequent effects on wing motions has rarely been carried out in a systematic comparative sense. The small size of most insects unfortunately precludes most such investigations at the present time.

Wing elevation in all insect orders is primarily attained through action of indirect dorsoventral muscles that connect the notum to the sternum as well as to proximal leg segments (fig. 2.3B). Dorsal oblique muscles that run from the notum to the base of the cuticular phragmata (partitions) between segments may also function as wing elevators in some cases (e.g., Barber and Pringle, 1966). Indirect elevation of wings is attained by muscle action that lowers the notum relative to the pleural wing process, thereby rotating the wing upwards about the axillary hinge. Muscles that elevate the wings thus act antagonistically to the dorsolongitudinal muscles (and, in some cases, to direct dorsoventral muscles) that function in wing depression.

The phylogenetic diversity of direct and indirect muscle action has not been systematically explored, but the major differences even among pterygote orders in muscle number and configuration (see Matsuda, 1970) must be reflected in alternative strategies of thoracic deformation and active wing control. For any given taxon, only a minority of flight muscles has been studied electrophysiologically on tethered experimental preparations (i.e., stationary insects flapping their wings in fictive flight). One general result that emerges, however, is that phylogenetically more derived orders decouple more effectively the powerful muscles involved in dorsoventral wing movement from those muscles controlling wing orientation. In dipterans, hymenopterans, and coleopterans, for example, the indirect flight muscles that effect notal bending and thoracic deformation are physiologically and mechanically distinct from the wing control muscles (see Nachtigall and Wilson, 1967; Kammer, 1985; Dickinson and Tu, 1997; Wisser, 1997; Nachtigall et al., 1998). Such a decoupling may be a necessary feature of any evolutionary increase in wingbeat frequency, as contractile dynamics become more difficult to regulate actively for a rapidly oscillating muscle. For the asynchronous flight muscles characterized by multiple contractions for a single nervous impulse (see section 4.2.2), wingbeat-by-wingbeat control is impossible to attain neuronally. Instead, wing orientation and other kinematic features not deriving from dorsoventral movements are under the active control of additional muscles that act at the wing base.

Dipterans, hymenopterans, and coleopterans are in fact all characterized by asynchronous indirect flight muscles (section 4.2.2), whereas the direct control muscles remain synchronous, contracting phasically at a usually high frequency but producing little power. A remarkable total of eighteen such control muscles can be found in the dipteran mesothoracic segment (Dickinson and Tu, 1997). These direct muscles (particularly the basalar and pleurosternal muscles) receive neural activation at the wingbeat frequency and impinge both tonically (at constant contraction strength) and phasically (with temporally fluctuating contraction strength) on wing motions (Heide, 1968, 1971a,b; Dickinson et al., 1993). The first basalar muscle in particular appears to influence wing base mechanics and thus wing orientation through changes in stiffness induced by variable timing of neural activation (Tu and Dickinson, 1994). Action of the first and second basalars together with that of additional axillary muscles influences both stroke amplitude and wing-tip paths (Heide and Götz, 1996; Lehmann and Götz, 1996; Tu and Dickinson, 1996). Flies thus exemplify the physiological and functional divergence between indirect power-producing muscles and direct control muscles (Dickinson and Tu, 1997). Evolution of this functional dichotomy must presumably parallel the evolution of asynchronous muscle in many taxa, but the relative contributions of indirect and direct flight muscles remain to be determined in comparative contexts.

2.1.3 Diversity of Thoracic Design

A general theme of thoracic evolution is that segment size can be dramatically modified or reduced in accordance with evolution of wing function. The prothoracic segment of winged insects, bearing a pair of legs but no wings, is much smaller in volume than either pterothoracic segment. Schwanwitsch (1943, 1958) used the terminology of anteromotorism, posteromotorism, and bimotorism to describe relatively enlarged forewings, enlarged hindwings, or equivalent wing size, respectively (see table 2.1). Here these terms are used to indicate not only relative size of wings on pterothoracic segments but also the likely extent of aerodynamic contributions during flight. For most taxa, the relative roles of wings in force production has not been assessed quantitatively, although the relative sizes of the meso- and metathorax tend to follow trends of wing specialization and thus locomotor dedication of the segment in question. For example, the coleopteran mesothorax

TABLE 2.1
Wing Number and Pterothoracic Specialization among Extant Pterygote Insect Orders

Order	Alary Mode (wing number)	Locomotor Mode	Tegminization/ Elytrization	Wing Miniaturization
Ephemeroptera	Tetraptery[a]	Anteromotorism	None	None
Odonata	Tetraptery	Bimotorism[b]	None	None
Plecoptera	Tetraptery	Posteromotorism[c]	None	None
Embioptera	Tetraptery	Bimotorism	None	None
Orthoptera	Tetraptery	Posteromotorism	Tegmina	None
Phasmatodea	Tetraptery	Posteromotorism	Tegmina	None
Grylloblattodea	Aptery	n/a	n/a	n/a
Dermaptera	Tetraptery	Posteromotorism	Tegmina	None
Isoptera	Tetraptery	Bimotorism[d]	None	None
Mantodea	Tetraptery	Posteromotorism	Tegmina	None
Blattaria	Tetraptery	Posteromotorism	Tegmina	None
Zoraptera	Tetraptery[e]	Anteromotorism	None	None
Thysanoptera	Tetraptery	Anteromotorism	None	None
Hemiptera	Tetraptery	Anteromotorism	Hemelytra	None
Homoptera	Tetraptery	Anteromotorism	None[f]	None[g]
Psocoptera	Tetraptery	Anteromotorism	None	None
Phthiraptera	Aptery	n/a	n/a	n/a
Coleoptera	Tetraptery	Posteromotorism	Elytra	None[h]
Neuroptera	Tetraptery	Bimotorism[i]	None	None
Megaloptera	Tetraptery	Bimotorism	None	None
Raphidioptera	Tetraptery	Bimotorism	None	None
Hymenoptera	Tetraptery[j]	Anteromotorism	None	None
Trichoptera	Tetraptery[k]	Anteromotorism[l]	None	None
Lepidoptera	Tetraptery[m]	Anteromotorism[n]	None	None
Strepsiptera	Diptery	Posteromotorism	None	Halteres (forewing)[o]
Diptera	Diptery	Anteromotorism	None	Halteres (hindwing)
Siphonaptera	Aptery	n/a	n/a	n/a
Mecoptera	Tetraptery	Bimotorism	None	None

Notes: Alary mode - the number of wings (tetra-, di-, or aptery indicate 4, 2, and 0 wings, respectively). Locomotor mode - anteromotorism (enlarged forewings), posteromotorism (enlarged hindwings), bimotorism (equivalent aerodynamic use of homonomous wing pairs). Details of wing tegminization/elytrization and wing miniaturization are given in the text. Because flightlessness has evolved in approximately half of the extant orders (fig. 6.2), these characterizations represent the general trend only of winged taxa within each order.

[a] Hindwings are reduced in Ephemeroptera and are absent in Caenidae and in certain Baetidae and Tricorythidae.

[b] Hindwings are generally larger than forewings in the suborder Anisoptera.

[c] Wings are homonomous in some Plecoptera.

[d] The family Mastotermitidae exhibits enlarged hindwings.

[e] Hindwings are vestigial in some species.

[f] Forewings are tegminized in some Auchenorrhyncha (e.g., Fulgoridae).

[g] Miniaturized hindwings of Coccidae may act as halteres.

is substantially smaller than the metathorax (Matsuda, 1970; Schneider, 1978a; Crowson, 1981), consistent with reduced aerodynamic
roles of the forewing (section 2.2.3). By contrast, the mesothorax predominates in both the Diptera (Young, 1921) and the Hymenoptera
(Snodgrass, 1910), as well as in many Hemiptera and Homoptera.
Other orders with alary (wing) specialization show a similar tendency
(fig. 2.4). Mesothoracic musculature in the Orthoptera is, for example,
reduced relative to that of the metathorax (Tiegs, 1955), mirroring the
relative aerodynamic roles of wings on the two segments (see section
3.2.1). In insects with high wingbeat frequencies (e.g., various Diptera
and Hymenoptera), reduction in the number of indirect muscles
within the mesothoracic segment is evident, suggesting functional consolidation of muscle action (see Brodsky, 1994). More generally, evolutionary trends in thoracic musculature have been described by Matsuda (1963a, b, 1970), who also emphasized homology of pterygote
flight muscles with locomotor muscles in apterygote taxa.

One recurrent biomechanical theme in thoracic design is that use of
indirect flight muscle to effect wing motions necessitates substantial
cuticular bending. Dorsoventral muscles pull and deform the notum
ventrally, whereas action of the opposing dorsolongitudinal muscles
reverses this action at frequencies often exceeding 100 Hz (see chapter 3). These muscle-induced deformations of the thorax are usually
constrained morphologically. Various notal grooves (particularly the
suture between scutum and scutellum; see fig. 2.1B) facilitate bending
along prefixed lines, whereas the internally strengthened pleural suture prevents excessive segmental deformation in the transverse axis.
Notal grooves are most evident in orders with pronounced notal flexion during flight (e.g., Diptera and Hymenoptera; see Janet, 1899;
Brodsky, 1994). Thoracic deformations are, by contrast, limited by the
phragmata that occur at intersegmental boundaries of adjacent notal
sclerites. These planar structures run dorsoventrally, strengthening the
pterothoracic segment in the transverse plane as well as serving as
attachment points for dorsolongitudinal and dorsal oblique muscles.

TABLE 2.1 (cont.)

[h] Elytra of some beetles are miniaturized and may function gyroscopically.

[i] Anteromotorism characterizes the family Nemopteridae.

[j] Hindwings may be absent in mymarid and mymarommatid Hymenoptera.

[k] Hindwings are vestigial in some Hydroptilidae.

[l] Wings are homonomous in some Trichoptera.

[m] Hindwings may be absent in ctenuchid (= syntomid) moths.

[n] Forewings are generally enlarged, but wings are homonomous in some Lepidoptera;
hindwings can be enlarged relative to forewings, particularly among butterflies.

[o] Male strepsipterans have miniaturized forewings; females are wingless.

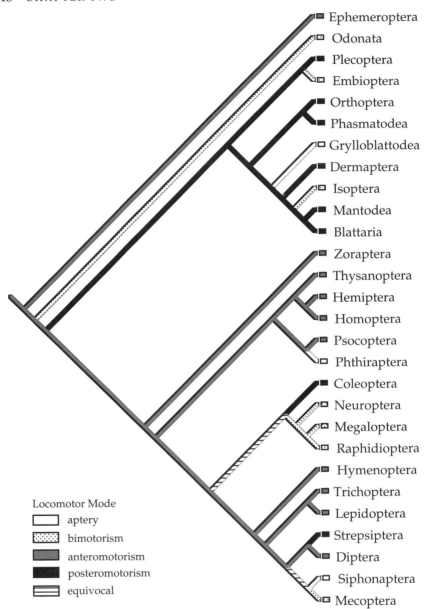

FIG. 2.4. Phylogenetic distribution of locomotor modes among pterygote orders. Note that the most parsimonious reconstruction using the pterygote phylogeny of fig. 1.1 identifies the ancestral locomotor mode as anteromotorism rather than as homonomous bimotorism (MacClade 3.0; see Maddison and Maddison, 1992).

Segmental rigidity in all taxa may be augmented by internal apodemes (chitinous projections) that also act as attachment points for muscles. Such apodemes, particularly pleural and sternal projections (e.g., the ventral furca of fig. 2.3B), are often joined by pleurosternal muscles. Contraction of these muscles can tighten the pleuron and wing base relative to the sternum (e.g., Nachtigall and Wilson, 1967; Kutsch and Hug, 1981), thereby altering the stiffness of the mechanically driven pterothoracic system. In odonates, the use of dorsoventral muscles to act as both wing elevators and depressors necessitates additional measures to prevent thoracic compression, for which purpose a pronounced scalariform (ladder-like) apodeme strengthens the pleuron dorsoventrally (Sargent, 1937; Russenberger and Russenberger, 1959, 1960; see fig. 2.3A).

Mechanically driven systems of given stiffness and damping (internal resistance to motion) exhibit resonant oscillatory frequencies at which advantageous energetic expenditure is highest. Any mechanical system can be driven at a frequency different from the resonant frequency, but the driving force is then out of phase with motion of the structure or object in question. The amplitude of oscillation is reduced considerably and substantial energy is dissipated in nonuseful work. The presence of elastic recoil within the thorax and antagonistic action between wing depressors and elevators suggest that the insect flight thorax is a mechanically resonant system driven periodically through muscular contraction (Pringle, 1949; Roeder, 1951; Danzer, 1956; Russenberger and Russenberger, 1959, 1960; Greenewalt, 1960b). Muscles contract against the inertial load of the aerodynamically active wings and the mass of the muscles themselves. Intrinsic stiffness of the axillary apparatus derives from the geometry of the wing articulation, the pterothoracic cuticle, and activity of direct flight muscles at the wing base. Tonic sternal and tergal (i.e., originating on the notum) muscles that insert on the pleural ridge are particularly well suited for this latter purpose. Restorative elastic elements in the flight apparatus include the thoracic cuticle as well as specific anatomical components of the flight muscle (section 3.3.2). The pterothorax thus contains those elements necessary for effective mechanical resonance. Wingbeat frequencies of free-flying insects are characterized by fairly low coefficients of variation, and both interspecific and experimentally induced variation in wingbeat frequency is consistent with the hypothesis of mechanical resonance within the thorax (see section 3.1.2).

The presence of elastic elements within the pterothorax is of particular importance for flight energetics. Such elements stretch at the end of a down- or upstroke (i.e., at the end of any given half-stroke) and store wing kinetic energy as strain energy. This energy is subsequently released and contributes to the generation of subsequent half-strokes (Weis-Fogh, 1959, 1965, 1972). Such action is energetically important because a major component of power expenditure during flight is the inertial energy required to accelerate the wings during each half-stroke (section 3.3.2). The actual magnitude of elastic storage during free flight is unknown for any insect, but elastic components within the locust thorax have been systematically investigated (see section 3.3.2). Elastic storage by biological materials is generally frequency dependent (i.e., viscoelastic). Although time-dependent measurements of elastic storage by the integral flight apparatus have not been made, some energetic recovery of wing inertial energy (particularly by elastic components inherent to the flight musculature) is likely in all winged insects. The inertial energy of oscillating appendages increases with the square of oscillation frequency, and insects with high wingbeat frequencies likely exhibit substantial elastic storage within the pterothorax (section 3.3.2).

Although the insect flight apparatus might superficially seem less complicated than comparable vertebrate musculoskeletal systems, mechanical operation of the pterothorax is not understood in detail for any insect group. The basic action of wing elevators (indirect muscles) and depressors (direct and indirect muscles) is not disputed, but detailed action of muscles and their interaction with the axillary sclerites to produce wing motions are much less clear. Complex pterothoracic mechanics arise from a high number of interacting elements, as exemplified by functional intricacies of the dipteran thorax. Sixteen sclerites articulate at the wing base between the notum and the wing, with as many direct control muscles (in addition to the indirect flight musculature) influencing sclerite position and axillary deformation (Ritter, 1911; Heide, 1971a,b; Dickinson and Tu, 1997). The dipteran thorax has been the target of classic investigations in insect flight mechanics. Using anesthetized flies, Boettiger and Furshpan (1952) described a click mechanism whereby wing articulation and pterothoracic musculature could stably position the wings only at either dorsal or ventral extremes of the wingbeat, with intermediate positions being elastically unstable due to the specific configuration of axillary sclerites. Wings were accelerated and then decelerated rapidly between these two extreme positions, with a waveform very different from the sinusoidal motion generally characteristic of simple resonant systems. This explanation of wing movement, with two stable extreme positions, persisted for many years in the specialized flight literature as well as in entomology textbooks.

Recent work suggests, however, that the click mechanism is likely an experimental artifact. Miyan and Ewing (1985b) showed that wing motions of tethered flies were inconsistent with the click mechanism. Carbon tetrachloride was used on flies by Boettiger and Furshpan (1952) as an anesthetic, and action of this chemical may have induced abnormal contraction of the pleurosternal muscles, altering the normal axillary configuration. The issue of normal operating configuration is in fact central to the resolution of conflicting views on the functional action of the dipteran thorax. For example, diverse anatomical studies of the dipteran wing base have suggested the presence of a mechanical stop on the pleural wing process that constrains ventral motion of the wing through physical contact with a projection at the base of the radial vein (e.g., Pfau, 1973, 1987; Wisser and Nachtigall, 1984, 1997a; Miyan and Ewing, 1985a, 1988; Ennos, 1987; Wisser, 1987, 1988). In stroboscopic video studies of tethered flies, however, Nalbach (1989) rarely saw contact between the base of the radial vein and the proposed endstop of the pleural wing process. The kinematic results of Ennos (1987) also implicated rotational mobility of the lateral edge of the scutum in control of wing movements, although no evidence for such rotation was found in tethered flies (Miyan and Ewing, 1988).

Such contrasting explanations of pterothoracic operation cannot be systematically compared for at least two general reasons. Different fly taxa under variable tethering and experimental regimes evidently activate their wings differently, and wingbeat kinematics during free flight may differ substantially from those in tethered flight (see section 3.1.2). For example, wing-tip motions of flies in free flight closely approximate simple harmonic motion (Ellington, 1984c; Ennos, 1989c), a waveform not found in aforementioned studies of tethered flies. Unitary explanations of pterothoracic operation are also unlikely because force output required from the flight motor can vary dramatically with airspeed and other dynamic demands, necessitating changes in kinematic parameters such as wingbeat frequency and amplitude (see chapter 3). Because axillary sclerites are three-dimensional (albeit small and flattened) structures, quantitative analysis of wing base motions requires image reconstruction comparable to that now available for wing-tip motions (section 3.1.2). The best approach to evaluating pterothoracic and axillary movements during flight would be to implement high-speed studies of wing-base motions in free-flying insects; such a goal would be attainable using a high degree of optical magnification and remote optomotor control of insect position in space (see chapter 5). Unfortunately, the intense illumination required for such studies can substantially disrupt the optomotor responses and natural flight behavior of subject insects under such circumstances.

2.2 WINGS

The thoracic structures responsible for generating useful aerodynamic forces are the wings, the often cambered surfaces that project laterally from pterothoracic segments. Venation serves to strengthen these surfaces against deleterious deformations and to facilitate advantageous wing geometries. Wing modification and functional differentiation between pterothoracic segments define major features of ordinal-level insect diversity.

2.2.1 Descriptive Anatomy

Insect wings are thin cuticular structures that arise ontogenetically from specialized patches of cells (in many taxa, the imaginal disks) on developing pterothoracic segments. Alary buds, or external structures of differentiated pre-wing tissue, are evident in larval stages of many exo- and endopterygote taxa. These winglike but nonarticulated structures are retained through consecutive larval instars (molts) until the final molt (for hemimetabolous insects) or emergence from the pupa (in holometabolous insects). At this point, the incipient wings within the alary buds expand into their final configuration. Expansion involves hydrostatic use of the hemolymph (insect blood) within the veins, implying involvement of the insect's circulatory system. Physiological processes endogenous to the wing may also be involved, as even isolated wing buds expand autonomously (see Glaser and Vincent, 1979). The wings remain soft and flexible during expansion, but dry out upon reaching their final geometry over a period of up to several hours.

A fully expanded wing consists of membranous regions of epidermal bilayers supported by veins. Extracellular cuticle layers expressed dorsally and ventrally from the epidermis determine the structural characteristics of the wing membrane per se. Wing veins are typically hollow and circular in cross section, providing a conduit for nerves and hemolymph. Wootton (1992) emphasizes, however, that there are a large number of exceptions to this characterization; veins and other functional equivalents (e.g., cuticular thickenings on the wing) are best interpreted in terms of their structural roles rather than morphological origins. Some veins contain no nerves and others contain no hemolymph. Venational cross sections range from circular to oval and campanulate (bell shaped). Some veins are flattened and consist merely of thickened regions of cuticle with no intervening lumen between dorsal and ventral layers, whereas others are annulate structures of high flexibility. Substantial phylogenetic variation is also evident in vein morphology. In Hemiptera (Heteroptera), for example, so-called channel veins consist simply of valley-like basins expressed either convexly or concavely on the wing surface (Betts, 1986a). These channel veins may or may not be delineated by ridgelike edges. Veinlike thickened structures of nontracheal origin may also serve to strengthen wings in diverse orders (e.g., Wootton and Betts, 1986).

Given such morphological diversity of veins, associated mechanical properties are likely to be highly variable. The flexibility of wings also relies on cuticular hydration of the wings; wings of desiccated insect specimens are notoriously fragile. The mechanical properties of veins and of wings generally are sustained in part by circulating hemolymph (Arnold, 1964; Wasserthal, 1982; Wootton, 1992). Anteriorly within the wing, veins are connected to the hemocoel of the body and are exposed to low positive pressure generated by the insect heart. Apical motion of hemolymph within the wings may also be facilitated by centrifugal pressures induced by wing flapping (see Larimer and Dudley, 1994). Posteriorly within the wing, circulation is maintained by action of accessory pulsatile organs, structures of the insect circulatory system that are located at the wing base of all winged insects (see Krenn and Pass, 1994, 1994/95). These muscular pumps create suction that pulls hemolymph basally within the posterior wing veins. Such action by the accessory pulsatile organs may be physiologically necessary if the wing cuticle is to remain adequately hydrated.

Patterns of wing venation are often highly complex and divergent even at the ordinal level; the phylogenetically most ancestral pattern of venation must remain speculative in the absence of modern cladistic analysis (see Kukalová-Peck, 1991). A general scheme for wing venation is that of Wootton (1979; see fig. 2.5A), who emphasized the descriptive character of the scheme without implying historical directionality. The major veins originate at the axillary apparatus, running distally and, in some cases, toward the trailing edge of the wing (fig. 2.5A). Ordinal-level differences in wing venation are largely defined by repeated bifurcations, anastomoses (merging of veins), and often loss of one or more of the major veins. Cross-venation and cuticular thickenings between major veins and associated minor branches are widespread and morphologically diverse (see Hamilton, 1972; Wootton, 1992).

Most anteriorly, the leading edge of the wing is defined by the costa and subcosta, thickened veins that provide structural rigidity (fig. 2.5A). The radius, a vein typically thicker than the costa and subcosta, extends and branches behind the subcosta, defining terminally the wing apex. The medial, cubitus, and anal veins project from the

(A)

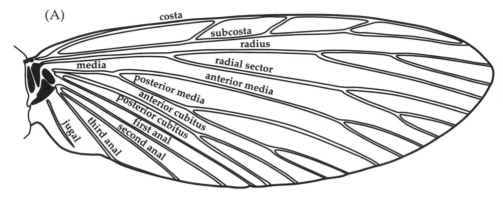

(B)

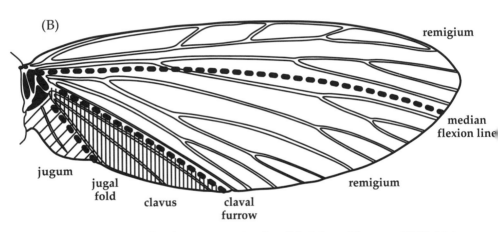

FIG. 2.5. (A) Generalized wing venation (modified from Wootton, 1979). Note that the medial vein bifurcates into the anterior media and the posterior media near the wing base, as does analogously the cubitus. (B) Major flexion lines and functional regions of the forewing.

wing base more posteriorly and largely define the trailing edge of the wing. The median flexion line represents a radial groove or region of increased flexibility along which the wing can deform and yield variable camber (see fig. 2.5B). The median flexion line generally runs anterior to the median vein, variably crossing secondary branches of the radial vein. A similar line of flexion, the claval furrow, lies radially between the cubitus and anal veins. Longitudinal bending about the claval furrow demonstrates functional partitioning between the claval region (or clavus) defined by the posterior anal veins, and the much larger anterior region termed the remigium (fig. 2.5B). A jugal fold separates the most posterior and basal region of the wing, the jugum, from the claval region. Some insect taxa, most notably the orthopterans, blattarians, mantodeans, and some plecopterans, exhibit basally expanded hindwings. This region is termed the vannus and an additional folding line, the vannal fold, then separates this region from the wing anteriorly. The jugal and vannal fold lines generally facilitate hindwing folding against the body.

Wing mass arises most immediately from the mass of wing venation; contributions of the membrane in nontegminized wings is negligible by comparison. The ratio of wing mass to body mass is highly variable in insects, ranging from 0.5–4% in typical dipterans and hymenopterans (Ellington, 1984b) to 3–10% in butterflies (Betts and Wootton, 1988; Dudley, 1990). Although relatively light, wings and particularly the longitudinal distribution of wing mass are of mechanical significance. Spanwise mass distribution determines the wing's moment of inertia and influences the potentially high inertial power expended during wing flapping (section 3.3.2). Mass tends thus to be concentrated near the wing base rather than at the wing tip. The center of wing mass in dipterans and hymenopterans typically lies at 0.3–0.4R, where R is the wing length (see Ellington, 1984b), but can be apically shifted as high as 0.44R in some butterflies (Dudley, 1990). Both span- and chordwise distributions of wing mass arise primarily from patterns in venational distribution. Wing mass, vein diameter, and particularly endocuticular thickness of veins (see Banerjee, 1988b) decrease from base to tip, indicating a gradient of wing stiffness. The concentration of veins at the wing base (fig. 2.5A) is primarily responsible for the greatest stiffness of the wing in this region. The spanwise stiffness gradient of wings is paralleled in chordwise transects, with the trailing edge of the wing being significantly more flexible than the leading edge. Chordwise mass distributions are much less studied than are spanwise mass distributions, but the greater number of larger veins at the leading edge of the wing (see fig. 2.5A) is primarily responsible for such variation in stiffness.

Parallel with venational patterns, wing shape (usually considered to be the two-dimensional projection of wing area, or planform) exhibits considerable geometrical diversity (plate 2). The total wing area S influences the magnitude of aerodynamic forces produced by the flapping wings (see eqs. 1.2 and 1.4); increased wing area yields increased forces in most cases of steady-state as well as unsteady flow. In nonaccelerating flight, the average pressure exerted on the surrounding air by the wings is given by the wing loading (p_w), the ratio of body weight ($m\mathbf{g}$) to total wing area. Wing area tends to increase with the square of linear body dimensions, whereas body mass is a general function of volume and increases with the cube of linear dimensions. The ratio of mass (or body weight) to wing area thus tends to increase linearly with body dimensions. Wing loading is therefore generally higher in insects of greater body mass, with consequent implications for wing aerodynamic pressures and airspeeds during flight (section 3.1.1). Also, insects of equivalent body mass can vary dramatically in wing loading because of differences in total wing area. For example, wing loadings in bumblebee workers are typically 15 N/m^2 (Dudley and Ellington, 1990a), whereas butterflies of comparable body mass exhibit wing loadings an order of magnitude lower (see Dudley, 1990). Wing loading is thus a general comparative parameter (much as is the Re) that is of greatest significance when used in conjunction with other measures of morphology.

A complete description of wing shape necessitates, in addition to knowledge of wing length R and wing area, two- and even three-dimensional analyses of wing geometry. The wing aspect ratio $\mathcal{R}$ (= $4R^2/S$) provides the simplest means of describing wing shape; a higher aspect ratio indicates a relatively more narrow wing. Aspect ratios of insect wings range from low values near 2 for a coupled wing pair of some butterflies (Dudley, 1990) to values of 10 or higher for certain odonate wings (see plate 2A). Aspect ratios of individual wings from otherwise coupled wing pairs (e.g., as in many Lepidoptera) can be near unity. The aspect ratio indicates the ratio of the wing length to the mean wing chord but does not address spanwise distribution of wing area (i.e., area could be concentrated basally or distally for two wings of equal aspect ratio). This parameter must therefore be used in conjunction with additional analyses of the distribution of wing area in both spanwise and chordwise dimensions.

For a flapping wing, the local velocity of any given wing section varies linearly with the radial distance from the wing base. More distal wing sections experience higher relative air velocities and thus generate greater aerodynamic force per unit area. The spanwise distributions of wing area have been well studied through the efforts of Ellington (1984b; see also Magnan, 1934; Weis-Fogh, 1972, 1973), who determined such distributions on diverse insect wings. The center of wing area (the first moment of distribution for the wing area) is typically located at 0.4–0.6R (e.g., Ellington, 1984b; Betts, 1986a; Ennos, 1989; Dudley, 1990). Although the mass of wing veins is high relative to that of the intervening membrane, distributions of wing mass and wing area are tightly coupled in many insects (see Ellington, 1984b). An additional morphological parameter derived from the wing area distribution is the virtual mass of the wing. As a wing section accelerates, a volume of air around the wing is simultaneously accelerated. The mass of this air volume, the virtual mass (v), is proportional to the chord length of the wing section and the air density (Ellington, 1984b). Wing area distributions determine the spanwise location of the mean wing chord and thus influence virtual mass distributions. The virtual wing mass can be comparable in magnitude to the wing mass itself at high wingbeat frequencies, and thus represents an additional component of wing inertia resisting acceleration by the flight muscle (see section 3.2.2.1).

Diverse ultrastructural features can be found on the surface of many wings (Wootton, 1992). Major veins may be supported by small membranous brackets projecting vertically from the wing surface. Sensory structures, particularly the mechanoreceptive campaniform sensillae (section 5.1.1), are found along vein surfaces. Various spines and trichiae (hairs) similarly relieve the smoothness of venation. Surface texture of some heteropteran hindwings resembles minute pointed cones, and is of no known functional significance (Betts, 1986a). Many insects, particularly Lepidoptera (Downey and Allyn, 1975) but also some Trichoptera (Huxley and Barnard, 1988) and to a lesser degree several other orders, possess flattened epidermal cells called wing scales. Most scales are utilized primarily in nonaerodynamic roles, particularly for coloration, but may subtly influence flow patterns and boundary layer structure over wings (see below). Finally, many insects possess discrete pigmented regions (pterostigmata) near the leading edge of the wing that are heavier than the wing membrane and the veins in the immediate vicinity of pigmentation. Pterostigmata are found on both fore- and hindwings in Odonata and Mecoptera, but also on the forewings of certain Homoptera, Hymenoptera, Neuroptera, and Psocoptera.

2.2.2 Functional Morphology

Wings are subjected to aerodynamic and inertial forces during flapping that induce deformation, bending, and torsion. Venational patterns correspondingly appear to reduce and control the extent of wing bending in both transverse and longitudinal planes (Wootton, 1992). Venational mass and vein density is highest at the base (see fig. 2.5A) where wing bending moments during flapping are greatest. This proximal concentration of wing mass also reduces the wing's moment of inertia and the associated energy required for angular acceleration (see section 3.3.2). The three-dimensional configuration of veins projecting from the membranous surface also serves a mechanical function. In a chordwise transect from leading to trailing edge of the wing, corrugation in the form of alternating furrows and ridges is evident as veins. Corrugation tends to be more pronounced toward the wing base and in anterior regions of the wing (Rees, 1975a, b). Veins alternately project dorsally and ventrally from the surface, yielding a pattern of convex and concave veins that transects from leading to trailing edge. The net result is a corrugated three-dimensional structure that is of higher flexural stiffness than a flat beam of equivalent mass (Rees, 1975b). Such corrugation, however, is effective, only if chordwise flattening and bending can be prevented by the presence of cross-veins and similar mechanical connections that run chordwise (see Newman and Wootton, 1986). Mechanically effective venation thus requires both spanwise and chordwise components. The presence of both components is most evident in wings with reticulate (meshlike) venation (e.g., wings of Odonata and Neuroptera).

Venational patterns may also mitigate damage during collisions of wings with their contralateral counterparts or with external objects. Typically, wings flex and yield in such collisions. For example, flattening of corrugations and chordwise bending characterizes impacts of dragonfly wings with enclosure walls (Newman and Wootton, 1986). Quantitative description of the particular strengthening afforded by vein geometry has recently been made possible through application of finite element analysis (FEM) to wing structures. This engineering method spatially decomposes three-dimensional structures into a large network of interacting elements, and enables predictions of wing deformation in response to applied force when elastic features of constituent elements are specified (e.g., Kesel, 1997). This method can be used not only to predict effects of particular veins and their geometry, but also to explore biomechanical consequences of venational designs not realized in nature. In addition to the veins, the wing membrane itself functions in mechanical stiffening, as evidenced by Wagner tension fields within cells of stressed dragonfly wings (Newman and Wootton, 1986). These fields are manifested as folds induced in the membrane that are parallel to the major axis of imposed tension. Such folding indicates force transmission through the wing membrane, demonstrating service in structural as well as aerodynamic roles.

The functional design of wings illustrates design based on structural hierarchy. The flapping wing itself is comprised of a venational network and associated membrane, whereas the particulars of vein geometry and cross-connections with other veins influence patterns of deformation on smaller spatial scales. This deformation in turn derives from the local stiffness of the constructional material, a function that ultimately emerges from molecular composition and organization. Relative to apterygote cuticle, a feature specific to pterygote wing veins is the presence of both helicoidal and parallel layers of cuticle (Neville, 1993). Within helicoidal layers, microfibrils exhibit continuous change in orientation across a cuticular transect, whereas parallel layers of cuticle are characterized by unidirectional orientation of chitin microfibrils. Such varying orientation of chitin fibrils in insect wing veins is attributed to the need to resist both bending and torsion of venation during flight (Neville, 1984; see also Banerjee, 1988a).

Such microstructural arrangements are probably widespread in wing design. The claval furrow of locusts, an axis of extensive bending during the upstroke, possesses larger chitin microfibrils than are found elsewhere on the wing, presumably in response to increased mechanical demands during the upstroke (Banerjee, 1988a). Biochemical features of cuticular construction may influence wing durability, as Jacobs (1985) showed that injection of a key cuticular amino acid (beta-alanine) into newly eclosed *Drosophila* increased the puncture resistance of wings. One of the less-appreciated points in flight mechanics is the number of cycles over which wings reciprocate in the course of an insect's lifetime. Mean forces comparable in magnitude to the body weight are applied with each cycle, and the high wingbeat frequencies of most insects ensure at least some alary degradation. If only 5% of a typical five-week life span of an adult *Drosophila* is utilized for flight at a wingbeat frequency of 200 Hz, then the wings undergo in excess of 30 million reciprocations. Interestingly, about 23 million wingbeats were obtained in a tethered simulation of long-duration flight using a single *Drosophila melanogaster* (Götz, 1987). Mechanical changes in wing structure under such intensive cycling are not known, but wing damage among insects, not surprisingly, does increase with age. The more flexible trailing margin is particularly susceptible to tearing. Flight performance of insects also declines with age, although this effect is matched in part by parallel deterioration of the flight muscle (see section 4.2.4).

In general, effective airfoils are moderately cambered structures that operate at low angles of attack relative to oncoming airflow (see section 3.2). The otherwise well-cambered design of most wings would superficially seem to be compromised by the extensive venational corrugation along chordwise transects. Air flowing around the wing can, however, be trapped in the venational folds, forming recirculating bubbles that alter the effective aerodynamic contour of the wing (see section 3.2.1.2). An irregular wing surface can thus be as functionally effective as is a smooth airfoil. If viewed end-on, wings from larger insects often exhibit a geometrical twist along the wing span, with the wing chord becoming more pitched ventrally toward the wing tip. Because the relative air velocity of flapping wings increases linearly with distance from the wing base, the direction of this vector also changes to become more coincident with the plane of the beating wings further along the wingspan. Intrinsic wing twist maintains an approximately constant chord orientation and angle of attack along the wingspan. Because lift and drag on wings can vary considerably with angle of attack (section 3.2.1.2), a constant wing chord relative to oncoming air results in advantageous aerodynamic characteristics along the length of the wing. In contrast to large wings, spanwise wing twisting is less pronounced in smaller wings that operate at fairly low Re, presumably because of a reduced orientational dependence of force production in more viscous flows (see section 3.2.1.2).

Wing corrugation, camber, and spanwise twist are not morphologically invariant but rather vary with inertial and aerodynamic loading. Wootton (1981b) has argued that wing structural features limit associated three-dimensional deformation and help to regulate aerodynamic shape; insect wings can thus act as adaptive airfoils. For example, camber is automatically induced in the vannus of the orthopteran hindwing as the downstroke proceeds (Wootton, 1995). The precise aerodynamic consequences of such camber induction are not known, but an increase in force production is likely, at least at higher Re. The corrugated design of wing venation appears predisposed to effect such shape changes without substantial bending. Using physical and theoretical models of wings, Ennos (1988a) showed that branching of corrugated spars from a thickened leading edge results in high resistance to bending but substantial torsional flexibility. Externally imposed aerodynamic and inertial forces may, via this mechanism, induce flexion about the wing base and facilitate advantageous changes in angular orientation of the wing.

The extent of such flexion depends on the torsional stiffness of the wing and the location of the wing's rotational axis relative to the point of an externally applied force. The rotational axis usually lies near the leading edge of the wing, whereas aerodynamic forces typically act behind both this axis and the chordwise center of wing mass. Aerodynamic forces are thus not coincident with the wing's rotational axis and tend not only to alter wing camber but also to twist the wing (Ennos, 1988a). If steady-state conditions apply, then aerodynamic torque is greatest in the middle of a half-stroke when the wing's relative velocity is highest. This torque will have the effect of inducing wing pronation during the downstroke, and provides for indirect aerodynamic regulation of wing orientation (Ennos, 1989b). Furthermore, a force on the concave side of a cambered plate will, in addition to bending the plate, increase camber if the point where the force is applied is not along the longitudinal axis of the structure (Ennos,

1995). Aerodynamic forces on flapping wings are imposed ventrally (on the concave wing surface) during the downstroke, and will thus enhance wing camber. This increased camber will at the same time resist further wing torsion and deformation, yielding an equilibrium (and perhaps aerodynamically advantageous) profile for a deformable wing.

By contrast, forces applied to the convex surface of a cambered wing can enhance both longitudinal bending and wing twisting (Ennos, 1995). Such forces are applied aerodynamically through the upstroke and during supinatory rotation. Wings should therefore exhibit more flexion during supination than during pronation, as has been confirmed by static bending tests on butterfly wings (Wootton, 1993). Both camber and angular orientation of flapping wings can thus be influenced by the force and direction of the instantaneous aerodynamic force vector. This method of aerodynamic control arises passively from the interaction of airflow with wing geometry, and is distinct from active muscular torques applied at the wing base. Not all insect wings conform to this model, however. The torsional resistance of dragonfly wings is higher than stiffness in bending (Zeng, Matsumoto, et al., 1996), and passively induced aerodynamic changes in wing profile are not likely to be pronounced. Wing profile in odonates is instead influenced by the action of a small muscle (the fulcroalar muscle) that is integral to the axillary complex at the wing base. This muscle acts as the base of the cubitus vein and is used to control wing profile (Newman and Wootton, 1988) in addition to the changes in wing contour caused by the direct downstroke muscles.

Whereas steady-state aerodynamic forces predominate in the middle of a half-stroke, inertial forces are greatest at the ends of half-strokes when the wing mass and virtual mass first decelerate and then reaccelerate in the opposite direction. Transverse inertial bending of wings is most likely at either end of the wingbeat when first angular deceleration and then acceleration are highest. At the ventral extremes of wingbeats, transverse bending associated with inertial deceleration is often pronounced and may be facilitated by a transverse flexion line. Flexion in the posterior region of the wing and bending at the wing tip may mitigate inertial bending at the wing base. In dipterans and hymenopterans, the extent of bending at the wing tip is generally less than 5% of total wing length (e.g., Ellington, 1984c), although other taxa (e.g., *Panorpa*) often exhibit a much more pronounced tip deflection (see Dalton, 1975; Brackenbury, 1992). Substantial spanwise deflection is uncommon at the end of the upstroke, at which point the positive camber and corrugated venation of the wing resist ventro-dorsal bending.

Not only wing bending but also twisting and rotation may be induced by wing inertial forces at the ends of each half-stroke. Norberg (1972c) first noted that the chordwise center of wing mass typically lies behind the rotational axis of the wing, and that wing deceleration at the end of half-strokes would tend to swing the wing mass chordwise about the longitudinal axis. Inertially induced rotation of different spanwise wing regions will vary according to the local section mass and to the distance between the rotational and inertial axes. Such rotation will be most pronounced for those regions closest to the torsional axis of the wing (e.g., the distal regions of fly wings; Ennos, 1988b). To alleviate potentially excessive inertial torques, the center of wing mass must be located more anteriorly and closer to the wing's rotational axis. This requirement is met by a concentrated mass along the leading edge of the wing that is located distally near the wing tip; the pterostigmata of wings from diverse taxa is of sufficient mass and in the appropriate location to passively regulate wing orientation (Norberg, 1972c). The relative size of hymenopteran pterostigmata increases with decreased body size, possibly in response to the greater torsional inertia associated with the higher wingbeat frequencies that come with small body size (Danforth, 1989; see also chapter 3). In beetles with folding hindwing apices, the pterostigma at the base of the fold may inertially augment wing supination (Brackenbury, 1994b). Many insect taxa have no obvious pterostigmata. For all winged insects, however, location of the center of wing mass behind the wing's rotational axis indicates that inertia will promote supination at the end of the downstroke.

Similarly, wing inertia will advantageously induce pronation at the end of the upstroke, although the location of the relevant mechanical axes is more difficult to specify for a wing already variably twisted about its base in supination. Wing orientation through the wingbeat and particularly timing and extent of rotation are also influenced by action of the pterothoracic musculature. Inertial regulation of wing orientation is, however, probably more pronounced in those insects with high wingbeat frequencies. Ennos (1988b, 1989b) suggested that wing inertia at stroke reversal was sufficient by itself to generate the rotations observed in flies. These calculations of wing inertia, however, involved formulations for the virtual wing mass that may not adequately incorporate the complex accelerations associated with rotating wings (see section 3.2.2).

Forward flapping flight necessitates force production that is asymmetric between down- and upstroke (section 3.1.2), and a variety of kinematic strategies are employed toward this goal. Direct modifications of wing geometry include changes in the angle of attack (i.e., wing twisting) and variation in wing camber and profile. Differences in wing orientation between the down- and upstroke are particularly marked in the flight of taxa with only one actual or functional wing pair (e.g., Diptera, Hymenoptera). One indirect consequence of wing supination is the reduction of the effective wing surface area during the upstroke. Twisting and reversal of wing orientation indicate that basal regions of the wing will be less effective aerodynamically during the upstroke. In hindwings, this effect can be enhanced by the presence of the vannus, which mechanically folds during the upstroke to reduce the total effective wing area (Wootton, 1992). Similarly, the sense of wing camber (i.e., convex or concave) often reverses from down- to upstroke. Bending along flexible cross-veins as well as along the median flexion line facilitates camber reversal at the transition between half-strokes. Because of the dorsoventral asymmetry in the design of biological wings and their basal connection, no animal wing is fully reversible between half-strokes in terms either of camber or effective surface area. As a consequence, the upstroke is inevitably less effective aerodynamically than the downstroke.

Variation in wing profile between half-strokes is largely effected via the radial flexion lines, most importantly the claval furrow and the median flexion line (Rohdendorf, 1958/59b; Wootton, 1979; Grodnitsky and Morozov, 1994; see fig. 2.5B). Radial bending along these lines at stroke reversal generates a "Z"-shaped cross-sectional profile during the upstroke, first described in the forewing of the locust (Jensen, 1956; see also Pfau, 1977, 1978; Nachtigall, 1981a) but characteristic of many other insects (Wootton, 1992). Note, however, that a diverse cross section of profiles is subsumed within the concept of a "Z" shape, and upstroke profiles likely vary according to spanwise location and the particular force output required under different flight conditions. For example, claval flexion can be extensive during wing supination in hovering hymenopterans (see Brackenbury, 1994a) but decreases at higher airspeeds as rotational angles decrease. Aerodynamic characteristics of the "Z"-shaped upstroke profile have been investigated only in locusts (Jensen, 1956; Nachtigall, 1981a,b), but in general such flexion and associated profile change render the wing less effective in lift production. Bending along the median flexion line may also result in advantageous vortex shedding from the wing in an unsteady flex mechanism (Ellington, 1984d; see section 3.2.2.3). Evolutionary change in patterns of upstroke flexion may be associated with enhanced maneuverability and flight performance. In a mecopteran, for example, the wing base assumes a "Z"-shaped profile through a longitudinal flexion about the claval furrow (Ennos and Wootton, 1989). By contrast, the more agile dipterans (the sister taxon of Mecoptera; Wootton and Ennos, 1989; see fig. 1.1) extensively supinate the wings and reverse wing camber in the upstroke, exhibiting neither claval flexion nor an irregular wing profile.

In contrast to forward flight, hovering demands a nearly complete reversal of wing camber and orientation if consecutive half-strokes are to be aerodynamically symmetric with no net thrust production. The magnitude of torsional reversal is constrained by the size of the wing articulation relative to the wing chord. Particularly for insects with linked fore- and hindwings, reversal of wing orientation between half-strokes is structurally impeded, as exemplified by many butterflies. Not surprisingly, hovering is usually limited to insects with highly flexible wings that effectively reverse camber and exhibit neither vannal folding of hindwings nor a pronounced claval region. Even so, reduced aerodynamic output in the upstroke is likely. In support of this hypothesis, wing elevator muscles tend to be smaller than depressor muscles in insects, although systematic data relating to this observation have not been collected in insects. It is noteworthy, however, that even in hummingbirds the main elevator muscle (the supracoracoideus) is approximately one-third the mass of depressor muscles (Greenewalt, 1962) in spite of the approximately symmetric wing motions and near-perfect reversal of wing camber and orientation exhibited by these birds during hovering (see section 7.5.2).

When at rest, paleopterous insects hold their wings either laterally outspread (Odonata: suborder Anisoptera) or obliquely above the body (Odonata: suborders Anisozygoptera and Zygoptera, Ephemeroptera). Neopterous insects at rest typically fold their wings on or over the abdomen, in some cases utilizing radial plication in the jugal and vannal regions (i.e., fanwise folding of the membrane along radial lines) to reduce exposed wing surface area against the body. The expanded vannus of the orthopteran hindwing provides the best example of such plication (Wootton, 1995; see fig. 2.6B), and such folding strategies are generally characteristic of orders with tegminized forewings. Radial pleating is also used by some Hymenoptera to fold the forewings when the insect is at rest (Danforth and Michener, 1988). The longitudinal wing span is unaffected by such folding, and in many neopterous insects the wings correspondingly lie above and extend substantially beyond the abdomen.

In some orders, however, transverse folding is used to dramatically reduce wing length. Elytrization in part demands such transverse folding if the hindwings are to be effectively concealed beneath the elytra. The Coleoptera, for example, have elaborate means for both radial and transverse wing folding that position the hindwings beneath the smaller elytral forewings (fig. 2.6E). Similar transverse folding characterizes the Blattaria and Dermaptera (Scudder, 1876; Kleinow, 1966; Haas and Wootton, 1996). Because wings lack the intrinsic musculature necessary to effect such transverse bending, indirect forces must instead be used to carry out such unfolding and subsequent refolding. The well-defined creases in the wing membrane act together with the inherent elasticity of the wing to move the wing tip outward when suitable torsion is applied at the wing base (Haas and Wootton, 1996). Some fully unfolded wings remain stably unfolded without active maintenance of tension, whereas others are unstable and must remain forced open. Some of the necessary tension in this latter mode may be provided by the aerodynamic and inertial forces of wing flapping (Haas and Wootton, 1996). Thus, elytrization of forewings is necessarily associated with evolution of sophisticated structural mechanisms in the hindwing, as exemplified by elaborate folding mechanisms in the beetles (see section 8.3.2).

One apparently unique structural modification for wing folding is found within the pterophorid (plume) moths, many of which, when at rest, roll ipsilateral fore- and hindwings together into a tubular structure of small diameter (Wasserthal, 1974). This peculiar manner of wing folding about the longitudinal axis may have evolved to escape detection by visually oriented predators. Interestingly, some pterophorid species as well as the unrelated orneodid moths exhibit highly lobed fore- and hindwings that often are fringed with fine hairs. No kinematic adaptations particular to the extraordinary wing structure of pterophorids are evident, and adjacent wing lobes appear to form a continuous surface (Norberg, 1972b). Wing fringing is also characteristic of many small dipterans, coleopterans, and hymenopterans (as well as thrips and microlepidopterans), and has the effect of creating an aerodynamically continuous wing (see section 7.1.3).

The aerodynamic implications of ornamentation on wing structure are largely uninvestigated. A tremendous diversity of scales, spines, protrusions, and varied surface textures can be found on wings of different taxa. Whatever their ultimate function, most of these structures may simply yield viscous drag on the wing surface. Some microstructural features may, however, subtly interact with a turbulent boundary layer to reduce drag. Lift characteristics may also be affected, and an unresolved issue in this field concerns the potentially advantageous aerodynamic presence of scales on wings of butterflies and other Lepidoptera (see section 3.2.1.2). The presence of scales is clearly not required to effect flight in Lepidoptera, as many moths (e.g., various Sesiidae; Kristensen, 1974) and some butterflies (e.g., the clearwing ithomiine genera, the satyrine genera *Cithaerias*, *Dulcedo*, and *Haetera*) exhibit only minimal scalation along the wing margins. Longitudinally striated wing scales can also be found in various Megaloptera, Psocoptera, Trichoptera (the sister taxon to Lepidoptera), and even in some Coleoptera and Diptera. Scales are also abundant on the bodies of both Trichoptera and Lepidoptera, but their aerodynamic significance, if any, is unknown.

2.2.3 Phylogenetic Diversity

The phylogenetically ancestral pterygote condition is usually assumed to have been meso- and metathoracic homonomous wings of approximately equivalent size, shape, and function. Using the present-day distribution of pterothoracic character states, however, the most parsimonious reconstruction of the ancestral wing configuration shows it to be anteromotoric i.e., flight derives predominantly from action of the forewings (fig. 2.4). Homonomy evidently characterized a number of Paleozoic orders (e.g., Paleodictyoptera, Protodonata), but homonomy of pterothoracic segments and associated bimotorism (equivalent action of fore- and hindwings) is infrequent among extant insects. Instead, modification of one pterothoracic segment results in either antero- or posteromotorism, and in some cases a transformation from tetraptery (expression of four wings) to either functional or actual diptery (table 2.1). Any presumed ancestral condition of homonomy has clearly been modified many times during the evolution of winged insects.

Antero- and posteromotorism occur at approximately equal frequencies among the insect orders, although posteromotorism is (with the exception of the Coleoptera and Strepsiptera) confined to the exopterygotes (fig. 2.4). In most extant insect orders, one wing pair is accordingly either reduced in relative size or is transformed morphologically to serve in various nonaerodynamic roles. Evolution of alary heteronomy (wings of different function) is thus a general trend in insect biology. Such transformations are associated with changes in the relative size and shape of ipsilateral wings if both wing pairs retain aerodynamic function. Brodsky (1994) noted that posteromotorism tends to be associated with broadening of the hindwings at the base, whereas those insects characterized by bimotorism tend to have wings with elliptical planform. Anteromotoric insects, by contrast, often display hindwing reduction (see below). Although a dominant theme in pterygote evolution, such trends in wing shape and size have not been rigorously quantified, nor have patterns of intraordinal wing heteronomy been systematically evaluated.

The most common such transformation is the use of a strengthened mesothoracic wing pair in defensive actions. Forewings of most insects tend to be slightly thickened relative to the hindwings (Rohdendorf, 1949), but this effect is most pronounced in tegminization and elytrization, and typically involves a reduction in area but an increase in cuticular strength and rigidity of the forewings. Tegmina (characteristic of the Orthoptera but also found in other orders; see table 2.1) tend to be thickened leathery structures with a reduced aerodynamic role relative to the hindwings (plate 2C, D). Elytra of Coleoptera are an autapomorphy (a unique derived character) of the taxon; elytra are generally much tougher than tegmina and in many cases are rigid (plate 2E). Wing venation tends to be reduced or essentially eliminated from both tegmina and elytra. Elytrization of the forewings pair is also associated with venational modifications of the metathoracic pair to enable folding either radially or transversely beneath the protective elytra.

Posteromotorism and locomotor specialization of the hindwings is a necessary consequence of such mesothoracic specialization (table 2.1). In the Coleoptera, the elytral forewings play a small aerodynamic role whereas the hindwings are the major force producers (see section 3.2.1.2). Elytra are typically well sclerotized and function primarily as sheathlike devices to protect the hindwings and abdomen. When the insect is at rest, the paired elytra meet dorsally at the midline and typically lock together at an elytral suture or fit into notal grooves. A similar modification characterizes some Mecoptera (Hlavac, 1974) as well as various Dermaptera and Hemiptera. Such linkage contributes to the overall mechanical rigidity of the pterothoracic complex when the insect is not in flight. Ventral surfaces of most beetles are also heavily sclerotized, rendering the entire body resistant to crushing. Few data exist on the mechanical protection afforded by elytra, although Hough-Goldstein et al. (1993) found that a chrysomelid beetle with these structures removed was less likely to survive pecking by domestic fowl (see also Thiele, 1977). Variation in microstructural design undoubtedly contributes to the wide range of elytral rigidity and toughness evident in different beetle taxa. Krzelj (1969) found that elytral flexibility varied primarily with thickness and relative size of the exo-to endocuticular layers, although variation in material stiffness (i.e., Young's modulus) may also play a role (Krzelj and Jeuniaux, 1968; see also Hepburn and Ball, 1973). The elytral conversion of mesothoracic beetle wings has been correlated with an anterior shift of the hindwing base, perhaps to compensate for reduction in torque about the center of body mass when the forewings no longer have a major aerodynamic function (Crowson, 1981). Also, the demands of pterothoracic cuticularization probably operate at crosspurposes to thoracic deformations induced by indirect flight muscles. The indirect dorsolongitudinal muscles are correspondingly reduced in many beetle families, and direct basalar muscles are the primary means of effecting wing depression (Larsen, 1966).

Elytriform wings of other orders also serve nonaerodynamic purposes. The hemelytra of Hemiptera (Heteroptera) represent intermediate conversion of the forewing from an aerodynamic to a protective structure. Interestingly, the membranous and coriaceous (leathery) regions of the hemelytron do not merge smoothly into one another but rather meet at a distinct junction (see plate 2B). Elytrization of hemipteran forewings probably represents a compromise between mechanical protection and aerodynamic force production, as hemelytra are linked in flight to the hindwings and remain aerodynamically functional (Puchkova, 1971; Betts, 1986a; Wootton and Betts, 1986; Wootton, 1996). In many heteropteran lineages with scutella that extend the full length of the abdomen (e.g., Scutelleridae, Thyreocoridae, some Pentatomidae), the hemelytra are essentially elytral in character and interface with the scutellum to result in a nearly continuous dorsal protective surface. Also contributing to mechanical protection are the partially or totally sclerotized forewings of Blattaria, Dermaptera, Orthoptera, Phasmatodea, and some Homoptera. Forewings are also slightly coriaceous in many Neuroptera, Mecoptera, and Megaloptera. In tetrigid orthopterans, the tegmina are vestigial but the pronotum (the notum of the prothoracic segment) extends posteriorly to cover the hindwings as well as much of the abdomen. The extinct orders Glosselytrodea and Protelytroptera were also characterized by robust elytra (Carpenter, 1992).

In addition to elytrization and tegminization, morphological strategies have evolved that enhance cuticular defense of the thorax and abdomen, most notably the various posterior extensions of the metathorax in the membracid homopterans, and a remarkable shieldlike projection of the scutellum in the dipteran family Celyphidae (the beetle-flies). Use of such static morphological defense against predators also has behavioral correlates. Many insects with tegmina or elytra are behaviorally more reluctant to fly than are other insects, or they tend to use wing flapping only as an adjunct to a jump (e.g., the saltatorial, or jumping, Orthoptera). Flight may in fact be energetically more costly in taxa with thickened forewings. The added cuticular weight of elytriform wings and associated body sclerotization (cuticular hardening) may increase substantially the mechanical power expended during flight, but the general theme of pterothoracic modification for defense has nonetheless been widespread in insect evolution (table 2.1).

Anteromotorism (i.e., hindwing reduction) represents the second form of pterothoracic differentiation. Flight using only one wing pair is true diptery, and at the ordinal level is characteristic only of the

Diptera and Strepsiptera (see table 2.1; plate 2G). More typically, the hindwing is reduced in size and is mechanically coupled to the forewing, a functional form of diptery (Chadwick, 1940; Grodnitsky, 1995) first evident historically in the Permian paleodictyopteran taxon Permothemistida (Brodsky, 1981). A variety of insect orders display functional diptery (see Table 2.1), as exemplified by the linked wings of contemporary Hymenoptera (Gauld and Bolton, 1988; see plate 2F). Quantitative data on relative masses and areas of hymenopteran wings are unfortunately lacking. Some Ephemeroptera (e.g., the Caenidae) exhibit extreme hindwing reduction (although the fore- and hindwings remain uncoupled), and in some cases the hindwings are completely lost. An extreme case of hindwing reduction is miniaturization of the wing pair, as in the dipteran halteres used for stabilization (section 5.1.1). The general trend in anteromotorism, however, tends to be overlapping or physical coupling of the relatively smaller hindwing to the forewing to yield one continuous aerodynamic surface (e.g., Lepidoptera).

Conversely, some insect orders demonstrate an increase in hindwing dimensions relative to the forewing, although this trend is numerically less significant than is relative expansion of the forewing (table 2.1). The phylogenetically basal order Odonata exhibits limited expansion of the hindwings within the suborder Anisoptera. Hindwings are expanded relative to forewings in the Orthoptera (see plate 2B) and the related Phasmida, in Blattaria and its sister taxon Mantodea, in some Plecoptera, and in the mastotermitid termites (Wootton, 1992). In general, the expanded basal region of such hindwings precludes substantial rotation about the longitudinal wing axis, and correspondingly prevents a reversal of wing orientation between down-and upstroke. Such insects are generally unable to hover and display only limited maneuverability in the air.

Diversity of wing coupling mechanisms parallels phylogenetic pathways in differentiation of pterothoracic segments. Nachtigall (1974a) discusses the general physical principles behind biological connectors and illustrates basic mechanisms for wings. If wings overlap and operate in phase (but with no specific mechanical connection), the coupling is termed amplexiform. Amplexiform wings are characteristic of mayflies (Ephemeroptera) and butterflies (Bourgogne, 1951). Moths typically possess a frenulum (set of fused setae or hairs) projecting from the hindwing that terminates in a small hook on the forewing, joining the two wings (Tillyard, 1918). Some moth taxa alternatively have a cuticular lobe on the forewing (the jugum) that hooks under the hindwing; these mechanisms of wing coupling render the majority of moths functionally dipterous. Some phylogenetically derived moths, particularly arctiids, sesiids, and sphingids, have evolved high wing-beat frequencies and hindwings that are much reduced in relative area. However, phylogenetically basal moths have an ineffective frenulum and are functionally tetrapterous with fore- and hindwings of comparable size (Grodnitsky and Kozlov, 1985). Mechanical coupling between wings thus appears to be correlated with relative wing reduction.

In many butterflies, wingbeat frequencies and concomitant inertial forces are low enough to obviate the need for extensive mechanical connections between the large and similarly sized fore- and hindwings (Grodnitsky and Kozlov, 1985, 1990, 1991). Similarly, low wingbeat frequencies and reduced inertial forces characterize such orders as Ephemeroptera, Mecoptera, Psocoptera, Raphidioptera, and Trichoptera, all of which have simple hooklike mechanisms connecting fore- and hindwings (Edmunds and Traver, 1954; Ivanov, 1985, 1990; Lawson and Chu, 1974; New, 1974). Fore- and hindwings of auchenorrhynchous Homoptera are connected by a claval fold on the forewing that links with a smaller fold, lobe, or hook (New, 1974; D'Urso and Ippolito, 1994). The claval fold and the hindwing do not lock together but rather permit the two wings to slide relative to each other both during flapping and when wings are being deployed for flight (Ossiannilsson, 1950). Mechanically constrained sliding between fore- and hindwing is evidently unnecessary given the reduced extent of relative wing motion that occurs during flight. When such a wing coupling device is absent, the taxon in question tends to be either brachypterous (with reduced wings) or saltatorial (D'Urso and Ippolito, 1994). New (1974) described additional projections near the pterostigma of the homopteran forewing that are used to lock fore- and hindwings when the insect is at rest.

By contrast, tetrapterous insect taxa with high wingbeat frequencies tend to have sophisticated wing connectors that link fore- and hindwings either tightly at one point or across a finite length of the contiguous wings (see Schneider and Schill, 1978; D'Urso, 1993; Brodsky, 1994). Hymenopteran wings are tightly linked by a line of small hooks, the hamuli, that vary in location along the anterior edge of the hindwing (Basibuyuk and Quicke, 1997). Similar structures characterize wing coupling mechanisms in the true Hemiptera (Heteroptera). The sternorrhynchous Homoptera (psyllids, aleyrodids, aphids and coccids) either possess hooks that couple fore- and hindwing, or alternatively exhibit hindwing reduction. Such tight connections between fore- and hindwing, together with substantial hindwing reduction, are almost certainly necessitated by the high wingbeat frequencies found in these taxa (see section 4.2.2).

Insect wings are used in a variety of functions supplemental to their principle role of force production during flight. Various hymenopterans use wing motions to create convection currents used in thermoregulation (see Neuhaus and Wohlgemuth, 1960; Herbst and Freund, 1962; Wohlgemuth, 1962; Stern and Dudley, 1991). The approximately planar structure of most nonelytral wings presents a convenient location upon which to display both two-dimensional patterns as well as coloration that may function in either natural or sexual selection. Such patterned wings appeared early in pterygote evolution and are well developed in some taxa by the Upper Carboniferous (Carpenter, 1971). Many extant insects are cryptic by virtue of pattern matching between the dorsal wing surface and the natural background. Cryptic patterns may be inherent to the cuticle itself or may be exogenous in character—certain weevil species (Curculionidae: Coleoptera) in New Guinea carry moss gardens on their elytra that apparently function in camouflage (Gressitt et al., 1968). Contrariwise, aposematic insects can use wings to display warning coloration. Deimatic (startle) displays on hindwings are also common among insects and are presumably used to frighten would-be predators (Edmunds, 1974). Parallel with natural selection, sexual selection is a force of comparable significance for the evolution of wing coloration. Intraspecific communication via wing patterns is particularly widespread among the butterflies (Silberglied, 1984; plate 2II), but can also be found in insect taxa as disparate as dragonflies, drosophilid flies, and euglossine bees (Thornhill and Alcock, 1983). Because wing appearance and aerodynamic function can essentially be decoupled, the superficial alary morphology is particularly well suited to modification for use in visual communication to predators and conspecifics.

Wings are also used by some insects to generate acoustic signals. The sounds of wing vibration, for example, are used for communication in honeybees and in drosophilid flies (Bennet-Clark and Ewing, 1968; Ewing, 1979). Beetles are particularly well known for sound production via stridulation between the elytra and the abdomen or hindwings (Dumortier, 1963). Among the Orthoptera, leg motion against the wings or stridulation of overlapping opposite wings are among the mechanisms used to produce the diverse sounds characteristic of this order. It is noteworthy that sound production in both the Coleoptera and Orthoptera is associated with forewing thickening (elytrization and tegminization, respectively). Sclerotization of the forewings in evolutionary time may thus interact synergistically with the evolution of sound production (see also Toms, 1986; Desutter-Grandcolas, 1995). Interestingly, stridulation between opposite wings has apparently evolved at least once in the Diptera (Petrunkevitch, 1956), whereas clicking of the forewing against the pronotum to produce sound has been recorded in an Old World cicada (Popov, 1981). In a noncommunicational context, collection of pollination by bees grasping flowers often involves high-frequency but low-amplitude wing motions that dislodge pollen from the anthers by the production of high-frequency sound (Buchmann, 1983; King et al., 1996). Thus, a variety of insect taxa use wing vibration to produce sound, but this behavior usually occurs only when the insect is at rest.

Intentional sound production during flight is much less common among insects. In the Neotropical butterfly genus *Hamadryas*, dorsal impact of opposite wings during flight generates pulses of audible clicking sounds that are used in territorial interactions (Otero, 1990; Monge-Nájera and Hernández, 1991; Monge-Nájera, 1992). Percussive use of wings during flight has also been described in agaristid (Bailey, 1982; Alcock et al., 1989) and noctuid moths (Dumortier, 1963; see also Hampson, 1892; McCrae, 1975). Sound production through opposite wing impact during flight may occur in some satyrid butterflies (Kane, 1982) and in some temperate-zone grasshoppers both during escape from predators and during apparent advertisement to conspecifics (pers. obs.). Finally, a novel use of wings has been described in the African cricket *Phaeophilacris spectrum*. Using wing motions, these cave crickets generate vortices that subsequently travel to conspecifics and perhaps communicate information (Heinzel and Dambach, 1987; see also Heidelbach et al., 1991). Such behavioral use of wing-generated vortices is perhaps unique in the Insecta.

Loss of wings is widespread on evolutionary (section 6.2), ecological (section 7.4), and behavioral timescales. Alary polymorphisms and facultative wing expression range from fully winged individuals to brachypterous and apterous morphs. Production of different wing morphs depends on local ecological circumstances and is usually correlated with a behavioral tendency to migrate. Some insects can also autotomize (voluntarily break off) their wings. Winged female reproductives of ants and termites similarly undergo dealation following mating flights, and at least one dipteran species (*Lipotena*) engages in this drastic behavior. Zorapterans lose their wings at sexual maturity, as do various Thysanoptera and even some Blattaria.

2.3 ANCILLARY STRUCTURES

Sensory functions such as vision and mechanoreception act prominently in the control of flight. The morphology and physiology of flight-related sensory structures (including eyes, antennae, and various mechanoreceptors) are discussed in chapter 5. More directly, insect legs and the insect abdomen play various minor but nonetheless significant roles in flight. The most direct contribution of insect legs is to enable the obligatory jumping takeoff subsequent to which the tarsal reflex (absence of contact with ground) initiates wing flapping (see section 5.2.1). Jumping may also have been an evolutionary prerequisite to the initial evolution of insect flight (section 6.1.4). In some saltatorial taxa (e.g., alticine chrysomelids, various homopterans, orthopterans), legs are hypertrophied and flight becomes an adjunct to locomotion by jumping, although wings are usually retained. Within the Diptera, Rohdendorf (1958/59a,c) has correlated taxonomic patterns of wing venation with leg structure and cursorial performance. In general, a phylogenetic trend toward heavier and more muscular legs was observed in the more derived families and superfamilies; this correlation presumably indicates more rapid jumping and takeoff abilities in the aerodynamically more agile flies. Insect legs are typically retracted during steady flight to reduce drag forces on the body. In some cases, laterally asymmetric leg movements during flight may contribute to maneuverability (section 5.2.2).

Aerodynamic drag of the insect abdomen and associated expenditure of power exert a major influence on performance during flight (section 3.3.2). Little work has been done on the nature of abdominal design toward reduced drag, but effective streamlining of the abdominal and the overall body profile is likely (see section 3.2.1). Abdominal cerci may ancestrally have contributed to stability in pitch and yaw (section 6.1.3), whereas abdominal motions are used by some insects to generate advantageous aerodynamic torques during maneuvers (section 5.2.2). Contrariwise, disadvantageous aerodynamic torques may ensue if the abdomen and the longitudinal body axis are not aligned parallel to the flight path. For insects with relatively short abdomina (e.g., Diptera and Hymenoptera), these torques are likely to be small in magnitude. By contrast, aerodynamic torque on the elongate abdomen of Odonata (required perhaps by the secondary genital pores of males and associated aerial copulatory habits) may preclude effective use of yawing and pitching movements during flight. Odonates therefore typically initiate maneuvers by rolling about the longitudinal body axis. In general, the location of the center of body mass must influence rotational maneuverability during flight (section 5.2.2). The abdomen may not strictly be necessary for flight in dragonflies: one living odonate specimen that had lost its abdomen escaped from its captors and flew off into the forest on Barro Colorado Island (Zotz and Dudley, pers. obs.)! Finally, differential allocation of mass to reproductive organs in the abdomen may account for substantial differences in wing morphology and flight performance between male and female insects, although such sexual dimorphism has not been systematically investigated from the perspective of flight biomechanics. Egg loads of gravid females, endogenous lipid reserves of migrants, and food stored in the crop can dramatically increase the power requirements of flight (section 3.3.3). A secondary reduction in maneuverability and an increased vulnerability to predation may also ensue.

2.4 SUMMARY

The pterothorax of insects consists of flight muscles, cuticular sclerites, a complex axillary articulation, and wings. Induction of wing movements through cuticular transmission of muscular force applied to the wing base is widespread among insects. Both direct and indirect muscles contribute to dorsoventral wing motions as well as to orientational control of the wing. Wings are cambered surfaces supported by networks of veins that act to control and limit deformations induced by aerodynamic and inertial forces. Wings can also function as adaptive airfoils that deform advantageously in response to applied forces (e.g., alteration or reversal of camber). Thoracic configuration is broadly similar in all insects, but wing modification and reduction in relative size have substantially altered the homonomous condition of equivalent wing size and function. Dominant evolutionary themes in pterothoracic modification include tegminization, elytrization, and/or relative reduction in size of one wing pair. Bimotorism is rare among extant winged insects relative to antero- and posteromotorism. Co-option of wings for secondary or supplemental nonaerodynamic purposes (e.g., elytra, halteres) largely defines patterns of interordinal insect diversity. Body shape, legs, and various sensory structures, although not directly contributing to flight aerodynamics, exert secondary influences on aerial performance.

Chapter Three

KINEMATICS AND AERODYNAMICS
OF FLIGHT

E FFECTIVE FLIGHT in insects cannot emerge solely from the morphological expression of wings. Instead, flight in animals derives from rhythmic wingbeat motions that generate aerodynamic forces. Vertical forces are necessary to support the body weight and to control altitude; a net thrust is used to effect forward propulsion. Subtle alterations to the motions of flapping wings transiently disrupt this balance and permit flying insects to alter flight speed and trajectory. Flight is energetically costly, and associated expenditure of mechanical power is influenced by wing and body morphology, wingbeat kinematics, and patterns of body movement in three-dimensional space. Morphological diversity of winged insects is matched by a corresponding kinematic diversity in wing and body motions. Although less well described than trends of morphological diversification, these motions largely determine the temporal patterns of force production and aerodynamic power expenditure during flight.

3.1 Wing and Body Motions

Fundamental to a biomechanical analysis of flight performance is the description of wing and body kinematics. Insect flight speeds are most generally referenced to the surrounding air volume (the airspeed of a flying insect), but can also refer to the speed relative to an external coordinate system fixed in the earth (the so-called groundspeed). Airspeed and groundspeed are not equivalent because of the effects of ambient winds. Of all wing and body motions, speed of the wings with respect to the surrounding air is the most important aerodynamically and reflects the combined influence of the body's velocity and the motions of the wings relative to the body. The latter kinematics of the wingbeat, including wingbeat frequency, stroke amplitude, and wing orientation, directly influence the magnitude of and temporal variation in aerodynamic forces and mechanical power expenditure. Methodology specific to the kinematic analysis of flying insects was reviewed by Dudley (1992).

3.1.1 Speed of Flight

Although one of the most fundamental of kinematic variables, insect airspeeds are also one of the least known features of flight performance. This unfortunate situation arises because flight velocities in nature reflect the vector combination of the ambient wind velocity and the insect's velocity with respect to the surrounding air (see fig. 3.1). Insects fly through air that is also potentially moving relative to the earth itself (e.g., wind). Groundspeed measurements thus combine insect air velocity with the motions of the air itself relative to the terrestrial coordinate system, and these latter motions are nontrivial with respect to insect flight speeds. Typical diurnal values for ambient wind speeds both within and above vegetational cover range from 0.1 to 10 m/s, and are often higher (Rumney, 1968; see section 7.4.1). Wind speeds can substantially exceed insect airspeeds and are, moreover, difficult if not impossible to measure in the immediate vicinity of flying insects.

One approach to dealing with the effects of ambient wind motions is simply to study flight indoors. Flight speeds within large wind-free enclosures or rooms, for example, can provide at least minimum estimates for insect airspeeds in forward flight. Demoll (1918), for example, used a stopwatch to time various insects in their flight across a room to a brightly lit window. A hawkmoth attained the highest speed (15 m/s) and a tabanid fly and a dragonfly (*Agrion* sp.) reached 14 m/s; most other insects flew in the range of 1–4 m/s. The values in excess of 10 m/s seem inordinately high and may be associated with timing errors for rapid flight over short distances. By contrast, Stevenson et al. (1995) obtained maximum airspeeds of 5.3 m/s for sphingid moths (*Manduca sexta*) flying freely in a large arena, whereas syrphid flies could attain airspeeds of 10 m/s in chases (Collett and Land, 1978). In a much-cited study, Lewis and Taylor (1967) determined flight speeds of 0.4–8 m/s for a group of mostly neopterous insects flying across a room. Body size of individual insects was presented as a product of wing and body length, and body mass was not reported. In general, detailed morphological data have not been taken on insects for which flight speed measurements have been made, although most such studies have been made on fairly large insects.

By contrast, remarkably few data exist on the flight speeds of small insects (e.g., 1–3 mm body length), in spite of their majority in the contemporary fauna (section 7.1). David (1978) recorded maximum speeds of 0.9 m/s for *Drosophila melanogaster* in experimental settings (see also Craig, 1986; Marden et al., 1997). Comparably small aphids

(A)

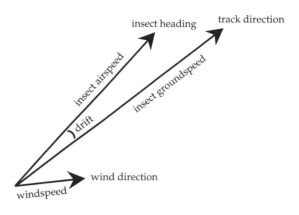

(B)

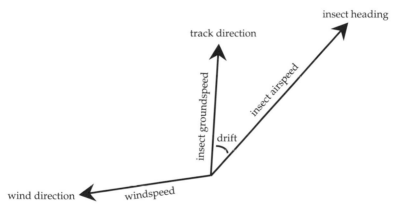

FIG. 3.1. Insect flight velocity measured with respect to a terrestrial coordinate system is the vector summation of the ambient wind velocity and the air velocity of the insect. (A) When ambient wind speed is small relative to the insect airspeed, insect drift with wind is small and the airspeed is similar to the groundspeed. (B) When wind speed is high and oriented against the insect's heading, groundspeed measurements can give misleading estimates of the true airspeed during flight.

fly at airspeeds less than 0.8 m/s (Kennedy and Thomas, 1974; David and Hardie, 1988; Hardie and Young, 1997). Tiny diaspidid homopterans are essentially incapable of flying at airspeeds greater than 0.5 m/s (Rice and Moreno, 1970). Airspeeds of small insects thus rarely exceed 1 m/s, a low value relative to the speed of most ambient winds (see section 7.4.2). Badly needed are further airspeed and morphological data for diverse insect orders flying under controlled experimental conditions (particularly with control of insect thermal regime). Information on variance in airspeeds within and among individuals would also be desirable. Such data should be analyzed using a comparative method to control for potentially nonrandom phylogenetic associations among species (see Harvey and Pagel, 1991). Unfortunately, chamber dimensions influence choice of airspeed, at least in butterflies (Dudley and Srygley, 1994) and in the sphingid moth *Manduca sexta* (Stevenson et al., 1995), so particular attention must be paid to the experimental context during such measurements. The effects of air temperature and illumination on insect flight speed are potentially substantial. For example, indoor flight speeds of the fly *Calliphora* increase with ambient temperature, whereas flight speeds of a hymenopteran and various dipterans and lepidopterans toward an illuminated window increase with the intensity of illumination (Schneider, 1965).

Relative to flight in confined spaces, insect airspeeds in nature are even less well known. Most such estimates are conjectural or are compromised by experimental artifacts such as entrainment in the flow field of moving vehicles (see Johnson, 1969). Generations of entomologists hanging out the windows of speeding cars or even trains have done little more than to confirm that substantial air volumes (and sometimes insects) are transported alongside moving vehicles. In a similar vein, suggestions of supersonic flight in deerflies (Townsend, 1927) are entertaining but clearly implausible on physical grounds alone (Langmuir, 1938). Townsend (1939, 1942) responded to this latter criticism by stating that the flies in question had "never been captured" and were "entirely invisible at top speed." Most estimates of insect airspeeds in natural contexts have decoupled airspeed from ground speed by assuming an average ambient wind speed and direction. This method has given airspeed values generally higher than those estimated for indoor settings (table 3.1). Because winds exhibit considerable spatial and temporal heterogeneity, assumptions of constant flight speed and direction may not be appropriate, particularly for insects flying at heights different from the height at which wind velocity was measured. None of the airspeed studies presented in table 3.1 measured ambient wind motions directly in the vicinity of a

TABLE 3.1

Insect Airspeeds Calculated Indirectly from Measurements of Ground Speed, Flight Direction, and the Wind Velocity (see fig. 3.1)

Order	Species	Airspeed (m/s)	Reference
Odonata	(Various)	2–5	Rüppell (1989)
		10 (maximum)	May (1991)
Orthoptera	*Schistocerca gregaria*	3–8	Rainey et al. (1957)
	Nomadacris septemfasciata	13 (maximum)	Waloff (1972a)
	(Migratory locusts)		Baker (1981)
Hymenoptera	*Apis mellifera*	6–8	von Frisch and
			Lindauer (1955)
			Wenner (1963)
Diptera	*Glossina* spp.	5	Brady (1991)
	(Tsetse flies)		

Note: Wind velocity is typically measured in the vicinity of the investigator and not of the flying insect.

flying insect, and logistical methods preclude such measurements under most circumstances. Sayter (1965) presented an ingenious photographic method that relies on multiple images of the same insect in different orientations to deduce locally the ambient wind velocity and insect airspeed. Major reorientation of the body axis in a short period of time is required to make such calculations, and this method is therefore not generally applicable to free-flying insects in nature.

However, direct airspeed measurements have been recently made on insects in natural free flight. Flight speeds of insects crossing a body of water can be measured by holding a unidirectional anemometer laterally from the bow of a boat moving parallel to and at the same speed (relative to a terrestrial coordinate system) as the flying insect (see Dudley, 1992). This approach is analogous to flying alongside the insect and obtaining a measurement of local airspeed. Because the anemometer is held outside of the flow field around the boat's hull, this speed measurement directly corresponds to the speed of the insect relative to the surrounding air. Flight direction of the insect is determined from a compass reading of boat orientation during such airspeed measurements. Wingbeat kinematics can also be obtained simultaneous with airspeed measurements using cameras mounted on the boat. Following capture of the insect (plate 3), the boat can immediately be stopped to permit measurements of ambient wind speed and direction, thus uniquely resolving the vector triangle of insect air velocity, velocity of the insect with respect to the ground, and the wind velocity (see fig. 3.1).

To date, this free-flight method has been used only on butterflies and moths that are fairly large and thus easily tracked during their flights over a large body of water. For example, an average airspeed of 3.9 m/s was determined for the diurnal uraniid moth *Urania fulgens* during migratory flight (DeVries and Dudley, 1990; Dudley and DeVries, 1990). Similarly, airspeeds for sixty-two Neotropical butterfly species in natural free flight ranged from 0.7 to 7.5 m/s (Srygley and Dudley, 1993; Dudley and Srygley, 1994). Direct measurements of airspeed can potentially be implemented on diverse pterygote taxa other than the aforementioned Lepidoptera, as insects can also be taken to the center of bodies of water, released to elicit flight, and then subsequently followed. Aforementioned data for both constrained and natural free flight airspeeds indicate, however, that insects generally fly at airspeeds between 0.5 and 10 m/s. Insect airspeeds in excess of 10 m/s are rare and require careful documentation.

Much of the variation in insect flight speeds is a consequence of body size, as larger insects tend to fly faster (e.g., Lewis and Taylor, 1967). If larger insects are similar in shape to smaller insects and vary only in size, then both wing length R and wing area S will scale isometrically with body mass m (i.e., $R \propto m^{1/3}$ and $S \propto m^{2/3}$ for geometrically similar organisms). The general dependence of steady-state aerodynamic forces on wing area and on the square of the relative velocity of airfoils (eqs. 1.2 and 1.4) permits the body weight of a flying insect to be equated with the mean aerodynamic lift on the flapping wings (see Lighthill, 1977; Norberg, 1990). Variation of forward airspeed with wing loading p_w and with body mass can then be estimated for isometrically scaled organisms:

$$V \propto p_w^{\frac{1}{2}}$$
(3.1)

$$V \propto m^{\frac{1}{6}}.$$
(3.2)

These equations strictly refer to conditions in which the forward airspeed is high relative to the mean flapping speed of the wings, but provide a useful initial basis for comparison with empirical results. Because insect airspeed data are so sparse, the generality of equations (3.1) and (3.2) has not been evaluated for a wide diversity of taxa. Johnson (1969) compiled data relating insect airspeed to body length for a variety of orders; these data refer to some chosen airspeed in a laboratory context, and are derived primarily from results of Lewis and Taylor (1967). If isometry in body length is assumed, the interspecific analysis of Johnson (1969) indicates a dependence of airspeed on $m^{0.29}$ (see Dudley, 1994). This exponent differs significantly from the value of

0.167 indicated by equation (3.2), although the underlying assumption of isometric body design is unlikely to be met in such interordinal comparisons.

Allometric studies of insect flight speeds in nature are limited in terms of taxonomic coverage to the butterflies (Lepidoptera: Rhopalocera). Butterfly airspeeds in natural free flight exhibit interspecifically a stronger dependence on wing loading ($V \propto p_w^{1.11}$) and on body mass ($V \propto m^{0.56}$) than is predicted by equations (3.1) and (3.2) (see Srygley and Dudley, 1993; Dudley and Srygley, 1994). Such relatively increased flight speeds at higher body masses may reflect the positive allometry of wing loading in butterflies and the corresponding deviation from isometric design (see Chai and Srygley, 1990; Dudley, 1990; Srygley and Dudley, 1993). No specific prediction exists on aerodynamic grounds for the relationship between flight speed and wing aspect ratio (Norberg and Rayner, 1987; Norberg, 1990), and only a weak inverse relationship exists between aspect ratio and natural airspeeds of butterflies (Dudley and Srygley, 1994). In contrast to butterflies, bats demonstrate a positive correlation between flight speed and wing aspect ratio (Norberg and Rayner, 1987). The underlying reasons for such a difference between these taxa are not known.

Because average insect body size is low by anthropomorphic standards (4–5 mm; section 7.1), only relatively few insect species (e.g., those with body masses approximately >250 mg) are likely to fly at airspeeds exceeding 5 m/s (Dudley, 1994; see section 7.4.2). Exactly such insects, however, have been the primary focus of biomechanical as well as ecological investigation. Natural airspeeds as well as aerodynamic mechanisms of the smaller but far more numerous insect taxa are little known (section 7.1). Also, consequences of intraspecific morphological variation for airspeed selection have not been comprehensively examined. Much of such variation is, as with interspecific variation, influenced directly by body size. However, the implications of sexual dimorphism for choice of airspeed may be more subtle and likely derive from the varying ecological roles of flight between the sexes (e.g., McLachlan, 1986a; Wickman, 1992).

Intraspecific variation in flight speeds is also evident on evolutionary timescales. Flies of the genus *Drosophila* respond positively to artificial selection for increased flight speeds (Weber, 1988, 1996; Marden et al., 1997), indicating substantial genetic variation in flight performance. When coupled with biomechanical assays of maximum performance (section 5.4.1), such selection experiments could provide substantial insight into functional constraints on flight capacity as well as evolutionary responsiveness of the flight apparatus to variable selective regimes. Particularly with *Drosophila*, genetic variation in flight performance is consistent with the widespread suggestion that long-term laboratory culture indirectly selects on fly strains for reduced flight capacity. For example, cultures of *Drosophila melanogaster* respond rapidly to a selection regime for enhanced upwind flight performance, and also decline substantially in performance once selection is relaxed (see Weber, 1996). Evolutionary lability in flight capacity is therefore considerable, and phenotypic interactions between flight ability and fecundity (see section 6.2.4) may be particularly pronounced within the artificially constrained environment of a laboratory culture bottle. This hypothesis has not been systematically investigated, but is clearly relevant for assays of flight performance in this important genus.

Airspeed is but one of many features of performance that delineate the flight envelope of insects. Flight behaviors that entail curved trajectories and often complex three-dimensional maneuvers are described in section 5.3. Rectilinear flight may involve components of vertical ascent and descent as well as horizontal motion. Few data are available on the kinematic correlates of ascending or descending flight. Larsen (1934) evaluated the flight of honeybees through an experimental tube that could be tilted horizontally; wingbeat frequency increased whereas upwards flight speed decreased as the tube was progressively tilted toward vertical. Also, foraging honeybees fly more quickly to downhill feeders than to uphill feeders (Heran, 1956). These changes in flight speed are consistent with increased flight costs associated with gain in potential energy (see section 5.4.2), but further investigation can be envisioned. For example, controlled flight in a variable-tilt wind tunnel could be used to evaluate interactions between horizontal flight speed and flight angle over the entire range of flight speeds, not just at the preferred cruising speed.

If natural insect airspeeds are little known, capacities for acceleration are even less well described. Success in aerial chasing and evasion may be critically dependent on acceleration and directional changes (section 5.3), but few comparative data exist on such transient features of flight performance. In general, estimates of acceleration from positional data must be treated cautiously because the results are sensitive to timing and digitization errors; such studies must be carried out with high filming frequencies and adequate spatial resolution of film images if sufficient accuracy is to be attained (see Harper and Blake, 1989; Walker, 1998). Termier (1970a) suggested that hoverflies could attain accelerations of up to 18 **g** when startled, whereas accelerations up to 9 **g** have been proposed in corduliid dragonflies (Ryazanova, 1966). More realistically, male hoverflies in chase attain accelerations of 3 **g** (Collett and Land, 1978). Lacewings escaping from bats accelerate up to 2.6 **g** (Miller and Olsen, 1979), whereas odonates in the suborder Anisoptera have greater capacity for acceleration (up to 2.5 **g**) than do those in the suborder Zygoptera (Rüppell, 1989; May, 1991; Wakeling and Ellington, 1997b). Higher-order temporal derivatives of body position (e.g., the rate of change of acceleration) are uninvestigated in insects but may be important when capturing prey, when escaping from predators, and during mating encounters.

Gliding flight characterizes very few insects, most notably some dragonflies (Hankin, 1921a, b; Wakeling and Ellington, 1997a) and butterflies. Slow glides with intermittent flapping are, for example, characteristic of many unpalatable ithomiine and heliconiine butterflies (e.g., Benson et al., 1989). During such glides, the fore- and hindwings of gliding butterflies are often nonoverlapping (see Betts and Wootton, 1988). Nemopterid neuropterans also glide occasionally (Picker, 1987), and some mayflies engage in transient parachuting with rapid vertical descent (Brodsky, 1994). Many insects engaged in long-distance migration may glide intermittently, albeit for periods usually less than a second (e.g., dragonflies: Corbet, 1962; locusts: Baker and Cooter, 1979b; the moth *Urania fulgens*: Dudley and DeVries, 1990). Similarly, the sphingid moth *Manduca sexta* glides for periods of several wingbeats when flying in large enclosures (Stevenson et al., 1995). Best studied of insect gliders is the monarch butterfly *Danaus plexippus*, which utilizes upward convection and prevailing winds to facilitate long-distance migration (Gibo and Pallett, 1979; Gibo, 1981a–c). Glides of several centimeters have been recorded in secondarily flightless beetles; wings are held outstretched in such glides but are not flapped (Bilton, 1994). Similarly, parachuting and in some cases gliding behavior has been recorded in both larvae and adults of several Southeast Asian leaf-mimicking species of stick insects that are flattened dorsoventrally (Siedlecki, 1917).

Selection of a forward flight speed near or equal to zero is termed either stationary flight (Magnan and Sainte-Laguë, 1933) or hovering flight. Hovering is aerodynamically challenging because the absence of a forward velocity vector demands that all requisite forces must be produced solely by wing flapping. Because wing area relative to body mass generally increases for smaller animals, the ability to hover increases with reduced body size. Hovering may have facilitated the diversification of small dipteran and hymenopteran parasitoids (section 7.1), and is also an important characteristic of many insect pollinators (section 7.2). Most insect hoverers can also fly backward, if only for transient periods, and in general are highly maneuverable. For example, many insect hoverers are capable of flying sideways (e.g., Collett and Land, 1975). Hovering is also an important feature of many dipteran and odonate mating systems. Although most common among the neopterous insects, the ability to hover also characterizes termites in their nuptial flights, Thysanoptera, and many Homoptera. At least one orthopteran (*Dissosteira carolina*; see Pierce, 1948) hovers during a courtship display. Among flying vertebrates, only hummingbirds and some nectarivorous bats can engage in hovering for other than transient intervals.

3.1.2 Wing Kinematics

Relative to the study of translational body motions, quantitative analysis of the rapid movements of beating insect wings requires substantially higher temporal resolution from filming or other experimental methodology. Moreover, most evaluations of wingbeat kinematics have been carried out in the laboratory because of the difficulties of studying insect flight performance in nature. Free flight in enclosed chambers and insectaries as well as controlled free flight in wind tunnels have been widely used to this end. Unfortunately, insects are psychologically resistant to anthropogenic suggestions of directed flight performance at a constant velocity. The study of free flight thus usually relies on analysis of volitional flight rather than of a performance regime determined by the experimenter, and behavioral variation in kinematics may accordingly be high. Studies of free flight at low or zero airspeed alleviate many methodological problems associated with the evaluation of forward flight, and much of our knowledge and parameter standardization of wingbeat kinematics derives from the comprehensive hovering study of Ellington (1984c).

Much experimental work on wing kinematics has involved attaching insects by the thorax or abdomen to rigid mounts or tethers. Wing beating can then be elicited using a variety of sensory cues; this experimental preparation is best termed a tethered simulation of flight as the phrase "tethered flight" is an oxymoron. Classic examples of this approach include work on locusts (Weis-Fogh, 1956a; Zarnack, 1972), calliphorid flies (Hollick, 1940; Nachtigall, 1966, 1973), and *Drosophila* (Vogel, 1966, 1967a; Götz, 1968). Potential artifactual complications of tethering are not well understood, although kinematics of free flight and of tethered preparations can differ substantially. It was, in fact, first suggested by Marey (1891) that tethering could induce abnormal wing movements (see also Voss, 1914; Schmidt, 1938). To cite Magnan and Planiol (1933b) on tethered preparations: "En effet la plupart des insectes sont doués d'un tempérament qui semble à l'opérateur essentiellement capricieux, leurs actions étant généralement imprévues, fantasques et décourageantes" [In effect, the majority of insects are endowed with a temperament that seems to the operator to be essentially capricious, their actions being generally unforeseen, weird and discouraging].

This observation has rarely been evaluated quantitatively, although Baker et al. (1981) determined that wingbeat frequencies of free-flying locusts are significantly greater than those of tethered animals (see also Kutsch and Stevenson, 1981; Schneider, 1981a). Wingbeat kinematics of tethered insects can also vary substantially with elapsed flapping time and with the particular sensory regime experienced by the restrained animal (e.g., Weis-Fogh, 1956a; Vogel, 1967a; Gewecke and Kutsch, 1979; Curtsinger and Laurie-Ahlberg, 1981; Kutsch and Hug, 1981; Ward and Baker, 1982; Grodnitsky and Kozlov, 1987; Grodnitsky and Morozov, 1993; Lehmann and Dickinson, 1997; Nachtigall, 1997b). Furthermore, the force balance of tethered insects may differ substantially from that in free flight (section 3.2.2). Tethered simulations of flight are thus heuristically informative but cannot evaluate quantitatively the extent to which associated kinematics and aerodynamic mechanisms are those actually used by free-flying insects. On the other hand, a major limitation of free-flight studies is the often high illumination required to visualize wingbeat kinematics—intense lights likely disrupt normal optomotor responses and may substantially alter natural flight behavior. Also, most investigations of the muscle actions and neurophysiology underlying wing motions (section 4.2) are logistically impossible unless the insect can be fixed to an experimental mount. Progress in understanding insect flight mechanics has thus derived historically from a combination of tethered and free-flight approaches.

The longitudinal axis of insect wings moves in an approximate plane largely constrained by the geometry of the axillary apparatus and the line of action of the dorsoventral musculature (figs. 2.1B, 2.3). This plane is oriented at an angle β (the stroke plane angle) relative to horizontal (see fig. 3.2A). The orientation of the longitudinal body axis relative to the line of action of the dorsoventral musculature is constrained anatomically, whereas the body axis oriented relative to horizontal defines the body angle χ when the insect is viewed laterally (fig. 3.2A). A wingbeat consists of two consecutive half-strokes in the stroke plane that are termed the down- and upstroke. In fact, these terms are misnomers as the wingbeat is most appropriately referenced to the body coordinate system and not to gravity; dorsoventral and ventrodorsal strokes, respectively, would be more accurate, although I will follow the conventional usage here.

Insects in flight move their wings rhythmically within the stroke plane (fig. 3.2B). The mean number of oscillatory cycles per second (i.e., Hz) is termed the wingbeat frequency (n). Because wing motions

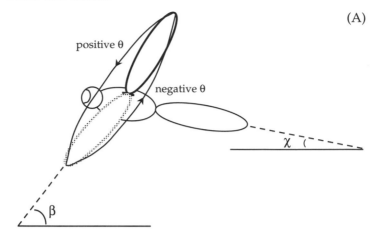

(A)

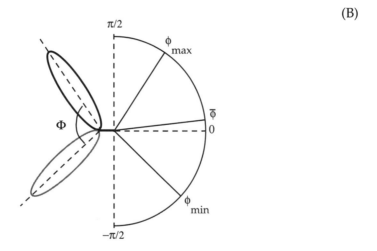

(B)

FIG. 3.2. Wingbeat kinematics and body orientation for a flying insect. (A) The longitudinal axis of the body forms an angle χ with respect to horizontal when the insect is viewed laterally. Wing motions define a stroke plane oriented at the stroke plane angle β. The wing tip can be elevated above or below the stroke plane angle (positive and negative θ, respectively). (B) The angular position φ describes the location of the wing tip projected onto the stroke plane. The wing moves through an arc of amplitude Φ (= $\varphi_{max} - \varphi_{min}$) during a wingbeat. The mean positional angle of the wingbeat, $\bar{\varphi}$, is equal to (φ_{max} + φ_{min})/2.

produce sound waves at audible frequencies, many early studies of insect flight kinematics involved acoustic determination of wingbeat frequency. The Finnish entomologist Olavi Sotavalta provided what are now classic contributions to this literature (Sotavalta, 1941, 1947, 1952a, b, 1953, 1954) based on his ability of absolute pitch (identification of the dominant frequency of any acoustic tone). The fundamental frequency of acoustic waves generated by insects usually corresponds to the wingbeat frequency, although higher harmonics may also be present and sometimes form the dominant frequency (e.g., Williams and Galambos, 1950; Webb et al., 1976). In spite of the often loud sounds of flying insects, actual acoustic power is generally on the order of microwatts (see, e.g., Haskell, 1957; Belton, 1986).

In addition to acoustic measurement, other methods can be used to determine wingbeat frequency. Cinematography or videography can be used to count the number of frames necessary to complete a wingbeat. Filming frequency must be high relative to wingbeat frequency to avoid aliasing (interaction between signal frequency and sampling frequency to produce an apparent but falsely complete beat). Radar echoes of flying insects often encode wingbeat frequencies (see Schaefer, 1976) and can be used to obtain such data on distant insects of albeit unknown identity. Kutsch et al. (1993) devised instrumentation for remote telemetry of flight muscle action potentials and, by extension, of wingbeat frequencies for large free-flying insects (see section 4.2.2). Optical methods can be used to determine wingbeat frequencies based on light reflected from wings of free-flying or tethered insects (Unwin and Ellington, 1979). Similarly, automated photodiode arrangements can be used to determine both wingbeat frequency and stroke amplitude of tethered insects over large numbers of wingbeats (Götz, 1987). Miniature induction coils can also be mounted on the wings of tethered insects to measure wingbeat frequency and additional wing kinematic parameters (e.g., Koch, 1977; Zarnack, 1978a; Schwenne and Zarnack, 1987). Automated methods in particular provide the opportunity to gather extensive data on the kinematic correlates of variable force production when tethered insects are exposed to different sensory regimes.

Extensive interspecific comparisons indicate that wingbeat frequency ranges over three orders of magnitude and varies in inverse proportion to body size, whereas wing area scales isometrically in proportion to $m^{0.66}$ (Sotavalta, 1947; Greenewalt, 1962; see fig. 3.3A, B). Some of the lowest recorded wingbeat frequencies for a free-flying insect belong to large male helicopter damselflies (5.5 Hz in the species *Megaloprepus coerulatus*; Rüppell and Fincke, 1989). Comparably low

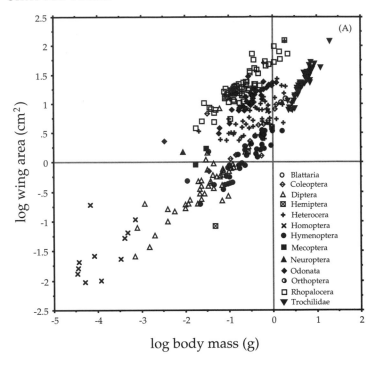

FIG. 3.3A. Allometry of wing area for insects and for hummingbirds (Trochilidae). Insect wing area scales with body mass$^{0.71}$ ($r^2 = 0.6$, $P < .001$; 95% confidence limits for exponent: 0.63–0.78); for hummingbirds, area scales with mass$^{1.16}$ (95% confidence limits for exponent: 1.03–1.29).

values characterize free-flying female birdwing butterflies (*Ornithoptera priamus*, Papilionidae; Dudley, unpublished data). The highest wingbeat frequency recorded on an unmanipulated insect is 1046 Hz for a diminutive ceratopogonid fly (*Forcipomyia* sp.; Sotavalta, 1947); this and other exceptionally high values for wingbeat frequency require corroboration as the potentially confounding presence of harmonics cannot be excluded. Wingbeat frequencies above 100 Hz are typically associated with the presence of asynchronous muscle, a muscle type unique to certain winged insects that exhibits repeated stretch-activated contractions in response to a single nervous impulse (section 4.2.2). Note that, for a given body mass, insect taxa can vary dramatically both in wing area (fig. 3.3A) and in wingbeat frequency (fig. 3.3B). Within the Lepidoptera, for example, butterflies have much higher relative wing area and lower wingbeat frequencies relative to other taxa in the order.

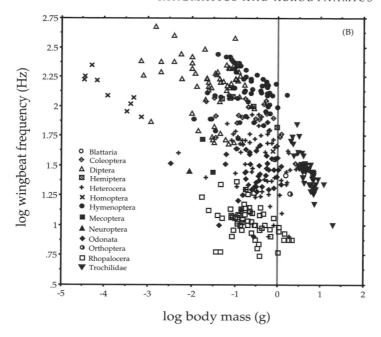

FIG. 3.3B. Allometry of insect and hummingbird wingbeat frequencies. For all insects, wingbeat frequency scales with mass$^{-0.24}$ (r^2 = .17, P < .001; 95% confidence limits for exponent: −0.18 to −0.29); wingbeat frequency of hummingbirds scales with mass$^{-0.61}$ (95% confidence limits for exponent: −0.49 to −0.73). Data from Magnan (1934), Sotavalta (1947, 1952b), Greenewalt (1962), Weis-Fogh (1972, 1973), Bartholomew and Heinrich (1973), Bartholomew and Casey (1978), Casey (1981b), May (1981b), Ahmad (1984), Ellington (1984c), Casey et al. (1985), Betts (1986b), Betts and Wootton (1988), Byrne et al. (1988), Ennos (1989), Rüppell (1989), Dudley (1990; unpublished data).

A variety of physical constraints impinge on the allometry of wingbeat frequency (see Lighthill, 1977; Weis-Fogh, 1977; Rayner, 1996). Because wing area and wing mass increase with body mass, larger insects must oscillate heavier wings. For mechanically resonant systems generally, oscillatory frequency varies inversely with the inertial load on the system. Wing amputation as well as loading experiments have established that wingbeat frequency varies inversely with wing inertia (Sotavalta, 1952a, b, 1954). Such effects are most pronounced in taxa with asynchronous flight muscle, a muscle type for which contractile activity is dependent on mechanical loading (see section 4.2.2). Reduced-density experiments also implicate wing inertia as a major determinant of wingbeat frequency, although the quantitative character of this relationship is both taxon- and context-specific (Chadwick and Williams, 1949; Chadwick, 1951; Sotavalta, 1952a, b).

Additional experimental evidence is consistent with the theory that wings are driven by a mechanically resonant oscillator. Positive correlations between wingbeat frequency and body temperature (e.g., Sotavalta, 1947; Sotavalta and Laulajainen, 1961; Schneider, 1980a; May, 1981b; Oertli, 1989; Joos et al., 1991) can be interpreted in part as simple thermal dependence of muscular power output driving the resonant system (see section 4.2.4). Smaller insects exhibit a greater dependence of wingbeat frequency on ambient temperature than do larger insects (Unwin and Corbet, 1984), a result consistent with a higher surface area : volume ratio and reduced thermoregulatory capacity at smaller body mass (section 4.3.2). Variation in the elastic modulus of flight muscle (Machin et al., 1962), in thoracic stiffness (Heide, 1971b), and in neural activation of the flight muscle (Hanegan and Heath, 1970; Pfau et al., 1989; Janssen, 1992; Xu and Robertson, 1994) all potentially underlie thermal sensitivity of wingbeat frequency. By comparing the relative thermal sensitivity of these different parameters, Oertli (1989) concluded for beetles that most temperature-related variation in wingbeat frequency arises from changes in resonant properties of the flight apparatus. Such an analysis has not been extended to other orders but is likely to be a general result for insects.

In addition to the wing mass, the virtual mass of the surrounding air must also influence wing acceleration and mechanical loading of the pterothorax. This complication has been omitted from many physical analyses of the allometry of wingbeat frequency (e.g., Reed et al., 1942; Rashevsky, 1944; Attila, 1947; Sotavalta, 1952a, b; Pennycuick, 1975). Hill (1950) argued that wingbeat frequency should vary with $m^{-1/3}$ for animals that vary isometrically in flight muscle mass. This theoretical prediction is not corroborated empirically for insects (see fig. 3.3B), although the allometry of flight muscle mass has not been systematically determined. Neither have phylogenetic associations of wingbeat frequency been adequately evaluated, although effects of among-clade variation may be substantial. For example, hummingbirds deviate significantly from the insect allometry by expressing relatively greater wing areas and a correspondingly more negative allometry of wingbeat frequency (fig. 3.3A, B).

Within an approximately unchanging behavioral context, wingbeat frequencies of individual insects are fairly constant. For example, coefficients of variation in wingbeat frequency are typically 5% or less for neopterous insects in standard hovering flight (Ellington, 1984c; see also Weis-Fogh, 1956a, for tethered locusts). One likely explanation for this relative invariance is that energetic expenditure is minimized at the resonant frequency of oscillation, and that insects incur substantial costs for any deviation from this preferred value. To summarize, both empirical and theoretical results are consistent with a theory of resonant pterothoracic oscillation (see also Greenewalt, 1960b). Direct mass addition or subtraction from the wing alters the moment of inertia of the resonant system, whereas air density manipulations change the wing virtual mass (and wing inertia) as well as the external damping on the wing. Thermal manipulations alter activation forces, stiffness, and damping characteristics of the pterothorax. Experimental perturbations of wing inertia have yielded predicted changes in resonant frequency, whereas interspecific variation demonstrates a large-scale inverse relationship between body mass and wingbeat frequency.

Within the stroke plane, wings reciprocate between well-defined angular limits for any particular dynamic situation (fig. 3.2B). The dorsal extreme of angular movement is termed the maximum positional angle (φ_{max}), and its ventral counterpart is the minimum positional angle (φ_{min}; see fig. 3.2B). Stroke amplitude Φ, the angular extent of motion in the stroke plane, is given by $\varphi_{max} - \varphi_{min}$. Values of Φ can range from 66° (hovering syrphid flies; Ellington, 1984c) to 180° or even slightly higher in some beetles and moths (see Pringle, 1957; Atkins, 1960; Termier, 1970b; Wilkin, 1991). Values of Φ in excess of 180° can only occur when the wing-base separation permits bilateral wing movement without contralateral wing interference or impact (see fig. 3.2B). Because the wing-base separation is generally small relative to wing length in insects, the capacity to increase maximum and minimum positional angles beyond ±90°, respectively, is very limited. Maximum stroke amplitudes exceeding 180° are thus rare, and this limit poses a geometric constraint for all flying animals that flap bilaterally paired appendages (see section 5.4.3). Maximum stroke amplitudes for many insects are usually smaller than 180° and likely reflect anatomical limits to the wing motion imposed by the axillary apparatus. For example, maximum stroke amplitudes of euglossine orchid bees under conditions of maximal load lifting are approximately 150° (Dillon and Dudley, in prep.). The maximum positional angle usually reaches 90° in such cases (see fig. 3.2B), whereas the absolute value of the minimum positional angle is usually lower. This observation suggests that the angular extent of the ventral half-stroke is limited by physical interaction between the wing base and the pleural wing process.

Rhythmic oscillation of wings within the stroke plane is often sinusoidal in character. For example, wing-tip motions of hovering dipterans and hymenopterans are nearly equivalent to simple harmonic motion (Ellington, 1984c; Ennos, 1989c; Dudley and Ellington, 1990a), although Vanderplank (1950) reported deviation from this pattern in the tsetse fly *Glossina palpalis*. The suggestion by Vanderplank (1950) that this taxon is characterized by substantial stationary periods at the ends of half-strokes particularly requires corroboration. One feature of simple harmonic wing motion is equal duration of the down- and up-stroke. The downstroke : upstroke ratio is approximately unity in hovering of many taxa (Ellington, 1984c) and for some insects in free forward flight (e.g., bumblebees: Dudley and Ellington, 1990a). Lepidoptera and Odonata often deviate, however, from this pattern of similar half-stroke duration (see Betts and Wootton, 1988; Dudley, 1990; Wakeling and Ellington, 1997b; Willmott and Ellington, 1997b). The downstroke : upstroke ratio may also vary with aerodynamic demand; for example, values of this ratio varied in free-flying migratory locusts with the rate of ascent or descent (Baker and Cooter, 1979a). Magnan (1934) recorded a more rapid downstroke relative to the upstroke for a variety of tethered insects; this result may have derived from artifactual effects of tethering. For example, Weis-Fogh (1956a) described relative half-stroke durations from tethered locusts (*Schistocerca gregaria*) that differ substantially from those of migratory locusts (*Locusta migratoria*) in natural free flight (see Baker and Cooter, 1979a), although taxonomic differences may also be involved.

Insects with synchronous muscle may demonstrate greater kinematic irregularity and more divergence from resonant behavior in the pterothorax than do insects with the more regular mechanical loading of asynchronous flight muscle (see section 4.2.2). Insects with this latter muscle type may also exhibit greater autocorrelation in kinematic parameters within any given sequence of stretch-induced contractions. A higher degree of autocorrelation would indicate reduced kinematic variance (e.g., reduced coefficients of variation in wingbeat frequency), although systematic comparisons of flight muscle type and kinematic variance have not been carried out. In locusts, a taxon with synchronous flight muscle, Zarnack (1972) and Baker and Cooter (1979a) noted substantial variation among consecutive wingbeats in stroke amplitude, stroke plane angle, and the downstroke : upstroke ratio (see also section 7.3.2). Temporal variation in wing kinematics of Lepidoptera (e.g., butterflies) can also be substantial. By contrast, the regularity in wingbeats of dipterans and hymenopterans is striking; coleopterans and other asynchronous fliers have not been investigated in this regard.

Wing chords are oriented variably with respect to the stroke plane and thus with respect to the horizontal. Methodologically, the wing angle of attack has proven to be difficult to quantify for arbitrary and possibly bilaterally asymmetric wing motions of free-flying animals. Three-dimensional reconstructions of wing orientation are possible using two synchronized film images taken from different but known orientations (stereophotogrammetry). If bilateral kinematic symmetry of the wingbeat can be assumed, a single camera view provides sufficient spatial information to reconstruct wing orientation (see Ellington, 1984c). Many features of insect flight performance rely, however, on asymmetric wing kinematics (see section 5.2). A recently developed method provides highly accurate reconstructions of arbitrary wing configurations in three-dimensional space using a single two-dimensional image (Willmott and Ellington, 1997a). This method relies on operator-guided visual matching of spanwise wing sections given an orthogonal view of the wing planform as the starting perspective. Using a computer depiction of a virtual wing, individual wing sections are rotated about the longitudinal wing axis until the observed two-dimensional projection of the wing section is replicated for any arbitrary three-dimensional position of the wing tip. This procedure is repeated consecutively along the length of the wing for a complete reconstruction and is highly robust to digitization and operator errors (Willmott and Ellington, 1997a).

Wing chord orientation relative to the stroke plane is termed the geometrical angle of attack (Ellington, 1984d) and typically ranges from 20° to 40° during the downstroke (e.g., Ellington, 1984c; Dudley and Ellington, 1990a; Willmott and Ellington, 1997c). Wing orientation is often reversed during the upstroke (as is the wing camber), although the absolute value of the geometrical angle of attack in the upstroke is often less than that of the downstroke. Wing-tip motions also deviate from the average stroke plane to describe a three-dimensional and often elliptical curve, although the angular extent of such deviations (sometimes termed the elevational angle θ; see fig. 3.2) is usually small relative to angular displacement within the stroke plane. The three-dimensional trajectories of wings are in general not highly repeatable from wingbeat to wingbeat, and both tethered (e.g., Nachtigall, 1966) and free-flying insects (e.g., Baker and Cooter, 1979a; Ellington, 1984c) exhibit substantial variation in wing-tip paths. The magnitude of wing-tip deviation from the stroke plane is under active neuromuscular control in flies (Tu and Dickinson, 1996), whereas free-flying bumblebees (Dudley and Ellington, 1990a) and the moth *Manduca sexta* (Willmott and Ellington, 1997b) show systematic variation in wing-tip paths with forward airspeed. Such small variation in wing motions is correlated with the varying aerodynamic demands of flight at different speeds (section 3.2.2.1).

Wing trajectories in three-dimensional space are defined not only by motion relative to the stroke plane but also by the translational velocity of the insect. Because of wing reciprocation about the base, a velocity gradient extends from the wing base (zero flapping velocity) to wing tip (maximum flapping velocity). The relative magnitude of flapping to translational motion was first nondimensionalized by Walker (1925, 1927) as the reduced frequency parameter. Since Vanderplank (1950), the inverse of the reduced frequency, termed the advance ratio (J), has been widely used in studies of insect flight. The advance ratio, as defined by Ellington (1984c), incorporates both stroke amplitude and wingbeat frequency in the estimate of flapping velocity:

$$J = V/2\Phi nR, \tag{3.3}$$

where V is the forward airspeed. The advance ratio is zero in hovering flight (no body translation) and increases with forward airspeed. Ellington (1984c) proposed a definition of hovering as flight with $J < 0.1$. In gliding flight with no wing oscillation, the advance ratio is effectively infinity ($n = 0$). The advance ratio is indicative of the relative magnitude of steady-state to unsteady effects in flight aerodynamics. Steady flows predominate at high advance ratios (e.g., $J > 10$, or in the extreme case, gliding flight). Oscillatory and unsteady aerodynamic flows are present to a variable extent for $J < 10$, although the relative magnitude of such effects is highly context-specific (Spedding, 1992).

Vanderplank (1950) suggested that an approximately constant advance ratio characterizes the forward flight of differently sized insect taxa. This hypothesis can be evaluated from empirically determined allometries of wingbeat frequency ($n \propto m^{-0.24}$; fig. 3.3B) and of airspeed ($V \propto m^{0.29}$; see Dudley, 1994) among insect species, and by assuming isometry of body dimensions as well as size invariance in stroke amplitude. Contrary to the prediction of Vanderplank (1950), the advance ratio will then scale with $m^{0.20}$, indicating that larger insects fly at absolutely higher advance ratios. Even for small insects in slow forward flight, however, advance ratios are well above the values typical of hovering flight. For example, a small *Drosophila* fly ($V \sim 0.5$ m/s; $R \sim 2$ mm; $n \sim 250$ Hz) is characterized by an advance ratio near 0.25. Although flight speeds are not available for most insects, forward airspeeds will usually be greater than the mean flapping speed of the wings (with the exception of slow-flying taxa such as craneflies; Ellington, 1984c). The standard cruising flight of insects nonetheless lies in a range of advance ratios for which flapping and translational velocities are within the same order of magnitude, and for which unsteady aerodynamic effects cannot be neglected.

Flapping velocity of the wing, the forward airspeed of the insect, and the induced velocity sum vectorially to produce variable wind speed and direction across spanwise sections during the wingbeat. The effects of translational forward velocities are most easily appreciated by considering motions of the wing tip. In hovering motions with a

(A)

(B)

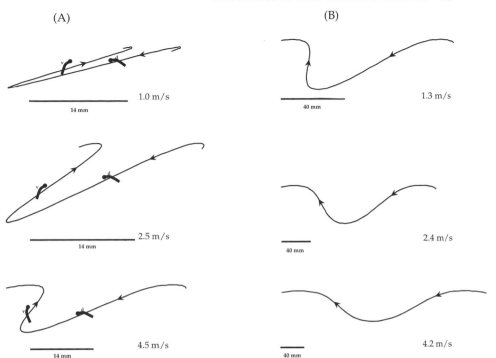

1.0 m/s

14 mm

2.5 m/s

14 mm

4.5 m/s

14 mm

1.3 m/s

40 mm

2.4 m/s

40 mm

4.2 m/s

40 mm

FIG. 3.4. Wing-tip path viewed laterally at different forward airspeeds: (A) the bumblebee *Bombus terrestris*, and (B) the moth *Urania fulgens*. Wing motions are assumed to be confined to the stroke plane and to follow simple harmonic motion; mean kinematic data are from Dudley and Ellington (1990a) and Dudley and DeVries (1990). For *B. terrestris*, wing chord orientation during down-stroke and upstroke are indicated; *d* and *v* indicate dorsal and ventral wing surfaces, respectively. Constant stroke amplitude Φ and wingbeat frequency *n* are assumed for the two species (*B. terrestris*: $\Phi = 120°$, $n = 145$ Hz; *U. fulgens*: $\Phi = 125°$, $n = 14$ Hz. Note that the upstroke becomes more vertical and travels less distance (i.e., lower mean velocity) as airspeed increases in (A), whereas in (B) the upstroke becomes more horizontal and increases in mean velocity over the same airspeed range.

near-zero stroke plane angle, half-strokes are approximately sym-metric and wing translational velocities are comparable between the down- and upstroke (Ellington, 1984c). Addition of a forward velocity vector induces an asymmetry between wing half-strokes that may in-crease the mean wing speed during the downstroke and decrease that during the upstroke (fig. 3.4A). Wing speed may, however, increase during the upstroke if the advance ratio is sufficiently high (e.g., fig. 3.4B). In general, the wing velocity of the downstroke relative to that of the upstroke varies as a function of both the stroke plane angle

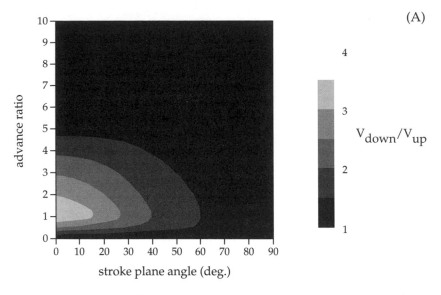

FIG. 3.5A. Ratio between mean downstroke and upstroke velocities as a function of the advance ratio J and the stroke plane angle β. The down- and upstroke progressively differ in mean velocity at lower advance ratios and lower stroke plane angles, although these two parameters interact to determine the ratio of mean half-stroke velocities.

and the advance ratio (see fig. 3.5). Such sensitivity of velocity and thus aerodynamic force partitioning between down- and upstrokes provides a useful means for regulating net force output over a wide range of airspeeds (see section 3.2.2.1).

The transition between half-strokes is characterized by often extensive deformation and by wing rotation that is typically 90° out of phase with wing oscillation in the stroke plane. Both the timing (e.g., Ellington, 1984c; Dickinson et al., 1993) and extent (e.g., Dudley and Ellington, 1990a; Willmott and Ellington, 1997c) of wing rotation can be varied by individual insects in accordance with demands of force production. Dorsal pronation and ventral supination occur at remarkably high angular velocities in some insects (e.g., $10^{5°}$/s in *Drosophila*; Zanker, 1990a), and arise via inertial torques as well as through the action of muscles at the wing base (chapter 2). The extent of wing rotation and deformation is constrained by venation, by the relative size of different wing regions, and by the extent of wing coupling if present. Insects with a pronounced claval or vannal wing region (e.g., hindwings of Orthoptera) or with broad wings generally (e.g., Lepidoptera) exhibit reduced pronation and supination. Odonates, dipterans,

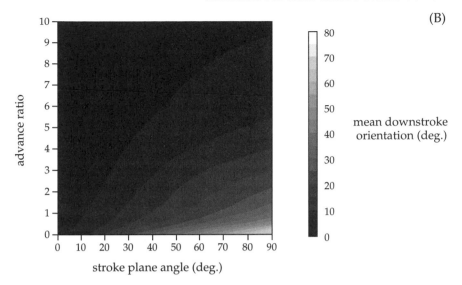

Fig. 3.5B. Mean downstroke orientation relative to horizontal as a function of the advance ratio J and the stroke plane angle β. A mean downstroke orientation of 0° indicates that the wing moves horizontally in the downstroke. The downstroke becomes progressively more vertical at higher stroke plane angles, but this effect is reduced at higher advance ratios. Simple harmonic motion of the wing tip in the stroke plane is assumed in (A) and (B). Upstroke orientation is much more variable than that of the downstroke and is usually negative (i.e., the wing tip moves on average backwards relative to the downstroke; see fig. 3.4A). Quantitative results in (A) and (B) refer to a stroke amplitude of two radians and a mean positional angle of zero, but qualitative results are independent of these kinematic parameters.

and hymenopterans, by contrast, possess more flexible wings that rotate and deform extensively. Torque imposed at the wing base as well as wing inertia may generate a torsional wave that spreads along the trailing wing edge from tip to base in many holometabolous insects (see Ellington, 1984c; Ennos, 1987, 1988a, b) and in odonates (Norberg, 1975; Brodsky, 1994), among other taxa. This wave begins distally and translates proximally along the wing because inertial effects are most pronounced near the wing tip, where section velocities and accelerations are greater. Extensive deformation of the entire wing surface may occur simultaneously with a torsional wave, although flexion about a longitudinal axis (see section 2.2.2) is the most general correlate of this inertially induced kinematic feature.

Half-stroke transition may also be characterized by proximity and even contact of contralateral wings or wing pairs. Weis-Fogh (1973)

(A)   (B)

(C)   (D)

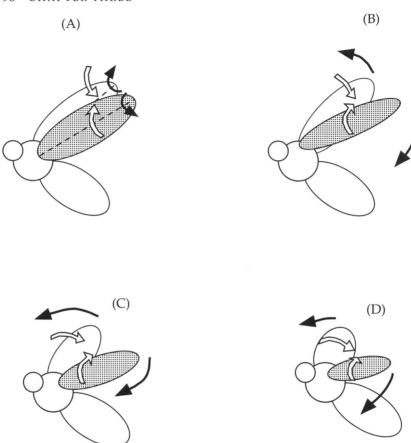

FIG. 3.6. Clap-fling mechanism of Weis-Fogh (1973). (A) Wings clap together dorsally at the end of upstroke and initiate pronation about the longitudinal wing axis. Air flows into the gap between the wings prior to wing translation. (B) Wings initiate downstroke and translation while wing rotation is slowed. Air continues to flow into the increasing gap between the wings. (C) Wings continue to separate as translational circulation augments that established by wing rotation. (D) Circulation is fully established as downstroke continues. Shading indicates the ventral surface of the nearside wing.

described a clap-fling mechanism in which contralateral wings close together dorsally and then separate by rotation about the adjacent trailing wing edges prior to the translational motion of the downstroke (see fig. 3.6). Although first observed in the small chalcid hymenopteran *Encarsia formosa* (Weis-Fogh, 1973), such contralateral wing proximity also occurs in damselflies (Wakeling and Ellington, 1997b) and occasionally in free-flying *Drosophila melanogaster* (Ennos, 1989c). Particular wing shapes may enhance the performance of this kinematic mechanism. For example, damselflies have more petiolate (narrowly stemmed) wings than do dragonflies in the closely related odonate suborder Anisoptera, likely facilitating wing rotation during the clap-fling typical of this lineage (Wakeling, 1997). A more distal location of the center of wing mass in anisopterous dragonflies, by contrast, appears to be associated with the typically higher airspeeds and advance ratios characterizing flight in this taxon (see May, 1981; Rüppell, 1989; Wootton, 1991; Grabow and Rüppell, 1995).

The concept of the clap-fling subsumes a diversity of kinematic possibilities relating to the extent of wing deformation, to the relative proximity of contralateral wings, and to the timing of wing rotation relative to the ventral flapping motion. Dorsal separation of contralateral wings or wing pairs may involve not rigid rotation but rather a peeling apart of the leading edges of the two opposed wings. Such peeling or partial peeling (Ellington, 1984c) often characterizes insects with broad wings and pronounced vannal regions (such as *Locusta migratoria* in natural free flight; Cooter and Baker, 1977). Butterfly wing separation at the onset of the downstroke involves a peeling apart of contralateral wings (Brackenbury, J. H., 1991), and such motions also occasionally occur in tethered *Drosophila* (see Vogel, 1967a; Götz, 1987). Ellington (1984c) noted that a continuum of potential separation characterizes contralateral wing proximity at the end of the upstroke; when wings approximate the clap motion but do not actually contact each other, the motion is best described as a near clap. The important aerodynamic consequences of dorsal wing proximity and the clap-fling mechanism are treated in section 3.2.2.

Evolutionary trends in wingbeat kinematics have not been systematically investigated, although an increase in wingbeat frequency has clearly been important in many of the holometabolous orders (Voss, 1913; Dudley, 1991b; Brodsky, 1994). Brodsky (1994) also suggested that a parallel phylogenetic decrease has characterized stroke amplitude, although the salient example of high stroke amplitudes in many Coleoptera (see above) suggests involvement of additional factors. Little is known of free-flight kinematics in the anteromotoric and most basal pterygote order (Ephemeroptera). Instead, the Odonata have received disproportionate attention of the two extant paleopterous orders, and may serve as a representative tetrapterous order.

The maneuverability of extant dragonflies is well known (in Chinese mythology, the dragonfly is symbolic of instability and rapid change), and is consistent with a well-developed flight capacity in the Protodonata and other Carboniferous pterygotes. Stroke amplitudes of fore-

and hindwings are generally similar in odonates (e.g., Wakeling and Ellington, 1997b), although stroke plane angles are not necessarily equivalent between the two wing pairs. Ipsilateral wing motions are generally out of phase in free-flying odonates, although in-phase synchronized wing strokes are used in display to conspecifics (e.g., Alexander, 1984; Rüppell, 1985), and transiently to maximize aerodynamic force production (Rüppell, 1989; Wakeling and Ellington, 1997b). Grodnitsky and Morozov (1993) advanced the intriguing hypothesis that out-of-phase wingbeating by odonates is an adaptation for reducing body oscillations and apparency to potential targets. In part, such kinematic variability is facilitated by the direct musculature acting at each wing base separately during the downstroke. Because these muscles can act independently for each wing, contralateral as well as ipsilateral wings become decoupled kinematically, enhancing wingbeat asymmetries during maneuvers.

By contrast, use of indirect flight musculature to effect the downstroke promotes enhanced kinematic coordination between adjacent thoracic segments. In such cases, the meso- and metathoracic segments share a common dorsoventral phragma upon which the longitudinal muscles insert. This mechanical linkage between adjacent wing pairs permits intersegmental transfer of kinetic energy anteroposteriorly through the pterothorax (Brodsky, 1994). In parallel with such transfer, increased downstroke synchronization between ipsilateral wings would ultimately promote dominance by one wing pair and either loss of the hindwing or the evolution of effective coupling mechanisms. Either the meso- or metathoracic wing pair can become dominant, leading to antero- or posteromotorism, respectively. Transformation for nonaerodynamic purposes or reduction of the other wing pair necessarily follows. Such reduced aerodynamic contributions require compensation by the dominant pair, usually by means of an increase in wing area or frequency. For example, evolution of functional or actual diptery (as in true flies) usually involves an increase in wingbeat frequency. If, as in the Hymenoptera, one wing pair remains, but in diminished form, an effective wing coupling mechanism typically evolves. Higher rotational velocities during pronation and supination will follow as a consequence of increased wingbeat frequency, and both inertially induced rotation as well as pronounced torsional waves are typical of insects with high wingbeat frequencies.

No explicit evaluation of these kinematic trends has been conducted across an ordinal-level phylogeny of winged insects, but such effects are readily evident when considering posteromotoric orders with tegmina or elytra. If elytrization or tegminization is not well developed (e.g., Orthoptera), the reduced-amplitude flapping of the forewings (see Weis-Fogh, 1956a; Brodsky, 1985b) is compensated for by hind-wing areal expansion and presence of a vannus. Elytrization of forewings is matched by faster hindwing motions enabled by asynchronous flight muscle (see section 7.1.2). The virtually complete posteromotorism of beetles is ensured by often negligible or very low-amplitude flapping of the elytra (Nachtigall, 1964; Schneider, 1975a; Schneider and Hermes, 1976). Some beetle taxa do, however, retain flapping motions of the forewings (Schneider and Meurer, 1975; Schneider, 1978a). Both an increase in wing area (as evidenced by the elaborate patterns of hindwing folding beneath the elytra) and higher wingbeat frequencies characterize the Coleoptera. In sum, transformation of one wing pair necessitates either kinematic or morphological compensation by the remaining wing pair. Increased wingbeat frequency appears to be the most common compensatory pathway, and acquisition of asynchronous flight muscle is a common physiological underpinning of this trend (see section 7.1.1).

In contrast to major orders with often high wingbeat frequencies, diverse endopterygote orders (e.g., the two clades [Neuroptera + Megaloptera + Raphidioptera] and [Trichoptera + Lepidoptera]) exhibit wingbeat frequencies generally lower than those of sister taxa. Even within these orders, however, kinematic patterns can evolve divergently. Kozlov et al. (1986), for example, pointed out that ancestrally homonomous wing pairs of Lepidoptera have evolved toward higher flapping frequencies with concomitant hindwing reduction (e.g., Sphingidae), or toward lower frequencies with chordwise broadening of both fore- and hindwing. Mapping of these kinematic and morphological traits onto a lepidopteran phylogeny is clearly required to evaluate these trends quantitatively. One interesting consequence of such frequency reduction is a less pronounced emphasis on ipsilateral wing connection (e.g., the amplexiform coupling of butterflies), which in turn reduces rotational capacity of the wing pair.

Wingbeat kinematics are not invariant to particular insects but vary according to aerodynamic demand. Control of flight speed, for example, necessitates alteration of kinematics to attain the requisite balance of vertical and horizontal forces (section 3.2.1). Wingbeat frequency and stroke amplitude are obvious candidates for parameters involved in airspeed regulation, although substantial changes in frequency are unlikely for reasons discussed above. Burst acceleration may, however, involve a transient increase in wingbeat frequency. For logistical reasons, variation in wingbeat kinematics with forward airspeed has been difficult to determine on individual insects in free flight. The issue of relevant sampling interval is also germane, as analyses of only several wingbeats may obscure time-averaged trends if variance in kinematic parameters is high. However, available data indicate that wingbeat frequency tends to be independent of airspeed in bumblebees, the sphingid moth *Manduca sexta*, and most other insects as well as in volant vertebrate taxa (see Dudley and Ellington, 1990b; Willmott and Ellington, 1997b). However, slower flight in some papilionid butterflies is characterized by a small decrease in the wingbeat frequency (Betts and Wootton, 1988). Among individual migratory locusts in free flight, wingbeat frequency and flight speed are positively correlated among individuals (Baker et al., 1981; see also Gewecke, 1975, for tethered locusts), although kinematic correlates of variable flight speed were not determined on individual insects. Results of these latter studies with butterflies and locusts may also reflect the greater kinematic variance associated with synchronous flight muscle (see section 4.2.2).

Variation in stroke amplitude with forward airspeed has been evaluated in detail for four insect taxa (dragonflies: Azuma and Watanabe, 1988, Wakeling and Ellington, 1997b; bumblebees: Dudley and Ellington, 1990a; the uraniid moth *Urania fulgens*: Dudley and DeVries, 1990; the sphingid moth *Manduca sexta*: Willmott and Ellington, 1997b). In none of these taxa is there a systematic trend relating stroke amplitude to airspeed. Stroke amplitude in *M. sexta* does, however, decline over the airspeed range of 0–3 m/s, but not between 3–5 m/s (Willmott and Ellington, 1997b). Contrariwise, two genera of papilionid butterflies show an increase in stroke amplitude at lower flight speeds (Betts and Wootton, 1988). Similar studies with tethered insects show systematic but conflicting trends in stroke amplitude variation. Tethered beetles showed an increase in wingbeat amplitude with increased airstream velocity (Schneider, 1982), whereas stroke amplitude in tethered dragonflies and locusts decreased as air velocity increased (Gewecke, 1970; Gewecke et al., 1974). These studies involve different insect taxa as well as different tethering methods, and are thus difficult to compare. In general, no systematic trend characterizes variation in stroke amplitude with airspeed among insects, although some particular taxa have demonstrated consistent trends. In contrast to wingbeat frequency, diverse patterns in the variation of stroke amplitude might be expected for insects of different sizes, shapes, and taxonomic affiliation.

Although less well studied, subtle variation in wing kinematics is also evident at different airspeeds. The mean positional angle of the wings tended to increase slightly with forward airspeed in free-flying bumblebees (Dudley and Ellington, 1990b) and hawkmoths (Willmott and Ellington, 1997b). Such dorsal and posterior motion of the mean position of the wings may correlate with the increase in nose-down dorsal pitching moments required for a reduction in body angle over the same speed range (see below). In bumblebees, other kinematic variables that showed consistent change with airspeed were wing rotational velocities and the extent of wing-tip deviation from the stroke plane (Dudley and Ellington, 1990b). Rotational angles during pronation and supination also showed systematic variation with airspeed in free-flying *Manduca sexta* (Willmott and Ellington, 1997b), whereas consistent variation in wing-tip paths was observed by Hollick (1940) for a dipteran tethered in variable-speed airstreams. Because the unsteady aerodynamic mechanisms used by insects are highly sensitive to the finer details of wingbeat motions (section 3.2.2), many different kinematic patterns might be expected among insect taxa in response to variable force requirements, including flight at different airspeeds.

3.1.3 Body Orientation and Oscillation

The insect body is a necessary underpinning to aerodynamic force production by flapping wings. Because the center of insect body mass lies posterior to the wing base, the longitudinal body axis tends to rotate vertically about the wing base axis under the influence of gravity. However, net forces on the wings act at some distance from the wing base and generate a net torque that yields a nonvertical and potentially variable body orientation relative to horizontal (section 3.2.1.1). Body angle tends to be reduced as insects fly more rapidly, whereas the stroke plane angle concomitantly increases (e.g., Nachtigall et al., 1971; Dudley and Ellington, 1990a; Willmott and Ellington, 1997b). In part, changes in body orientation function indirectly to alter the stroke plane angle and thus to regulate mean wing velocities and orientation during the downstroke and upstroke (fig. 3.5). Body orientations in fast forward flight are approximately horizontal and wing motions approximate a vertical stroke plane (for classic illustrations, see Magnan 1934 and Séguy 1950).

An additional consequence of lower body angles at high airspeeds is reduced body drag. Because of the dependence of drag on the square of airspeed (eq. 1.2), the disproportionate increase in body drag at higher airspeeds is partially offset by a reduced projected area of the body at lower angles relative to flow. Conversely, body drag is of little significance for slow-flying insects. In hovering and low-speed flight, many neopterous taxa exhibit body angles between 40° and 60°. Certain hoverflies, however, hover with inclined stroke planes and body angles that are correspondingly close to horizontal (see Ellington, 1984c). At the other end of the spectrum, lucanid beetles fly with bodies oriented vertically (Poujade, 1884), and many other large beetles seem to fly in similar fashion (see Crowson, 1981; Schneider,

1981b). Reduction of body drag is unlikely to be important at the low airspeeds characteristic of these taxa.

The aerodynamic and inertial forces produced by repetitive wing motions act through the wing base to result in coupled oscillations of body position and body angle (Demoll, 1918; Magnan, 1934; Ellington, 1984c). Inertial lag of the body can damp these oscillations considerably, particularly when wingbeat frequencies are high. In hovering dipterans and hymenopterans, for example, vertical and horizontal oscillations of the body occur at the wingbeat frequency but are typically low in amplitude (i.e., less than 5% of the wing length; Ellington, 1984c). Low-amplitude oscillations in body angle are similarly coupled with force production by the wings, although such angular displacement is typically only on the order of several degrees (see Ellington, 1984c). Body oscillations are more pronounced for insects with larger wings and relatively greater moment arms between the wing base and the center of action for either aerodynamic or wing inertial forces. For example, ascending flight of a large saturniid moth (n = 14.5 Hz) is characterized by large changes in body angle through the wingbeat (Wilkin, 1991). Similar rotation and displacement of the body characterizes the wingbeats of palatable butterflies (Dudley, 1990). Such oscillations, when coupled with variability in the wingbeat frequency, may be an important component of the erratic and unpredictable flight paths in these insects (see section 7.3.2).

Ancillary flight structures are usually placed in a stereotypical posture during flight. Flying insects typically point their antennae forward and upward. Fore- and midlegs are often held against the thorax with the hindlegs trailing posteriorly, apparently to reduce drag. For example, Weis-Fogh (1956a) showed that locust hindlegs oriented disadvantageously could increase total body drag by 48%. Many dipterans (Hocking, 1953) and orthopterans (Weis-Fogh, 1956a) fly with both meso- and metathoracic leg pairs deployed posteriorly along the abdomen (see also Atkins, 1958; Coggshall, 1972). The abdomen is generally regarded as a passive structure during flight. However, the abdominal tip rotates dorsally during the downstroke of pierid and other butterflies in takeoff (Ellington, 1984a; Grodnitsky, 1993; Sunada et al., 1993b). Lateral abdominal deflection is also used by some insects to effect yawing maneuvers (see section 5.2.2).

Most insects fly as solitary individuals, but mating swarms (section 5.3.3) and mass migrations (section 7.4) are also common. Little is known about synchronization of wingbeat and body kinematics in such aggregations, but behavioral mechanisms are clearly necessary to maintain cohesiveness as well as to prevent collisions (see section 5.3.3). Aerial aggregations also offer the possibility of aerodynamic interactions among individuals. In tethered locusts, individuals volitionally phase-lock their wingbeats with those of neighbors (Kutsch et al., 1994; Camhi et al., 1995), possibly utilizing useful vortex upwash as postulated for birds in formation (see Norberg, 1990). Aerodynamic consequences of swarming in flies and other taxa have not been studied, although the short interindividual distances are certainly suggestive of interaction among the vortex wakes produced by flying neighbors.

3.2 Aerodynamics

The motion of insect wings and bodies relative to the surrounding air generates aerodynamic forces on these structures. Steady-state lift and drag on insect bodies can be measured at airspeeds corresponding to those typical of forward flight. Substantial data also exist on the aerodynamic properties of insect wings exposed to time-invariant flow. By contrast, forces on oscillating insect wings are more complex and vary substantially in space and time. Unsteady aerodynamic effects are found in virtually all situations of interest in the aerial locomotion of insects. Independent of the character of aerodynamic mechanisms, an overall force and moment balance characterizes hovering and forward flight in insects. Alteration of this force and moment balance is utilized to regulate the speed of forward flight as well as to engage transiently in maneuvers.

3.2.1 Steady-State Aerodynamic Analysis

Aerodynamic forces parallel to relative flow are termed drag, whereas orthogonal forces are termed lift (see chapter 1). Knowledge of steady-state lift and drag coefficients on insect bodies enables analysis of the overall force balance acting on the insect body during flight. The net forces on flapping wings can be estimated through use of the so-called quasi-steady assumption that mathematically represents the dynamics of oscillating wings as a succession of independent conditions of steady-state force production. Although this assumption is unrealistic for most cases of wing flapping, such estimates provide a baseline comparison for more detailed unsteady aerodynamic models.

3.2.1.1 BODY FORCES AND MOMENTS

Bodies of insects in steady flight experience both lift and parasite drag. From empirical measurements of lift and drag on isolated bodies, nondimensional lift and body drag coefficients are easily calculated (eqs. 1.2 and 1.4). These coefficients can then be compared among taxa over a range of *Re* and body orientations relative to flow. Drag co-efficients of insect bodies range from 0.4 to 1.5, depending on flow conditions and taxonomic identity of the body (Hocking, 1953; Nachtigall, 1964; Weis-Fogh, 1956a; Vogel, 1966, 1981; Dudley and Ellington, 1990b; Nachtigall, 1991). Drag coefficients of insect bodies tend to decrease with increased *Re*, as indicated by variation of body drag on air velocity to some power less than two (e.g., flies: Hocking, 1953; locusts: Weis-Fogh, 1956a; *Drosophila*: Vogel, 1966; moths: Chance, 1975; Willmott and Ellington, 1997c; bees: Dudley and Ellington, 1990b; Nachtigall and Hanauer-Thieser, 1992; dragonflies: May, 1991; Wakeling and Ellington, 1997a).

Body drag coefficients also tend to decrease with decreased body angle, a general kinematic correlate of increased flight speed. Because of the increasing influence of viscous drag relative to pressure drag at lower *Re*, the effects of body orientation on parasite drag are, however, less pronounced for smaller insects (e.g., *Drosophila*: Vogel, 1966). Given that faster flight is necessarily associated with an increase in *Re* of the body (eq. 1.1), body drag of all insects exhibits a relative decrease at higher airspeeds for three reasons: the drag coefficient declines with *Re* and with body angle, and the cross-sectional area of the body relative to flow also decreases (see also Rayner, 1979a). An absolute reduction in body drag is not possible because of the general dependence of this quantity on the square of air velocity, but the substantial relative reduction in body drag and associated parasite power (see section 3.3.2) is nonetheless advantageous.

Streamlining to reduce overall pressure drag likely plays some role in insect body design, although a highly streamlined design would be unlikely given the diverse biological roles of insect bodies. In fact, body drag coefficients of insects tend to be more comparable to those of spheres than of streamlined objects moving at comparable *Re* (Dudley and Ellington, 1990b). Effects of streamlining are more pronounced at higher *Re*, and the generally elongated bodies of larger insects (e.g., dragonflies) might be expected to exhibit a more streamlined design appropriate for their more rapid flight. At normal cruising speeds, however, the ratio of body drag to weight is generally small (see below). Drag minimization at such speeds will probably not have substantial energetic consequences. Near maximal airspeeds, however, drag forces and associated power expenditure are likely to be mechanically limiting. The rapid increase in parasite drag and parasite power with air velocity may well limit top speed during flight, as was suggested by May (1991) for various dragonflies. Knowledge of drag forces and maximum flight speeds in other insect taxa is unfortunately too sparse to analyze the generality of such limits. The hawkmoth

Manduca sexta, however, exhibits very low body drag coefficients (Willmott and Ellington, 1997c) and may be, together with other sphingids, among the fastest-flying insects. Neither cruising nor maximum flight speeds of sphingids have been measured under natural operating conditions, but this taxon is a suitable candidate for airspeed measurements over a lake as described previously (section 3.1.1).

One indirect method of assessing the relative extent of streamlining is to measure drag for a given body orientation and then to repeat the measurement with the body reversed by 180° relative to oncoming flow. If drag is comparable in the two orientations, streamlining is likely to be of little significance (Vogel, 1981). Contrariwise, a high degree of asymmetry in parasite drag is consistent with effective streamlining. Such manipulations have revealed substantial unidirectional streamlining in bumblebees (Dudley and Ellington, 1990b) as well as in honeybees and large dipterans (Nachtigall and Hanauer-Thieser, 1992). No such orientational asymmetry in drag characterizes bodies of *Drosophila*, which fly at much lower *Re* and for which streamlining is of less importance in drag reduction (Vogel, 1981). Pilosity (hairs covering the body) may also contribute to the orientational dependence of body drag in bumblebees and other taxa (see Dudley and Ellington, 1990b). Additional structures such as legs and antennae may also influence relative streamlining. The presence of legs in normal flight orientation neither augments nor subtracts from total body drag in the honeybee (Nachtigall and Hanauer-Thieser, 1992) and in *Drosophila* (Vogel, 1981). This results suggests that leg placement is judiciously selected so as to minimize disruption of flow over the body, and is consistent with selection for drag reduction on the fuselage as a whole. Elongated or plumose (feathery) antennae (e.g., Vogel, 1983) may incidentally augment total body drag. For example, the energetic costs associated with drag on the extraordinarily long antennae of many longicorn beetles (Cerambycidae) could be substantial, although these insects typically fly rather slowly.

Lift forces on insect bodies are typically less than 10% of the body weight at the airspeeds and body orientations typical of forward flight. A historic claim that up to one-third of the body weight of a flying honeybee could be supported by body lift (Hocking, 1953) was shown to be a considerable overestimate of actual lift support (see Dudley and Ellington, 1990b). Additional studies support the conclusion that body lift rarely exceeds 10% of the body weight (e.g., cockchafer beetle: Nachtigall, 1964; the fly *Phormia regina*: Wood, 1970; noctuid moths: Chance, 1975; dragonflies: Okamoto et al., 1996; Wakeling and Ellington, 1997a; hymenopterans: Dudley and Ellington, 1990b; Nachtigall, 1992; Nachtigall and Hanauer-Thieser, 1992). Lift coefficients also show, in common with drag coefficients, a tendency to decrease at higher *Re* (see Dudley and Ellington, 1990b; Nachtigall and Hanauer-Thieser, 1992; Wakeling and Ellington, 1997a). Overall, the influence of body lift on the force balance of flying insects is small and of biomechanical importance only for fast-flying insects. As with parasite drag, body lift may be of significance for the overall force balance at energetically limiting airspeeds; this possibility has not been systematically investigated.

Empirical measurement of forces on insect bodies can be time consuming because variation with body orientation relative to flow and with *Re* must be measured independently. As an adjunct to direct measurements of lift and drag at variable orientations, a resolved flow analysis has recently been used to predict aerodynamic force coefficients (Wakeling and Ellington, 1997a). This method requires only knowledge of the drag force with the body oriented parallel to as well as orthogonal to flow. For other body angles, the normal and tangential components of air velocity relative to the body are first estimated trigonometrically. Potential interaction between the two components of velocity is assumed to be insignificant, and any lift (i.e., circulation) generated on the body is ignored. The drag forces corresponding to normal and tangential components of flow are estimated using equation (1.2) and the empirically determined drag coefficients along each flow dimension. Net force on the body is the vector sum of the normal and tangential forces, and can then be composed into components orthogonal and parallel to flow. This approach is best suited for long, narrow insects such as dragonflies, and in general can adequately predict drag but not lift coefficients. In odonates, the resolved flow method is, however, sufficient to reconstruct body drag and the overall force balance during flight (Wakeling and Ellington, 1997a). A general description of flight forces nonetheless requires empirical measurements of lift on insect bodies for the range of *Re* and body angles relevant to free flight.

Because gravitational acceleration substantially exceeds deceleration by body drag at most speeds, the body weight of insects is the dominant force acting in forward flight (see fig. 1.5). Except at very high forward speeds, thrust is small relative to vertical force production, and the magnitude of the resultant aerodynamic vector remains approximately constant to offset gravity. This conclusion indicates that regulation of forward airspeed can be attained primarily by changes in the direction but not magnitude of the net aerodynamic forces on the wings. Reorientation of the resultant force vector is most simply attained by altering the stroke plane angle, as first suggested by Bull

(1910). Only a small forward tilt of the force vector opposing gravity will be sufficient to provide horizontal components for thrust generation. For example, stroke plane angles of free-flying bumblebees change from 5° in hovering to 40° in forward flight at 4.5 m/s, but the forward tilt of the resultant force vector produced by the wings is only 8° (Dudley and Ellington, 1990b). Similar changes in direction but not magnitude of the output force vector characterize both tethered and free-flying *Drosophila* (see Vogel, 1966; David, 1978). In free-flying houseflies, ascent angle and body angle are positively correlated, consistent with simple reorientation of the net force vector (Wagner, 1986a). Flight at different airspeeds and ascent angles thus involves use of an approximately constant force vector acting at variable inclinations. This conclusion does not indicate, however, that wing aerodynamics and associated power expenditure are equivalent under all flight conditions. Wing relative velocities and orientations can change dramatically with airspeed (e.g., fig. 3.4), and substantial differences in wing lift and drag may be associated with otherwise subtle reorientation of the output force vector. Also, slight alterations in the magnitude of this vector are required to maintain a horizontal trajectory, as ascent or descent would otherwise ensue along with airspeed changes. Because insect wings are bilaterally paired appendages, net roll and yaw torques on the body are balanced when contralateral wings or wing pairs beat symmetrically.

Given that a major correlate of change in forward airspeed is variation in the stroke plane angle, regulatory mechanisms underlying this latter parameter are aerodynamically of critical importance. Dorsoventral wing motions within the stroke plane are largely but not entirely determined by the orientation of the axillary apparatus relative to the pleural wing process (see fig. 2.2). Substantial decoupling of the longitudinal body axis and the stroke plane angle is evident in some taxa, particularly among the Diptera (e.g., Ennos, 1989c) and Hymenoptera (e.g., Dudley and Ellington, 1990a). Such kinematic flexibility may be of particular significance for maneuvers. In forward flight, however, changes in thoracic inclination relative to horizontal, and thus changes in body angle, serve as the primary mechanism to alter the stroke plane angle. Pitch of the body relative to horizontal thus acts to partition indirectly the output force vector between vertical and horizontal components. Nose-down pitching moments will, in part, arise passively during forward flight because the point of application of resultant aerodynamic forces (body lift and drag) is, in general, posterior to the center of body mass. At least in bumblebees, however, the magnitude of this passive pitching moment is inadequate to maintain the body angles observed in free flight (Dudley and Ellington, 1990b). Active production of body pitching moments is instead indicated in forward flight, as well as for stationary hovering (see Ellington, 1984b).

The only active mechanism available to regulate body angle is aerodynamic force generated by the wings and imparted to the body as a pitching moment about the wing base axis. The center of body mass is located posterior to the thoraco-abdominal junction in many if not all insect taxa (e.g., Ellington, 1984b). A greater distance between the wing base and the center of body mass increases the nose-up gravitational pitching moment that must be offset by aerodynamic torque. Conversely, body responsiveness to aerodynamic torques is enhanced if the distance between the wing base and the center of body mass is reduced (see Ellington, 1984b). For example, the evolutionary tendency within Diptera toward reduction in the number of abdominal segments may increase body responsiveness in pitching while concomitantly decreasing parasite drag (Komárek, 1929). In Neotropical butterflies, such reduced rotational inertia of the body is correlated with enhanced evasive capacities (Srygley and Dudley, 1993). Body design in other taxa may, by contrast, obviate the need for substantial pitching moments. The inclined stroke plane angles of some hovering syrphid flies, for example, enable rapid airspeed changes by facilitating reorientation of the net aerodynamic force vector with little or no change in body angle (Ellington, 1984f; Ennos, 1989c). Other taxa, such as odonates, exhibit a center of mass located more posteriorly within the abdomen; such elongated bodies preclude rapid response in pitch or yaw. Alternative methods of effecting speed changes and maneuvers are instead used by odonates and by morphologically similar taxa such as ascalaphid neuropterans (see section 5.2.2).

Variation in mean positional angle of the wings provides the simplest means of generating nose-down pitching moments that influence body angle (Hollick, 1940; Ellington, 1984c). By changing the mean position of aerodynamic force production within the stroke plane, the moment arm of net aerodynamic forces about the wing base is altered in length, and the pitching torque is correspondingly changed. Positive mean positional angles indicate a dorsal location for the aerodynamic force vector if steady-state forces are generated. A decrease in the mean positional angle will shift the net aerodynamic force anteriorly and reduce the nose-downwards torque generated by the wings (e.g., Hollick, 1940; Schneider, 1980b). Changes in wing promotion and remotion relative to the mean stroke-plane angle also potentially yield dorsoventral shifts of the aerodynamic force vector relative to the center of body mass. Systematic changes in wing-tip paths with airspeed may thus contribute to control of body angle (e.g., Zanker, 1988b; Dudley and Ellington, 1990a).

Modulation of unsteady aerodynamic forces either during translation or at moments of wing rotation cannot, however, be excluded as potential mechanisms of body pitch control. For example, subtle kinematic modifications to the clap-fling mechanism may substantially alter pitching torque, particularly as the extreme dorsal location of the clap-fling lengthens the moment arm through which associated forces would act on the wing base (Ellington, 1984c). Similar arguments apply to wing supination at the end of the downstroke, although minimum positional angles are generally less in absolute magnitude than are maximum positional angles, reducing the length of the moment arm through which aerodynamic torque will act. Finally, passive aerodynamic pitch on a translating body may be altered via active positioning of the abdomen. Contraction of abdominal musculature in a pierid butterfly alters the angle between thorax and abdomen, and may contribute to regulation of body pitching moments (Obara, 1975). Similarly in *Drosophila*, dorsoventral rotation of the abdomen relative to the thorax generates aerodynamic torque that may alter the body angle in flight (Zanker, 1988b).

One final means of demonstrating the overall force and moment balance during flight is to examine performance at reduced or zero gravity. Generation of predominantly vertical forces that act against a gravitational field suggest that removal of this field will result in unbalanced forces during flight. May et al. (1980) showed that *Manduca sexta* in zero gravity could not hover relative to the local coordinate system but rather translated in space and collided with the surrounding container walls (see also Oosterveld and Greven, 1975, for similar experiments with pigeons). Roll and yaw were stable for *M. sexta* in zero gravity, although pitching was not well controlled. Because the aerodynamic torque from the wings was not counterbalanced by gravitational torque, the normal control of body angle was disrupted (May et al., 1980). Insect flight at low gravity thus demonstrates the predominance of vertical force production in earth-based flight as well as the aerodynamic regulation of body angle. Flight experiments over a graded range of hypogravity would test these trends quantitatively, and additionally could be used to generalize the results over a range of insect body sizes. Enhanced gravity experiments would also be informative, as such manipulations would provide a non-invasive means of increasing vertical force requirements from the flapping wings without the intrusive attachment of weights to the insect. No such experiments have been conducted with free-flying insects, although hypergravity can be applied chronically in centrifuges (Smith, 1992) as well as transiently during aircraft pullouts from steep dives.

3.2.1.2 STEADY-STATE WING FORCES

Wing size and speed are of paramount importance in steady-state lift production. Lift on a moving wing is proportional to the relative velocity of the air, the air density, and the circulatory strength of the bound vortex. In steady-state flow, wing circulation and lift typically increase with angle of attack of the wing until a plateau is attained (fig. 3.7A). Flow separates from the dorsal surface of the wing at higher angles of attack, forming a turbulent wake and in some cases then reattaching to the wing. Lift may remain substantially elevated at even higher angles of attack, but ultimately all flow separates from the wing, and stall ensues. Because circulatory lift relies on the inertial forces generated by vortical flow, relative lift performance improves at higher Re. Maximum values of the lift coefficient under steady-state conditions are near 1 for most insects and show little variation with Re among different taxa (fig. 3.7B). By contrast, the increased effects of viscous drag at lower Re result in higher profile drag, which varies in approximate inverse proportion to the square of the Re (Ellington, 1984c). The ratio of maximum lift to drag thus increases at higher Re (fig. 3.7B). The magnitude of induced drag influences the quantitative extent of this increase, but aerodynamic performance in general improves for larger wings moving more rapidly.

As with the ratio of lift to drag, the characteristics of stalling behavior are also Re dependent. At higher Re, lift tends to plateau at lower angles of attack, and the onset of stall is more pronounced. Lift maxima are thus more tightly constrained relative to angle of attack at higher Re, and are followed by a progressive decline in post-stall circulation and lift production (fig. 3.7A). The rate of post-stall decline in lift is more pronounced at higher Re, indicating a greater sensitivity of lift production to wing orientation. Aerodynamic effects of this orientational sensitivity are most pronounced in the presence of the spanwise velocity gradient generated during wing flapping. Wing twist to maintain a relatively constant angle of attack along the wing span might thus be more important at higher Re. Resting wing twist seems to be more pronounced in larger insects, but this trend has not been systematically evaluated. Neither has the relationship between resting wing twist and that actually manifested during the course of a wingbeat been determined. The importance of wing inclination relative to flow may, however, be of reduced significance in smaller insects. For

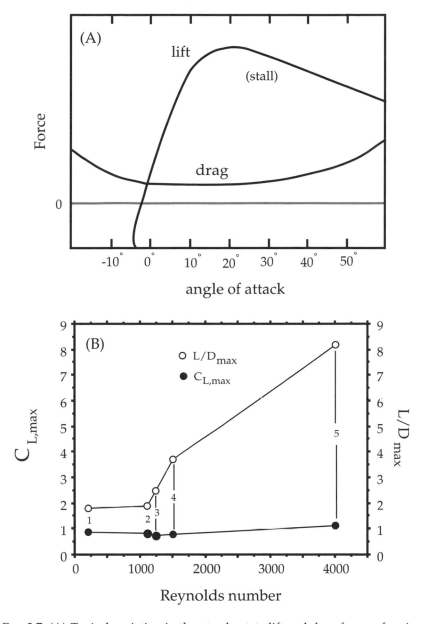

FIG. 3.7. (A) Typical variation in the steady-state lift and drag forces of an insect wing as a function of angle of attack. (B) Maximum lift coefficient ($C_{L,max}$) and maximum lift : drag ratio (L/D_{max}) versus the Reynolds number for various insect wings under conditions of steady-state flow. Data from (1) Vogel (1967b); (2) Nachtigall (1964); (3) Dudley and Ellington (1990b); (4) Nachtigall (1977); (5) Jensen (1956). For non-zero lift production, all values incorporate effects of the wing induced velocity. Data from Jensen (1956) used wings mounted in the boundary layer of a wind tunnel to mimic radial variation in flapping velocity; other studies used a constant airspeed across the wingspan.

example, little spanwise twist characterizes *Drosophila* wings (Vogel, 1967a; Weis-Fogh, 1972), suggesting that the angle of attack is more variable along the wingspan during flapping motions. Such orientational variability will have, however, reduced aerodynamic impact given the relative insensitivity of lift on low-*Re* wings to post-stall orientation.

The three-dimensional surface contour of wings often has important aerodynamic consequences. Effective airfoils are often cambered structures, and increased camber enhances lift production of many insect wings (see Vogel, 1967b; Ellington, 1984d). Wings of most large insects, even the seemingly planar wings of many Lepidoptera, are cambered to some extent at rest. Imposition of aerodynamic and inertial forces alters camber and results in conformational changes during the wingbeat, the most pronounced of which is camber reversal between down-and upstrokes. The potential variability in wing cross-sectional geometries provided by radial flexion lines as well as by transverse bending may also convey aerodynamic advantage. The best-studied example of radial bending is the "Z"-shaped upstroke profile of the locust forewing (section 2.2.2). Relative to a flattened profile of the same wing, the "Z"-shaped cross section exhibits increased lift production in steady-state flow at fairly high *Re* (Nachtigall, 1981a,b; see also table 1.2). At lower *Re*, however, the effects of camber on lift forces are much reduced (Dickinson and Götz, 1993).

Modification of wing curvature in either half-stroke in response to inertial and aerodynamic forces indicates that insect wings may act as reactive airfoils that are morphologically responsive to variable dynamic circumstances. Radial flexion combined with camber changes can dynamically yield a great diversity of wing profiles and configurations. Variable coupling of and bending between the fore- and hindwing of a coupled wing pair may yield similar aerodynamic flexibility. Many of these deformations and wing bendings are potentially rapid and aerodynamically adaptive changes that meet variable requirements of force production. For example, transiently excessive force applied to one region of the wing might induce a camber change that locally modifies or disrupts wing circulation.

Similar dynamic responsiveness can be postulated for entrainment and modulation of airflow within the irregular contours of the wing surface. The dorsoventral corrugation of wing venation provides local three-dimensional troughs within which trapped air may form recirculating flows (Brodsky, 1970; Ellington, 1984d). Most anteriorly on the wing, formation of a leading edge recirculation bubble may enhance overall lifting performance, particularly at the *Re* characteristic of much of insect flight (10^2–10^4). Recirculation bubbles within venational corrugation facilitate laminar reattachment of flow from the wing following initial separation from the leading edge (see Newman et al., 1977; Buckholz, 1986; Okamoto et al., 1996). Consistent with these observations, Rees (1975a) found that models of corrugated insect wings exhibited aerodynamic performance comparable with that of the same models covered with a continuous membrane delineating the dorsoventral extremes of corrugation. Recirculation of air between venational peaks can thus alter the effective aerodynamic contour of the wing. Rudolph (1978) concluded that such corrugational effects were more pronounced at higher Re, consistent with lift production dominated by circulating airflows. Under conditions of steady-state flow, abrupt changes in the geometry of trapped vortices in adjacent corrugations may also contribute to irregular jumps in otherwise smooth plots of lift versus drag coefficients (see Newman et al., 1977). Insects may potentially utilize similar transients in force production to rapidly alter magnitude and direction of the instantaneous force, although detailed three-dimensional analysis of a series of wingbeats and associated body accelerations would probably be necessary to establish this possibility.

Although not relevant to flapping wings, steady-state lift production is potentially significant for the usually stationary elytra of beetles and for other heavily sclerotized or tegminized forewings. Contributions of elytra and tegmina to aerodynamic force production are of particular interest because of the apparent conflict between structural and aerodynamic roles. Flapping of heavy structures designed for physical protection is unlikely to be biomechanically advantageous, and elytra in most beetles are held laterally and stationary relative to the body during flight. Beetles usually fly slowly, and the ensuing elytral contribution to body weight support is likely to be small. Flow visualizations with elytra at low angles of attack reveal patterns of airflow that are consistent with limited lift production (Schneider and Hermes, 1977). Burton and Sandeman (1961), however, concluded that optimally oriented elytra of a scarabaeid beetle could offset up to 40% of body weight at a speed of 5 m/s. The potential aerodynamic role of stationary elytra clearly must depend on the airspeed during flight, but virtually no relevant data are available for free-flying beetles.

Lift generation by elytra differs substantially from that produced by the flapping hindwings in that the former produce no net thrust. Profile drag and the small induced drag associated with elytral lift production must in fact be offset by the hindwings. By contrast, hindwing flapping generates both vertical forces as well as a horizontal force that overcomes parasite drag and produces thrust (see below). In some beetle species, the elytra do make low-amplitude flapping motions in phase with the hindwings. Even such flapping elytra typically produce much less lift than do the beating hindwings (e.g., Nachtigall, 1964; Schneider and Hermes, 1976). Elytral size and elytral kinematics during flight have not been mapped onto a phylogeny of beetles, but clearly would be of interest to understand further the origins of this important forewing modification.

No aerodynamic studies have been made on the hemelytra of hemipterans nor, with the exception of some Orthoptera, on tegminized or similarly modified forewings (see table 2.1). In orthopterans, however, the tegmina appear to provide aerodynamic contributions intermediate to those of elytrized and nonelytrized wings. Flapping forewings of tethered migratory locusts produce only about 30% of total aerodynamic forces (Jensen, 1956). The fore- and hindwings beat at similar frequency, but, not surprisingly, both area and stroke amplitude of the forewing are substantially reduced relative to its ipsilateral counterpart (Weis-Fogh, 1956a). Potential aerodynamic interactions between fore- and hindwings were not evaluated in the study of Jensen (1956), and in general are not well understood. Such interactions are a potentially important influence in the evolution of both tegmina and elytra. Rigidly held elytra, for example, may interfere aerodynamically with flapping hindwings; the generally high stroke amplitude of beetles (section 3.1.2) may partially mitigate such interference. As with elytra, little is known of the phylogenetic origins or evolutionary diversification of either hemelytra or tegmina.

Some of the most diverse wing shapes among insects occur in the Lepidoptera. For example, a salient feature of many papilionid butterflies (particularly those in the tribes Papilionini and Leptocircini) is the elongated tail of the hindwings. Aerodynamic consequences of variable papilionid hindwing shapes are not known, although Demoll (1918) manipulated wings of the swallowtail *Papilio machaon* and found no change in gliding performance following unilateral removal of the tail. Elongated hindwings are widespread in the Lepidoptera (e.g., various Hesperiidae, Saturniidae, Uraniidae, Zygaenidae) and also appear in other insect orders, most notably the nemopterid Neuroptera (plate 1D). Such elongation, when sexually dimorphic, is most probably the consequence of sexual selection (as in bird tails; see Andersson, 1994). Preference by female insects for elongated hindwings remains, however, to be demonstrated. In steady-state airflow, the elongated hindwings of some butterflies appear qualitatively to stabilize vortices shed from the wing (Martin and Carpenter, 1977). As with bird tails, elongated hindwings may invoke some aerodynamic cost if only by increasing both pressure and viscous drag (see Balmford et al., 1993; Cuervo et al., 1996). The possible use of elongated hindwings to generate advantageous aerodynamic torque during maneuvers is also uninvestigated in insects (see Oehme, 1965, and Norberg, 1994, for compelling avian examples).

Surface texture of insect wings may also influence aerodynamic performance. In dragonflies, large spines projecting from major wing veins potentially modify boundary-layer characteristics and lift production (D'Andrea and Carfì, 1988, 1989). Such effects have not yet been demonstrated experimentally, although lift and drag measurements prior to and following spine extirpation are experimentally straightforward. The setae (hairlike projections) and fringing characteristic of many insect wings can obviously increase effective surface area if present in sufficient quantity. Scales, as modified hairs, represents the most extreme modification of wing texture. In the Lepidoptera, scales normally overlap one another to produce a shingled pattern on the wing. Moreover, these scales are usually characterized by parallel ribs that run the length of the scale. Extensive fluting and transverse cross-ribs between these longitudinal ribs contribute to the architectonic complexity of the scale (Downey and Allyn, 1975). Overlapping of scales on the wing surface, in conjunction with the braced composition of scales at finer levels of structural resolution, may contribute to mechanical integrity and rigidity of the wing. However, direct selection for the structural support provided by scales (as opposed to that of veins) is difficult to envision.

Most interest in the aerodynamics of wing surface structure has focused on potentially advantageous effects of butterfly wing scales. Nachtigall (1965, 1967; see also Brodsky and Vorobjov, 1990) measured lift characteristics of entire butterflies as well as on isolated wings in the presence and absence of wing scales. Scale removal slightly reduced lift, consistent with lift enhancement by scales when present in their natural configuration. Total drag was surprisingly unaffected by scale removal, however, and correlated changes might be expected in lift and drag if scales exerted an aerodynamic influence. Another possibility is that conformational changes in wing geometry associated with the process of scale removal influenced subsequent aerodynamic measurements. Aerodynamic consequences of scale removal are thus not fully resolved. The vertical projection of butterfly scales is low relative to the thickness of wing boundary layers (Martin and Carpenter, 1977), and their aerodynamic properties might also be expected to be minimal.

However, microstructures sometimes have unanticipated effects on the characteristics of turbulent boundary layers and on drag. Walsh (1980), for example, found that parallel ribbing on a surface reduced drag in steady flow by as much as 7%; such ribbing reduced lateral velocity fluctuations in the turbulent boundary layer over the surface. Shark skin, the scales of which possess fine longitudinal ridges, exhibits a similar effect (Bechert et al. 1985). The three-dimensional shape of such scales as well as the interscale distance appear to influence flow within the viscous sublayer of a turbulent boundary layer; these effects occur at Re characteristic of many butterfly wings (see Dudley, 1990). At somewhat higher Re, substantial drag reduction can also be attained by microstructures arranged semirandomly within turbulent boundary layers (see Bushnell and Moore, 1991; Sirovich and Karlsson, 1997). Both the striations on individual scales as well as the sequential arrangement of scales on the wing may thus influence flow patterns over butterfly wings. If such effects do exist, technological emulation of the diverse microstructural components described from lepidopteran scales might be constructive. The presence of scales may, however, impose a direct energetic cost via an increased inertial power to accelerate the mass of wing scales during flapping. For example, scales represent approximately one-third of the total wing mass in heliconiine butterflies (Gilbert and Dudley, unpublished data). The actual magnitude of inertial power expenditure depends, however, upon the extent of elastic energy storage within the pterothorax (see section 3.3.2).

The most relevant entomological context for applying steady-state aerodynamic analysis is in gliding flight. Steady gliding occurs at a constant airspeed and at a constant angle θ relative to the horizontal (fig. 3.8). Lift forces on the wing and body act perpendicularly to the flight path, whereas drag forces (the induced and profile drags on the wings together with parasite drag on the body) act parallel to the flight path. Vertical components of lift and drag on the wings and body must balance body weight in nonaccelerational gliding. Similarly, horizontal components of lift and drag balance each other out. Under these conditions, the ratio of drag to lift is then given by arctan θ (fig. 3.8). A higher lift to drag ratio thus yields, not surprisingly, a more shallow glide.

Because induced wing drag effectively acts to decrease the glide angle, gliding in general is more effective with wings of higher aspect ratio. Profile drag also increases with aspect ratio, however, and wings of intermediate aspect-ratio may yield the lowest glide angle (Ennos, 1989a). No insect wing approaches the high aspect ratios typical of well-known avian gliders, and relatively low aspect-ratio wings may be most appropriate for gliding in the Re range typical of insect fliers (Ennos, 1989a). As mentioned previously, gliding flight is not widespread among insects, although jump-initiated gliding may have played an important role in the initial evolution of flight (section 6.1.5).

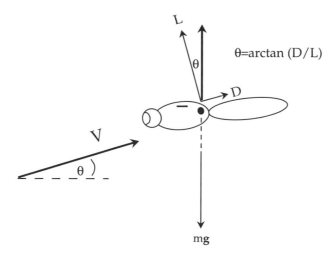

FIG. 3.8. Mechanics of gliding at constant airspeed V and glide angle θ. Lift forces (L) are generated on the wing and body, whereas the drag forces (D) include induced and profile drag on the wings as well as parasite drag. Summed vertical components of lift and drag offset the body weight $m\mathbf{g}$. Note that not only forces but also rotational moments must balance in stable gliding (see fig. 1.5).

This latter form of gliding likely involved both vertical and horizontal accelerations rather than flight at a constant velocity. The interconvertibility of kinetic and potential energy permits horizontal as well as upward glides if deceleration occurs simultaneously (see Wakeling and Ellington, 1997a).

3.2.1.3 QUASI-STEADY ANALYSIS

The applicability of steady-state aerodynamics to force production by flapping animal wings has historically been the subject of great interest. To what extent do the aerodynamics of flying animals deviate from or transcend the maximum performance obtained from wings in steady flow? Generation and regulation of unsteady aerodynamic forces has likely been fundamental to insect diversification, but only a quantitative description of underlying mechanisms will permit an evaluation of their evolutionary role. The identification and description of novel unsteady mechanisms also hold the possibility of technological application. Fixed-wing aircraft as well as helicopters with their continuously rotating blades reflect only faintly the wide range of aerodynamic possibilities afforded by unsteady aerodynamic mechanisms. Given the increased significance of wing flapping and temporally variable relative velocities at low advance ratios (eq. 3.1), unsteady mechanisms are most likely to predominate during slow forward flight and hovering.

One approach to estimating the relative importance of unsteady aerodynamics in animal flight has been to calculate the average lift performance from flapping wings that would be sufficient to offset the body weight. This method conceptually reduces wing flapping to a series of consecutive conditions of steady-state flow over the wing or over a set of spanwise wing sections. Spanwise flow along the wing is neglected; because individual spanwise sections of the wing are evaluated independently, this method is also termed a quasi-steady blade element analysis. Using available kinematic and morphological data, aerodynamic forces are calculated assuming steady-state wing lift and drag production (eqs. 1.2 and 1.4) for wing sections at appropriate instantaneous angles of attack and relative velocity. These forces are summed temporally through the wingbeat cycle, and a mean lift coefficient is then calculated (following eq. 1.4) to satisfy the requirement that vertical forces equal the body weight. If such estimated lift coefficients exceed a maximum measured directly on actual wings in steady-state flow, then unsteady mechanisms are implicated in a so-called proof-by-contradiction (i.e., estimated lift coefficients contradict the empirical maxima). Estimated coefficients that are equal to or lower than maximum values in steady-state flow do not, however, unequivocally preclude unsteady mechanisms (Ellington, 1984a; Spedding, 1992). The quasi-steady approach is of limited utility, therefore, but represents a useful first approximation to the overall question of mechanisms of force production.

Using a quasi-steady approach, Hoff (1919) evaluated likely aerodynamic performance of flapping insect wings during forward flight. The flapping velocity of the wings was excluded from calculations of wing relative velocity, and mean lift coefficients were thus substantially overestimated. Von Holst and Küchemann (1941, 1942; see also Garrick, 1936) constructed a generalized two-dimensional quasi-steady model for flapping flight that ignored the induced velocity in estimates of wing relative velocity, and that moreover was limited to flight at high advance ratios. Osborne (1951) implemented the first detailed quasi-steady approach to insect flight whereby mean lift and drag coefficients required to support the body weight were estimated from species-specific kinematic and morphological data. Relative velocities of wing sections were estimated from the wing flapping velocity, the forward velocity of the insects, and the induced airflow associated with the tip vortex. Ensuing aerodynamic forces were integrated along the wingspan and through the wingbeat to derive total forces.

For flight at low advance ratios, quasi-steady studies of wing relative velocities necessarily rely on meaningful estimates of the induced velocity. Osborne (1951) and many subsequent studies have estimated the induced velocity on flapping wings using a Rankine-Froude propeller model. This model presumes that a downward momentum impulse is applied continuously to the air across a theoretical two-dimensional region termed the actuator disk. The impulsive generation of a net momentum flux across the actuator disk offsets the body weight in flight and results in a downward jet of air. The actuator disk may be assumed to be either a circular disk (with radius equal to the wing length), or the actual area swept out by the wings (i.e., ΦR^2; see Ellington, 1984e). Based on the momentum of the jet and the area of the actuator disk, the downward air velocity in the plane of the beating wings can be calculated (see eq. 3.7); this airflow is conceptually identical to the induced velocity of the wings. In the simplest formulation of actuator disk theory, the induced velocity is assumed to be constant across the wing and through the wingbeat. Temporal and spanwise variation in lift production and associated induced flow can, however, be substantial during wing flapping. Nonetheless, use by Osborne (1951) and others of an invariant induced velocity provided a first approximation that proved useful for indicating qualitative trends in estimates of force coefficients.

For a variety of insects in forward flight, Osborne (1951) calculated high mean quasi-steady force coefficients (e.g., mostly >1) that exceeded steady-state values for airfoils at comparable Re. This work was thus the first detailed evaluation of the relative importance of unsteady aerodynamic mechanisms in insect flight. Although the kinematic data (and particularly the wingbeat frequencies) used by Osborne (1951) were probably inaccurate for the insect taxa under consideration (see Weis-Fogh and Jensen, 1956), the general conclusion of substantial unsteady effects during slow forward flight and hovering was probably robust. More detailed quasi-steady studies of forward flight in insects have reached similar conclusions as to the relative importance of unsteady effects. A detailed quasi-steady analysis applied to the free forward flight of bumblebees concluded that unsteady mechanisms characterized force production over a range of airspeeds from hovering up to 4.5 m/s (Dudley and Ellington, 1990b). Mean lift coefficients for flapping bumblebee wings tended to decrease with increased airspeed, an observation consistent with that of Osborne (1951), who similarly concluded that lift coefficients necessary for forward flight were inversely correlated with the advance ratio. At higher advance ratios, effects of wing flapping become less significant, and the temporally and spatially invariant forward airspeed progressively dominates flow over the wings.

In contrast to these estimates suggesting substantial unsteady effects in wing flapping, a classic study of tethered locusts (Jensen, 1956; see also Nachtigall, 1981a, b; Zarnack, 1975) concluded that steady-state mechanisms were sufficient for both weight support and horizontal thrust production. These studies considered locusts flapping their wings in airstream velocities of about 3 m/s, a value corresponding to slow forward flight in this taxon. A detailed series of analyses (Weis-Fogh, 1972, 1973, 1975, 1976; see also Norberg, 1975) also concluded that steady-state mechanisms were sufficient to sustain hovering flight in *Drosophila* and other insects, as well as in hummingbirds. These analyses used morphological and kinematic data that were more accurate and detailed than those used by Osborne (1951). In the migratory moth *Urania fulgens* (Dudley and DeVries, 1990) as well as in the hawkmoth *Manduca sexta* (Willmott and Ellington, 1997c), mean lift performance from the wings exceeds maximum steady-state values only at low airspeeds and, for the hawkmoth, during hovering flight. Thus, a number of different studies suggest that steady-state mechanisms of force production are, at least in quasi-steady principle, sufficient to sustain flight for certain insect taxa flying at low airspeeds and even when hovering.

Use of the proof-by-contradiction method to validate steady-state lift performance does not, however, unequivocally eliminate the possibility of unsteady effects acting at particular moments during the wingbeat. Transient unsteady peaks of high force production may be subsumed within a mean coefficient for the wingbeat that is consistent with steady-state flow. As noted by Spedding (1992), unsteady effects, no matter how small their relative magnitude, must be incorporated into any general model of flapping flight. This observation is best illustrated by a study that directly measured the instantaneous forces produced by a tethered locust flapping its wings in an airstream (Cloupeau et al., 1979; see also Wilkin, 1990). Inertial forces associated with wing flapping were subtracted from total forces to yield the instantaneous aerodynamic force as a function of time through the wingbeat. Peak values of the aerodynamic forces were found to be approximately twice the magnitude predicted by the analysis of Jensen (1956), although these forces averaged through the wingbeat were consistent with steady-state lift production. Similarly, direct force measurements on tethered *Drosophila* revealed that both the magnitude and direction of aerodynamic forces were incompatible with steady-state mechanisms (Zanker, 1990b; Zanker and Götz, 1990). These results shows that the proof-by-contradiction method, while heuristically informative, can yield insight into neither the temporal dynamics nor precise mechanisms of force production.

In contrast to the aforementioned actuator disk model of induced velocity, vortex wake models specify more precisely the shape and magnitude of the velocity flow field shed by flapping wings. These methods thus provide a more accurate estimate of temporal and spatial variation in the induced velocity, and are of particular importance for studies of hovering flight. In this condition, the absence of a forward velocity vector requires that momentum flux arise solely through action of the flapping wings; induced velocities are correspondingly highest in hovering. Accurate estimates of the induced velocity are thus essential for meaningful calculations of wing relative velocity and lift performance during hovering. One approach to estimating induced velocity is to treat wing circulation as bound linear element of vorticity that stretches across the wingspan. Such lifting line models have been extensively applied to flapping and rotating wings in technological contexts (e.g., Betteridge and Archer, 1974; Archer et al., 1979; Jones, 1980; DeLaurier; 1993a, b), but typically assume low-amplitude flapping motions and high advance ratios that preclude any influence of shed vorticity on subsequent changes in wing circulation. Similar models that assume high aspect ratio wings and a constrained vortex wake geometry have also been developed for flapping wings (see Phlips et al., 1981; Ahmadi and Widnall, 1986; Guermond and Sellier, 1991). Flying insects violate at least two basic assumptions of these aerodynamic models given generally high amplitudes of wing flapping and associated components of spanwise airflow. Shed wake vorticity is also of sufficiently high magnitude and proximity at low advance ratios to alter substantially the dynamics of circulation growth on flapping wings.

For hovering flight, Ellington (1984e, f) developed a detailed vortex wake theory to analyze more accurately the induced velocities on flapping wings. The area of the Rankine-Froude actuator disk across which downward momentum is generated was assumed to be the horizontal projection of the actual area in space through which the flapping wings sweep and shed vorticity. The otherwise constant momentum flux produced by a Rankine-Froude actuator disk was then modified to reflect the fact that momentum pulses are periodically applied across the area of the disk as the wings sweep back and forth within the stroke plane. Spatial variance in the applied pressure impulse was also incorporated given that wing circulation and the induced velocity vary with radial position along the wingspan. Addition of these spatial and temporal fluctuations to the wake increased estimates of the induced velocity by 15% on average for hovering with a horizontal stroke plane, and by a much higher value if the stroke plane was inclined relative to horizontal. Contrary to the results of Weis-Fogh (1972, 1973), Ellington (1984f) concluded that hovering flight was incompatible with steady-state mechanisms, particularly in those animals that hover with an inclined stroke plane. Bumblebees in forward flight (Dudley and Ellington, 1990b), tethered *Drosophila* (Zanker and Götz, 1990), and various hovering insects (Ellington, 1984f) unequivocally rely on unsteady mechanisms, whereas other insects may at least transiently deviate from steady-state force production during both hovering and forward flight. The associated unsteady aerodynamic mechanisms are discussed in section 3.2.2.

Independent of the character of associated aerodynamic mechanisms, both hovering and steady forward flight require an equilibrium of net forces and torques acting on the center of body mass (fig. 1.5). In hovering flight with a horizontal stroke plane, down- and upstroke motions are approximately symmetric (Ellington, 1984c). Vertical forces derive primarily from lift on the wing and are approximately equivalent in magnitude between the down- and upstrokes. Drag forces on the wings are primarily horizontal and offset one another between half-strokes. As the advance ratio increases above the values characteristic of hovering flight, progressively greater aerodynamic asymmetry characterizes the two half-strokes. At higher advance ratios, the velocity of the downstroke increases with respect to that of the upstroke (fig. 3.5A). The relative orientation of the two half-strokes also changes, with the downstroke becoming more horizontal and the upstroke more vertical (figs. 3.4, 3.5B).

These orientational effects combine to direct wing lift forces during the downstroke progressively more vertically at higher advance ratios. The increased relative velocity of the downstroke also contributes to enhanced lift production offsetting the body weight; insect taxa that cannot hover are constrained by a minimum flight speed below which the wing relative velocity is insufficient to generate the forces required to offset the body weight. Thrust must be generated to overcome body drag in forward flight, and the vertical motions of the upstroke correspondingly orient the wing lift vector more horizontally. Thrust production thus predominates during the upstroke. Because lift-to-drag ratios are generally high for insect wings, the influence of wing drag on the overall force balance is small in forward flight. Contributions of wing drag to the vertical and horizontal force balance may, however, be more pronounced for smaller insects flying at lower Re. Quasi-steady analyses of locusts tethered in an airstream (Jensen, 1956) and of free forward flight in bumblebees (Dudley and Ellington, 1990b) confirm these observations; most vertical force is produced during the downstroke. This effect arises most directly from the use of a tilted stroke plane in forward flight.

Because wing orientation relative to the stroke plane may differ between down- and upstrokes, variation in the wing's angle of attack also potentially contributes to half-stroke asymmetry. Such variation may, depending on the orientational dependence of lift and drag (fig. 3.7A), lead to substantial asymmetry in force production. If pronation and supination are substantial, the sense of air circulation about the wing likely reverses direction during the upstroke (Lighthill, 1977; Dudley and Ellington, 1990b). Such circulation reversal is matched in many insects by wing camber reversal during the upstroke. Because camber cannot be perfectly reversible in such wings, however, upstroke lift is likely to be even further reduced relative to that of the downstroke. The effective area of the wing during the upstroke may be partially reduced because of the basal wing twisting required for stroke reversal, and because of potentially variable overlap between ipsilateral wing pairs. At least the latter effect is, however, unlikely to be substantial. For example, no significant difference in effective wing area between half-strokes characterizes the moth *Manduca sexta* over a range of forward airspeeds (Willmott and Ellington, 1997b). In contrast to flying vertebrates that exhibit often considerable spanwise wing bending during the upstroke (Norberg, 1990), no reduction in effective wing area has been documented for insects. The various asymmetries introduced kinematically are, however, more than sufficient to partition the down- and upstroke into the particular vertical and horizontal components of force required for steady forward flight.

3.2.2 Unsteady Aerodynamic Mechanisms

Quasi-steady assumptions are clearly violated by insects over a range of forward airspeeds and for hovering flight. What exactly are the unsteady aerodynamic mechanisms that contribute to force and moment balance in flight? Reviews for flying animals generally (Spedding, 1992, 1993; Spedding and DeLaurier, 1996; Dickinson, 1996), insects (Nachtigall, 1979a; Maxworthy, 1981; Ellington, 1984d; Ellington, 1995), and dragonflies in particular (Wakeling, 1993) have all outlined the possible patterns and dynamics of unsteady flows utilized in flight. More broadly, unsteady aerodynamic mechanisms were reviewed by Sarpkaya (1996). It should be emphasized, however, that for no insect is even a complete qualitative description of unsteady flow patterns and associated forces available to describe the dynamics of a full wingbeat. Most quantitative descriptions of unsteady mechanisms, moreover, are confined to simplified theoretical or experimental contexts that do not necessarily correspond to biological reality. From an entomological perspective, even these complex models are remarkably constrained in their representation of actual wing motions. Nonetheless, several broad categories of unsteady forces are likely to be relevant to insect flight.

3.2.2.1 VIRTUAL MASS FORCES

Insect wing motions are multioscillatory in that wings flap about their base and also rotate about their longitudinal axis. Wing flapping and rotation are approximately 90° out of phase. It is important to distinguish conceptually between the concept of a plunging airfoil that moves up and down uniformly within an airstream, and that of a flapping airfoil. The former situation refers to vertical airfoil motions in a fluid with no spanwise flow along the wing, whereas the latter refers to periodic reciprocation of an airfoil hinged about a fixed axis, as characterizes insect wings. The relative phase and extent of flapping and rotation determine the magnitude of circulation shed from the wing, the structure of the vortex wake, and associated lift and drag on the wing. Spanwise flow arising from flapping, and three-dimensional effects more generally, must accordingly be an important feature in insect flight. Oscillatory motion also implies continuous acceleration and deceleration of the wing, with concomitant forces associated both with wing inertia and with the virtual mass of air accelerated around the wing.

The virtual (or added) mass may be comparable to the wing mass in some insects (Vogel, 1962; Ellington, 1984b) and represents a potential unsteady contribution to total forces on flapping animal wings, as recognized by Oehmichen (1920) and Osborne (1951). Although the force associated with acceleration of the wing virtual mass has been termed "aerodynamic inertial force" (Weis-Fogh and Jensen, 1956), this force is best viewed as an aerodynamic force contingent upon wing acceleration. The virtual mass of flapping wings has empirically been shown to vary with wing geometry and with wing orientation relative to the axis of flapping (Jones, 1941; Scruton, 1941). Also, the advance ratio influences the virtual mass of flat plates in periodically oscillating fluids (Keulegan and Carpenter, 1958). Both wing morphology and specific kinematic features therefore contribute to the magnitude of the added mass. To date, no experimental determinations of added mass for the wing kinematics and Re specific to insect flight have been carried out. Such measurements must address simultaneously the influence of the forward velocity vector (Otzen and Kuiper, 1983), wing geometry and orientation, and the effects of wing rotation. The kinematic complexity and number of degrees of freedom that characterize this problem in dynamics mandate that biological applications use simplified representations for the added mass of wings.

Ellington (1984b) pointed out that wing accelerations for many insects are greatest at the ends of wing strokes, at which point wing chords are approximately perpendicular to the direction of travel. In this case, the added mass about a wing chord is given by $\pi\rho c^2/4$, where ρ is air density and c is the chord length; this chord-specific mass may be summed along the wingspan to determine the total added mass. This formulation for virtual mass has been used widely in studies of insect flight, although such phenomena as rotation and acceleration nonorthogonally to the axis of wing translation clearly complicate this generalized formulation. Moreover, subtle changes in wing angular position through the wingbeat are magnified in velocity and acceleration profiles, and slight variation in wingbeat kinematics may thus have major aerodynamic implications. Inertial power to accelerate the wing mass and virtual mass, for example, varies with the square of the maximum angular velocity. Even small deviations from simple harmonic motion can substantially alter wing angular velocities and result in amplified energetic consequences. Temporal variation in acceleration (as characterizes flapping wings) may also influence the magnitude of the virtual mass. At present, neither physical theory nor empirical data are available to estimate the magnitude of these effects for flapping insect wings.

Because wingbeat kinematics are usually symmetric, inertial forces associated with the wing mass and virtual mass will alter only maximum transient rather than mean forces during flight (Ellington, 1984d). Houghton (1964) in fact suggested that such inertial forces were predominant in insect flight, but reached this conclusion using unlikely values for wing lift coefficients and for the inertia of the wing mass and virtual mass. In some cases, however, instantaneous values for wing inertial forces are high (e.g., Cloupeau et al., 1979), and the relative contribution of the virtual mass to total wing inertia can be considerable. Significant kinematic differences between the down- and upstroke may result in substantial half-stroke asymmetry in the inertial forces, potentially assisting weight support. The role of inertial forces in maneuvers is also unclear but may be important given the associated asymmetries in ipsilateral and contralateral wing motions (see section 5.2.2). Understanding of these inertial effects is hindered by the absence of a detailed formulation for the wing virtual mass under different kinematic regimes.

3.2.2.2 WAGNER EFFECT

Time-dependent effects also characterize aerodynamic forces on wings moving at constant airspeed. Most important of these is the temporal evolution of circulation and lift on an impulsively started airfoil. This phenomenon was first noted by Wagner (1925) and is eponymously termed the Wagner effect. As confirmed empirically (Walker, 1931) and theoretically (Chow and Huang, 1982; Graham, 1983), an airfoil must translate 7–8 chord lengths before circulation attains the steady-state value predicted by the Kutta-Joukowski law (eq. 1.3). This delay derives from interference of already-shed circulation with subsequent development of circulation about the wing (see fig. 1.4). The Wagner effect is particularly critical for reciprocating animal wings, as substantial chord travel will be necessary each time a wing is accelerated in order to attain maximum lift production. The Wagner effect also functions during wing deceleration, so that circulation declines only slowly during the second half of a half-stroke relative to the half-stroke maximum. If the sense of circulation changes at wing reversal, lift production is even further impeded as the Wagner effect applies first in the decay of existing circulation on the wing, and then during buildup of circulation in the opposite direction. In fast forward flight, the high rates of chord travel during the downstroke (fig. 3.4) suggest that the Wagner effect will be of minimal significance. In hovering flight, however, the absence of horizontal translation minimizes chord travel during each half-stroke, and time-dependent effects of lift buildup can be substantial (Ellington, 1984d).

Few experimental data exist on the magnitude of the Wagner effect for wings at the Re and kinematic profiles appropriate to insect flight. One general theoretical prediction is that the dynamics of circulatory change over time are Re dependent (Childress, 1981). Dickinson and Götz (1993) evaluated transient aerodynamic characteristics of linearly translating insect wing models, and concluded that delay in lift buildup is less pronounced at the lower Re relevant to flight of small-to medium-sized insects (i.e., $10^2 < Re < 10^3$). Because of the positive allometry of advance ratios (section 3.1.2), the reduced chord travel per half-stroke that characterizes smaller insects will be, in part, offset by a less pronounced Wagner effect. Conversely, the hovering flight of large insects and hummingbirds will be characterized by substantial delays in the buildup of wing circulation. Furthermore, the shedding of wing circulation at the end of each half-stroke deposits vorticity in the wake that may interfere with circulation growth in the subsequent half-stroke. This interference will depend on the local wake geometry and on the magnitude of the forward velocity, again being most significant in hovering flight.

3.2.2.3 WING ROTATION AND THE CLAP-FLING

Effects of rotation often interact with wing flapping to yield complex aerodynamic outcomes. In some cases, the coupling of wing rotation with translation can be aerodynamically advantageous. An increasing angle of attack during airfoil translation may transiently increase lift above steady-state values (Kramer, 1932; Bennett, 1970; Ellington, 1984d). This so-called Kramer's effect is closely related to the delayed or dynamic stall that can occur transiently at angles of attack above the stalling value (Francis and Cohen, 1933). Under such conditions, the standard dropoff of lift is delayed and the airfoil experiences enhanced lift production. This dynamic stall derives from unsteady separation of flow from the leading edge of the wing; leading edge vortices appear to be responsible for the concomitantly high lift forces (Mehta and Lavan, 1975; Izumi and Kuwahara, 1983; Wu et al., 1991; see section 3.2.2.4). The Re dependence of dynamic stall has not been systematically investigated, but this mechanism clearly is of relevance to insect flight at low Re (e.g., Dickinson and Götz, 1993). Attached leading-edge vortices may further be stabilized by the radial components of airflow associated with wing flapping (Maxworthy, 1979; Ellington et al., 1996). The wide range of kinematic possibilities afforded by variable wing rotation during the wingbeat suggests that an equivalent aerodynamic diversity may characterize lift production under post-stall conditions.

Wing rotation at the ends of half-strokes may also result in substantial rotational airflow about the wing, augmenting the circulation associated with wing translation. If rotational velocities are sufficiently high, the quasi-steady circulation associated with pronation and supination can be comparable to that generated during the wingbeat (see Ellington, 1984d). Few empirical results are available on this point for the low Re characteristic of insect flight, although models of insect wings rotated about their trailing edge demonstrate transiently enhanced circulation and lift production (Dickinson, 1994). Both chordwise location of the rotational axis as well as the location of previously shed vortices influence the magnitude of lift produced in this manner. A general finding of this and related empirical studies on unsteady force production is that lift forces are highly sensitive to particular kinematic details. The actual wing kinematics of free-flying insects must therefore be accurately replicated if unsteady aerodynamic forces are to be reliably estimated in biological contexts. The presence of wing flexibility reinforces this conclusion; bending about flexion lines may subtly alter shedding of vorticity from the wing. Ellington (1984d) proposed that wing flexion at the end of the downstroke might shed circulation from the leading rather than the trailing edge of the wing, indirectly facilitating rapid growth of circulation in the following half-stroke. Both the timing and magnitude of wing rotation relative to translation may therefore be of major importance for force production (Ellington, 1995). The effects of variable camber and possible camber reversal between half-strokes further indicate the relevance of kinematic details to patterns of unsteady force production.

Of all unsteady aerodynamic mechanisms, potential interaction between contralateral wings or wing pairs is the most biologically immediate given the morphological condition of bilaterally paired appendages. Wing pronation and supination occur at the dorsal and ventral extremes of half-strokes, respectively, and thus potentially involve aerodynamic interaction between opposite wings or wing pairs. Physical proximity of contralateral wings is usually greatest during pronation, as maximum positional angles often approach 90° (see fig. 3.2B). In many insects, the opposite wings or wing pairs actually contact each other physically at the top of the wingbeat. Wings then rotate apart while beginning the translatory motion of the downstroke; the phase relationship between rotation and translation is then of critical importance in determining associated patterns of airflow.

In particular, the rotational separation of contralateral wings raises the possibility of circulation being simultaneously generated about each rotating wing. In the unsteady clap-fling mechanism first described by Weis-Fogh (1973) for the chalcid wasp *Encarsia formosa*, contralateral wing pairs first become juxtaposed at the end of the upstroke; the wings are in essence clapped together. Each wing pair then rotates about the trailing edge of the corresponding hindwing, and air flows into the region between the wing pairs (fig. 3.6A). This rotational air motion is of opposite senses for the two wings, ensuring that the net angular momentum of air remains zero. Circulation begins to develop independently, however, about each of the two wing pairs. This process continues as the wings rotate further apart and initiate translational movements within the stroke plane (fig. 3.6B/C). When the wing pairs are completely separated, circulation is substantial about each pair although there has been no associated vortex shedding (fig. 3.6D). Most importantly, the Wagner effect does not apply in the clap-fling mechanism because no net circulation is generated. The clap-fling thus provides a rapid means of establishing circulation and producing lift prior to wing translation.

The qualitative patterns of airflow associated with the clap-fling (fig. 3.6) have been corroborated by a variety of theoretical models (Lighthill, 1973; Belotserkovsky et al., 1974; Haussling, 1979; Edwards and Cheng, 1982), all of which suggest that substantial vorticity can be generated about the leading edges of two separating airfoils. Empirically, the experimental work by Bennett (1977) demonstrated the feasibility of the clap-fling for lift generation at Re near 10^5. Flow visualization on three-dimensional flinging wing models furthermore illustrates enhanced circulation at Re of 32 and 13,000 (Maxworthy, 1979, 1981). The clap-fling thus is effective over a wide range of Re and is likely to be a general feature of flight in many insects. Because the inflow of air associated with wing rotational separation occurs independently of viscous effects, the clap-fling may function even at very low Re for which circulatory lift is otherwise inhibited (Lighthill, 1973, 1978). Viscous effects may in fact contribute to the formation of separation bubbles attached to the leading edge of the wings during rotation and translation, enhancing overall lift production.

Kinematic characteristics of the clap-fling have not been determined for taxa other than *Encarsia formosa* and the fly *Drosophila melanogaster* (Götz, 1987; Zanker, 1990b), and a wide range of kinematic and morphological variables may potentially interact to influence lift production in such small insects. For example, theoretical modeling of the clap-fling (Lighthill, 1973; Weis-Fogh, 1973) and related empirical studies (Spedding and Maxworthy, 1986) suggest that wing circulation is generated in direct proportion to the angular velocity of wing rotation. Insects may explicitly vary the rate of wing opening during the clap-fling to meet variable force requirements, although no kinematic data evaluate this hypothesis directly. To test theoretical models of the clap-fling, lift forces predicted from observed kinematics should be explicitly compared with empirically determined body accelerations. For example, a clap-fling should generate large pitching moments that tend to tip the body nose-down (Bennett, 1977). Curiously, no such body rotation is evident in film images of *Encarsia formosa*, although the very high lift associated with the clap-fling is in fact sufficient to offset body weight in this species (Maxworthy, 1979).

This discussion of the clap-fling mechanisms presupposes fairly rigid wings rotating about an axis defined by their juxtaposed trailing edges. This motion is but one of a great diversity of kinematic possibilities if wings are variably flexible and can bend during the rotational and/or translational process. Ellington (1980, 1984c) described in butterflies a kinematic variant on the clap-fling (the clap-peel) that involves a continuous peeling apart of juxtaposed wings or wing pairs. The clap-peel is widespread among butterflies and also characterizes the takeoff in mantids (Brackenbury, J., 1991) and other insects with enlarged hind wings (e.g., locusts: Cooter and Baker, 1977). Also, wings of many insects during flight are not clapped dorsally but rather approach one another in near proximity, and then separate. Ellington (1984c) proposed that this event could be termed a near clap-fling; the extent of contralateral wing proximity is best described as a continuum. Flow visualization as well as vortex modeling (Sunada et al. 1993a, b) confirm that wing circulation can be substantially enhanced during a near clap-fling.

The high maximum positional angles typical of many insects suggest that the near clap-fling is a general feature of insect flight. Dorsal wing proximity and particularly wing contact may also facilitate production of a propulsive jet of air directed parallel to the wing chord (e.g., Götz, 1987). A jetlike effect may similarly characterize dorsal wing contact in butterflies and moths (Bocharova-Messner and Aksyuk, 1981; Grodnitsky, 1992, 1993; Willmott et al., 1997). Furthermore, Ellington (1984c) suggested that the degree of dorsal wing proximity could be varied to regulate the magnitude of wing circulation. For example, *Drosophila melanogaster* tethered in an airstream reduce the extent of dorsal wing contact as the speed of the airstream is increased (Götz, 1987; Zanker, 1990b). The clap-fling of hovering in this taxon is progressively converted to a near clap-fling and finally to isolated pronation of contralateral wings. Comparable reduction of contralateral wing contact has not been documented in free-flying *D. melanogaster* over a range of airspeeds, but would represent a straightforward method for regulating overall direction and magnitude of aerodynamic forces.

3.2.2.4 NET FORCE PRODUCTION BY WING FLAPPING

One experimental approach to the aerodynamic problem of wing flapping is to measure the total forces generated through a wingbeat either by tethered insects or by flapping wing models. In fact, the biologically inspired experiments of Katzmayr (1922) represent the first such empirical work in unsteady aerodynamics. This study evaluated the total forces on airfoils that were undergoing plunging motions and simultaneous changes in angle of attack; there was no spanwise flow in these experiments. Studies on flapping model wings date at least to the work by von Holst (1943) on swinging and rotating airfoils. More recently, Bennett (1966) showed that flapping of model wings augmented force production above values associated with continuous rotation. Archer et al. (1979) constructed a flapping device to measure total forces on wings at high advance ratios and flapping amplitudes up to 30°. Neither of these latter studies used wing morphologies or kinematics specifically relevant to the flight of particular insect taxa. The forces measured in these experiments also reflected the combined sum of aerodynamic and inertial forces on the wings; inertial forces must be calculated and subtracted from the total measured force if wing lift and drag are to be determined. If dynamic similarity is maintained through constancy of the *Re*, models of insect wings can be used in mechanical flapping devices to measure total force production. Such flapping models have been used in visualization of unsteady vortex wakes (Ellington et al., 1996; see section 3.2.2.5), but to date have not been used in estimates of instantaneous lift and drag on moving wings.

Measurement of the unsteady forces generated by flapping wings can also be made on tethered insects. Tethering of insects is characterized experimentally, however, by a biomechanical condition that renders generalization of results uncertain. The total force produced by tethered insects invariably contains both vertical and horizontal components. Even in simulations of hovering flight (i.e., tethering in still air), the restrained insect typically produces a net thrust (as in forward flight) as well as vertical forces (e.g., Hollick, 1940; Vogel, 1966; Götz, 1968; Wilkin, 1991). No forward velocity vector is present in this experimental context, however, and the flow situation around the wings must deviate substantially from flight at an airspeed that would correspond to the horizontal thrust generated by the insect (Weis-Fogh and Jensen, 1956). The positive thrust produced by the tethered insect also indicates that the situation deviates dynamically from true hovering. A similar situation applies to tethered insects flapping in an airstream but producing a thrust different from that required to translate at the corresponding airspeed. Unless the thrust exactly balances the horizontal component of drag, the insect is in aerodynamic disequilibrium.

The force balance of tethered insects may thus correspond neither to that of hovering nor to that of forward flight, and the general relevance of this method to flight mechanics is correspondingly unclear. Simulated flight of insects tethered to the arm of a rotating mill is similarly characterized by a force balance different from that of free flight (see Riley et al., 1997). Force production by tethered insects can furthermore vary with elapsed flapping time (e.g., Kutsch and Gewecke, 1979), whereas pitching moments generated by tethered insects can differ from those in free flight (e.g., Hollick, 1940). Bennett (1975) has also drawn attention to possibly confounding effects of viscoelastic coupling between wing forces and the force transducer. As with studies of the kinematics of tethered insects (section 3.1.2), few direct assessments of the effects of tethering on force production have been carried out. However, the maximum loads that *Drosophila melanogaster* flies can carry aloft during takeoff exceed by approximately 20% the maximum force supplemental to body weight that can be produced by tethered flies (Lehmann and Dickinson, 1997). This result suggests that tethering may artifactually interfere with the capacity for maximum force production. Such objections notwithstanding, tethering has been widely used to evaluate the possible range of forces produced by insects flapping their wings. These studies address biomechanical feasibility of particular aerodynamic mechanisms but cannot determine either quantitative similarity or likelihood of use of these mechanisms during free flight.

Most such studies have evaluated vertical force production, as body weight is the dominant force that must be overcome by the flapping wings. Vertical force production is occasionally termed "lift" in the biomechanics literature. This usage is, however, misleading in that aerodynamic lift on wings and bodies is not necessarily directed vertically; use of the phrase "vertical force" avoids potential confusion. Vertical forces produced by tethered insects often show substantial variation, but have generally been found to be proportional to the wingbeat frequency (e.g., odonates: Zeng et al., 1996f; locusts: Weis-Fogh, 1956a; honeybees: Esch et al., 1975; Esch, 1976; a calliphorid fly: Nachtigall and Roth, 1983). As with wingbeat frequency, vertical force production by tethered insects can also illustrate temperature dependence (e.g., Coelho, 1991; Marden, 1995). Additional kinematic variables implicated in vertical force production include stroke amplitude, angle of attack, and pronational and supinational velocities during wing rotation (Weis-Fogh, 1956a; Gettrup and Wilson, 1964; Luttges, 1989). These kinematic variables are similar to those identified in the regulation of forward airspeed in free-flying insects, and likely contribute to variable partitioning of aerodynamic force output into vertical and horizontal components. Similarly, externally imposed changes in the body orientation of tethered insects are invariably associated with changes in vertical force production (Gewecke and Niehaus, 1981; Zarnack and Wortmann, 1989; Wortmann and Zarnack, 1993; Zarnack, 1997). Such changes in body angle necessarily change the stroke plane angle, and thus alter the relative magnitudes of vertical and horizontal forces produced by the flapping wings.

Measurement of vertical forces alone does not constitute a complete description of tethered flight, as three translational and three rotational degrees of kinematic freedom characterize a flying animal (fig. 1.5). Few studies of tethered insects have examined the full dimensionality of aerodynamic force production. However, Buckholz (1981) measured both vertical and horizontal forces on a calliphorid fly tethered in airstreams of variable speed. The downstroke generated primarily vertical forces whereas the upstroke generated thrust, consistent with quasi-steady inferences drawn for forward flight in various insect taxa (section 3.2.1.3). In an exemplary study, Blondeau (1981) described dynamic measurements for six degrees of kinematic freedom on a tethered fly. Neither the work of Blondeau (1981) nor that of Buckholz (1981) evaluated potential similarity of tethered wingbeat kinematics to those of free flight, nor did these studies differentiate inertial from aerodynamic forces. By contrast, demonstration of unsteady force production by tethered insects requires subtraction of wing inertial forces from total forces acting on the insect. In locusts as well as in saturniid and sphingid moths (Cloupeau et al., 1979; Wilkin, 1990, 1991, 1993), this approach has demonstrated that noninertial aerodynamic forces exceed quasi-steady estimates at numerous points through the wingbeat. This informative conclusion does not, however, address the issue of kinematic or dynamic similarity of the tethered preparation to actual free flight.

An alternative approach to estimating total force production is to measure whole-body accelerations of a free-flying animal during the course of a wingbeat. The temporal patterns of force acting on the center of body mass can then be indirectly inferred. This technique has been successfully applied to free-flying pigeons using implanted acceleration transducers (Bilo et al., 1982; Bilo et al., 1985), but for reasons of scale is not at present feasible for application to insects. High-speed cinematography could, however, be used to derive body accelerations from positional data, enabling subsequent estimates of force production through a wingbeat. For example, Sunada et al. (1993) presented a detailed acceleration profile for a butterfly in takeoff flight. Net body forces estimated from such accelerational data will reflect contributions from both inertial and aerodynamic forces on the wings. In addition to body positional data, detailed wing positional data are therefore necessary in order to estimate and then subtract wing inertial forces from the total predicted force acting on the center of body mass. Accurate measurements of wing kinematics are, however, difficult to obtain for free-flying animals. The contribution of wing virtual mass also presents problems in this context because there is presently no accurate formulation for the added mass of wings undergoing variable acceleration and rotation.

3.2.2.5 VORTEX WAKES AND UNSTEADY FLOW MODELS

Because lift production involves generation of a vorticity field in the surrounding air, measurements of velocity and pressure in the wake of a flying animal potentially allow quantification of momentum transfer from flapping wings to the surrounding air. Time-averaged mapping of air pressure (Magnan, 1934) and air velocity (Hocking, 1953; Wood, 1970, 1972; Bennett, 1975, 1976; Chance, 1975) around tethered insects confirm that wing flapping generates a net pressure gradient across the plane of the beating wings. Because such measurements cannot resolve the temporal dynamics of the velocity flow field, however, this approach can evaluate neither underlying aerodynamic mechanisms nor the mechanical costs of vortex production. Instead, visualization and quantification of the vortex wake in space and time are required.

Early attempts at insect wake visualization were carried out by Demoll (1918) and Magnan (1934). Chance (1975) published apparently the first clear photograph of the vortices produced by wing flapping of tethered insects (see also Ellington, 1978, for the vortex produced by a butterfly in takeoff). Extensive work in the 1980s by Brodsky and co-workers (Brodsky and Ivanov, 1983b, 1984, 1985; Brodsky, 1985a, 1986a, 1994; Brodsky and Grodnitsky, 1985; Ivanov, 1990) established the generality of vortex production by tethered insects. For example, butterflies and skippers tethered in an airstream produce a discrete vortex per half-stroke. The vortex produced during the downstroke is oriented ventro-posteriorly and contributes primarily to vertical force production, whereas that of the upstroke is oriented more vertically and yields horizontal thrust (Brodsky, 1991; see fig. 3.9A). Vorticity of opposite sense is shed in each consecutive half-stroke, and force production is partitioned between half-strokes into predominantly vertical and horizontal components. The down- and upstroke vortices are linked ventrally, as has similarly been described in tethered *Manduca sexta* hawkmoths (Willmott et al., 1997). The most likely structure of the extended wake under such circumstances is thus a linked chain of vortex rings (fig. 3.9B).

By contrast, Grodnitsky and Morozov (1992) observed for tethered lacewings (Neuroptera) that both the down- and upstroke generate separate "U"-shaped vortex tubes that then fuse to produce a single vortex ring for the complete wingbeat (see fig. 3.9C). This vortex ring incorporates both vertical and horizontal components of momentum flux in the wake of the insect. The Neuroptera are tetrapterous insects, and vorticity shed from ipsilateral wings merges during translational motion to create a single vortex loop. Vortex loops on contralateral wings then link near the wing bases at the end of the downstroke, producing a single dorsal vortex loop. During the upstroke, the tips of the dorsal vortex loop merge in the region of the body if the upstroke sheds no further vorticity, or otherwise become loosely fused beneath the body, creating a closed vortex tube (fig. 3.9C). Similar vortex patterns have been observed in the wakes of various functionally two-winged insects in tethered flight (e.g., Hemiptera, Diptera, Lepidoptera: Grodnitsky and Morozov, 1993; *Drosophila*: Dickinson and Götz, 1996), and in free-flying butterflies (Grodnitsky and Dudley, 1996).

(A)

(B)

(C)

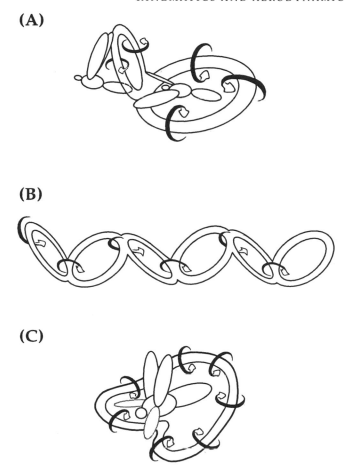

FIG. 3.9. (A) Distinct vortices produced in the down- and upstroke by various tethered Lepidoptera (modified from Brodsky, 1991; Willmott et al., 1997). (B) Possible free-flight wake structure arising from such vortices. (C) Single closed vortex tube produced by a complete wingbeat of various tethered insects (modified from Grodnitsky and Morozov, 1992; Dickinson and Götz, 1996).

Grodnitsky and Morozov (1993) concluded that only a single vortex ring is generated per wingbeat in functionally two-winged insects, and proposed that the vortex wake in such taxa is a series of unlinked rings. This conclusion based on vortex visualizations with tethered insects must be tempered by the twin restrictions of tethering and the absence of a translational airstream and associated body drag corresponding in magnitude to the actual thrust generated by the tethered insect. For example, the two connected vortex rings generated in the wingbeat of *Manduca sexta* (Willmott and Ellington, 1997c) suggest that a wide range of wake geometries is possible in morphologically as well as functionally two-winged insects. The often reduced weight support by tethered insects and the potentially variable influences of the surrounding free-stream velocity on wake structure suggest that further experimental work should be implemented on free-flying as well as tethered insects. The pattern of a single vortex per wingbeat is in fact characteristic of an unloaded upstroke (i.e., lift and associated circulation are minimal), with the downstroke assuming full weight support and thrust production. Situations intermediate to this and to the condition of a fully loaded upstroke (yielding lift and thrust) are clearly possible, and flight at different airspeeds probably involves changes in the structure of the vortex wake. As documented for flying vertebrates (Norberg, 1990), variable roles of the upstroke over a range of forward airspeeds may result in distinct gaits or patterns of shed vorticity, although this issue has not been evaluated for free-flying insects.

Evolutionary transformation of vortex wakes is also possible. Brodsky (1985a, 1986b, 1991) postulated that the ancestral vortex wake consisted of vortex loops linked between contralateral wings, whereas subsequent evolution of high wingbeat frequencies and low stroke amplitudes resulted in separated and distinct vortex wakes on each side of a flying insect. The former vortex pattern, however, appears to be typical of phylogenetically derived insect taxa (Grodnitsky and Morozov, 1993). Definitive conclusions concerning the evolution of insect vortex wakes await further data collection in a rigorous comparative context. Most empirical results are consistent with the presence of only a single vortex ring that subsumes the combined shed vorticity from contralateral wings or wing pairs. Dathe et al. (1984), however, concluded that a distinct vortex ring would be generated beneath each wing of a contralateral wing pair; different combinations of kinematics and wing morphology may therefore result in substantial geometric diversity of vortex wakes. The great diversity in insect wingbeat kinematics and the wide range of Re represented in insect flight suggests that further comparative investigation is warranted before a double vortex wake is deemed impossible.

A recent series of papers by Ellington and coworkers (Ellington et al., 1996; Van den Berg and Ellington, 1997a, b; Willmott et al., 1997) has substantially extended flow visualization methodology to evaluate quantitatively those unsteady aerodynamic mechanisms used by the hawkmoth *Manduca sexta* to effect high lift production. On the wings of moths tethered in a constant airstream, a large leading edge vortex is formed during the downstroke that is sufficient in circulation to produce the vertical forces generated by the tethered insect (Ellington et al., 1996). The leading edge vortex is produced solely via dynamic stall as the wing translates downwards; no capture of previously shed vorticity (i.e., via wing flexion during rotation) is observed. The leading edge vortex also moves radially toward the wing tip during the downstroke. Stability of this vortex is enhanced by the spanwise flow associated with wing flapping, and possibly by centrifugal acceleration and by presence of the tip vortex. As the airstream velocity around tethered *M. sexta* increases, the size of the leading edge vortex during the downstroke also increases (Willmott et al., 1996).

In contrast to the downstroke, the upstroke of *M. sexta* produces much less aerodynamic force; no leading edge vortex is evident on the wing. At the end of either half-stroke, bound vortices on opposite wing pairs are shed to yield a single vortex ring. The vortex wake in forward flight thus consists of alternating vortices shed at the end of each half-stroke, suggesting reversal of wing circulation between half-strokes (see fig. 3.9A). At high forward airspeeds, down- and upstroke circulation may, however, be of the same sense (Willmott et al., 1996; Willmott and Ellington, 1997c). These findings on tethered insects were replicated using a dynamically scaled mechanical flapping model, which provided further geometric and aerodynamic details both for the vortex wake (Van den Berg and Ellington, 1996a) and for the leading edge vortex (Van den Berg and Ellington, 1996b). In aggregate, these studies unequivocally demonstrate that delayed stall via a leading edge vortex is the predominant mechanism of unsteady force production in large insects. This mechanism may characterize much of insect flight at lower Re given the prevalence of kinematics appropriate for dynamic stall and unsteady flow separation from flapping insect wings.

Similar conclusions apply to visualization studies of flapping model wings at the upper end of Re relevant to insect flight. In the Re range of 10^4–10^5, dynamic stall is a common feature of plunging airfoils (Maresca et al., 1979; Favier et al., 1982; Freymuth, 1989, 1990). At Re around 10^4 (Gursul and Ho, 1992; Panda and Zaman, 1992, 1994) and as low as 1500 (Ohmi et al., 1990, 1991), dynamic stall and enhanced lift on flapping wings also depend on the presence of a leading edge vortex. Patterns of vortex formation and shed vorticity for oscillating plates and wings depend, not surprisingly, on the advance ratio, oscillatory amplitude and frequency, and on the shape of the structures involved (e.g., Savage et al., 1979; Robinson and Luttges, 1983; Ashworth and Luttges, 1986; Schreck and Luttges, 1988). A general conclusion from these and related physical studies of flapping airfoils is that vortex production and unsteady wing forces are profoundly sensitive to kinematic details. One basic implication of this result is that biologically relevant simulations of flapping flight must accurately replicate wing motions. The usually low frequency variation in the mechanically tuned insect pterothorax is unlikely to exert a major influence on vortex wake structure. Stroke amplitude and variable timing of wing rotation, however, are two major kinematic parameters subject to wingbeat-by-wingbeat variation and potentially involved in control of vortex shedding. Further, reactive modulation of wingbeat kinematics in response to prior vortex shedding may permit maintenance of an energetically advantageous wake structure. This possibility can be addressed only through simultaneous studies of wingbeat kinematics and vortex structure over a series of consecutive wingbeats.

The dramatic consequences of variable phase relationships between translation and rotation of flapping wings have been qualitatively demonstrated with plunging flat plates (Kliss et al., 1989) as well as with a mechanical model of a dragonfly wing (Saharon and Luttges, 1987, 1988). Both geometry and stability of the vortex wake can change dramatically in these systems according to choice of kinematic parameters. Plunging airfoils have also been investigated theoretically (Yates, 1986) and experimentally (Triantafyllou and Triantafyllou, 1991; Send, 1992; Triantafyllou et al., 1993) to determine values of kinematic parameters that result in optimal force production. Interestingly, these studies demonstrate that location of the wing's rotational axis can exert a strong influence on power expenditure. For a plunging and rotating airfoil, the optimal location of the rotational axis is in the anterior region of the wing (Yates, 1986), as characterizes all winged insects (see section 2.2.2). With the exception of the flapping wing models studied by Saharon and Luttges (1987, 1988), three-dimensional effects are unfortunately ignored in these aerodynamic studies. Given this restriction as well as the absence of wing flexibility in these studies, related conclusions concerning optimal kinematic parameters are not yet directly relevant for analysis of the flapping and rotation of insect wings.

An additional complication relates to unsteady interaction between the uncoupled ipsilateral wings of many tetrapterous taxa. Empirical results show that variable phase relationships between uncoupled fore- and hindwings can yield dramatically different patterns of airflow (e.g., Somps and Luttges, 1985; Reavis and Luttges, 1988; Luttges, 1989) and of vertical force production by tethered insects (e.g., Zarnack, 1983; Wortmann and Zarnack, 1993). This effect has a technological counterpart in the form of an oscillatory canard, a small aerodynamic surface mounted in front of the principal airfoil that serves in flight control (Ashworth et al., 1988). Both dramatic disruption as well as enhancement of hindwing airflow are possible consequences of variable forewing movements in tetrapterous taxa.

Extending models of a continuous momentum jet and the pulsed actuator disk during flight, a variety of theoretical approaches to wing flapping have been employed to more accurately estimate induced flow fields during flight. The local circulation method of Azuma and Kawachi (1979; see also Kawachi, 1981; Azuma et al., 1983) estimates aerodynamic forces by using steady-state lift polars at the Re and angles of attack appropriate for flapping wings. The downwash of the tip vortex is estimated by conceptually decomposing the wing into a series of consecutively shorter wings with elliptical distributions of circulation along the wingspan, and thus with known induced velocity distributions. Shed vortices were ignored in an initial analysis of dragonfly flight using the local circulation method (Azuma et al., 1985), but were subsequently incorporated in a more detailed but still static wake geometry (Azuma and Watanabe, 1988; see also Azuma, 1992; Sato and Azuma, 1997). Because the local circulation method relies on lift polars that are determined empirically on wings in steady-state flow, unsteady flow separation from wings and effects of spanwise flow cannot be incorporated in these models. Absence of wake deformation also renders the local circulation method most appropriate for modeling insect flight at high advance ratios. These models nonetheless predict, for forward flight, net vertical forces in the downstroke and net horizontal forces in the upstroke that are consistent with steady-state lift production.

The quasi-steady blade element analysis of flapping flight (section 3.2.1.3) represents bound circulation on the wing as a single linear element of concentrated vorticity along the wingspan. A two-dimensional extension of this lifting line is the lifting surface model that envisions a network or lattice of vorticity distributed over the wing surface (Albano and Rodden, 1969). The induced flows from bound elements of vorticity are then summed over the wing to determine the total induced velocity field (e.g., Chopra, 1974, 1976; Chopra and Kambe, 1977). Additional versions of the lifting surface include the unsteady vortex lattice models of Lan (1979), Möllenstädt (1980), and Sunada et al. (1993b). These lifting surface models incorporate neither effects of shed vorticity nor unsteady wake deformation on wing lift generation, and are only relevant for flight at high advance ratios. Lifting surface models assuming high aspect ratio structures and low-amplitude oscillations have been developed by Cheng and Murillo (1984) and Karpouzian et al. (1990), but similarly are not of general applicability to insect flight. Specifically with reference to insects, lifting surface models were elaborated with application to flapping wings by Smith (1996), Smith et al. (1996), and Vest and Katz (1996). All such models assume that vorticity is confined to the wing surface and cannot evaluate such effects as viscous flow separation from the wing nor the presence of leading edge vortices. Recent empirical results describing leading edge vortices on hawkmoth wings (Ellington et al., 1996) suggest that lifting surface models are probably of limited relevance for the Re range relevant to insect flight.

An alternative theoretical approach is to analyze flapping flight from first principles by using the Navier-Stokes equations that govern the mechanics of fluid flow. This approach is generally termed computational fluid dynamics (CFD), and typically imposes either a two- or three-dimensional grid onto the spatial configuration of a structure moving within a fluid. Temporal interactions between adjacent fluid elements are then evaluated assuming local conservation of mass and momentum. Presence of a wing imposes a physical boundary across which fluid elements cannot move; wing motions correspondingly impose force on adjacent fluid elements. The global pressure and velocity field of the fluid around the wing then emerges from spatial averaging and temporal tracking of individual fluid elements. Adequate spatial resolution for the analysis grid can require on the order of millions of individual elements even for simple flow situations. Once an appropriate CFD model is developed, however, sensitivity of results to variation in Re, wing shape, and kinematic parameters is easily determined by altering setup parameters. Use of CFD thus provides a convenient method for evaluating the consequences of particular wingbeat kinematics and morphologies, both real and imaginary.

Two-dimensional CFD studies of rotating and plunging airfoils at Re near 10^4 (e.g., Gürgey and Thiele, 1991; Gustafson and Leben, 1991; Gustafson et al., 1992) generally confirm patterns of vortex production described empirically using flow visualization. Until recently, no comparable CFD model had been implemented for realistic kinematics of an insect wingbeat. However, a three-dimensional CFD model has been recently applied to the kinematics of wing flapping in hawkmoths (Liu and Kawachi, 1998; Liu et al., 1998; see also Zeng, Liu, and Kawachi, 1996). To facilitate comparison with empirical results, this computer model used morphological and kinematic data for a robotic wing flapper that simulated hawkmoth hovering (Ellington et al., 1996; Van den Berg and Ellington, 1997a, b). Computationally, the hawkmoth downstroke was shown to generate a leading edge vortex that moved axially along the wingspan as the half-stroke proceeds, and that persisted in diminutive form during both supination and subsequent pronation. The upstroke was characterized by a similar but smaller leading edge vortex. In general, calculated flow structures were similar to those observed empirically on the flapping wing model and on tethered moths themselves (see Ellington et al., 1996; Willmott et al., 1997). Substantial spanwise flows are characteristic of both the computational model and empirical visualizations; the explicit neglect of spanwise flow in quasi-steady models of insect flight (section 3.2.1.3) is a major shortcoming of such models.

Spedding (1992) concluded that unsteady effects of wing flapping are typical of most flight phenomena in insects, and further suggested that the relevant unsteady aerodynamic effects associated with wing flapping and three-dimensional flows could be explicitly calculated. This claim is only true, though, under various combinations of simplifying assumptions such as an inviscid fluid, static wake geometries, and inflexibility of wings. Precisely those complicating features of real wings and fluids, especially flow separation, wing flexibility, and opposite wing interaction make such predictions of limited utility. Given the existing complexity of CFD models even for simplified wake geometries and wing motions, biologists may well speculate as to the future computational power necessary for accurate simulation of the flapping of biological airfoils. There is, moreover, every reason to suspect that unsteady effects will be nonlinearly additive, necessitating accurate representation of wing motions and possibly dynamic wing morphology if computational results are to be biologically meaningful. Simplified computational models can, however, provide valuable qualitative insight into the likely character of unsteady flows. Alternatively, flying animals can themselves be used in experimental manipulations and flow visualizations that elucidate unsteady aerodynamic mechanisms.

Clearly, one of the most pressing experimental needs is for further flow visualization both of insects in free flight as well as of flapping wing models at the *Re* relevant to insect flight (Ellington, 1995). Analysis of vortical signatures for free-flying insects is important because the variable force balance of tethered insects may influence the geometry and translational dynamics of shed vortices. For wing models engaged in flapping motions similar to those of free-flying insects, replication of and subsequent variation in biologically realistic kinematic profiles can evaluate the implications of particular wing motions (e.g., wing proximity effects, accelerations, etc.) for the ensuing vortex wake. Such correlational analyses of wingbeat kinematics and associated force production will contribute to an understanding of unsteady aerodynamic forces as utilized within their biological context.

3.3 MECHANICAL POWER REQUIREMENTS

Knowledge of wingbeat kinematics allows calculation of the associated energetic costs of flapping flight. Flight is energetically expensive, and estimates of the mechanical power expended in wing flapping permit comparative studies of biomechanical design and performance. The variation in mechanical power requirements with forward airspeed, the so-called power curve, is also of interest because of implications for the metabolic costs of flight and optimal strategies of airspeed selection. Aerodynamic theories that evaluate the power curve quantitatively have been extensively applied to the kinematics of flying birds and bats (e.g., Pennycuick, 1968, 1975; Rayner, 1979a, b; Norberg, 1990). By contrast, fewer such analyses have been carried out for insects in forward flight, although the conceptual and aerodynamic approaches are identical to those used for vertebrates.

3.3.1 Power and Efficiency in Flight

Mechanical power requirements and metabolic power input are related by the overall efficiency of the flight muscle. If muscle efficiency does not vary with airspeed, the shape of the curve relating metabolic power input to airspeed will be the same as that for mechanical power output. Metabolic rates in relation to airspeed have been determined for numerous free-flying vertebrates, yielding power curves that deviate often dramatically from quantitative aerodynamic estimates (see Norberg, 1990; Walsberg, 1990; Ellington, 1991b). The metabolic power curve has only been determined once on insects (Ellington et al., 1990; see section 4.1.1), and comparative analysis of power expenditure during flight awaits further data collection on a variety of insects.

3.3.2 Components of Mechanical Power Expenditure

In steady forward flight, four distinct avenues exist for expenditure of mechanical power: profile power P_{pro}, parasite power P_{par}, induced power P_{ind}, and inertial power during the first half of a half-stroke, P_{acc}. Additionally, mechanical power may be required to maintain oscillations in the center of body mass (P_{cm}). Mechanical power requirements are frequently expressed in mass-specific form (i.e., power/kg body mass), and such usage is conventionally indicated by an asterisk (e.g., P^*_{acc}). The total mechanical power required for flight can be determined by evaluating the individual components of parasite power, induced power, profile power, and inertial power. By comparing the maximum power output of striated muscle with calculated power requirements assuming variable inertial power requirements, the likelihood of elastic energy storage during the wingbeat can also be evaluated.

Profile power is the power required to overcome profile drag forces on the wings. For flying animals generally, a variety of approaches have been used to estimate profile power. At intermediate flight speeds, profile power for flapping wings has been treated as a constant fraction of the minimum power required to fly (which occurs at the speed that maximizes flight time per unit energy; see Pennycuick, 1975). Profile power requirements have also been evaluated assuming that aerodynamic forces of the upstroke are negligible (Rayner, 1979a, b). Neither of these methods is necessarily appropriate for analysis of flapping flight in insects. Instead, a quasi-steady approach can be used that estimates instantaneous values of wing drag on individual wing sections. The relative velocity of each section is estimated from the vector sum of the forward, flapping, and induced velocities, and a profile drag coefficient ($C_{D,\mathrm{pro}}$) is taken from empirical data for the appropriate Re and section orientation. Values of $C_{D,\mathrm{pro}}$ include the combined effects of pressure drag and skin friction on the wing but not the induced drag, which is evaluated separately (see below). Because unsteady drag coefficients are unavailable for the wing motions and Re characteristic of insect flight, $C_{D,\mathrm{pro}}$ is instead evaluated for wing sections based on steady-state drag measurements and the mean Re during the wingbeat (see Ellington, 1984f). Use of a mean Re rather than the instantaneous Re is appropriate here because so little is known empirically about profile drag on wings undergoing unsteady motions. This steady-state approximation nonetheless represents the best available representation of wing profile drag given existing empirical results. Following identification of the appropriate drag coefficient, equation (1.2) is used to estimate instantaneous profile drag on any given wing section; associated profile power is then calculated as the product of the section drag and the relative velocity.

In this quasi-steady formulation, total profile power is integrated along the wingspan to yield an instantaneous value for the wing as a whole:

$$P_{\mathrm{pro}} = \rho \int_{0}^{R} C_{D,\mathrm{pro}}\, c(r)\, V_R(r)^3 dr, \qquad (3.5)$$

where ρ is the mass density of air, $c(r)$ the wing chord at radial distance r from the wing base, and $V_R(r)$ the relative velocity of the wing at distance r. Equation (3.5) tacitly incorporates a factor of two to account for profile power requirements of both wings (or wing pairs). If the forward airspeed is augmented by sideslip, modifications to equation (3.5) are necessary to account for directional asymmetry in the translational velocity vector between contralateral wings or wing pairs (see Wakeling and Ellington, 1997c). Instantaneous profile power for individual wing sections is then averaged through the wingbeat to derive a mean rate of energetic expenditure necessary to overcome profile drag (see Norberg, 1976; Dudley and Ellington, 1990b; Willmott and Ellington, 1997c). This method thus addresses both spanwise and temporal variation in wing drag, albeit using steady-state assumptions.

Parasite power P_{par} is the power required to overcome drag forces on the body and is equal to the product of the body drag D_b and the forward airspeed:

$$P_{\text{par}} = D_b V. \tag{3.6}$$

Body drag can either be measured directly for the body angles and Re appropriate for the flying insect, or it can be estimated using equation (1.2) and a reference area for the body. In the latter approach, a body drag coefficient must be taken from the literature for the same or similar taxa at comparable Re and body orientations. Because body drag is usually small relative to the weight of an insect, mechanical power estimates will not be in most cases overly sensitive to the choice of drag coefficient. At high velocities, however, parasite power predominates aerodynamic power requirements and the influence of body drag becomes more significant. The general dependence of body drag on the square of air velocity (eq. 1.2), when combined with equation (3.6), indicates that parasite power varies with the cube of forward airspeed. Flight at very high speeds is thus likely to be energetically constrained by available power. By contrast, parasite power in hovering flight is assumed to be zero because there is no translational velocity and thus no net drag force on the body; effects of the induced velocity on body drag are typically ignored even though such velocities may be substantial (e.g., Ellington, 1984f).

Induced power (P_{ind}) is the power required to overcome induced drag forces on the wing and represents the cost of imparting sufficient downwards momentum to the surrounding air so as to offset the body weight. In the Rankine-Froude propeller model of induced velocities (section 3.2.1.3), this momentum flux is modeled by assuming that an actuator disk applies a constant pressure impulse to the surrounding air. The downward movement of air that ensues is assumed to be of constant velocity (the induced velocity V_i), which in forward flight is given by:

$$V_i = ((mg - L_b)/2\rho A_0(V^2 + V_i^2))^{1/2}, \tag{3.7}$$

where L_b is lift on the body, V is the forward airspeed, A_0 is the area of the actuator disk (see Dudley and Ellington, 1990b, and Norberg, 1990, for discussion of the appropriate disk area for forward flight). Note that V_i appears on both sides of eq. (3.7) and must be computed iteratively. This Rankine-Froude estimate of V_i represents a theoretical minimum for the power required to sustain lift production. Spatial fluctuations in applied pressure and wing circulation will increase the induced velocity above this minimum estimate, as will the application of periodic pressure impulses which occur during wing reciprocation. The combined effects of spatial and temporal variance in the applied pressure impulse have been explicitly calculated for hovering flight by Ellington (1984e). No such corrections have been estimated for the case of forward flight, however, and equation (3.7) is the most appropriate model to apply when advance ratios exceed the typical hovering maximum of 0.1. In accelerating flight, the induced velocity can be further modified to include the total mass flux of air across the actuator disk that generates the observed force balance (Wakeling and Ellington, 1997c). If whole-body accelerations are high, induced velocities may be substantially increased relative to values in hovering or in forward flight at constant airspeeds. Because induced power directly reflects the costs of lift production by three-dimensional airfoils, these costs can be reduced through the use of high aspect ratio wings that produce relatively smaller tip vortices and associated induced velocities.

Once induced velocity is known, the induced power P_{ind} can then be calculated by the product of this quantity and the body weight (less any body lift in forward flight, L_b):

$$P_{ind} = (mg - L_b)V_i. \tag{3.8}$$

One modification to this estimate of the induced power concerns potentially deleterious effects of wing profile drag in vertical force production. If profile drag is directed downward (as may occur during the upstroke, for example), additional vertical force will be required from wing lift as averaged through the wingbeat to maintain force equilibrium. In the moth *Manduca sexta* (Willmott and Ellington, 1997c) and in dragonflies (Wakeling and Ellington, 1997c), downwards contributions of wing profile drag to the overall force balance significantly increase the induced power expenditure.

The expenditure of induced power during flight can be substantially reduced via the so-called ground effect. This effect occurs through interactions between the vortex wake of a flying animal and an underlying physical surface (Lighthill, 1979; Rayner, 1991). Upwash from the surface reduces the downward momentum flux required to offset the body weight, and the ground effect is therefore most effective in hovering flight when induced velocities are highest. No insect is known to utilize the ground effect specifically for energetic reasons, although many satyrine butterflies fly close to the ground and may accordingly derive some aerodynamic advantage. In particular, male butterflies of the Neotropical genera *Cithaerias*, *Haetera*, and *Pierella* (see DeVries, 1987) regularly patrol the forest floor at heights appropriate for the ground effect (1–3 *R*). Empidid flies that fly close to the surface of water to find prey may also utilize the ground effect (see Zeil et al., 1989), as may adult caddisflies, stoneflies, and mayflies in some circumstances. More generally, all surfaces surrounding flying animals can potentially affect patterns of aerodynamic flow (Rayner and Thomas, 1991). The flight of insects from foliage, for example, may be influenced by potentially advantageous backwash that facilitates speed of takeoff. In laboratory contexts, vertical walls as well as horizontal surfaces above and below a flying animal can substantially influence vortex wake formation and energetic expenditure during flight. These effects may be particularly pronounced in small chambers that are constrained by size for respirometric purposes. Aerodynamically significant boundary interactions may also characterize forward flight of animals in wind tunnels (Rayner, 1994). Particular care must therefore be exercised to avoid such confounding effects in laboratory studies of flight performance.

In addition to the aerodynamic avenues of profile, parasite, and induced power expenditure, inertial power is required to accelerate the wing mass and virtual mass during flapping motions. Inertial power during the first half of a half-stroke, P_{acc}, is given by:

$$P_{acc} = 2nI \, (d\varphi/dt)_{max}^2, \qquad (3.9)$$

where I is the moment of inertia of the wing mass and virtual mass, and $(d\varphi/dt)_{max}$ is the maximum angular velocity attained by the wing during the half-stroke (Ellington, 1984f). Values of $(d\varphi/dt)_{max}$ can either be determined from actual kinematic data or estimated by assuming simple harmonic motion of the wings. If the downstroke and upstroke are of unequal duration, maximum angular velocities and associated inertial power must be determined separately for each half-stroke. Inertial costs may also be associated with wing movements during pronation and supination, although the moment of inertia of the wing mass and virtual mass about the longitudinal wing axis is typically much less than that about the wing base. The rotation of low aspect ratio wings about their longitudinal axis may, however, entail nontrivial inertial costs; this effect has not yet been investigated but may be relevant for Lepidoptera and Orthoptera, among other taxa.

During wing deceleration in the second half of a half-stroke, elastic structures in the thorax may allow kinetic energy of the wing mass and virtual mass to be stored transiently as strain energy. If this energy can subsequently be released in the following half-stroke to reaccelerate the wing, then net costs of accelerating the wing in either half-stroke will be minimal, and inertial power requirements as averaged through the wingbeat will be zero. Mechanical power requirements for this case of perfect elastic energy storage (P_{per}) simply equal the aerodynamic power, P_{aero} (= $P_{pro} + P_{par} + P_{ind}$; see Ellington, 1984f). This estimate of aerodynamic power represents a lower bound on the total possible power expenditure during flapping flight.

Alternatively, if there is no elastic energy storage of wing inertial energy, then supplemental power P_{acc} will be required to accelerate the wing during the first half of a half-stroke. During the second half of this half-stroke, however, essentially no energy will be required to decelerate the wing because the negative power requirements associated with tension generation during muscle stretching are small relative to energetic expenditures associated with positive work (Margaria, 1968; Ellington, 1984f). Aerodynamic power requirements over the same period, which are in many cases low relative to the inertial power, will be supplied by the kinetic energy of the decelerating wings. Remaining wing inertial energy can be dissipated by thoracic structures as the wing comes to a full stop. In this case, power output as averaged over the half-stroke will then equal one-half the sum of the aerodynamic power requirements and the inertial power during the first half of the half-stroke (Ellington, 1984f). Total mechanical power requirements given zero elastic energy storage, P_{zero}, are thus given by:

$$P_{zero} = (P_{aero} + P_{acc})/2. \tag{3.10}$$

Depending on the extent of elastic energy storage, actual power requirements lie somewhere between P_{per} and P_{zero}.

The absence of elastic energy storage does not necessarily increase total power expenditure, however. If P_{acc} is less than P_{aero}, then elastically stored energy derived from wing inertia will be insufficient to satisfy aerodynamic power requirements, and the mechanical costs of flight will always be equal to P_{aero}. In flight of many medium- to large-sized insects, however, inertial power requirements will substantially exceed the aerodynamic power requirements if elastic energy storage is negligible (e.g., Weis-Fogh, 1972, 1973; Ellington, 1984f; Dudley, 1995b). The relative magnitude of inertial and aerodynamic power requirements bears no necessary relationship to body mass, however.

Both in the moth *Urania fulgens* (200 mg $< m <$ 600 mg) and in the fly *Drosophila melanogaster* ($m <$ 5 mg), values of P_{acc} are less than P_{aero}, indicating that elastic energy storage cannot act to minimize the energetic costs of flight (see Dudley and DeVries, 1990; Dickinson and Lighton, 1995). Specific interactions between wingbeat kinematics and wing and body morphology thus influence relative partitioning of mechanical power expenditure between inertial and aerodynamic components.

Various pterothoracic structures that can potentially store wing inertial energy have been identified morphologically and in some cases investigated biomechanically. For the downstroke of the locusts, most of the energy necessary to accelerate the wings can be supplied via pterothoracic elasticity (Weis-Fogh, 1961). The cuticle of the locust thorax is fairly elastic but fails at strains greater than 1% (Jensen and Weis-Fogh, 1962). As a result, only one-quarter of the necessary strain energy can be stored in the thoracic exoskeleton. High indentation hardness of the locust thorax is, however, consistent with substantial elastic storage in this structure (Hillerton et al., 1982). Another 35% of energy required for the locust downstroke is stored in elastic components of the flight muscle. The capacity of insect flight muscle to store elastic strain energy is high (1–3 J/kg; see Alexander and Bennet-Clark, 1977), and may in fact be sufficient to store all wing inertial energy in insects with synchronous as well as asynchronous flight muscle. Specific molecular components of striated muscle that may deform elastically include the motor proteins myosin and actin as well as very large structural proteins (e.g., titin; see section 4.2.2).

In addition to the strain energy stored within the cuticle and flight muscle, elastic storage may also occur within a rubberlike protein called resilin. Resilin exhibits a highly elastic restorative response to applied deformations (Weis-Fogh, 1960; Weis-Fogh, 1961a, b; Alexander, 1988; Lombardi and Kaplan, 1993), and thus conforms to any a priori expectation that the flight apparatus should express structures that minimize energetic expenditure during flight. In locusts as well as in a variety of other insect taxa, a pad of resilin is located at the wing base (Weis-Fogh, 1960, 1965; Anderson and Weis-Fogh, 1964). Alternating compression and extension of this structure at the corresponding wingbeat frequency provides for nearly complete recovery of stored strain energy. The general absence of tendons in the flight apparatus (section 2.1.2) precludes widespread elastic storage in such structures as is characteristic of many other locomotor systems (see Alexander, 1988). Resilin is, however, present in the tendons of odonate wing depressors (Weis-Fogh, 1960, 1965), and likely fulfills an important role in elastic energy storage for this paleopterous taxon.

Direct experimental demonstrations of elastic storage in the ptero-thorax are remarkably few given the potential significance of this mechanism for energy conservation during flight. One approach to this problem is simply to measure the restorative torque acting on wings when depressed or elevated from their resting position in ex-perimental preparations. Weis-Fogh (1972) measured such elastic torque as a function of wing positional angle for an aeshnid dragonfly, locusts, and a hawkmoth. Torque was measured first following extir-pation of thoracic ganglia (thereby precluding muscle contraction), and then again upon excision of flight musculature. At mean posi-tional angles near 0°, the restorative elastic torque was minimal as would be expected for wings in a resting position. Increasing angular deviations from 0° produced progressively greater elastic resistance, consistent with a high degree of pterothoracic elasticity. Measurement of elastic torque prior to and following excision of the flight muscula-ture revealed that most of this elastic response was associated with nonmuscular components of the thorax. Muscular contributions to elasticity were, however, significant in all taxa, and in dragonflies likely represented contributions both of the elastic tendons as well as of the muscle per se.

The measurement of static elastic torques in the pterothorax yields results consistent with substantial elastic energy storage, but cannot evaluate potentially time-dependent elastic responses of pterothoracic structures at the oscillation frequencies relevant to flight. The method of Weis-Fogh (1972) could be modified, however, to evaluate wing os-cillations over a range of frequencies and to measure restorative torques dynamically. Existing data on static elastic torques, however, as well as on elastic features of flight muscle itself, strongly suggest that elastic storage is substantial in a diversity of insect taxa with syn-chronous muscles (e.g., locusts, dragonflies, hawkmoths). Similar ar-guments apply for insects that possess asynchronous flight muscle. The asynchronous flight muscles of various dipterans and hymenop-terans probably store most if not all kinetic energy of the oscillating wing mass and virtual mass (see Ellington, 1984f; Dudley and Elling-ton, 1990b). Estimates of elastic energy storage in tethered *Drosophila* are also consistent with a high degree of pterothoracic elasticity in much smaller insects (Dickinson and Lighton, 1995). In sum, much evidence indicates that elastic energy storage both within the flight muscle and in other thoracic structures is extensive (i.e., >50% of wing inertial energy).

Unfortunately, neither the partitioning of this storage among spe-cific pterothoracic structures nor allometry of elastic storage among different insect taxa have been evaluated quantitatively for the wingbeat frequencies typical of flight. One general consequence of extensive elastic storage, however, is that the flight muscles of insects appear to do work against primarily aerodynamic rather than inertial loads. For muscles that contract primarily against noninertial loads, muscle series compliance (i.e., the inverse of stiffness) should be low and contraction velocities should be high if metabolic power expenditure is to be minimized (Alexander, 1997). Morphologically, this prediction is consistent with the general absence of elastic tendons connecting insect flight muscles to the thoracic cuticle (see section 2.1.2). Physiologically, such minimization of the metabolic costs of contraction may have promoted indirect selection for high muscle stiffness as well as for high wingbeat frequencies. Both factors are potentially involved in the evolution of asynchronous flight muscle (see section 4.2.2).

3.3.3 Total Mechanical Power

Knowledge of the individual components of mechanical power at different flight speeds enables construction of a power curve relating total energetic expenditure to forward airspeed. Relative to numerous such efforts for flying vertebrates (see Norberg, 1990), far fewer power curves have been estimated for insects. Hocking (1953) published the first such power curve for insects, evaluating flight energetics for various dipterans using simplified aerodynamic models. More recently, steady-state aerodynamic theory has been used to estimate energetic expenditure for diverse insect taxa flying over a range of forward airspeeds (bumblebees: Dudley and Ellington, 1990b; a diurnal migratory moth: Dudley and DeVries, 1990; various odonates: Wakeling and Ellington, 1997c; the hawkmoth *Manduca sexta*: Willmott and Ellington, 1997c). The studies of Dudley and Devries (1990) and Wakeling and Ellington (1997c) derived species-specific power curves from individual insects, each flying at a different airspeed; the other studies evaluated power expenditure for individual insects flying over a range of airspeeds. Some insects (as well as hummingbirds) can also fly backwards and sideways, although concomitant expenditure of mechanical power during these behaviors has not been evaluated. Power curves can potentially be evaluated by measuring directly the power produced by flight muscles over a range of flight speeds (e.g., Dial et al., 1997). This invasive method is most appropriate for application to those volant vertebrates for which a single large muscle with a small region of insertion predominates during flight. For reasons of scale and the typically broad insertion of flight muscles on the thoracic cuticle, this method has not proven technologically feasible for insects.

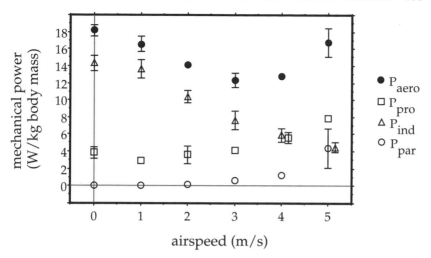

FIG. 3.10. Body mass-specific aerodynamic power curve and individual power components for free flight of the sphingid *Manduca sexta* (modified from Willmott and Ellington, 1997c). P_{par}, parasite power; P_{pro}, profile power; P_{ind}, induced power; P_{aero}, aerodynamic power. Data points refer to means (s.d.) derived from one female and two male moths; error bars within symbol dimensions are omitted. Estimates of profile power use a profile drag coefficient based on the mean wing Re during the wingbeat; sinusoidal wing motion is assumed (see Willmott and Ellington, 1997c). Values of forward airspeed for individual insects were in all cases within 0.2 m/s of the integer value indicated graphically.

For flying animals generally, the classic quasi-steady model for the overall shape of power curves is that of Pennycuick (1975). This model predicts a "U"-shaped power curve characterized by high values of energetic expenditure in hovering flight, minimum power expenditure at intermediate flight speeds, and high costs of flight again at higher airspeeds. The limited range of airspeeds studied in flying insects precludes comprehensive assessment of this prediction. However, interesting interspecific differences do characterize the shape of mechanical power curves. Bumblebees exhibit approximately constant power requirements over the speed range of hovering to 4.5 m/s (Dudley and Ellington, 1990b). By contrast, the moths *Manduca sexta* (Willmott and Ellington, 1997c; see fig. 3.10) and *Urania fulgens* (Dudley and DeVries, 1990), as well as odonates (Wakeling and Ellington, 1997c), exhibit substantial variation in mechanical power expenditure with airspeed. None of the insect power curves conform quantitatively to predictions of the general model proposed by Pennycuick (1975), although the lowest values of power expenditure in all studied taxa do indeed occur at intermediate airspeeds. Ellington (1991b) in fact argued that the power curve for both vertebrate and insect fliers is best described as "J-shaped," for which power requirements at hovering and intermediate airspeeds are roughly similar but increase substantially at higher speeds. The insect energetic data available to date suggest that such a positively curved shape is probably the most appropriate qualitative description of insect power curves.

The shape of the power curve derives from trends in individual components of power expenditure over the airspeed range under consideration. Expenditure of power to accelerate the wings is potentially high given its dependence on the square of flapping velocity (eq. 3.9), but inertial power requirements are partially or totally mitigated by elastic storage. Fairly constant wingbeat frequencies during flight (section 3.1.2) also suggest that inertial power is approximately invariant with respect to forward airspeed. The relative magnitudes of parasite, induced, and profile power do show, however, systematic variation with forward airspeed (fig. 3.10). Parasite power is zero in hovering but then increases approximately with the cube of forward airspeed (eqs. 1.2 and 3.6). Induced power requirements are maximal in hovering flight and typically exceed profile power requirements, but then decline as forward airspeed increases (eqs. 3.7 and 3.8). Profile power requirements increase with the cube of the wing relative velocity (eq. 3.5), but the relative magnitude of profile power varies according to the advance ratio of the wings (Dudley and DeVries, 1990). For the relatively higher flapping velocities characteristic of flight at low advance ratios, changes in forward airspeed have a reduced influence on the relative air velocity experienced by wing sections (see fig. 3.4A). The flapping velocity instead dominates airflow over the wing, and profile power requirements are relatively insensitive to variation in forward airspeed.

At higher advance ratios, however, a change in the forward airspeed will significantly alter the relative air velocities experienced by wing sections (see fig. 3.4). Profile power then increases substantially with forward airspeed, together with parasite power contributing to a steep overall rise in aerodynamic power requirements at higher airspeeds (fig. 3.10). For example, advance ratios of the uraniid moth *U. fulgens* are approximately an order of magnitude higher than those of bumblebees at equivalent airspeeds, reflecting the approximately 10-fold higher wingbeat frequency of the latter taxon. As a consequence, profile and total power requirements of flight in moths are much more sensitive to changes in airspeed (see Dudley and DeVries, 1990). The considerations suggest that power curves might be advantageously referenced with respect to the nondimensional advance ratio rather than to the forward airspeed. Certainly, interspecific comparisons of mechanical power expenditure will be most informative aerodynamically when made at equivalent advance ratios during forward flight.

Whereas assumptions of quasi-steady modeling must be quantitatively inaccurate in some respects, this qualitative trend relating advance ratio to shape of the power curve is unlikely to be different in the presence of unsteady effects. A single generalized model for the mechanical power requirements of flight is, however, probably inappropriate for application to all insect groups. The great diversity in morphology, wingbeat kinematics, and airspeeds of flying insects suggests a corresponding diversity in the relative magnitude of the components of the power curve. Wingbeat kinematics may also vary with airspeed (see section 3.1.2), further confounding simplified analyses of mechanical power requirements. At extreme airspeeds, however, total power must increase dramatically in reflection of the dependence of parasite power on the cube of forward speed. A similar effect applies to profile power when airspeeds and the advance ratio are sufficiently high. Independent of the costs of flight at intermediate airspeeds, insect power curves must increase substantially at higher airspeeds.

Systematic comparisons of power curves among different insect taxa represent a potentially rich area for investigation, particularly in ecological contexts involving dispersal and long-distance flight. Quantitative knowledge of the power curve can be used to determine those particular speeds that optimize energetic expenditure given fixed energy reserves (see Pennycuick, 1975). The maximum range speed of flying animals is the airspeed that minimizes energetic expenditure per unit distance traveled, thereby maximizing the distance flown. Analogously, the minimum power speed is the airspeed that minimizes energy expenditure per unit time, maximizing the time spent aloft. Optimal selection of flight speeds might be expected in a variety of flight-related contexts because the energetic costs of flight are high, typically several orders of magnitude higher than metabolic costs at rest (section 4.1.1).

Depending on the particular behavior under question (e.g., foraging, food or mate transport, migration), minimization of either energetic or temporal costs during flight may be important (Hedenström and Alerstam, 1995). For example, flight at minimum power expenditure might be important for female insects searching for oviposition sites. Location of an appropriate host plant often requires extensive sampling over a long duration of time, and the success of this behavior is more likely to be dependent on the total time spent searching rather than on the total energetic expenditure during flight. Alternatively, long-distance insect migrants might be expected to maximize the distance flown per available energy reserves. Such theoretical predictions of optimal flight speeds may be altered, however, if endogenous energetic reserves can be continuously supplemented during flight. For example, many migrating insects obtain nectar at flowers or otherwise engage in feeding behavior (see section 7.4.3).

Although this treatment of flight energetics emphasizes costs of hovering and steady forward flight, many other features of insect flight performance invoke energetic expenditure. Additional costs during flight may also be associated with internal mass loading or with the carrying of external objects (see section 5.4.2). One interesting and important avenue for supplemental power expenditure is ascending flight, either as directly vertical flight or as climbing flight with a constant vertical component of total velocity. The additional power required for ascending flight in this case is given by the product of body mass, gravitational acceleration, and the vertical component of velocity. This power represents the rate of increase in potential energy associated with height changes of the center of body mass.

A related component of power expenditure during flight derives from oscillatory or accelerational movements of the center of body mass. In hovering flight, oscillations of the center of body mass are minimal (e.g., Ellington, 1984c), and the concomitant power required to maintain the associated small fluctuations in kinetic and potential energy of the body mass is negligible. In forward flight, however, many insects display erratic movements and sudden vertical and lateral accelerational components in otherwise approximately straight trajectories. Many such behaviors are probably protean devices that serve to confound potential predators (Driver and Humphries, 1988; see section 7.3.2). To estimate the power associated with these flight path oscillations, the position of the center of body mass as a function of time is used to derive fluctuations in potential energy and in the vertical and horizontal components of kinetic energy (Dudley, 1991a). Over any given period of time, a positive change in the total energy of the center of mass requires energetic input from the flight motor. The summed positive increments in total body energy divided by the corresponding time period represents the average power required to maintain such fluctuations (P_{cm}). Values of P_{cm} are added to either P_{per} or P_{zero} to obtain total mechanical power requirements during flight.

Although there are few empirical studies of erratic flight paths in insects, aerial protean display can entail substantial energetic expenditure. For example, flight path oscillations in erratically flying Neotropical butterflies increase total mechanical power requirements (assuming perfect elastic storage of wing inertial energy) by an average of 43% (Dudley, 1991a). Other flight behaviors also may involve changes

PLATES

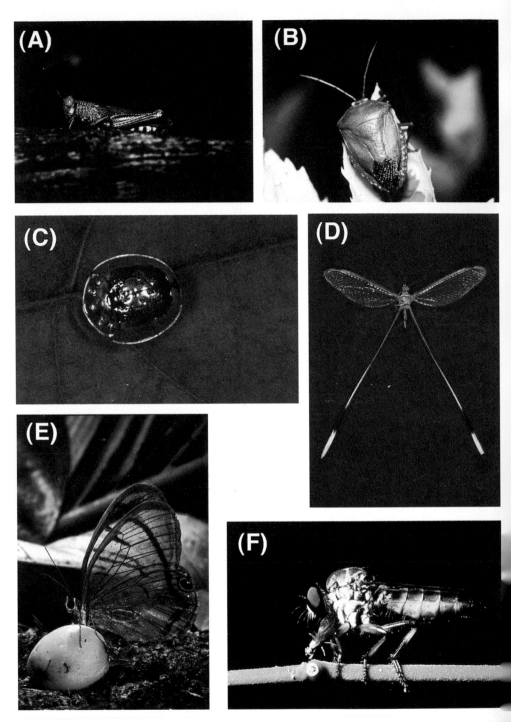

PLATE 1. Diverse insect body forms: (A) acridid orthopteran (photo courtesy of P. Chai); (B) pentatomid hemipteran; (C) cassidine chrysomelid (Coleoptera); (D) nemopterid neuropteran (*Nemopterella* sp.; specimen courtesy of M. Picker); (E) satyrine nymphalid (*Dulcedo polita*; photo courtesy of P. DeVries); (F) asilid dipteran (photo courtesy of L. Gilbert).

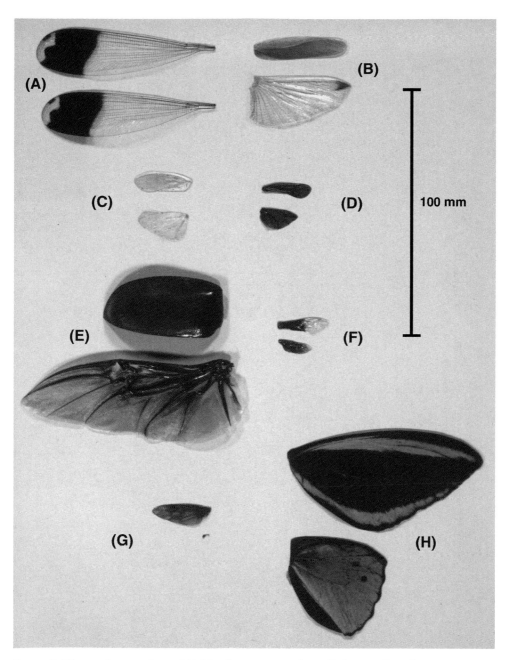

PLATE 2. Diverse insect wings: (A) *Megaloprepus coerulatus* (Odonata: Pseudostigmatidae, a Neotropical helicopter damselfly), forewing length $R = 71.9$ mm; (B) *Euceria insignis* (Orthoptera: Tettigoniidae), forewing length $R = 37.5$ mm; (C) *Enchophora floreps* (Homoptera: Fulgoridae), forewing length $R = 23.4$ mm; (D) *Edessa* sp. (Hemiptera: Pentatomidae), forewing length $R = 19.8$ mm; (E) *Megasoma elephas* (Coleoptera: Scarabaeidae), hindwing length $R = 88.7$ mm; (F) *Exaerete frontalis* (Hymenoptera: Apidae, a euglossine orchid bee), forewing length $R = 20.8$ mm; (G) *Pantophthalmus tabaninus* (Diptera: Pantophthalmidae); forewing length $R = 24$ mm (note the miniaturized hindwing, the haltere); (H) *Ornithoptera priamus* (Lepidoptera: Papilionidae, a Papuan birdwing butterfly), forewing length $R = 79.5$ mm.

PLATE 3. Capture of the butterfly *Morpho amathonte* over Lake Gatún, Republic of Panama.

PLATE 4. Rufous-tailed jacamar (*Galbula ruficauda*) handling a butterfly (*Morpho amathonte*) under experimental conditions in Corcovado National Park, Costa Rica (photo courtesy of P. Chai; see Chai, 1996).

PLATE 5. Dorsal perspectives of the heliconiine nymphalid butterfly *Dryas iulia* (top) and the nymphaline nymphalids *Marpesia petreus* (middle) and *M. berania* (bottom). Ventral wing coloration (a more uniform and lighter orange with no black lineation) is more similar than the dorsal coloration among the three species.

PLATE 6. Euglossine bee (*Euglossa imperialis*) hovering at an orchid. (Courtesy of P. DeVries.)

in total body energy and thus increase the total costs of flight. One important example of unsteady locomotion is aerial maneuvering (chapter 5), success of which may be dependent upon an animal's ability to alter rapidly the rate of power production by the flight muscle. Startle escapes, for example, depend critically upon the initial generation of aerodynamic forces and subsequent acceleration from a resting start.

The instantaneous mechanical power required for any particular body acceleration is given by the product of body mass, the acceleration, and the instantaneous velocity. The associated temporal derivative of this quantity, the power flux, is proportional to the square of the acceleration. Energetic demands of rapid maneuvers with vertical or lateral accelerations can therefore be particularly high. For example, May (1991) showed in dragonflies that power required for acceleration was substantial relative to other components of energetic expenditure in maneuvering forward flight. Additional power may be required to accelerate the added mass of air about the body. Systematic comparisons among insects exist neither for maximum power flux nor for the maximum capacity for acceleration. These biomechanical parameters potentially influence the kinematic characteristics of rapid aerial courtship displays, and may be of vital importance in antagonistic aerial interactions.

3.4 SUMMARY

Relative to the morphological diversity of insects, wing and body motions have received much less descriptive attention. The biomechanical consequences of particular wing motions are even less well elucidated. Some kinematic parameters such as characteristic flight speed and wingbeat frequency exhibit strong patterns of allometric variation. Other parameters such as stroke amplitude and stroke plane angle appear to be more constrained phylogenetically. Wingbeat kinematics reflect the complex combination of flapping motions about the wing base and rotation about the longitudinal wing axis. The use of time-averaged kinematic parameters to describe rhythmic wing flapping may conceal short-term autocorrelation of wing motions over a series of wingbeats. Various components of flight performance (e.g., hovering, forward flight, maneuvers) require precise modulation of wingbeat kinematics to regulate the requisite balance or transient imbalance of aerodynamic and gravitational forces. Biomechanical studies of flight kinematics and aerodynamics are most realistically implemented using free-flying and otherwise unrestrained insects.

4.1.1 Aerobic Metabolism during Flight

Insects lack lungs or otherwise centralized structures dedicated to the transport of respiratory gases. Instead, oxygen is supplied to flight muscles by the tracheal respiratory system, a branching network of cuticular tubes that projects throughout the insect body. The tracheal system functions in respiration primarily through gaseous diffusion rather than by convective air movements (Kestler, 1985; Mill, 1985). Convective pumping motions partially facilitate movement of oxygen and carbon dioxide molecules, but diffusion of these two gases along concentration gradients represents the primary means of respiratory gas transport. Oxygen influx and efflux of carbon dioxide occur almost exclusively via tracheal pathways, as diffusion of these molecules directly across the insect cuticle is very low in magnitude. Passive removal of carbon dioxide through diffusion is particularly enhanced by its low present-day atmospheric concentration (0.03%), although carbon dioxide concentrations may have been substantially higher during early evolution of winged insects (see Berner, 1991, 1994, 1998; Mora et al., 1996).

As with all respiratory systems, supply of oxygen and release of carbon dioxide in the insect tracheal system is regulated either partially or completely to match physiological requirements. At the interface between insect body and the atmosphere, small openings termed spiracles control the flux of respiratory gases between the tracheal system and the external environment. Air diffuses freely across the spiracles, and these structures therefore represent the predominant means of gas entry into and exit from the insect (Harrison, 1997). This flux is subject to behavioral control through the action of small muscles that can open or close individual spiracles. Internally, cuticular invaginations from the spiracles form tracheae, a network of cuticular tubing that extends throughout the body. Iterative branching of tracheal segments coupled with an ever-decreasing tracheal diameter yields a pervasive respiratory tract of small but continuous air-filled tubing. In some regions of the body, enlarged portions of tracheae also form internal air sacs often positioned closely to the flight muscles (e.g., Weis-Fogh, 1964c).

Although not centralized, the insect respiratory system can nonetheless represent a significant fraction of the internal body volume. For example, the relative tracheal area in flight muscle varies from 0.1% to 10% of the total cross-sectional area, depending on the insect taxon and the order of tracheal branching under consideration (see Weis-Fogh, 1964b). Tracheae can thus occupy a considerable portion of the total muscle, suggesting that substantial costs may be associated with tracheal construction as well as with the occupation of muscle volume that could be otherwise devoted to contractile myofibrils and/or mitochondria. Relative tracheal volumes are likely to be higher in insect taxa with higher rates of oxygen consumption. For example, the generally higher metabolic demands of asynchronous flight muscle (section 4.2.2) are matched by a higher tracheal density relative to that of synchronous muscle (e.g., Smith, 1961a, b).

The branching network of tracheae terminates in minute structures termed tracheoles, the closest region of contact between the respiratory system and other tissues. The blind-ended tracheoles often penetrate into surrounding tissue, including muscles, to yield a much-reduced distance between the tracheolar lumen and the actual regions of respiratory demand. Particularly with muscle fibers, tracheoles typically indent the cell membrane and terminate near individual mitochondria. Tracheolar diameters are comparable to the mean free-path length of oxygen (Weis-Fogh, 1964b), the average distance traveled by an oxygen molecule before colliding with another molecule. Diameters smaller than this distance would actually impede diffusion of oxygen (Pickard, 1974), so that structural design would appear in this instance to closely approximate the optimal value for effective oxygen transport.

In resting insects, tracheoles are typically filled with a liquid that is secreted by surrounding cells. Most of this liquid is apparently resorbed during flight, yielding a nearly continuous pathway of air throughout the tracheal system when the metabolic demand is substantially elevated. Because gas diffusion is much faster in air than in water, such resorption of tracheolar fluid enhances diffusion for what might otherwise be a rate-limiting step (see Wigglesworth and Lee, 1982; Wigglesworth, 1983; Kestler, 1985). One important issue relating to insect respiration concerns the maximum possible size of insects given diffusive capacity of the tracheal system and likely metabolic demands during flight. This problem can be reduced to analysis of the maximum oxygen flux potentially provided by tracheal diffusion, given the likely concentration of this gas within respiring tissue and in the surrounding atmosphere. In dragonflies, studies of tracheal density and geometry suggest that a thoracic radius of 0.5 cm represents an upper limit to the distance over which passive diffusion of oxygen can meet metabolic demands of flight (Weis-Fogh, 1964b). The taxonomic generality of this result has not been established, however, and many contemporary insects have thoracic radii in excess of 0.5 cm. Note also that the estimates of Weis-Fogh (1964b) did not incorporate oxygen diffusion through the final and partially liquid tracheolar stage, and thus potentially overestimated maximum thoracic size sufficient for adequate supply of oxygen through diffusive means alone.

In addition to the diffusion of oxygen and carbon dioxide along concentration gradients, convective motion of air may substantially increase gas flux within the tracheal system. One effect of thoracic muscle contraction is to alternately compress and expand various branches of the tracheal system as well as any internal air sacs. Such repetitive motions will induce convective air movements within the tracheal system, and potentially supplement via bulk gas transport the diffusive supply of oxygen and removal of carbon dioxide. This effect has been termed autoventilation, and is a direct consequence of muscular contraction. Empirically, direct measurements on tethered insects demonstrate that cuticular deformation induced by wing flapping can contribute significantly to thoracic ventilation (Weis-Fogh, 1967b; Bartholomew and Barnhart, 1984; Komai, 1998). Enhanced abdominal pumping may also enhance tracheal ventilation in flight, particularly in dipterans and hymenopterans (Weis-Fogh, 1967b), but also in beetles (Krogh and Zeuthen, 1941) and orthopterans (Miller, 1960; Ramirez and Pearson, 1989).

The convective airflow induced by such pumping motions ventilates not only the central (first-order branching) tracheae but also secondary tracheal branches. Higher-order branches within the tracheolar network are unlikely to experience such bulk airflow, although tracheoles that invaginate muscle-fiber membranes may potentially be deformed during muscular contraction to produce convective movements of air (Weis-Fogh, 1956d). Both diffusion and the capacity to generate tracheal convection represent potentially rate-limiting steps in insect respiration. The much reduced thoracic diameters of smaller insects (e.g., *Drosophila*) probably alleviate any metabolically based need for active ventilation of the tracheal system. In larger insects, diffusive limits to maximum body size likely predominate given the limited extent of convection that is induced through much of the tracheal network. In an extant dragonfly species, flight metabolic rates vary directly with ambient oxygen concentration (Harrison and Lighton, 1998), a result consistent with diffusion-limited oxygen delivery in the tracheal system. Indirect support for ultimately diffusive limits is also provided by the presence of widespread gigantism in Paleozoic pterygotes at times of elevated atmospheric oxygen concentration (see section 6.1.1). A final mechanism of respiration in some insect taxa is Bernoulli entrainment of air within the tracheal system. According to this mechanism, higher external airflow velocities near posterior spiracles generate a pressure gradient that pulls air into the anterior spiracles and yields bulk flow through the central tracheae. Although described at present only from beetles (Stride, 1958; Amos and Miller, 1965; Miller,

1966), Bernoulli entrainment may nonetheless be widespread, particularly for fast-flying insects.

For insects, diffusion and convection within the tracheal system are the only potentially rate-limiting steps for oxygen flux from the atmosphere into respiring tissue. By contrast, vertebrate respiration includes intervention of a circulatory system that imposes additional convective and diffusive resistances to flow. A circulatory system that incorporates convective transport of respiratory gases would also disadvantageously increase the body mass of a flying animal. The presence of such a circulatory step within the chain of respiratory gas exchange may, in fact, have imposed minimum size limits on vertebrates that substantially exceed those characteristic of insects (Schmidt-Nielsen, 1984). Diffusion of respiratory gases within insect tracheoles also facilitates the rapid changes in gas flux necessary to shift from resting to active levels of metabolism, although this latter capacity is also found in volant vertebrates. The sole use of a tracheal system for respiration does indicate, however, that the insect circulatory system potentially limits only transport of fuel and not of respiratory gases during flight. Consistent with this observation, the most important indirect consequence of muscular contraction may be convectively to move hemolymph and not tracheal air volumes (Weis-Fogh, 1964b).

The most striking feature of insect flight metabolism is the extremely high relative as well as absolute rate of oxygen consumption attained during repetitive muscular contraction. Relative to the standard resting metabolic rates of insects, flight represents an elevation of metabolism typically by a factor of 50–200 (Kammer and Heinrich, 1978; Casey, 1981c, 1989). Such high factorial scopes (the ratio of active to standard resting metabolic rate) are unprecedented either for vertebrate flight or for other forms of arthropod locomotion (see Norberg, 1990; Full, 1997). The correspondingly high rates of energetic expenditure during insect flight are afforded solely by aerobic (oxygen-dependent) activity of the flight muscles. Anaerobic metabolic pathways are present biochemically but are of negligible quantitative consequence for insect flight muscle (see Beenakkers et al., 1985).

Mass-specific rates of oxygen consumption in flying insects represent an extreme design in animal locomotor physiology, substantially exceeding maximal values recorded from other exercising invertebrates as well as vertebrates. For example, flying hummingbirds may represent a physiological extreme for exercising vertebrates (see Hochachka, 1987, 1994; Chai and Dudley, 1995), yet insect flight metabolism exceeds the metabolic turnover rate of hovering hummingbirds by up to a factor of 2–3. Such high rates of obligately aerobic metabolism must ultimately be powered energetically through provision of ATP by the mitochondria, and considerable attention has been focused subcellularly on mitochondrial structure and design in insects. Relative to most vertebrate muscles, mitochondrial densities of insect flight muscle are high, ranging as high as 45% (e.g., Casey et al., 1992). In parallel with whole-body respiration, mitochondrial respiration rates of insect flight muscle substantially exceed maxima determined on muscle from a variety of exercising vertebrates, including hummingbirds (see Casey and Ellington, 1989; Suarez and Moyes, 1992; Wells and Ellington, 1992). Differences in mitochondrial structure among insects and vertebrates might contribute to such large differences in metabolic rates. The relative volume of mitochondria within flight muscles is, however, comparable among insects and hummingbirds, as is the extent of cristal (inner membrane) surface density within mitochondria (see section 7.5.2). These structural similarities among flight muscle of insect and hummingbird mitochondria suggest that insect-specific molecular adaptations of respiratory enzymes may facilitate higher rates of metabolic turnover in this taxon.

As with other aspects of flight performance, activity metabolism of insects in tethered simulations of flight can differ substantially from that in free-flight (e.g., Heinrich, 1971; Heinrich, 1980; Nicolson and Louw, 1982). For example, the comparison of multiple studies of honeybee metabolism suggests that tethered honeybees exhibit rates of oxygen consumption substantially lower than those of their free-flying conspecifics (see Nachtigall et al., 1989; Nachtigall et al., 1995). Fortunately, extensive data exist on hovering metabolism for a variety of insects (Kammer and Heinrich, 1978; Casey, 1989). The highest recorded mass-specific rates of oxygen consumption are found among free-flying Diptera and Hymenoptera, although some particularly high values reported in early studies (e.g., Davis and Fraenkel, 1940) require corroboration using modern methods. Most interspecific variation in flight metabolism is allometrically correlated. Among closely related species, mass-specific rates of oxygen consumption in hovering flight tend to vary inversely with body mass (e.g., Casey, 1981a; Casey et al., 1985; Chappell, 1982, 1984). Morphological correlates of this allometry are not well studied, although flight muscles of smaller bee species have greater relative mitochondrial volumes that enable a greater mass-specific oxygen flux (Casey et al., 1992). Within a given species, allometric variation in hovering flight metabolism mirrors the interspecific trend of higher mass-specific metabolic rates at smaller body mass. Intraspecific differences in metabolic rates also derive from genetically based variation in metabolic enzymes (e.g., Laurie-Ahlberg et al., 1985; Coelho and Mitton, 1988; Harrison and Hall,

1993; Harrison, Nielsen, and Page, 1996). Such genetic bases for flight metabolism indicate that natural selection can potentially exert a strong influence on the molecular underpinnings of flight performance. For example, intraspecific variation in metabolic enzymes has direct implications both for the extent of flight activity and ultimately for female fecundity in the butterfly genus *Colias* (Watt et al., 1983; Watt, 1992).

Oxygen consumption of flying insects also varies according to aerodynamic demands on the flight system. For example, metabolic rates of hovering queen bumblebees (Heinrich, 1975b) and of honeybees in slow forward flight (Wolf et al., 1989) increase linearly with increased body mass due to nectar loading in the abdomen. Relative to flight in normodense air ($\sim$ 1.21 kg/m^3), oxygen consumption of actively flying honeybees declines with increased total pressure (and air density), but is also reduced at air densities below about 1 kg/m^3 (Withers, 1981). This latter response is surprising because induced power requirements typically predominate during hovering flight and should increase at lower air densities (eqs. 3.7 and 3.8). Because of the extended time period (5 minutes) over which metabolic measurements were averaged in the study of Withers (1981), these results may also reflect variable behavioral responses to changes in air density. For example, honeybees may not have been strictly hovering during the entire period; periods of hovering may have been interrupted by periods of slow flight at non-zero translational velocities together with variable accelerations or changes in direction. Measurements over shorter time intervals may clarify the metabolic responses of insects to flight in low-density gas mixtures. Some indication as to the likely magnitude of physiological responses to low-density air may be obtained from studies of hummingbirds. For example, oxygen consumption of hummingbirds hovering at feeders increases substantially in normobaric density reduction trials using much shorter time intervals for the measurement of flight metabolism (Chai and Dudley, 1996). Similar experiments with insect taxa in controlled hovering flight would clearly be desirable.

In addition to air density, other abiotic factors that potentially affect rates of oxygen consumption in free flight include ambient oxygen concentration and air temperature. Hypoxic conditions (reduced oxygen availability) as well as hyperoxia (increased oxygen) may alter the extent of gas diffusion in the tracheal system and influence flight performance accordingly. Few experimental data evaluate this possibility, although wingbeat frequencies and metabolic rates of honeybees flying in normobaric (standard pressure) hypoxia decline only at oxygen concentrations below 8% (Joos et al., 1997). The increased rates of water loss measured under these conditions suggest increased and possibly limiting rates of convective gas exchange at such low oxygen tensions. Conversely, hyperoxia may augment maximum flight performance in insects if diffusion-based oxygen availability ultimately limits metabolic capacity of thoracic muscle. Flight metabolism under hyperoxic conditions has not yet been investigated for free-flying insects, although Chai et al. (1996) found no effect of hyperoxia on metabolic rates of hummingbirds in either normal hovering conditions or during maximum performance. This latter result is consistent with nondiffusive constraints on hummingbird flight performance, and suggests either ventilatory or circulatory system limitations on maximal oxygen uptake in this taxon (and potentially in other vertebrates, including humans; see Wagner, 1996). A comparable hyperoxic manipulation with flying insects would test for similar effects on potentially diffusion-limited tracheal systems.

Oxygen consumption by hovering insects is generally independent of air temperature (Casey, 1989; see, however, Harrison, Fewell, et al., 1996; Roberts and Harrison, 1998; Roberts et al., 1998), a result suggesting that mechanical power output and muscle efficiencies are similarly constant under different thermal regimes. For insects in forward flight, such constancy of mechanical and metabolic power expenditure may not apply because of speed-dependent variation in convective cooling and in aerodynamic power requirements (section 3.3.2). No empirical results are available to address this issue, nor have thermal variables other than air temperature (e.g., radiant heat load) been manipulated in studies of flight energetics. In studies of hovering metabolism, thermally correlated variation in behavioral repertoire (e.g., slightly faster flight at higher temperatures) may alter convective heat loss and the overall heat balance. This variation in heat flux may induce compensatory metabolic responses for endothermic animals that regulate body temperature during flight. The analysis of thermal variation in metabolic rates during flight therefore requires careful interpretation. For example, Schuchmann (1979) reported that metabolic costs of flight in hummingbirds declined substantially at higher air temperatures. However, no such decline was found in the hummingbird study of Chai et al. (1998), which carried out more controlled measurements of flight metabolism that recorded rates of oxygen consumption only when birds were hovering at feeders. For flying insects, an analogous set of measurements that controlled for potentially variable behavioral responses to different air temperatures would be informative. One interesting if speculative implication of reduced flight costs at higher temperatures is that global warming may indirectly enhance pathogen transmission by volant insect vectors. This possibility is not generally

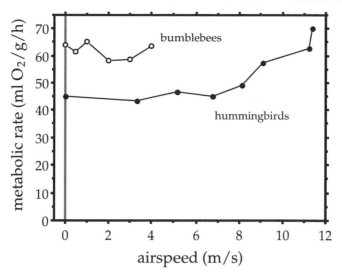

FIG. 4.1. Body mass-specific rates of oxygen consumption for bumblebees and hummingbirds in forward flight over a range of airspeeds. Bumblebee data are from Ellington et al. (1990) and represent means obtained from three queen bumblebees of *Bombus lucorum*, one *B. lucorum* worker, and two *B. pascuorum* workers. Hummingbird data are taken from Berger (1985) and represent means obtained from one individual *Colibri coruscans* and two *C. thalassinus*.

incorporated in analyses of climate change and vector biology (e.g., Marten et al., 1995; Martin and Lefebvre, 1995; Jetten et al., 1996), but clearly deserves further evaluation.

In contrast to the extensive data now available on hovering flight energetics, oxygen consumption of insects flying at different airspeeds has been measured only once. Using a closed-circuit wind tunnel, Ellington et al. (1990) determined energetic expenditure for bumblebees flying at airspeeds over a range of 0–4 m/s (fig. 4.1). Oxygen consumption showed no significant variation with forward airspeed, matching the relatively invariant mechanical power expenditure estimated for bumblebees over the same speed range (see Dudley and Ellington, 1990b). The relatively constant ratio of mechanical to metabolic power requirements indicates that flight muscle efficiency is also constant for these airspeeds (Casey and Ellington, 1989). Large changes in muscle efficiency would in fact be surprising given invariance in wingbeat frequency and stroke amplitude over the same speed range (see Dudley and Ellington, 1990a). For bumblebees flying at forward velocities above 4 m/s, metabolic expenditure probably rises in parallel with projected increases in mechanical power requirements, and particularly in parasite power (see section 3.3.3). Insects tethered in airstreams do in fact show an increase in metabolic power expenditure at the upper extremes of investigated flow velocities (e.g., Weis-Fogh, 1952; Hocking, 1953), although metabolic rates have never been determined for insects in free flight near their maximum airspeed.

Energetic costs of flight other than those associated with contraction of the flight muscles are poorly understood, but are likely to be relatively small. Clearly, some energy must be expended to ventilate both the abdomen and thorax, to mobilize fuels from the fat body, to circulate hemolymph for delivery of metabolic fuels, and to sustain increased activity of the nervous system (e.g., regeneration of ionic gradients used in signal transmission). For example, heart pulsation frequency increases above the resting rate in free-flying moths (Wasserthal, 1996), and additional power will be necessary to sustain such enhanced hemolymph circulation. Small but long-lasting oxygen debts immediately following flight activity (e.g., Chadwick and Gilmour, 1940; Krogh and Weis-Fogh, 1951) may reflect slowly diminishing energetic demands of such metabolic support activities once wing flapping ceases.

4.1.2 Respiratory Quotients and Metabolic Fuels

Biochemical pathways that sustain flight metabolism can be determined through identification of metabolic enzymes and analysis of the respiratory quotient (RQ), the ratio of respired carbon dioxide to consumed oxygen. As in animal metabolism generally, equal amounts of evolved carbon dioxide and consumed oxygen indicate oxidation of carbohydrates ($RQ = 1$); values of RQ less than than one indicate use of lipid or protein. Carbohydrate is utilized by most Diptera and Hymenoptera during flight (Sacktor, 1965; Crabtree and Newsholme, 1975; Beenakkers et al., 1984, 1985; Bertsch, 1984; Rothe and Nachtigall, 1989; Wheeler, 1989). Trehalose, a carbohydrate disaccharide, is the principal metabolic substrate in such taxa (Wyatt and Kalf, 1957). Other carbohydrates such as glycogen present in the flight muscle are used transiently by many insects at the outset of flight (see Downer and Matthews, 1976; Rowan and Newsholme, 1979; Burkhardt and Wegener, 1994). High-energy phosphate compounds (e.g., arginine phosphate) may also be sufficiently abundant within the muscle so as to offset rates of ATP consumption for several seconds (Candy, 1989). After such short periods, however, trehalose becomes the dominant fuel type. High trehalose concentrations in the hemolymph (insect blood) ensure ready availability for uptake and use by the flight muscles (Steele, 1981). The resting pool of trehalose also represents a substantial stored energetic capacity that can be used prior to mobilization of supplemental reserves from the fat body (see Auerswald and Gäde, 1995). High resting concentrations of carbohydrates also mitigate potential limits to fuel supply associated with slow convective movements of hemolymph within the open and low-pressure insect circulatory system.

Some insects utilize more than one type of metabolic fuel during flight. Migratory locusts shift from carbohydrate (trehalose and glycogen) to lipid fuels during flights that last more than 30 minutes (Weis-Fogh, 1952a). Similar patterns of fuel use characterize some Homoptera and Coleoptera. Many Lepidoptera (Zebe, 1954; Beenakkers et al., 1981) rely primarily on lipid metabolism, but some lepidopteran species use both fat and carbohydrate (see Stevenson, 1968; Van Handel and Nayar, 1972). Lipids are also used by various taxa in the orders Hemiptera and Odonata. Not surprisingly, lipase activity is high in the muscles of those insects that metabolize fat during flight (see George et al., 1958; George and Bhakthan, 1960; Crabtree and Newsholme, 1972). Few insect species oxidize amino acids in flight, although metabolic pathways based on the amino acid proline are found in the hematophagous tsetse fly and also in some plant-eating Coleoptera (Bursell, 1981). Various beetle species also combine carbohydrate and proline metabolism during flight (see Zebe and Gäde, 1993; Auerswald and Gäde, 1995). Amino acid concentrations in insect hemolymph tend to be high relative those in vertebrate blood, and those insect taxa that base flight metabolism on amino acids may have partially co-opted this resting pool of potential metabolites.

Although the biochemical diversity of flight metabolism is impressive, behavioral and ecological correlates of fuel utilization are less clear. Duve (1975) suggested that carbohydrate utilization is associated with high wingbeat frequencies and asynchronous flight muscle, whereas lipid metabolism is more characteristic of insects with synchronous muscle. Short-duration flights appear to be associated with utilization of carbohydrates (Wheeler, 1989), although feeding frequency, regularity of flight behavior, and ecological context can all influence the characteristics of flight metabolism (e.g., Yuval et al., 1994). One extreme case of metabolic specialization occurs in insects that utilize lipid during migration. Because of its higher energy density per unit mass relative to that of carbohydrates, lipid is the optimal fuel for long-distance migrants. Furthermore, in contrast to the anhydrous deposition of fats, glycogen molecules are invariably stored with water. This obligatory storage condition further decreases the energy density of carbohydrate reserves and renders them even less desirable for migrating animals (Weis-Fogh, 1952b, 1967a).

Alternatively, metabolic release of water stored with carbohydrates may contribute to hydric balance during flight. One disadvantage of a tracheal respiratory system is that not only respiratory gases but also water vapor pass freely through open spiracles. Water retention is particularly difficult for small insects with relatively high surface area : volume ratios. In flight, however, respiratory water loss from the tracheal system is generally offset by water production through metabolic oxidation (e.g., Cockbain, 1961; Weis-Fogh, 1952a, 1967b). Similar equilibrium between water loss and gain characterizes bumblebees (Bertsch, 1984) and carpenter bees in free flight (Nicolson and Louw, 1982), as well as honeybees in tethered simulations of forward flight (Louw and Hadley, 1985). Certain thermal conditions (e.g., high radiative heat load and high air temperature, low relative humidity) may impose water deficits on long-distance fliers and on xerophilic insects (e.g., Willmer, 1986). Also, thermoregulatory adaptations to xeric environments (see below) suggest that flight in many subtropical and temperate regions will necessitate evaporative water loss in order to avoid overheating. In many cases, however, evaporative cooling by water loss either via spiracles or through the cuticle directly is probably small relative to convective heat flux (Church, 1960a, b; Casey, 1980; Tsuji et al., 1986). Further studies of free-flying insects exposed to environmentally demanding conditions, particularly those that simulate xeric and tropical conditions, will be necessary to evaluate the generality of this conclusion.

Extended periods of flight require replenishment of fuels firstly from immediate muscle stores and secondly from the hemolymph as primary metabolites are progressively depleted. In long-duration flights, metabolic fuels are mobilized from the abdominal fat body and are then transported to the muscle via the hemolymph (Beenakkers et al., 1985; Wheeler, 1989). Mobilization of carbohydrate occurs via conversion of glycogen stored in the fat body to free-circulating trehalose. Lipid is mobilized as diglycerides from the fat body; specialized lipoproteins in the hemolymph then transport these diglycerides to the flight muscle (Chino, 1985). Fuel mobilization is in part under hormonal control that is specific to the metabolic substrate in question (Beenakkers et al., 1985). For example, adipokinetic hormone released from the corpora cardiaca specifically regulates lipid mobilization from the fat body (Stone et al., 1976). Similar effects derive from release of the biogenic amine octopamine (Goosey and Candy, 1980; Orchard et al., 1993). Hormonal regulation of fuel supply clearly provides for more precise temporal regulation of fuel supply than otherwise might be available through simple diffusion from source (fat body) to sink (respiring muscle). In prolonged flights, however, hemolymph concentrations of principal metabolic substrates typically show a slow decline. This decline may reflect long-term substrate depletion at the fat body and limiting supply via diffusion from the circulating hemolymph that convectively supplies fuels to the muscles. Weis-Fogh (1964b) in fact suggested that the relatively high trehalose and lipid concentrations of insect hemolymph were necessary to overcome diffusive limits on fuel supply at the level of the flight muscle.

Flight can be immediately initiated by most insects from a resting condition. This capacity is consistent with standing supplies of oxygen and energetic substrate within the flight muscle that are sufficient to support a rapid elevation in metabolic demands. In tethered locusts, muscle concentrations of ATP remain approximately constant across the transition from resting to active flapping metabolism, even though turnover of all energy-rich phosphate compounds in the muscle occurs in less then a second (Wegener, 1990; Wegener et al., 1991). This result indicates a fine-tuned biochemical balance between supply and demand, and suggests that metabolite supply during flight is very well regulated in spite of the 100-fold change in metabolic rate (Wegener, 1996). Of all forms of locomotion, the high metabolic rates of flight might be expected to yield a close match between functional capacity of metabolic pathways and maximal physiological performance at the organismal level. For example, maximal rates of oxygen consumption in flying honeybees are generally congruent with the maximal activity of various enzymes involved in flight metabolism (Suarez et al., 1996; Staples and Suarez, 1997; Suarez, 1998). Such matching is likely to be more pronounced in high-flux systems, for which animal flight is exemplary given high associated rates of oxygen consumption and rapid onset to active metabolism (Suarez et al., 1997). In this sense, animal flight metabolism is more representative of a transition between two distinct physiological states (rest and activity) rather than a continuum of power output, as is characteristic of other locomotor muscles.

Finally, an intriguing but conjectural metabolic hypothesis is that the high standing pools of ATP required for flight may bias nucleotide ratios of the genome in volant animals (Pettigrew, 1994). Enhanced ATP concentrations are probably required to meet the overall demands of flight, but this effect has not been systematically demonstrated for insects relative to other arthropods. Genomes of volant vertebrates do tend to exhibit an elevated adenine-thymine/guanine-cytosine ratio relative to genomes of nonvolant vertebrates, although no strict phylogenetic test of this observation has been carried out (Pettigrew, 1994). Similarly, Hughes and Hughes (1995) have suggested that both genome and cell size of volant vertebrates will be reduced relative to nonvolant vertebrates. This hypothesis proposes that the greatly elevated metabolic rates of flight require much-enhanced rates of transcellular metabolic exchange, which in turn are facilitated by higher surface-volume ratios and thus by smaller cells. Neither this hypothesis nor the interesting suggestion of Pettigrew (1994) has been evaluated for winged insects.

4.2 Muscle Physiology

Motor neurons projecting from the thoracic ganglia and, in some cases, abdominal ganglia regulate contractile activity of the flight musculature. In synchronous flight muscle, contraction frequency and amplitude are under direct neural control. The asynchronous flight muscle unique to some pterygote lineages permits repeated muscular contraction per nervous impulse and enables contraction frequencies typically in excess of 100 Hz. All muscular contraction involves generation of heat that, particularly in many large insects, can be used in active physiological regulation of body temperature. Flight muscle morphology and histochemical composition exhibit often dramatic ontogenetic changes. Dissolution of muscle fibers and decline of flight ability are the final stages in the developmental trajectory of the winged insect.

4.2.1 Innervation of Flight Muscle

The contraction of flight muscle fibers is regulated by motor neurons that originate within thoracic ganglia. These ganglia, although homologous across insect taxa (Schmitt, 1962; Horridge, 1965), are evolutionarily labile in terms of relative size and segmental association. Most insect taxa retain three separate thoracic ganglia. However, partial or complete fusion of meso- and metathoracic ganglia characterizes the more derived Diptera (Melin, 1941; Power, 1948; Vater, 1962), some Hemiptera (Johansson, 1957), and some Hymenoptera (Markl, 1966). Fusion of abdominal with thoracic ganglia has also occurred in some lineages, particularly in the Orthoptera, Diptera, and Coleoptera (Horridge, 1965; Robertson et al., 1982). Derived dipteran families (e.g., Muscidae) represent perhaps an extreme of ganglial consolidation, with the three thoracic ganglia and all abdominal ganglia fused into a single mesothoracic structure. Such centralization is rivaled only by the coccoid homopterans (Horridge, 1965), a group perhaps not coincidentally also characterized by haltere-like structures on the metathorax (see section 4.2.2). The functional implications of ganglial fusion within the pterothorax are unclear but may be related to progressive synchronization of muscle activity between segments and to use of only one functional or morphological wing pair. Apart from the aforementioned studies of Diptera, however, neuroanatomy of the thoracic ganglia has been examined in detail only for a few species from individual insect orders (see Horridge, 1965).

Within pterothoracic ganglia, the position of motor neuron somata (cell bodies) and the branching patterns of major nerve processes tend to be phylogenetically conservative (see Simmons, 1977a; Breidbach and Kutsch, 1990; Kutsch and Breidbach, 1994). Dorsolongitudinal muscles of the pterothoracic segments are typically innervated by the corresponding segmental ganglion, and in some cases also by the prothoracic ganglion as well as by abdominal ganglia (Kammer, 1985; Burrows, 1996). Such shared neural control between adjacent segments likely facilitates the coordination of ipsilateral wing movements as well as the control of potentially variable intersegmental phase relationships during flapping. In contrast to the dorsolongitudinal muscles, dorsoventral muscles together with the basalar and subalar muscles are innervated by motor neurons that originate only from the corresponding segmental ganglion (see Neville, 1963; Tyrer and Altman, 1974; Simmons, 1977a; Rind, 1983; Ritzmann et al., 1983; Kammer, 1985).

Any given flight muscle comprises a variable number of distinct motor units, groups of muscle fibers (i.e., muscle cells) that are of approximately identical histochemical composition and physiology and that tend to be activated simultaneously by a single motor neuron (Kammer, 1985; Stokes, 1987). In synchronous flight muscle, each contraction of a muscle fiber is initiated either by single or multiple impulses from a motor neuron that result in muscle depolarization and ultimately in an action potential (see Hoyle, 1983; Aidley, 1985, 1989). Variation in number, amplitude, and timing of impulses all potentially serve to regulate the extent of muscle contraction during a wingbeat (Moran and Ewer, 1966; Kammer and Heinrich, 1978; Kammer and Rheuben, 1981; Kammer, 1985). Some phylogenetic variation in impulse number is also evident (see Kammer, 1985). The presence of multiple motor units within flight muscles raises problems of synchronization among these units to ensure the general low-variance repetitive movements typical of flight. In *Drosophila*, for example, motor units are activated at fairly constant phase relationships that promote regularity of muscular contraction (see Harcombe and Wyman, 1978; Tanouye and Wyman, 1981; Wyman and Tanouye, 1982). In contrast to this dipteran example, however, phasic coupling between motor units appears to be less pronounced in other orders (e.g., Kendig, 1968; Mulloney, 1970b), possibly yielding greater variance in wingbeat kinematics.

Innervation of individual muscle fibers by more than one motor neuron has also been reported from several insect taxa (see Darwin and Pringle, 1959; Ikeda and Boettiger, 1965a, b; Kutsch and Underwood, 1970). Such multiple innervation may facilitate graded activation of muscle fibers and permit more precise control of both the timing and extent (i.e., amplitude) of muscular contraction.

Wilson (1961), in a classic paper of neurophysiology, demonstrated that the basic activational rhythm underlying the locust flight motor resides within the central nervous system. Patterned firing of flight motor neurons originates in the thoracic and (in some cases) abdominal ganglia, and is regulated by a complex network of ganglial interneurons (Burrows, 1996). Both excitatory and inhibitory interneuronal inputs to motor neurons influence rhythmic firing of the latter (Simmons, 1977b; Burrows, 1973, 1975a, b, 1977, 1996; Robertson and Pearson, 1982, 1985). Among different taxa, the total number of interneurons involved in generation of the flight pattern may vary substantially. The flight motor pattern is likely an emergent property of the entire interneuronal network rather than determined by a unique subset of these cells (Burrows, 1996). For example, large differences in muscle activation patterns between preflight warm-up and actual flight are inconsistent with a unitary oscillator underlying driving motor output in Lepidoptera (see Kammer, 1967, 1968). Those neural oscillators underlying the flight rhythm may also reside within multiple ganglia rather than being confined to the ganglion of a single thoracic segment (e.g., Robertson and Pearson, 1983, 1984; Wolf and Pearson, 1987; Ronacher et al., 1988; Robertson, 1989). Flight interneurons are furthermore found within abdominal ganglia in some taxa, a result consistent with flapping appendages being ancestrally arranged along both abdomen and thorax (Robertson and Pearson, 1982; see section 6.1.2). Identification of the full range of connectivity and extent of electrical coupling among those interneurons involved in flight-pattern generation remains a formidable task for neurophysiology (Burrows, 1996).

4.2.2 Contraction Dynamics

Most generally, rhythmic contraction within the thorax alternates between wing depressor and elevator muscles. Differences in phase relationships between activation of muscle antagonists may then influence wingbeat kinematics and ultimately aerodynamic output. For example, flies may use differences in stimulation rate between elevator and depressor muscles to regulate the downstroke : upstroke ratio of the wings (Mulloney, 1970a, b). Such effects may be displaced in time, persisting over the course of several wingbeats. Similarly in locusts, temporal variability in activation of the first basalar muscle induces correlated variation in the activity of wing elevator muscles during subsequent wingbeats (Möhl and Nachtigall, 1978). As with all muscles, electrical activation of flight muscles precedes contraction and, in the case of the pterothorax, wing depression and elevation (e.g., Wilson and Weis-Fogh, 1962). Electromyographic inferences of the timing of muscle contraction must accordingly incorporate potentially variable onset times for different flight muscles.

In a landmark paper of twentieth-century physiology, Pringle (1949; see also Roeder, 1951) identified a type of insect flight muscle that contracts repeatedly in response to only one nervous impulse and ensuing membrane depolarization. This muscle type is termed asynchronous muscle because of the temporal decoupling between contraction frequency and stimulation rate, although these two variables are often monotonically correlated (Nachtigall and Wilson, 1967; Bastian, 1972; Esch et al., 1975). Asynchronous muscle is also known as myogenic or fibrillar muscle, and was originally described by Pringle (1949) in dipteran haltere muscle. Wingbeat frequency in insects with asynchronous flight muscle is determined primarily by the mechanical characteristics of the pterothoracic apparatus, and secondarily by behavioral modulation (e.g., Nachtigall and Wilson, 1967; Kutsch and Hug, 1981). By contrast, the phylogenetically more ancestral synchronous muscle type is characterized by a one-to-one correspondence between nervous impulses and muscle contraction (see Usherwood 1975, Tregear 1977, and Pringle 1981 for general reviews of insect muscle). As with vertebrate striated muscle, frequency regulation of synchronous flight muscle is attained simply by varying the rate of neural stimulation. For example, wing clipping of synchronous fliers produces only small increases in wingbeat frequency consistent with neuronal regulation of contraction frequency (Roeder, 1951). Tiegs (1955), however, found that wing clipping of some synchronous fliers could produce substantial increases in wingbeat frequency, possibly indicating compensatory sensory reflexes as well as artifactual consequences of wing amputation.

Asynchronous muscle is advantageous in that evolution of this muscle type permits acquisition of high wingbeat frequencies and associated changes in aerodynamic force output (Dudley, 1991b). Constraints on the activation rates of synchronous muscle impose an upper limit on the operating frequency of about 100 Hz, although there are some important exceptions found among nonlocomotor muscles. For example, the synchronous tymbal muscle of some cicadas can reach contraction rates as high as 550 Hz during sound production

(Josephson and Young, 1985). In general, however, the intriguing physiological characteristics of asynchronous flight muscle permit much higher muscle contraction rates to be attained than would otherwise be possible with synchronous muscle. Wingbeat frequencies are correspondingly elevated, with typical values for such asynchronous fliers as bees and flies in the range of 100–300 Hz (Greenewalt, 1962). Rates of contraction as high as 1000 Hz have been inferred acoustically from the flight of some insects with asynchronous flight muscle (Sotavalta, 1953), whereas wingbeat frequencies of many synchronous fliers such as odonates and orthopterans are typically less than 50 Hz (e.g., Greenewalt, 1962; section 3.1.2). The high wingbeat frequencies enabled by asynchronous muscle are, in general, correlated with small body size (section 7.1). However, possession of asynchronous flight muscle does not necessarily mandate wingbeat frequencies in excess of 100 Hz. Large beetles and hemipterans, among others, exhibit markedly lower wingbeat frequencies. The flight muscle in such insects is clearly asynchronous (Pringle, 1957), but the characteristics of pterothoracic resonance necessitate reduced wingbeat frequencies given the increased inertial loading of larger wings (see below).

In addition to the salient parameter of contraction frequency, synchronous and asynchronous flight muscle also differ in a number of other morphological and physiological characteristics (see table 4.1). For example, contraction velocity of muscles is proportional to the product of the contraction frequency and the muscle strain, the relative change in length during contraction. Biophysical limits to the speed of muscle contraction generally indicate that high contraction frequencies can be attained only at low values of muscle strain. Muscle strain correspondingly tends to be high in synchronous relative to that in asynchronous flight muscle (table 4.1). As an extreme example, Wensler (1977) estimated a strain of 17% for the synchronous dorsolongitudinal muscle of monarch butterflies during the downstroke. By contrast, direct observations of dorsolongitudinal muscles in tethered bumblebees suggest mean strains of only about 2% during wing flapping (Gilmour and Ellington, 1993a; Josephson and Ellington, 1997). Similarly, strains average about 3.5% in the indirect flight muscles of *Drosophila virilis* (Chan and Dickinson, 1996). Difference in strain between synchronous and asynchronous flight muscle can thus be as high as an order of magnitude, although the extent to which this difference persists at equivalent operating frequencies has not yet been examined.

Important anatomical features also distinguish the two types of muscle (table 4.1). Asynchronous flight muscles tend to have myofibrillar diameters substantially larger than those of synchronous muscle (hence the traditional designations of fibrillar and close-packed

TABLE 4.1

Morphological and Physiological Characteristics of Synchronous and Asynchronous Flight Muscle

Trait	Synchronous Muscle	Asynchronous Muscle
Evolutionary condition	Ancestral	Derived
Sarcoplasmic reticulum	Abundant	Sparse
Stiffness	Lower	Higher
Operating frequency	Lower[a]	Higher
Contractions/nervous impulse	Single	Multiple
Operating strain	Higher	Lower[b]
Force/contraction	Higher	Lower[c]
Mechanical energy/contraction	Higher[d]	Lower
Oxygen consumption/contraction	Higher[d]	Lower

Notes: These qualitative characterizations only generally differentiate the two muscle types; considerable overlap if not reversal of characteristics may occur in comparison of particular taxa. Relevant literature citations can be found in the text.

[a] Operating frequencies of synchronous tymbal muscle in many cicadas exceed 100 Hz.

[b] Strain may be comparable at equivalent operating frequencies; the higher frequencies characteristic of asynchronous flight muscle are correlated with lower muscle strain.

[c] Isometric force production is lower in asynchronous flight muscle; force under oscillatory conditions may be comparable to that of synchronous muscle at equal contraction frequency.

[d] Mechanical and metabolic power expended per contraction is higher in synchronous muscle given the generally lower operating frequency; values may be comparable at equivalent frequencies.

muscle, respectively). Considerable overlap in fiber diameters can be found, however, between the two muscle types. Coupled with the low strain of asynchronous muscle is a nearly complete overlap of the contractile filaments (actin and myosin) along the length of myofibril subunits, the sarcomeres. Such a minimized range of filament motions permits a greater number of attachment points between contractile filaments and yields optimum power output for low-amplitude contractions (see Otten, 1987; Van Leeuwen, 1991). Interspecifically, the ratio of actin to myosin filaments appears to increase at the lower wingbeat frequencies characteristic of synchronous muscles (Auber, J., 1967; Auber, M. J., 1967; Elder, 1975; Candia Carnevali and Reger, 1982), although data that evaluate this trend phylogenetically are not available.

The diagnostic anatomical feature of asynchronous flight muscle is reduction in the sarcoplasmic reticulum, a membrane-bound network surrounding the myofibrils that is used for release and uptake of calcium during muscle excitation. In asynchronous muscle, the transverse tubules located near the ends of sarcomeres are widespread, but the sarcoplasmic reticulum that permits calcium excitation of the myofibrils is much reduced relative to that of synchronous flight muscle (see the extensive electron microscopy of Smith, 1961a, b, 1962, 1965, 1966a). Physiologically, the reduced network of sarcoplasmic reticulum yields a decreased dependence on the diffusion and binding of calcium to initiate and sustain myofibrillar contraction. In turn, myofibrillar diameters can be substantially larger in many asynchronous muscles given this relaxation of potential diffusional constraints on calcium-based activation (Smith, 1966b). In addition to varying in the extent of sarcoplasmic reticulum, flight muscles of different insect taxa potentially vary in terms of the actual geometry and three-dimensional structure of this network within the myofibrils. Particularly for extant insects that are evolutionarily transitional between synchronous and asynchronous muscle (see below), quantitative stereological evaluation of the sarcoplasmic reticulum may provide further morphological characters of physiological relevance to the designation of muscle type.

Reduced calcium cycling within asynchronous flight muscle yields an important physiological consequence, namely lower total costs of calcium transport and thus of muscle contraction as a whole (Josephson and Young, 1985). In vertebrate striated muscle, for example, calcium cycling can represent as much as 50% of total energy expended in isometric (constant-length) exertion of force (see Homsher and Kean, 1978; Rall, 1982). No comparable estimate exists for synchronous insect flight muscle, but the cumulative costs of calcium activation may be substantial. If such costs are high and are obviated by the particular contractile mechanism of asynchronous muscle, then efficiencies of contraction may well be higher in this muscle type. Evolution of asynchronous muscle may thus entail energetic savings in addition to selection for high wingbeat frequencies and enhanced flight performance. Maximum power output has not been systematically compared between synchronous and asynchronous muscle, although mass-specific mechanical and metabolic power expenditures are generally higher in asynchronous muscle given their usually much higher contraction frequencies (see section 5.4.2).

Oscillatory frequencies of asynchronous muscle are determined primarily by the inertial load acting on the muscle; neural impulses electrically activate the muscle but do not usually influence the frequency of contraction (Machin and Pringle, 1959, 1960). Asynchronous muscle typically oscillates through 5–25 contractions per activational neural impulse. Consistent with the hypothesis of mechanical resonance, the wingbeat frequency of asynchronous fliers invariably increases when wings are artificially shortened in spite of a concomitant decrease in muscle activation frequency (Roeder, 1951). High precision of wingbeat frequency is to be expected in the mechanically tuned pterothorax, and the low variation characteristic of wingbeat frequencies for insects with asynchronous flight muscle is consistent with this interpretation. During flight, asynchronous muscle can be viewed as in a state of continuous electrical activation relative to the rest condition. Multiterminal innervation may in fact be more extensive in asynchronous relative to synchronous muscle, yielding physiologically a more uniform state of activation (McCann and Boettiger, 1961). An increased frequency of activational impulses may marginally increase the contraction frequency of asynchronous muscle in some cases (e.g., Heide, 1974), although neither the linearity of this response nor its potential transient use to elevate wingbeat frequency and aerodynamic force output have been evaluated. When activated electrically, the contraction amplitude of asynchronous muscles is largely regulated by the independent activity of distinct steering and control muscles (section 2.1.2). Below a frequency threshold of neural stimulation, however, contraction amplitude and output force of asynchronous muscle decline.

The repeated contractions of asynchronous muscle ultimately derive from a biophysical mechanism of stretch activation. Detailed descriptions of the molecular physiology underlying stretch-activated contraction can be found in Pringle (1967), Usherwood (1975), Tregear (1977, 1983), and Pringle (1978). After an asynchronous muscle fiber contracts, internal calcium concentrations remain elevated and the muscle remains physiologically activated. Within the pterothorax, elevator and depressor muscles act antagonistically. Such phasically active muscles are, during periods of relaxation, necessarily stretched by action of their antagonistic counterpart. This stretching induces, after some delay, an opposing tension at a rate directly proportional to the operating frequency during flight (Molloy et al., 1987; Peckham et al., 1990). Ultimately this tension is sufficient to initiate a compensatory contraction. Thus, paired asynchronous and antagonistic muscles can repetitively induce and sustain reciprocal contraction if the development of tension is both stretch-activated and of sufficiently high magnitude.

Molecular mechanisms that generate this delayed tension following stretching must reside within the myofibrils themselves. For example, the cell membrane and its associated calcium ion gradient are not strictly necessary for stretch activation. Demembranated asynchronous muscle fibers from the belostomatid hemipteran genera *Hydrocy-rius* and *Lethocerus* exhibit stretch-activated contraction, and when working against an inertial load can oscillate for up to 23 hours given appropriate calcium activation and supply of ATP substrate (Jewell and Rüegg, 1966). Stretching of an asynchronous muscle sarcomere may in part induce conformational changes in the binding between contractile proteins that yields delayed tension. Geometrically, period matching between contractile filaments may facilitate both stretch activation as well as synchronized cross-bridge connection and detachment within myofibrils (Wray, 1979; see also Reedy, 1967). Physiologically, the total number of attached sites between myosin heads and actin filaments increases with stretch in asynchronous muscle (Thorson and White, 1969, 1983). Patterns of oxygen production during ATP hydrolysis are also consistent with helical period matching between actin and myosin in some asynchronous flight muscles (White et al., 1988). By contrast, stretch activation does not characterize the synchronous flight muscles of odonates, an order for which such helical matching is absent (Peckham and White, 1991).

However, precise helical matching between actin and myosin structures characterizes neither the asynchronous muscle of the hemipteran *Lethocerus* (Squire, 1992) nor that of bumblebees (Tregear et al., 1993), and additional biophysical features may be indicated. Of these, a high intrinsic stiffness is salient. For example, the mathematical model developed by Sicilia and Smith (1991) for asynchronous muscle suggests that continuous oscillation will occur only when a particular muscle stiffness is exceeded. Stretch-activated contraction also emerges from various nonlinear models of cross-bridge kinetics and force development within the sarcomere (Thomas and Thornhill, 1995a, b). Also contributing to stretch-activated responsiveness are nonlinear dependencies of inherent muscle stiffness (Pringle and Tregear, 1969; Josephson, 1997a) and induced tension (Abbott, 1973) on myofibrillar extension. Unfortunately, the potential evolutionary pathways for acquisition of these mechanisms is entirely unresolved. Most striated muscle does exhibit some degree of stretch activation and enhanced force production (potentiation) through prestretch (see Pringle, 1981; Ettema et al., 1992).

Stretch activation may in fact predominate in the asynchronous muscle of winged insects only because of specific molecular adaptations for increased stiffness. A causal association between muscle stiffness and stretch-activated contraction is strengthened by the observation that inactive asynchronous muscles have a high modulus of elasticity (Buchthal and Weis-Fogh, 1956; Machin and Pringle, 1959; Tregear, 1977). Such stiffness derives from neither active cross-bridge connections nor specific calcium ion concentrations (White, 1983), but probably reflects the presence of continuous structural elements that run parallel to contractile filaments. For striated muscle generally, inherent elastic stiffness derives from large endosarcomeric proteins (e.g., connectin, nebulin, projectin, and titin) that form an extensible but elastic matrix along the length of the sarcomere and possibly across sarcomeres (see Trombitas and Tigyi-Sebes, 1979; Bullard, 1983; Wang, 1985; Horowits et al., 1986; Trinick, 1991; Granzier and Wang, 1993; Ziegler, 1994). These and similar proteins may be particularly abundant or mechanically specialized in asynchronous flight muscle. Troponin components unique to asynchronous flight muscle may also contribute to stretch activation (Bullard et al., 1988; Peckham et al., 1992). Further biophysical understanding of asynchronous muscle is likely to emerge from specific molecular perturbations of muscle function. For example, various muscle mutants in *Drosophila* manifest physiological dysfunction as well as behavioral impairment during locomotion (e.g., O'Donnell and Bernstein, 1988; Sparrow et al., 1991; Warmke et al., 1992; Tohtong et al., 1995; Dickinson, Hyatt, et al., 1997; Vigoreaux et al., 1998). Study of such mutants can provide a mechanistic link between protein biophysics and free-flight performance.

The evolution of asynchronous flight muscle may also be linked with increased pterothoracic elasticity and consequent storage of wing inertial energy in the first half of a half-stroke. The magnitude of this quantity increases with the square of wingbeat frequency (section 3.3.2) and thus will be inordinately high for small insects with the elevated wingbeat frequencies associated with asynchronous flight muscle. Because selection for elastic recovery of wing inertial energy has probably been substantial in flight evolution, comparative analyses of the structural and physiological characteristics of such sarcomeric proteins (e.g., Wang et al., 1991, 1993; Labeit and Kolmerer, 1995) would directly link consequences of molecular design for organismal performance. Studies of the genesis and molecular evolution of elastic proteins (e.g., resilin) would be of particular interest given their direct role in maximizing energetic efficiency during flight. In this context, hummingbirds would be a particularly illuminating comparison relative both to insects as well as to other volant vertebrates (see section 7.5.2).

Asynchronous muscle is phylogenetically derived relative to synchronous flight muscle and is widely distributed among both endo- and exopterygote orders (fig. 4.2). Although asynchronous flight muscle has been historically interpreted as polyphyletic in origin (see Cullen, 1974; Pringle, 1981), a number of phylogenetic scenarios are possible. For example, assignment of asynchronous muscle to the equivocal branches of fig. 4.2 yields a monophyletic origin of

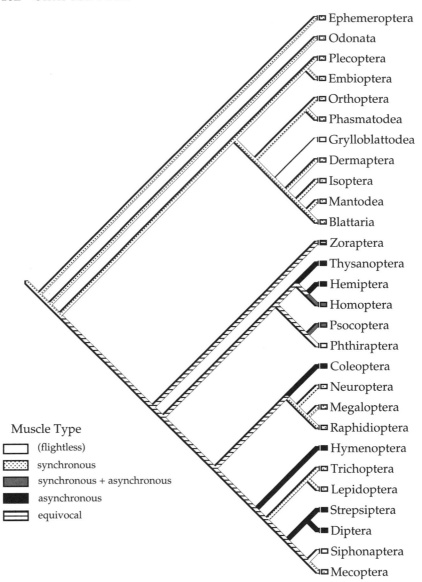

FIG. 4.2. Phylogenetic distribution of asynchronous flight muscle following Boettiger (1960), Cullen (1974), Smith (1984), and Smith and Kathirithamby (1984). MacClade 3.0 was used to generate the most parsimonious reconstruction of character states; equivocal branch designations indicate either an unknown (e.g., Zoraptera) or an unresolved state.

asynchronous flight muscle but subsequent reversion to the synchronous type in at least five lineages. Reversion from asynchronous to synchronous flight muscle is physiologically unlikely, but a systematic analysis of character distribution and evolution is not at present possible because muscle types are not well resolved for key taxa. For example, the flight muscle type of the key order Zoraptera (see fig. 4.2) has not been investigated but may well be asynchronous given the small body size of this taxon. Most muscle type designations have also been made on the grounds of morphological inference rather than by direct physiological measurement.

For example, histological studies of flight muscle in diverse Hymenoptera have revealed the presence of large muscle fibers and possibly asynchronous flight muscle in all but two families of sawfly (suborder Symphyta; see Daly, 1963). The basal hymenopteran family Xyelidae, however, expresses muscle that appears on morphological grounds to be asynchronous. This finding suggests that Hymenoptera ancestrally possessed asynchronous flight muscles, and that the two apparent reversions within the sawflies actually indicate the limits of inference from myofibrillar diameter alone without incorporation of physiological data. For example, a sample of ten symphytan and basal hymenopteran species examined by Sotavalta (1947) exhibited wingbeat frequencies ranging from 58 to 123 Hz; frequencies of three species were greater than 100 Hz and consistent with the presence of asynchronous muscle. Likewise, morphological criteria were used to infer presence of both synchronous and asynchronous muscle within the Psocoptera (Cullen, 1974). A more likely possibility is that this taxon is exclusively asynchronous given the small body sizes within the order and the vestigial hindwings of some species (see below). Physiological evaluation of these diverse flight muscles is clearly required, particularly as the quantitative extent to which muscular contraction is decoupled from activational impulses will not always be evident morphologically.

Asynchronous muscle may be present in yet additional taxa not generally thought to express this muscle type. Most suggestive are associations between high wingbeat frequency and small body size (section 7.1). Hindwing reduction and small body size together combine to demand ever increasing wingbeat frequency (Dudley, 1991b); extant taxa potentially satisfying these two morphological criteria include the small coniopterygid neuropterans (1–3 mm body length), some psocopterans, and the bucculatricid, gracillariid, lyonetiid, and nepticulid moths. Also within the Lepidoptera, physiological assessment of the Sphingidae would prove informative. Small body size together with extensive hindwing reduction appears to be common in some sphingid lineages (D'Abrera, 1986). Even if such taxa do not possess asynchronous muscle, stretch-activated contraction may be more pronounced than in synchronous sphingids, indicating possible intermediate morphological and physiological pathways for evolution of high frequencies. For example, certain sesiid moths have high wingbeat frequencies near 100 Hz (Sotavalta, 1947) that may indicate asynchronous contraction. Finally, the baetid, caenid, and tricorythid ephemeropterans are highly viable candidates for possession of asynchronous flight muscle. These mayflies have small forewings (2–10 mm wing length) and exhibit either greatly reduced or absent hindwings. Most intriguingly, at least one caenid species (*Caenis horaria*) has a wingbeat frequency exceeding 150 Hz (see Sotavalta, 1947). Such insects would be excellent choices for characterization of muscle type, particularly as asynchronous muscle has never been recorded from the paleopterous orders.

The distribution of flight muscle types within the order Homoptera poses further questions of evolutionary importance. This traditional order has recently been shown to be a paraphyletic assemblage (Campbell et al., 1994; von Dohlen and Moran, 1995; Campbell et al., 1995; Sorensen et al., 1995; Schaefer, 1996; see fig. 4.3A). Furthermore, the primarily morphological data of Tiegs (1955), Boettiger (1960), and Cullen (1974) indicate a complex distribution of muscle type within the Hemiptera/Homoptera clade (see fig. 4.3B). The family Cicadidae is the predominant synchronous representative within the Homoptera (the other member of the superfamily Cicadoidea is the monogeneric family Tettigarctidae that has only two species and an unknown type of flight muscle). Abdominal musculature in various cicadas has been modified to yield asynchronous tymbal muscles, although synchronous flight muscles with well-developed sarcoplasmic reticulum are retained (Pringle, 1954; Smith, 1966b; Josephson and Young, 1981; Young and Josephson, 1985). The suggestion that flight muscles of the cicada *Tibicen linnei* are asynchronous (Edwards et al., 1958) requires

---

On opposite page:
FIG. 4.3. (A) Phylogeny of the homopteran/hemipteran lineages following Campbell et al. (1994), von Dohlen and Moran (1995), Campbell et al. (1995), Sorensen et al. (1995), and Schaefer (1996). Suborder names (i.e., Sterno-, Clypeo-, Archaeo-, and Prosorrhyncha) follow Sorensen et al. (1995). Note that the classical taxon of the Homoptera is paraphyletic. Monophyly of the Prosorrhyncha is well established (e.g., Wheeler et al., 1993). (B) Distribution of flight muscle type among the homopteran/hemipteran lineages. Worldwide species counts for suborders are given in parentheses and follow Richards and Davies (1977) and Hodkinson and Casson (1991). Assignment of muscle type (see section 4.2.2) follows Tiegs (1955), Boettiger (1960), and Cullen (1974).

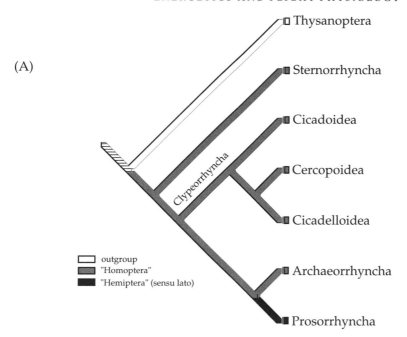

(A)

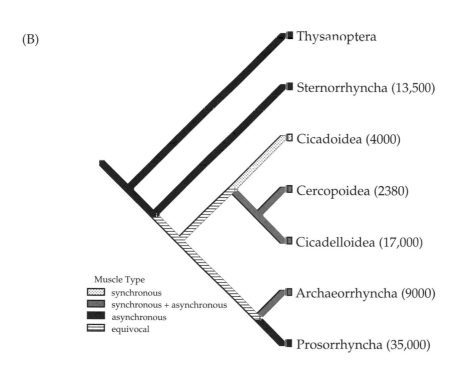

(B)

physiological evaluation. Evolution of flight and tymbal muscles within the Cicadidae represents potentially multiple origins of myogenic muscle rhythms, but this issue has not been evaluated phylogenetically. Of great importance is the ancestral condition of cicada flight muscle and of close relatives among other Hemiptera. Ancestral asynchronicity within the clade would indicate that synchronicity has been reacquired for some tymbal as well as for some flight muscles, a reversion in character state not yet documented within the winged insects. Another possibility is the existence of intermediate physiological states, as suggested by Pringle (1957) for the basal dipteran family Tipulidae. Particularly in large insects from generally asynchronous clades, reacquisition of synchronicity between nervous activation and contraction may reveal unique transitional forms of muscle physiology.

Asynchronous flight muscle likely characterizes all members of the hemipteran suborder Sternorrhyncha (comprising the superfamilies Aleyrodoidea, Aphidoidea, Coccoidea, and Psylloidea). Psyllids and aphids are clearly asynchronous fliers given their muscle histology (Cullen, 1974), small body size, and high wingbeat frequencies (Byrne et al., 1988). The families Aleyrodidae and Coccidae are presented in figure 4.3B as asynchronous even though morphological data of Cullen (1974) originally indicated otherwise. High wingbeat frequencies (143–181 Hz) of certain aleyrodids are consistent with asynchronous muscle (Wootton and Newman, 1979; Byrne et al., 1988), an observation corroborated by the finding of much reduced sarcoplasmic reticulum in this family (Smith, 1983). The Coccoidea are particularly curious in that hindwings of many males are miniaturized and are known as hamulohalteres (Kawecki, 1964; Gullan and Kosztarab, 1997). Hamulohalteres may well act physiologically like dipteran halteres (section 5.1.1), although this possibility has never been investigated empirically. Based on their small body size and reduced hindwings (see section 7.1), it is likely that coccids and, in fact, all sternorrhynchans possess asynchronous muscle.

Evaluation of this hypothesis merely requires simultaneous recording of wingbeat frequency and muscle action potentials in Coccidae and related taxa in the suborder. Examination of the relative extent of sarcoplasmic reticulum in the flight muscle would be similarly straightforward. Because contractile dynamics can now be determined on individual fibers (e.g., Peckham et al., 1990), comparative analysis of in vitro muscle physiology for the numerous small hemipteran lineages is also experimentally tractable. Most importantly, possession of asynchronous muscle is correlated with increased species diversity both among pterygote orders (fig. 7.1A) and among pterygote lineages if a comparative analysis is applied that controls statistically for phylogenetic relatedness (fig. 7.1B; see section 7.1.1). This finding suggests that aerodynamic and morphological correlates of increased wingbeat frequency have been historically associated with insect species diversification. Note finally that within the suborder Prosorrhyncha, the infraorder Peloridiomorpha consists of only twenty *Nothofagus*-associated species (Popov and Shcherbakov, 1996). Most species are flightless, and flight muscle type is unknown for the remaining species. This group is presently presumed to be asynchronous (fig. 4.3B), but if synchronous in muscle type would provide a remarkable contrast in species-level diversity with its sister group, the 35,000-strong Heteroptera.

Insect miniaturization is a key feature of the diversity-generating process associated with acquisition of asynchronous muscle (section 7.1). Pringle (1981) proposed that selection for small body size occurred during the initial evolution of asynchronous muscle, and was later followed by an increase in body size for some lineages. The low wingbeat frequencies of large extant Coleoptera, Cicadidae, Archaeorrhyncha, Prosorrhyncha, and Hymenoptera could then presumably have been derived from higher frequencies, with asynchronicity retained in these taxa even though muscle activation frequencies can be well below 100 Hz (Pringle, 1957). Functionally, the evolution of asynchronous muscle may have been associated with progressive dominance by one pterothoracic segment (either in antero- or posteromotorism) at higher operating frequencies, leading to wing coupling or loss of one pair. Because asynchronous muscles are not immediately responsive to changes in activational frequency, autoregulation of wing orientation must become a feature both of inertial oscillation and of control by high frequency but low-power synchronous steering muscles (section 2.1.2). In fact, some variance in wingbeat kinematics might be expected over consecutive wingbeats if potentially variable aerodynamic loading, wing inertia, and internal viscous damping combine to influence the timing of subsequent muscle contraction (Hocking, 1953). Such effects might be evident in time-series analysis of kinematics across many wingbeats of asynchronous fliers. An alternative hypothesis might propose greater variance in the dynamics of synchronous flight muscle given the physiological possibilities afforded by wingbeat-by-wingbeat modulation. For both muscle types, kinematic variance during well-defined flight modes (e.g., hovering) and in maneuvers (section 5.2.2) clearly deserves further attention.

As mentioned previously, the reduced costs of calcium cycling may render asynchronous flight muscle more energetically efficient than its synchronous counterpart. Higher contraction frequencies will also be facilitated by the reduced time necessary to activate and deactivate contractile proteins during each cycle (Josephson, 1993). Morphologically, a reduction in the volume fraction of sarcoplasmic reticulum permits an increased volume fraction of mitochondria and myofibrils, potentially allowing higher mass-specific force and power output to be attained. In an evolutionary context, Josephson and Young (1981) specifically proposed a pathway for acquisition of a myogenic rhythm in insect flight muscle. In the ancestrally synchronous muscle, cyclic activation is effected by a high-frequency train of neural impulses. Net work output of the muscle would then be increased by any physiological mechanism of stretch-activation and deactivation that enhanced contractile force during shortening and reduced this force during lengthening. Progressively increasing effectiveness of such a mechanism would enable the stretch-activated but time-delayed contractile forces associated with inherent stiffness to generate repeated oscillation. A subsequent reduction in activational frequency with the retention of high calcium concentrations would then permit asynchronous oscillation to be driven solely by stretch activation and subsequent deactivation on shortening (Josephson and Young, 1981). Reduction of sarcoplasmic reticulum would presumably occur after the intrinsic feature of continuous calcium excitation had been attained (Pringle, 1965). Intermediate stages of this postulated evolutionary trajectory may well exist among extant homopteran lineages (see fig. 4.3B).

4.2.3 Energetics and Efficiency

Flight muscles act as machinery that converts metabolic energy into useful mechanical work. For repetitively contracting muscles, total mechanical power is given by the product of the work per contraction and the contraction frequency. Work, in turn, is proportional to the product of the mean muscle strain (the effective displacement) and the mean myofibrillar force exerted during contraction. If the force of contraction and concomitant displacement of the muscle can be measured directly, then power generated during cyclic contraction can be estimated (Weis-Fogh and Alexander, 1977; Pennycuick and Rezende, 1984; Ellington, 1985). Efficiency of power generation as well as increased total power may both be reasonably assumed to be foci of selection for insects during flight. Considerable data are available on the mechanical characteristics of contracting flight muscle, although flight muscle efficiencies are less well known.

One classic approach to studying the mechanical characteristics of muscle is to allow muscles to contract against a constant inertial load (isotonic contraction). In such preparations, insect flight muscles exhibit an inverse relationship between contractile force and muscle strain. Power output is proportional to the product of these parameters and peaks at intermediate values of force production (e.g., Buchthal et al., 1957; Josephson, 1984; Malamud and Josephson, 1991). A single isotonic contraction does not replicate natural muscle mechanics, however, as flight muscles operate repetitively with temporally varying force production. More realistic assays of flight muscle physiology have instead relied on the work-loop technique originated by Machin and Pringle (1959, 1960) and subsequently elaborated by Josephson (1985a, b). In this method, sinusoidal length changes are imposed on a muscle that is either detached from the thorax and mounted on both ends in vitro, or that is partially detached in vivo (e.g., Josephson and Ellington, 1997). Repetitive electrical stimulation is applied to the muscle at a constant phase relationship relative to the sinusoidal function determining muscle length. By simultaneously monitoring both muscle length and instantaneous force during extension and contraction, the net work output per cycle of contraction can be determined (see fig. 4.4). Mechanical power is then given by the product of this work and the contraction frequency. Temporal patterns of force production and, by implication, power output of muscles or of muscle fibers can thus be studied over a range of relevant oscillatory frequencies. Systematic variation in oscillation frequency and in the phase of electrical stimulation allows the experimenter to search for those combinations of these two parameters that yield maximum power output (see Josephson, 1985a, 1993; Johnston and Altringham, 1988).

However, even such sophisticated work-loop studies do not necessarily replicate the biochemical and loading regimes experienced by muscles in vivo. For example, subtle but important features of thoracic configuration and muscle kinematics may be impossible to replicate under invasive experimental conditions. Slight deviations from simple harmonic motion in wing tip kinematics (e.g., Ellington, 1984c; Dudley and Ellington, 1990a) suggest parallel variation in temporal patterns of muscle contraction. Harmonics above the wingbeat frequency are substantial both in muscle motions of tethered preparations and in wing motions of free-flying bumblebees (Gilmour and Ellington, 1993a). In work-loop studies of bumblebee muscle fibers, inclusion of these higher harmonics typically decreased work output but could, on occasion, yield power outputs well in excess of average values (Gilmour and Ellington, 1993a). These technical considerations suggest that minimum estimates of power output can be reliably obtained in work-loop studies (see table 4.2), but also that the maximum power of such preparations probably falls well below that available to insects in flight.

(A)

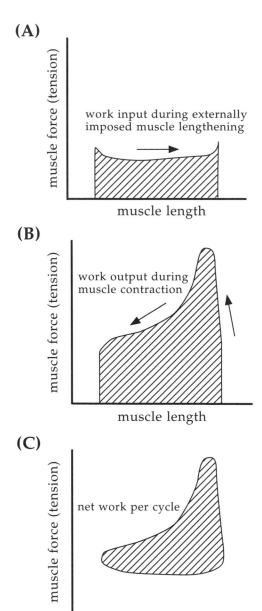

muscle force (tension)

work input during externally
imposed muscle lengthening

muscle length

(B)

muscle force (tension)

work output during
muscle contraction

muscle length

(C)

muscle force (tension)

net work per cycle

muscle length

FIG. 4.4. Work-loop method for determining work and power output of a repetitively contracting muscle (modified from Josephson, 1985a). (A) Mechanical work is required to extend the muscle during externally imposed stretching. Phasic electrical stimulation is applied near the time of maximum extension. (B) Contraction by the muscle results in positive work. (C) The sum of positive and negative work is the net work done by the muscle over the extension/contraction cycle. This value is multiplied by the contraction frequency to yield the mechanical power output.

TABLE 4.2
Maximal Mechanical Power Output Measured by the Work-Loop Method on Different
Flight Muscles

Muscle Identity	Muscle Type	Mechanical Power	Reference
Metathoracic tergocoxal of tettigoniid (Orthoptera)	Synchronous	76 W/kg [a]	Josephson (1985a)
Mesothoracic tergocoxal of tettigoniid (Orthoptera)	Synchronous	33 W/kg[b]	Josephson (1985b)
Metathoracic tergosternal of locust (Orthoptera)	Synchronous	37 W/kg	Josephson and Stevenson (1991)
Metathoracic tergocoxal of locust (Orthoptera)	Synchronous	73 W/kg [c]	Mizisin and Josephson (1987)
Mesothoracic dorsoventral of *Manduca sexta* (Lepidoptera)	Synchronous	90 W/kg [d]	Stevenson and Josephson (1990)
Metathoracic dorsolongitudinal of *Oryctes rhinocero* (Coleoptera)	Asynchronous	30 W/kg [e]	Machin and Pringle (1959)
Mesothoracic dorsolongitudinal of *Bombus* (Hymenoptera)	Asynchronous	60 W/kg [f]	Machin and Pringle (1959)
Mesothoracic dorsoventral of *Bombus* (Hymenoptera)	Asynchronous	110 W/kg [g]	Gilmour and Ellington (1993b)
Mesothoracic dorsoventral of *Bombus* (Hymenoptera)	Asynchronous	100 W/kg [h]	Josephson (1997b)

[a] Value obtained using multiple electrical stimuli per cycle at a stimulation frequency similar to the natural wingbeat frequency, and at a muscle temperature similar to the normal operating temperature during flight.

[b] Stimulation frequency similar to the natural wingbeat frequency; muscle temperature similar to the normal operating temperature. Note that the mesothoracic wing muscles also function in sound production and may not generate high power relative to dedicated flight muscles.

[c] Multiple electrical stimuli per cycle and stimulation frequency similar to the natural wingbeat frequency. Muscle evaluated at the normal operating temperature during flight.

[d] Stimulation frequency similar to the natural wingbeat frequency; muscle temperature similar to the normal operating temperature.

[e] Stimulation frequencies similar to those of free flight; muscle temperature (25–30°C) probably lower than the normal operating temperature during flight. Power output exhibited a strong temperature dependence (Machin et al., 1962).

[f] Muscle identity not specified in paper but likely as indicated here.

[g] Value obtained on demembranated muscle fibers at temperature of 40°C, and at a contraction frequency of 50 Hz and strain of 4–5%. These latter two values are lower and higher, respectively, than those characteristic of free-flying bumblebees. Power output exhibited a strong temperature dependence.

[h] Value obtained at operating frequencies (141–173 Hz) and muscle temperature (40°C) typical of free flight.

Indeed, more precise control of ATP availability, myofibrillar integrity, and pterothoracic configuration yields substantially increased power output from bumblebee flight muscle (see Gilmour and Ellington, 1993b; Josephson, 1997a, b). Further evidence indicates that the internal milieu of flight muscle will be difficult to mimic experimentally. For example, the neuromodulator octopamine plays a major role in modulating muscular contraction during flight (Candy, 1978; Orchard et al., 1993). Application of octopamine increases the force and power output of orthopteran flight muscles in isotonic contraction (see Evans and Siegler, 1982; Malamud et al., 1988; O'Gara and Drewes, 1990; Stevenson and Meuser, 1997); no comparable work-loop studies have been carried out. Accurate replication of the complex biochemical environment of working muscle is, however, likely to be a difficult experimental task.

Such restrictions notwithstanding, use of the work-loop method permits a direct measurement of the mechanical power produced by contracting muscles. Maximal values obtained by this method generally are 100 W/kg or less, although these measurements are strongly taxon and temperature dependent (table 4.2). For comparative purposes, rates of mechanical work done by free-flying insects can be estimated using aerodynamic models of wing flapping and knowledge of the flight muscle mass (section 3.3). Such estimates of power production during maximal flight performance often substantially exceed work-loop maxima (see section 5.4.2). Note also that power estimates for free-flying insects assume that all thoracic muscles are fully active during flight, a condition not necessarily met under all aerodynamic conditions. In hummingbirds, for example, the mass of the primary depressor muscle (the pectoralis major) is twice that of the primary elevator (the supracoracoideus), and is probably fully activated only during fast forward flight. During normal and maximal hovering performance, however, the down- and upstrokes are symmetric and serve equivalent aerodynamic roles, suggesting that the pectoralis major is only partially activated if it and the supracoracoideus are of similar physiological composition and biomechanical action. Comparable situations probably characterize the flight muscles of insects, although the relevant morphological and electromyographic data are not available.

Direct measurements of oxygen consumption by freely flying insects do, however, enable metabolic energy input to be determined parallel with estimates of mechanical work. Muscle efficiency during hovering or free forward flight can then be derived (see table 4.3). An alternative approach estimates total heat production (i.e., the muscle inefficiency) to derive total metabolic work (see Wakeling and Ellington, 1997c).

TABLE 4.3
Estimated Muscle Contraction Efficiencies for Various Flying Insects

Taxon	Muscle Type	Efficiency	Reference
Damselfly sp. 1 Dragonfly sp. 2 (Odonata)	Synchronous	9%[a] 13%[a]	Wakeling and Ellington (1997c)
Locust (Orthoptera)	Synchronous	4–10%[b]	Josephson and Stevenson (1991)
Sphingidae (Lepidoptera)	Synchronous	2–8% (perfect)[c] 4–12% (zero)[c]	Casey (1976, 1981a)
Euglossine bees (Hymenoptera)	Asynchronous	4–16% (perfect)[d] 8–34% (zero)[d]	Casey and Ellington (1989) Casey et al. (1992)
Apis, Bombus (Hymenoptera)	Asynchronous	5–6% (perfect)[e] 12–14% (zero)[e]	Ellington (1984f)
Bombus terrestris (Hymenoptera)	Asynchronous	12% (perfect)[f] 20% (zero)[f]	Casey and Ellington (1989)
Eristalis tenax (Diptera)	Asynchronous	8% (perfect)[e] 29% (zero)[e]	Ellington (1984f)
Drosophila hydei (Diptera)	Asynchronous	10–11%[g]	Dickinson and Lighton (1995)

Notes: Unless otherwise noted, values are based on aerodynamic estimates of mechanical work and direct measurements of metabolic rates during hovering flight. "Perfect" and "zero" following values of efficiency refer to the assumed extent of elastic storage and subsequent recovery of wing inertial energy.

[a] Aerodynamic power exceeded inertial power requirements, obviating the need for elastic energy storage. Metabolic rates estimated from measurements of heat production during flight.

[b] Estimate derived from work-loop measurements and simultaneous recording of oxygen consumption for a single muscle.

[c] Stroke amplitude of 120° assumed in estimates of induced power.

[d] Values refer to a multispecies sample of orchid bees ranging in body mass from 80 to 1000 mg. Mass-specific mechanical power was independent of body mass, whereas mass-specific rates of metabolism increased at lower body mass. Efficiency was thus lowest for the smallest bees.

[e] Values may be low because of underestimated thoracic muscle mass (see Dudley and Ellington, 1990b).

[f] Values refer to hovering and forward flight up to an airspeed of 4.5 m/s.

[g] Values refer to wing flapping by tethered insects. Aerodynamic power exceeded inertial power requirements, obviating the need for elastic energy storage.

Muscle efficiency reflects the total conversion efficiency of metabolic energy into repetitive muscle contraction, and subsumes a number of different processes. Molecules of ATP are hydrolyzed during both cross-bridge operation and pumping of calcium across the sarcoplasmic reticulum; only the former process yields useful mechanical work.

The cost of calcium pumping is more properly termed an informational cost associated with signal transduction, although both expenditures are conflated in a general parameter of muscle efficiency. This parameter is, however, probably the most relevant for any organismal perspective of total energy use during locomotion (Woledge, 1989). In general, flight muscle efficiencies of insect flight muscle tend to be fairly low (table 4.3). The limited data available to date preclude direct comparisons of efficiencies between synchronous and asynchronous muscle. In particular, size-dependent variation in muscle efficiency, uncertainties concerning elastic energy storage, and relative activation of available flight muscle must render premature any such comparative conclusion.

4.2.4 Ontogeny and Senescence

As hemimetabolous insects progress temporally to the winged adult stage via incomplete metamorphosis, marked changes are evident in the flight apparatus. Even though only the final adult stage of hemimetabolous insects is winged, flight-related structures are anatomically distinct and physiologically active throughout ontogeny. For example, pterothoracic muscle mass and mitochondrial density increase dramatically across larval stages in crickets (Ready and Najm, 1985); muscle contractile properties similarly exhibit ontogenetic variation (Ready and Josephson, 1982). Electromyographic activity similar to that during adult flight is also evident in the larval pterothoracic musculature of locusts (Stevenson and Kutsch, 1988; Kutsch and Hemmer, 1994b). Such patterned electrical activation of larval flight muscles indicates that ganglial centers of neural control are active during ontogeny. Indeed, various ganglial interneurons implicated in flight motor patterns demonstrate spiking activity in larval locust instars (Kutsch and Hemmer, 1994a), and motor patterns exhibited by the pupal muscles in holometabolous insects bear phasic resemblance to the motor patterns of free flight (e.g., Kammer and Rheuben, 1976; Kutsch et al., 1993). Interordinal variability in the ontogeny of motor rhythms has not been systematically evaluated, but likely mirrors phylogenetic transformation of the pterothoracic apparatus as a whole.

Following eclosion to the final winged stage, flight-related structures mature rapidly as physiological performance concomitantly increases. Following initial expansion to their final size, insect wings harden through desiccation in their final shape and size. For example, wing stiffness of freshly emerged butterflies increases asymptotically to the final value over a period of several days (Petersen et al., 1956). Flight muscles show similar maturational changes. Contractile ability increases, power production improves, and whole-animal locomotor capacity is enhanced (e.g., Petersen et al., 1956; Kutsch, 1989; Marden, 1995; Marden et al., 1996; Marden et al., 1998). As with senescent decline (see below), such temporal changes in flight ability are probably mediated via variable genomic expression of muscle proteins. For example, increased flight performance in dragonflies following eclosion is correlated with enhanced molecular sensitivity of the flight muscle to calcium activation (Fitzhugh and Marden, 1997). Also following eclosion, wingbeat frequencies increase to a plateau in a variety of insect taxa (see Kutsch, 1971, 1973; Altman, 1975; Gewecke and Kutsch, 1979; Curtsinger and Laurie-Ahlberg, 1981; Kutsch and Hug, 1981; see also Provine, 1994). At least for locusts, this increase may in part be mediated by increased conduction velocities of associated motor neurons (Gray and Robertson, 1994) as well as by ontogenetic changes in sensory receptors, particularly the wing stretch receptor (Gee and Robertson, 1994, 1996; see section 5.1.1). Wings, muscles, and elements of the nervous system involved in flight all potentially exhibit such posteclosion changes.

After peak locomotor performance is attained, senescence of physiological flight capacity and associated anatomical structures is readily evident. Metabolic enzymes decrease in activity, pathway turnover declines, and mitochondrial and myofibrillar structures exhibit age-dependent decline in insects in culture (e.g., Davies, 1974; Finlayson, 1975; Baker, 1976; Johnson, 1980; Ready and Josephson, 1982; Sohal, 1986; Collatz and Wilps, 1986; Shiga et al., 1991). These senescent effects are phylogenetically widespread among insects (see Johnson, 1976). Intriguingly, artificial dealation can also induce flight muscle atrophy and degeneration in adult crickets (Tanaka, 1991; Gomi et al., 1995). This finding suggests that programmed mechanisms of senescence may be facultatively activated, at least for the flight motor. In some taxa, metabolic pathways of flight clearly exhibit age-dependent shifts. For example, Harrison (1986) showed that both oxygen consumption rates in hovering and metabolic enzyme activities increase as honeybee workers shift ontogenetically from hive residence to active foraging. Such ontogenetic variation in metabolic pathways would be consistent with evolutionary theories of aging that predict at least partial genetic control of senescent decline (see Partridge and Barton, 1993).

In addition to these senescent changes in flight physiology, the external structures of flight may also exhibit age-dependent deterioration. Wing damage tends to increase with age in confined insects, although it is unclear if this effect is associated with actual use in flight or with mating attempts (Ragland and Sohal, 1973). Such decline in flight performance and in pterothoracic structures has not been well documented in natural insect populations, although abraded and frayed wings of older individuals are readily evident among many taxa and probably reduce life expectancy. For example, forager bumblebees with greater amounts of wing wear (either generated naturally or induced via wing clipping) experience higher mortality (Cartar, 1992). The direct aerodynamic consequences of such wing damage for flight performance are not known but may be presumed to be deleterious. Abrasion of and long-term impact damage within the axillary apparatus is an equally likely possibility with, at present, no empirical corroboration.

In a gerontological perspective, the extreme metabolic intensity of flight has given rise to speculation that insect life span varies in inverse proportion to cumulative flight activity and, more generally, to lifetime metabolic expenditure (e.g., McArthur and Sohal, 1982, Neukirch, 1982; Baret et al., 1994). Mechanistically, this effect may arise through the accumulation of deleterious superoxide radicals that are unavoidable consequences of the intense oxidative metabolism that characterizes flight (see Shigenaga et al., 1994; Sohal and Weindruch, 1996). Consistent with this hypothesis, overexpression within *Drosophila* of catalase and dismutase enzymes that break down superoxide radicals yields an increase of fly lifespan in cultures (Orr and Sohal, 1994). This interesting experimental result suggests that phylogenetic associations of lifetime flight intensity (perhaps evident in a comparison of insects with synchronous and asynchronous flight muscle) may be matched by underlying biochemical adaptations that mitigate deleterious consequences of oxidative metabolism.

4.3 THERMOREGULATION IN FLIGHT

Given that metabolic rates of flying insects are high and that contraction of flight muscles is inherently inefficient, endogenous heat generation during flight is substantial. In many taxa, this heat is secondarily co-opted to elevate and regulate muscle temperature (Heinrich, 1993). Such facultative endothermy is particularly common in large insects. Furthermore, increased power output of flight muscle provides a fundamental motivation for the evolution and elaboration of thermoregulatory mechanisms.

4.3.1 Power Output and Thermoregulatory Mechanisms

Many features of muscle physiology are strongly temperature dependent. Of these, power production is perhaps the most critical from the perspective of animal flight performance. Using work-loop methods,

thermal dependence of power generation has been documented for repetitively contracting synchronous (Stevenson and Josephson, 1990)
and asynchronous flight muscles (Machin et al., 1962; Jewell and
Rüegg, 1966; Steiger and Rüegg, 1969; Gilmour and Ellington, 1993b).
These experiments show that mechanical power output of muscle typically increases at higher temperatures, and is maximal near muscle
temperatures characteristic of the free-flying insect. Additional features of muscle performance such as tetanic force of contraction,
stretch-activated tension in asynchronous muscle, and contractile responsiveness also increase at higher temperatures (see, e.g., Weis-
Fogh, 1956c; Abbott and Steiger, 1977; Josephson, 1981). Such sensitivity suggests ample scope to respond to selective pressures that
demand elevated thoracic temperature during flight.

Equilibrium thoracic temperatures during flight represent a balance
between heat loss and heat gain. To elevate body temperature, insects
can rely either on endogenous (i.e., metabolic) or exogenous sources
(e.g., solar radiation). Existing estimates of muscle efficiency (table 4.3)
are consistent with substantial heat generation within the insect thorax
during flight. The dramatic amounts of heat produced by muscular
contraction are strikingly illustrated by the capacity of bumblebees
and certain noctuid moths to maintain thoracic temperatures exceeding 30°C when ambient air temperature is only 2–3°C (see Heinrich, 1975b; Heinrich and Mommsen, 1985). Further support of the link
between thermoregulation and flight is the use of pterothoracic muscles to engage in preflight warm-up. This activity elevates muscle
temperatures to those values optimal for power output and typically
necessary for takeoff (Heinrich, 1993). In contrast to existing studies
of power generation, no data are available on the temperature dependence of muscle contraction efficiency. However, the exemplary
work-loop study of Josephson and Stevenson (1991) measured oxygen consumption by contracting muscles (see table 4.3), and could be
readily adapted to study variation in muscle efficiency with operating
temperature.

Minimum thoracic temperatures required for takeoff and flight vary
widely among insect taxa, suggesting a wide diversity of physiological
strategies and thermal setpoints for optimal power output. The rapidity of takeoff in many small insects suggests that equilibrium temperature for the flight muscles is either reached very rapidly or is equal to
ambient air temperature. Larger insects, by contrast, are often more
reluctant to fly and require more time to attain physiologically optimal
temperatures. Some large beetles, for example, need several minutes of
preflight warm-up (Bartholomew and Casey, 1977; Bartholomew and
Heinrich, 1978). Ontogenetic variation in the temperature dependence

of muscular contraction can also be substantial. For example, thermal sensitivity of force production by dragonfly flight muscle is correlated with ontogenetically variable expression of myosin isoforms (Marden, 1995). This finding suggests that ample molecular capacity exists intraspecifically to match physiological features of flight performance with varying environmental conditions. Interspecifically, variation in the thermal tolerance of flight behavior is similarly diverse (Heinrich, 1993). The extent to which intra- and interspecific variation in thermoregulatory abilities derive from specific molecular adaptations will likely be a productive area for future research.

In addition to using the heat provided by metabolic activity, flying insects can thermoregulate behaviorally. A wide diversity of strategies is used by insects to alter body temperature both when at rest and during flight (Casey, 1988; Heinrich, 1993). Microhabitat selection in association with the complex thermal regime of vegetation is a common behavior, as is differential body orientation relative to ambient solar radiation and variation in cuticular reflectance to shortwave radiation (see Willmer and Unwin, 1981; Willmer, 1982, 1986b). With the exception of strongly endothermic taxa, foraging activity of nectarivorous insects is strongly correlated with solar radiation (e.g., Willmer, 1983). Sunshine necessarily augments the endogenous heat production of flying insects, and can be predominant under certain environmental conditions. In an extreme example, short- and longwave radiation dominate heat gain of honeybees foraging in the desert (Cooper et al., 1985). By contrast, flight metabolism can be expected to greatly exceed irradiant heat flux for flight in shaded microhabitats.

Heat loss from flying insects occurs under most environmental conditions via convective cooling. Convective heat flux from objects occurs at a rate proportional to airflow velocity and to the difference between object and air temperature. Airspeeds in forward flight are sufficiently high for most insects to result in net heat loss from the thorax at all but the highest environmental air temperatures. In flying locusts, for example, virtually all excess thoracic heat is removed via convection (Rainey et al., 1957; Rainey, 1974). Carpenter bees probably rely on convective cooling of the thorax to enable flight in hot desert conditions (Heinrich and Buchmann, 1986), whereas convection is high enough to produce net heat loss in free-flying honeybees except at high ambient temperatures (Cooper et al., 1985). The role of convection during hovering flight is unclear, although the induced velocity around the insect's body may be sufficient to cool the thorax. For example, Gilbert (1984; see also Heinrich, 1972) found that initiation of wing beating by resting hoverflies reduced their thoracic temperature, suggesting that the thermal losses associated with convection exceeded metabolic heat gain.

In addition to such external mechanisms of thermoregulation, many large insects control heat distributions internally via control of hemolymph circulation between the thorax and abdomen (Heinrich, 1993). If such heat flux to the abdomen (and secondarily to the head) can be assumed to be minimal (i.e., all heat produced through muscular contraction is either retained to elevate thoracic temperature or is lost via convection), then indirect estimates of metabolic heat production can be obtained from measurements of thoracic temperature before and during flight (e.g., Wakeling and Ellington, 1997c). Such estimates require an independent measurement of the Newtonian cooling constant of the body (rate of heat loss in convection as a function of airspeed) as well as knowledge of the specific heat of thoracic tissue. Metabolic heat thus determined can be added to aerodynamic estimates of work to provide an independent estimate of the metabolic rate during flight (Wakeling and Ellington, 1997c). This approach provides a minimum estimate of metabolic rate because any potential heat flux from thorax to abdomen represents supplemental heat production and, by implication, energetic expenditure by thoracic musculature.

Evaporative heat loss is in most cases a minor component of the overall energy balance during flight, although desert honeybees flying at air temperatures as high as 46°C can do so only because of water evaporation from regurgitated nectar (Heinrich, 1979, 1980; see also Willmer, 1986a). Evaporative heat loss is effective only when relative humidity is low, as in deserts. By contrast, the high relative humidities of the equatorial tropics combine with intense sunshine and high air temperatures to impose substantial thermal loads. Few data exist on thermoregulatory strategies of lowland tropical rain forest insects, although high body temperatures and the potential for overheating are clearly one consequence of flight in this ecosystem (e.g., May, 1976; DeVries and Dudley, 1990). In experimental contexts, flight tunnels enabling control of all relevant microclimatic variables simultaneous with video tracking of flying insects are now technologically feasible (see, e.g., Grace and Shipp, 1988). Use of such tunnels permits different components of the flight energy balance to be systematically decoupled and assessed quantitatively. The application of infrared thermal imaging cameras also permits remote assessment of body temperature with a high degree of spatial resolution (e.g., Stavenga et al., 1993). Such methods, when applied to insects flying under a variety of microclimatic regimes, will substantially enhance our knowledge of the physiological mechanisms of thermoregulation.

4.3.2 *Evolutionary Origins of Thermoregulation*

Selection for enhanced flight performance has apparently promoted the evolution of facultative endothermy in insects (Heinrich, 1977, 1993). With the exception of one study of warm-up rates in bees (Stone and Willmer, 1989), however, phylogenetically based comparative analysis of the evolution of thermoregulatory capacity does not exist for any clade of winged insects. Evolutionary acquisition of facultative endothermy in insects is likely to involve coordination between behavioral and physiological mechanisms that regulate high muscle temperatures. Temperature regulation is essential because proteins can become more specialized over narrower temperature ranges, and because physiological performance degrades beyond certain limits. For example, thermal tolerance of most insects is in the range of 40–45°C. An optimal temperature therefore exists to which many insects must warm themselves (using either endogenous or exogenous heat sources) before the flight muscles can generate sufficient power to fly (Heinrich, 1993).

Thermoregulatory capacity exhibits allometric dependence because convective cooling is more pronounced for the higher surface area : volume ratios of small insects. A phylogenetic study that traced evolution both of body size and of thermoregulatory capacity would be particularly informative in this regard. Many small insects are effectively isothermic during flight because rates of convective heat flux are high relative to any heat gained through muscle metabolism. Because the majority of insects are relatively small (section 7.1), convection likely dominates heat loss for all but a small percentage of the extant fauna. In larger insects, however, both heat gain via solar radiation and heat loss via convection tend to be small relative to metabolic heat production. As would be expected, large insects (e.g., bumblebees and sphingid moths) exhibit the most pronounced capacities for thermoregulation during flight (Heinrich, 1993). Muscle performance and efficiency in these taxa might correspondingly be expected to be maximal at such elevated and well-regulated body temperatures. Interestingly, muscle efficiencies in at least one taxon (orchid bees) also tend to increase at larger body size (Casey et al., 1992; see table 4.3), although the extent to which this effect is mediated thermally is not known.

Excessive rates of heat production associated with large body size may actually impose an upper limit to the elevation of body temperature. Some large insects may be unable to avoid overheating during continuous flapping flight because the endogenous and solar heat loads cannot be sufficiently offset by convective cooling. Behavioral avoidance of sun-lit areas as well as reduced wing flapping may then be necessary to avoid excessively high body temperatures. For example, alternation of wing flapping with periods of gliding is used by many tropical lowland dragonflies to avoid overheating during flight in full sunshine (May, 1981b, 1983). Some temperate-zone aeshnid dragonflies also increase gliding and decrease wingbeat frequency at higher ambient air temperatures, apparently to regulate body temperature (May, 1995). Interestingly, some members of the extinct Paleozoic dragonfly order Protodonata may have attained body masses on the order of 10–20 g, values much higher than the approximately 1 g of the largest contemporary odonates (May, 1982). Allometric extrapolation from extant odonate physiology suggests that the Protodonata may have been at considerable risk of overheating during flight (see May, 1982). These enormous aerial insectivores may thus have been predisposed to regular gliding as a means of limiting endogenous heat production. Increased crepuscular activity (as characterizes many large extant dragonflies in the tropics) may also have acted to reduce the exogenous heat load of giant Paleozoic dragonflies, as well as of other contemporaneous giant forms (see Dudley, 1998).

Thermoregulation and associated elevation of power output are likely to be advantageous under a variety of ecological circumstances. The success of such aerial behaviors as escape from predators, capture of mates, and acquisition of nutritional resources all potentially depend on flight capacity and duration. For example, Neotropical damselfly species that perch in brighter areas forage more frequently and for longer periods, presumably as a consequence of higher body temperatures (Shelly, 1982). In the context of sexual selection, elevated body temperatures can enhance mating success during aerial competitions (e.g., Heinrich and Bartholomew, 1979; Chappell, 1984; Srygley, 1994b). Such enhancement is often size dependent, with the more effective endothermic capacity of larger insects potentially yielding more effective flight performance and ultimately higher reproductive success (see Willmer, 1991). Phenotypic plasticity in the expression of flight-related morphology may also be induced by variation in the surrounding thermal environment. For example, total wing area within *Drosophila* is inversely correlated with ontogenetic, seasonal, and latitudinal changes in ambient air temperature (see Stalker, 1980; Starmer and Wolf, 1989). These changes in wing area and probably in wing loading may act to reduce induced and total power expenditure when muscle-operating temperatures are low and power output is correspondingly reduced. Finally, thermal strategies of tropical butterflies exhibit strong associations with antipredatory defenses and choice of microhabitat (Chai and Srygley, 1990; Srygley and Chai, 1990b; see section 7.3.2). These diverse ecophysiological examples indicate that further organismal studies of the interface between thermoregulatory capacity and flight performance will be highly informative.

4.4 SUMMARY

Exclusively aerobic metabolism characterizes the contractile activity of insect flight muscle. Metabolic fuels diffuse from the hemolymph surrounding muscle fibers to the point of oxidation within mitochondria, whereas bulk movement of hemolymph within the body cavity enables transport of fuels from the fat body to the flight muscles. Individual muscle contractions are associated with single neural impulses in the phylogenetically ancestral muscle type known as synchronous muscle. In the phylogenetically derived asynchronous muscle that is unique to some pterygote lineages, multiple contractions following initial neural stimulation derive from stretch activation of the muscle by an inertial load. Wingbeat frequencies well in excess of 100 Hz are typically associated with this muscle type. For both asynchronous and synchronous muscle, motor neurons emanating from thoracic ganglia convey oscillatory rhythms of excitation. Origins of these rhythms remain unclear but apparently represent emergent properties of interneuronal networks within the ganglia.

Muscle contraction and the generation of mechanical power require continuous oxidation of metabolic fuels. The efficiency of this conversion process in both synchronous and asynchronous muscles is low, typically between 5% and 10%. The inefficiency of muscle contraction combines with the elevated metabolic rates of flight to yield high rates of heat production during flight. Insect body temperatures represent an equilibrium between instantaneous rates of heat loss and gain. In small insects, most metabolic heat generated during flight is lost via convective cooling, and body temperature is close to if not equal with ambient air temperature. In larger insects, however, metabolic heat gain is high relative to convective loss, and body temperatures are correspondingly elevated. Regulation of rates of heat loss during flight enables partial or complete constancy of muscle temperature to be attained over a range of air temperatures. The evolution of thermoregulatory abilities in many insects is consistent with strong historical selection on muscle performance to meet the exacting energetic demands of flight.

Chapter Five

STABILITY, MANEUVERABILITY, AND MAXIMUM
FLIGHT PERFORMANCE

THE FULL RANGE of insect flight performance can readily be observed within a small sunlit gap of a tropical rain forest. Hovering syrphid flies rapidly chase other flies from their territories. Hesperiid and riodinid butterflies perch on vegetation and then engage in high-speed pursuits and spiral courtship displays. Asilid robberflies briskly intercept flying prey. An iridescent *Morpho* butterfly gracefully glides by, moments later to be the target of sallying insectivorous birds perched within the forest. At midheight in the forest, the heavy, labored flight of a large scarabaeid beetle contrasts sharply with that of far more rapid and agile insect taxa. Linear flight alone is insufficient to ensure aerial survival; regulation and near-instantaneous alteration of aerodynamic force output is essential. For an insect in midair, six degrees of freedom (translation along each of three orthogonal axes, and rotation about each of these axes) characterize movement in space. Continuous sensory feedback concerning wing and body positions enables equilibrium for each degree of freedom and the overall condition of stability to be monitored and regulated. A diversity of kinematic mechanisms is used to implement the dynamic stability that forms the essence of maneuverability. Simultaneous and sequential rotations and translations yield complex flight trajectories during episodes of aerial predation and mate selection, outcomes of which are critically dependent on three-dimensional agility.

5.1 STABILITY

Lift forces on wings are of no practical significance to insects unless both the direction and magnitude of these forces act to yield useful motions. Both gliding and flapping flight require a subtle balance of body forces and moments if a particular flight trajectory is to be maintained. Unintentional deviations from such an equilibrium condition are preferably met with restorative aerodynamic forces that provide for dynamic stability during flight. Such stability is required not only when flying through the usually turbulent air conditions of natural environments, but it is also a prerequisite for the initial evolution and subsequent elaboration of flight (Maynard Smith, 1952). Dynamic stability can be effected via either active changes in wing and body orientation, or by passive mechanisms deriving from transient lift and drag forces acting on an unchanged wing and body geometry. In insects, passive aerodynamic characteristics of the body provide partial compensation for unintended pitch, roll, and yaw torques during forward flight. Such passive mechanisms, however, can compensate only for a small subset of unintentional course deviations; hovering, for example, precludes the use of restorative torques generated by translational motions of the body.

Instead, active aerodynamic mechanisms must be used to compensate for unintentional rotations and translational motions of the body. Voss (1914) first emphasized the need for active stabilization during the flapping flight of insects. Rapid kinematic responses mediated by the nervous system lie at the core of such intentional stabilization. Remaining stable involves continuous transduction of aerodynamic and environmental cues, the central assessment of multiple sensory inputs, and finally the generation of locomotor output appropriate to changing biomechanical demands. Active recovery from transient disruption involves a diversity of bilaterally asymmetric wing motions that generate compensatory forces and torques; advantageous forces may also be produced through asymmetric abdominal and leg motions. Steady flight is perhaps best viewed as a sequence of consecutively unstable but controlled aerodynamic conditions; compensatory course correction must continuously characterize flight with flapping wings. Maneuverability can then be viewed conceptually as the intentional disruption of a transient force and moment balance.

5.1.1 Sensory Inputs

The analysis of body position and movement in space (proprioception) and the continuous transduction of environmental information (exteroception) occur via a number of different sensory pathways. Informational output of these pathways is processed primarily in the head and thoracic ganglia, influencing the timing and extent of compensatory motor output. A great diversity of sensory inputs impinges on the central mechanisms that ultimately drive muscle activation during flight (see reviews by Altman, 1983; Kammer, 1985; Möhl, 1989b; Reichert, 1993). This feedback to the central nervous system is a mandatory component of the normal operating rhythm of the flight motor. For example, deafferentation of sensory receptors (i.e., removal of feedback) reduces the frequency of phasic motor output generated autonomously by the thoracic ganglia (Wilson, 1961). Complete understanding of the insect flight motor correspondingly requires analysis of the relevant sensory inputs that interact with the central nervous system.

Of sensory modalities involved in flight control, visual cues exert the predominant influence on insect orientation and stability. The importance of vision for flight is suggested by the relative size of the brain region dedicated to processing afferent optical information. Of the three principal regions of the insect brain, the protocerebrum is the largest in volume relative to the other two brain regions, the deuto- and tritocerebrum (Hanström, 1928; Horridge, 1965). In all insects, the protocerebrum, and particularly the lobular plate, primarily comprises two optic lobes connected to the external compound eyes, physiologically serving to process visual information. The relative size of the protocerebrum thus correlates with eye size and potentially with rapidity of flight and maneuvers. The huge protocerebrum of dragonflies, for example, represents approximately 80% of total brain volume; dragonfly eyes are correspondingly exaggerated in size. Compound eyes are comprised of individual elements termed ommatidia, and are well configured morphologically for the rapid visual demands of flight. Many compound eyes obtain fields of view not only in front of but also above, below, and to some extent behind the insect. Such spatial coverage facilitates rapid accession of visual cues within a three-dimensional and structurally complex environment.

Forward flight for most insect taxa typically involves motion parallel to the longitudinal body axis. Optical flow, the motion of the visual field relative to the eyes, is then most rapid along the length of the body. Ommatidial coverage correspondingly exhibits horizontal extension and vertical compression (Land, 1989), facilitating spatial analysis of those portions of the visual field that are moving most quickly. Neuroanatomy of the visual system is also consistent with a major role of vision during flight. Numerous sensory neurons descend from the optic lobes, posterior protocerebrum, and additional protocerebral regions into motor centers of the pterothoracic ganglia. These descending pathways provide a direct link between afferent visual information and efferent motor output (e.g., Mimura, 1970; Rowell, 1988; Strausfeld and Gronenberg, 1990; Gronenberg and Strausfeld, 1990; Horridge and Marcelja, 1992; Strauss and Heisenberg, 1993; Homberg, 1994; Ilius et al., 1994). Neural pathways specifically associated with motion computation (e.g., that yield output proportional to image velocity) have also been identified within the protocerebrum (e.g., Rowell and Reichert, 1986; Egelhaaf et al., 1988; Hausen and Egelhaaf, 1989; Egelhaaf, 1990). Both anatomical and neurophysiological observations thus suggest a necessary requirement for the insect visual system to evaluate rapidly the dynamic spatial environment associated with flight.

Indeed, this requirement of flight has prompted the evolution of a phenomenal capacity for information processing by the insect visual system (Burkhardt, 1977). One measure of this capacity is the flicker fusion frequency, the maximum rate at which individual light impulses can be resolved temporally by the optical system and underlying neurophysiological processes. Flicker fusion frequencies are typically much higher in winged insects than in vertebrates, and in some dipterans and hymenopterans can reach 200–300 Hz (Autrum, 1950, 1958; Ruck, 1961; Srinivasan and Lehrer, 1984). By contrast, values of this parameter for humans rarely exceed 20 Hz. Flicker fusion frequencies of fast-flying dipterans, hymenopterans, and odonates are typically an order of magnitude higher than those of slower-flying Blattaria and Orthoptera (see Autrum, 1949, 1952; Autrum and Stoecker, 1950; Autrum and Gallwitz, 1951; Ruck, 1958; Miall, 1978). Fast-flying insect species also appear to have cephalization indices (ratio of brain mass to body mass) and oxygen consumption rates by brain tissue higher than those of other insect species (Kern, 1985). However, relationships among rapidity of visual responses, characteristic flight speed, and phylogenetic association have not been systematically investigated.

Movement of the visual field around an insect, due either to external perturbation or inadvertent movement of the insect itself, typically induces compensatory locomotor responses. This phenomenon is termed the optomotor response and can result in translation along or rotation about any of the three orthogonal body axes (see fig. 5.1). For a flying animal to maintain a particular flight direction and speed, constancy of lengthwise optical flow and absence of rotational optical flow about any body axis is preferred. Rotation of the visual field about a flying insect typically induces compensatory syndirectional (same-sense) body rotation, thereby reducing relative rotation and promoting visual balance in relation to the environment (e.g., Götz, 1968; Reichardt, 1973; Collett and King, 1975; Heisenberg and Wolf, 1984; Preiss and Spork, 1994). The speed of the optomotor torque response is proportional to the rotational velocity of the changing visual field about an insect (e.g., Kunze, 1961), yielding a dynamically responsive adjustment. At least in tethered *Drosophila*, the rapidity of rotational optomotor compensation is similar for roll, pitch, and yaw of the surrounding visual field (Blondeau and Heisenberg, 1982). Simultaneous operation of these compensatory responses enables maintenance of a constant flight trajectory and orientation with respect to the surround-

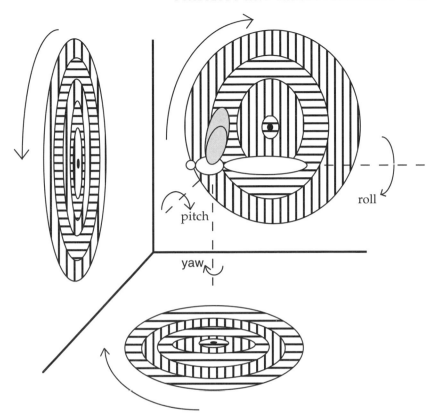

FIG. 5.1. Syndirectional optomotor torque induced by rotation of the visual field about the three orthogonal body axes of a flying insect. Various geometrical patterns can be used to provide moving visual contrast and thereby to induce optomotor responses.

ing environment in spite of transient three-dimensional disruptions induced by local air turbulence.

In contrast to syndirectional rotational responses, short-term translational motion of the optical flow field typically elicits a contradirectional response in both vertical and horizontal directions. Taxa as diverse as locusts (Spork and Preiss, 1993, 1994; Kennedy, 1951), flies (Kennedy, 1939; Schneider, 1965; Götz, 1968), and gypsy moths (Preiss, 1987, 1991) regulate thrust production to match translational speed of the visual field. Flying insects often tend to match their airspeed with instantaneously perceived groundspeed based on the translational velocity of the visual surround. Translational optomotor responses correspondingly tend to be most sensitive to motion of the ventral visual field and less responsive to lateral or dorsal motions. Compensation for translational motion of the visual field is attained by rapid changes in wingbeat kinematics, the best studied of which is wingbeat frequency. For example, tethered locusts momentarily increase wingbeat frequency in response to an increased flow field velocity (Baader et al., 1992). Tethered *Drosophila* similarly increase wingbeat frequencies in response to changes in speed of strips moving around them, as well as in response to approaching visual stimuli (Friedrich et al., 1994). However, the extent of coupling between visual cues and kinematic responses is not perfectly linear nor necessarily of the same sense, typically varying with the relative speed and other characteristics of the visual flow field.

In this context, use of the entire visual field for unequivocal differentiation between body rotation and translation is critical. The rotational and translational components of relative motion may be indistinguishable if only a small region of the visual space is used to assess apparent motion. Nonetheless, insects are fully capable of differentiating translational from rotational components of the visual flow field (e.g., Preiss and Spork, 1995; Kern and Varjú, 1998), whereas spatial sensitivity of the compound eye to moving cues derives from a complex nonlinear assessment of the visual field over the entire eye (e.g., Wehrhahn, 1978). Such information obtained at lower levels of visual processing then permits higher-order analysis of relative flow. For example, specific descending neurons of the visual pathway exhibit heightened sensitivity either to rotation about or to translation along particular body axes in response to whole-field stimulation (e.g., Krapp and Hengstenberg, 1996; Kern, 1998). Such neural capacity suggests a subtle sensitivity to the geometrical and dynamic characteristics of the insect's visual field. As a consequence, relative velocities in rotation and translation, the relative size of objects, spatial coverage, contrast, and even color composition of the visual field all potentially influence the intensity and rapidity of optomotor responses.

The temporal characteristics of visually induced wing and body movements are in fact critical for effective aerodynamic compensation in response to disturbance. Many optomotor studies have often considered only time-averaged values of compensatory force and moment production, whereas transient visual stimuli as well as variable stimulus composition and illumination can induce substantial temporal fluctuation in otherwise steady responses (e.g., McCann and MacGintie, 1965; Wehrhahn, 1986; Robert, 1988). McCann and MacGintie (1965) in particular drew attention to potentially significant differences between instantaneous and mean values of torque generation. Time-averaged measurements in general may conflate optomotor events with ambient saccadic (jumplike) reorientations of the body that characterize free flight (e.g., Collett, 1980a; Zeil, 1986). In tethered *Drosophila* exposed to a fixed visual field, for example, rotational yaw corrections are implemented at a frequency of about 5 Hz (Götz, 1987). Wolf and Heisenberg (1990) documented similar pulses of torque production in tethered *Drosophila*, and also found a strong correlation between instantaneous yaw torque and higher temporal derivatives of the moving visual field (e.g., acceleration), rather than between rotational velocity of the visual field and absolute mean torque. Similarly, saccadic head motions in the fly *Calliphora* provide stabilization of the visual field with respect to an imposed rotation of the body in yaw (Land, 1973, 1975; Geiger and Poggio, 1977). Such sensitivity to higher temporal derivatives of motion increases flight responsiveness under variable visual conditions, but also requires quantitative analysis focusing on instantaneous rather than time-averaged optomotor responses. Interestingly, the response time of visually induced vertical accelerations in *Drosophila* decreases with increasing ambient temperature (David, 1979a). Thermal effects are generally not evaluated in optomotor studies but could well influence dynamics of both visual and kinematic responses.

Furthermore, long-term responses by insects to moving visual cues may be different from transient patterns of compensatory torque or thrust production. In long-duration tethering experiments with desert locusts, optomotor torque in yaw exhibits a time-dependent response that is much reduced relative to that at the beginning or end of simulated flight (Preiss and Spork, 1993). Thrust production by tethered locusts exposed to moving patterns tends to fix on one of two values independent of pattern speed in experiments that range over several minutes in duration (Preiss, 1992). Comparable results are found in closed-loop optomotor experiments that use feedback to adjust pattern rotational velocities in response to the instantaneous torque generated by an insect. In flies and locusts, the intensity of such coupling between generated torque and pattern velocity can vary substantially over time (see Kirschfeld, 1989; Eggers et al., 1991). Kinematic responses in fictive maneuvers by tethered insects may also reflect short-term memory within flight neural circuitry as well as muscle sensitization for a quicker yaw response (Möhl, 1988). Certain *Drosophila* mutants exhibit similarly variable orientation responses for enhanced fixation on visual targets (Wolf and Heisenberg, 1986).

Given such flexibility in visual responsiveness, quantitative specification of the physiological and behavioral conditions under which tethered insects are constrained is necessary if useful generalities are to emerge. This conclusion is particularly appropriate in light of observations that, even in the absence of optomotor cues, instantaneous torque generation may fluctuate substantially because of variable phase activation between contralateral flight muscles (see Möhl, 1997). Most optomotor studies to date have investigated only responses in yaw by tethered insects. The visual responsiveness of free-flying insects may differ substantially, particularly when the exteroreceptive sensory cues normally associated with body rotation are precluded experimentally by a rigid tether. Potential interactions between rotational and translational optomotor responses are also poorly understood, particularly for free-flying insects.

Dynamics of the optomotor response have been best studied in the fly genus *Drosophila*. In free flight, the aerodynamic force vector generated by the flapping wings is of approximately constant magnitude and lies at a fixed orientation relative to the longitudinal body axis (Vogel, 1966, 1967a; Götz and Wandel, 1984a, b). Changes in body angle (possibly driven by modulation of stroke amplitude; see Götz et al., 1979) are used to regulate the relative vertical and horizontal components of total force output (David, 1978). As body angle is reduced, however, the force vector tilts forward and progressively greater thrust accelerates the fly until this component of force is balanced by the increased body drag. Body angle, in turn, is regulated by the horizontal component of the translational visual flow field. Total force output is thus partitioned by optomotor cues into a vertical component and the thrust necessary to maintain a constant groundspeed (David, 1985). Independent of horizontal thrust, *Drosophila* exhibits substantial capacity to vary vertical and total force production (e.g., Lehmann and Dickinson, 1997). Vertical forces must be held constant, however, if body weight is to be balanced over a range of airspeeds. Because thrust is usually small relative to body weight (section 3.2.1.1), however, only a small increase in total force production is required to fly horizontally relative to hovering flight. Unexpected vertical deviations from a horizontal flight trajectory are compensated by kinematic responsiveness to vertical components of the visual flow field (David, 1979a, b, 1984, 1985). These translational optomotor responses, when coupled with rotational responses about each body axis (Blondeau and Heisenberg, 1982), enable *Drosophila* flies to maintain a constant groundspeed and horizontal trajectory during flight.

Translational optomotor cues can be of considerable behavioral utility for free-flying insects. By flying at variable airspeeds but at a constant groundspeed based on visual cues in the environment, horizontal fluctuations in ambient wind velocity can be effectively buffered (e.g., Kennedy, 1951; Heran, 1955, 1956; David, 1982). Flight at a constant ground speed also provides a useful reference calibration for the absolute distance traveled; foraging honeybees integrate optical flow velocities over time to estimate their cumulative flight distance (Esch and Burns, 1996; Srinivasan et al., 1996, 1997). Translational optomotor responses are not, however, necessarily oriented parallel to the longitudinal body axis. In the sideways flight that is characteristic of some syrphid flies, optomotor responses are driven by image motion within the laterally oriented optical flow field (Collett, 1980b; see also Zeil, 1997). More generally, drift induced by air movement orthogonal to the direction of flight can be detected using the optical flow field, and used to correct for the corresponding lateral course deviations (e.g., David, 1986). Neural underpinnings of these translational optomotor responses may be tuned to species-specific behavioral characteristics. For example, insects that tend to hover have motion-sensitive neurons that are particularly responsive to low image velocities (O'Carroll et al., 1996). This sensitivity suggests a heightened optomotor capacity to detect and respond to small-scale displacements, and accordingly to maintain stationary hovering.

In addition to the role of translational cues in maintaining constant ground speed or in station keeping, the optomotor response indirectly facilitates distance assessment by flying insects. The apparent translational velocity of similarly sized background objects is a function of object distance—closer objects appear to move by more rapidly. If insects attempt to maintain a constant perceived groundspeed, a more distant visual background will necessitate faster flight. This effect is most evident when the earth's surface is used as the visual reference. Individual honeybees that fly higher above the ground tend to increase their flight speed, maintaining a constant translational velocity relative to their reference background (Steiner, 1953; Esch and Burns, 1996). Similarly, the apparent translational velocity of moving objects is used by both honeybees and hawkmoths to assess relative object or surface distance during flight (Kirchner and Srinivasan, 1989; Pfaff and Varjú, 1991; Srinivasan et al., 1991, 1996). This ability apparently underlies the capacity of many insects to maintain continuous flight trajectories inside relatively small volumes without colliding against constraining boundaries. Experimentally, the translational optomotor response can be used behaviorally to entrain free-flying insects within a small region of space. This method has been widely used to elicit steady flight by insects in both horizontal and vertical wind tunnels (see Young et al., 1993).

Optomotor responses also likely serve in the initiation and control of complex two- and three-dimensional maneuvers. For example, meliponine bees hovering in front of their nest entrance are capable of tracking nest oscillations and of following moving visual backgrounds (Zeil and Wittmann, 1989; Kelber and Zeil, 1990, 1997). Similar optomotor tracking during flight characterizes hawkmoths hovering in front of oscillating artificial flowers (Farina et al., 1994; Kern and Varjú, 1998). Particular neural algorithms decompose such complex moving visual fields into sets of orthogonal and noninteracting stimuli that elicit sensorimotor output along and about each body axis. For example, visual stimuli simulating roll, yaw, thrust, and ascent stimulated unique combinations of activity in different pleurodorsal steering muscles of tethered *M. sexta* (Wendler et al., 1993). Clearly, the omnipresence of dynamic visual cues available to free-flying insects provides a complex three-dimensional sensory substrate promoting the evolution of diverse optomotor mechanisms. Visual stimuli are not, however, strictly necessary for flight. Wagner (1986c) has shown that houseflies deprived of optical input can nonetheless fly freely and, moreover, implement turning dynamics similar to those of unmanipulated flies (see also Faust, 1952; Gettrup, 1966). As with wing-amputation experiments (see section 3.1.1), effects of sensory deprivation and extirpation have not been systematically investigated in free-flying insects, but would provide an informative perspective on potential redundancy built into the sensory control of flight.

In addition to optomotor responses mediated by the compound eyes, horizon-based pitch and roll detection may be functions served by insect ocelli, simple eyes that are generally located dorsally on the head (Mizunami, 1994). Morphological expression of ocelli is highly variable taxonomically, and, with several exceptions, has not been systematically correlated with flight performance. Ocelli do tend to be reduced, however, in wingless or flightless morphs of otherwise flying species (Kalmus, 1945), suggesting a physiological role in flight. For example, male homopteran coccids that hover with bodies oriented vertically tend to have ocelli located circumferentially around the head, consistent with visual monitoring of the horizon (Duelli, 1985). In locusts, the neural connections from the ocelli to the thoracic ganglia are consistent with an optomotor role in pitch and roll regulation (Rowell and Pearson, 1983; Reichert and Rowell, 1985). Behavioral and physiological evaluations of the function of ocelli generally suggest a role in flight control (Chen and Young, 1943; Goodman, 1965; Wilson, 1978; Stange, 1981; Stange and Howard, 1979; Taylor, 1981a, b; Kastberger, 1990; Kastberger and Schuhmann, 1993; see, however, Schuppe and Hengstenberg, 1993). Potential interactions between ocelli and the compound eyes are unclear, although the simpler optics of the ocelli suggest that horizon detection and monitoring of body orientation are the primary roles of this structure. Among apterygote insects, ocelli are always present in the saltatorial Archaeognatha but are absent in many Thysanura; true compound eyes among the apterygote wingless insects occur only in the jumping machilid archaeognathans (Paulus, 1975, 1979). Ocelli may thus ancestrally function in target guidance during jumps, and in this context could represent a sensory proto-adaptation for gliding flight (see section 6.1.4).

As with movement of the surrounding visual field, aerodynamically induced bending in such external structures as antennae and hairlike setae potentially provides information about the changing speed and direction of flight. Antennae are typically positioned within the moving airstream so as to bend in response to imposed aerodynamic drag. This bending is monitored by a specialized sensory structure (Johnston's organ) located basally within the second antennal segment of all winged insects. Mechanoreceptors within Johnston's organ convey afferent sensory signals that are proportional to the extent of antennal bending and thus to forward airspeed (see Burkhardt and Schneider, 1957; Heran, 1959; Schneider, 1965; Gewecke, 1967, 1970, 1972; Gewecke et al., 1974; Gewecke and Heinzel, 1980; Gewecke and Niehaus, 1981). Antennal deflection is thus used to monitor speed of the insect with respect to the surrounding air, whereas insect ground speed can be simultaneously inferred using optomotor cues. Lateral bending of antennae or mechanosensory setae may also indicate the presence of transient side winds relative to the insect's forward velocity (Heran, 1959; Neese, 1965). Behaviorally, tethered locusts respond to such lateral transients with compensatory asymmetric wingbeat kinematics (Gewecke, 1972), although the precise sensory transduction underlying this response has not been established.

In addition to undergoing steady deflection induced by the forward airspeed of a flying insect, bending of antennae can potentially track the temporal fluctuations in local airflow induced by wing flapping. Such phasic antennal bending appears to provide sensory input to the central control of the locust flight motor (Heinzel, 1983; Heinzel and Gewecke, 1987). Moreover, oscillation of the antennae can be actively induced through the activity of muscles at the base of the antenna. Interaction between such induced oscillation and that associated with wing flapping provides phasic sensory feedback that may regulate wingbeat frequency (see Gewecke and Schlegel, 1970). More posteriorly, an analogous sensory role characterizes the cerci, small paired appendages projecting from a terminal abdominal segment (see fig. 2.1). In the Blattaria and possibly in additional orders, aerodynamically induced bending of the abdominal cerci combines with active oscillation of these structures to monitor wingbeat frequency (Fraser, 1977; Heinzel, 1983; Libersat and Camhi, 1988; Libersat et al., 1989). As with antennal oscillations, phasic deflection of the cerci has moreover been implicated in the regulation of wingbeat frequency (Ritzmann et al., 1982; Libersat, 1992). Because both the antennae and cerci are large in mass relative to mechanosensory hairs, phasic sensory ability is perhaps amplified by monitoring the interaction frequency between actively and passively induced bending in these structures.

On a smaller spatial scale, hairs or setae are widely distributed over the insect head, thorax, and abdomen. Mechanoreception by individual hairs or, more typically, a field of sensory hairs, provides a high-frequency response to the local flow field. Such mechanosensitive setae are characteristic of all winged insects, and have been particularly well studied in the Orthoptera. Induced bending of wind-sensitive hairs on tethered locusts not only stimulates wing flapping and phasically entrains the motor rhythm, but also potentially indicates asymmetric airflow relative to the longitudinal body axis (Weis-Fogh, 1949, 1956b; Smola, 1970; Möhl, 1985a, b, c). Behaviorally, asymmetric stimulation of cephalic hair fields in locusts can induce wing and abdominal compensatory kinematics (Gewecke and Philippen, 1978; see section 5.2.2). Additional fields of deflection-sensitive hairs are located on sclerites of the neck. These hairs monitor head position relative to the body in Odonata (Mittelstaedt, 1950), Orthoptera (Goodman, 1959; Haskell, 1959), Diptera (Liske, 1977), and probably in all other orders. As with the antennae, active oscillation of the head combines with the induced flow field of flapping wings to modulate phasically the wingbeat frequency in locusts (Bacon and Möhl, 1983; Horsmann et al., 1983).

Optomotor responses may also be associated physiologically with the stimulation of cervical hair fields. As the head moves in response to visual cues, cervical sensory input rapidly evokes compensatory actions by the flight motor. For example, externally imposed head yaw elicits syndirectional body torque from flapping flies (Liske, 1977). Similarly, an imposed body roll is compensated by a head roll in the opposite direction and by wingbeat kinematics that generate a contra-directional restorative torque (Hengstenberg et al., 1986; Hengstenberg, 1988; Gilbert et al., 1995; see also Kien, 1977). However, intentional turns are associated with syndirectional head movements, at least in locusts (Robert and Rowell, 1992a, b). Proprioceptive feedback from cervical mechanoreceptors can thus be used either to mitigate unintentional rotations or to exacerbate intentional and at some times evasive movements. Such coupling of head rotation to compensatory

motor output provides an indirect but highly useful means of trans-
ducing moving visual stimuli.

In addition to monitoring antennal and setal deflection, mechanoreceptors are also used to determine deformation of wing and thoracic
structures. On wings, arrays of campaniform sensillae (dome-shaped
mechanoreceptors) monitor the rate and extent of local bending. These
sensory receptors form arrays concentrated anteriorly on the wing and
at the wing base where stiffness and bending moments are highest (see
Pringle, 1957). Nerves from the campaniform sensillae typically follow
wing veins and terminate within ipsilateral motor and sensory do-
mains of thoracic ganglia (Schmitt, 1962; Altman and Tyrer, 1974; Burrows, 1975a). The high surface density of campaniform sensillae and
their direct neural connections with central locations of flight control
suggest that these sensors monitor cyclical wing motions. Indeed, the
mechanical tuning of sensillae on fly wings corresponds approximately to the wingbeat frequency, and the spanwise topographical arrangement of these mechanoreceptors suggests a role in monitoring
progression of torsional waves during wing rotation (Dickinson,
1990a, b). The campaniform sensillae of fly wings also respond to deflections if camber is actively imposed to the wing either dorsally or
ventrally, as occurs during the down- and upstrokes, respectively
(Dickinson, 1992). Such physiological roles are likely widespread
among insect wing sensillae. Pringle (1957) proposed a general correlation across insect orders between the number of wing sensillae and
flight maneuverability, although this hypothesis remains qualitative in
character.

Not only passive but active involvement of the campaniform sensillae in the flight rhythm is possible if sensory feedback is used phasically to entrain the wingbeat. Mechanoreceptive cues from the wings
apparently play a major role in the control of flight motor patterns. In
the locust, sensory feedback from wing mechanoreceptors is necessary
for motor frequencies corresponding to those in free flight (Wilson and
Wyman, 1965; Gettrup, 1966). That this feedback derives exclusively
from wing motions can be shown by phasically applying torque to the
wings to simulate flapping motions. When thus driven by an external
force, wing oscillation elicits sensory feedback that entrains motor
rhythms generated by the thoracic ganglia (Wendler, 1974, 1978, 1983;
Wolf and Pearson, 1987, 1989; Pearson and Wolf, 1987; Möhl, 1991).
Phasic proprioception by wing sensory receptors is thus both necessary and sufficient to derive the normal oscillatory frequency of the
flight motor. Similarly in flies, externally supplied proprioceptive feedback from wing mechanoreceptors phase-locks thoracic motor neurons at the wing oscillation frequency (Ewing, 1979; Heide, 1979). Also

in flies, feedback from wing mechanoreceptors may influence wing-beat kinematics by phasically modulating contraction of the first basalar muscle at the forewing base (Dickinson, 1990a). Such indirect modulation of the output of the main asynchronous flight muscle is necessary as the stretch-activated contraction in this muscle type eliminates opportunities for direct phasic control.

Additional mechanoreceptors physiologically similar to the campaniform sensillae can be found elsewhere in the pterothorax. For example, stretch receptors are found at the wing base and within the pleural wall of many insects (e.g., locusts: Gettrup, 1962; Wilson and Gettrup, 1963; dipterans: Miyan and Ewing, 1984). In locusts and probably in all insects, activity of these stretch receptors phasically modulates output of elevator and depressor motor neurons, and also tonically modulates the wingbeat frequency (Pearson et al., 1983; Möhl, 1985b,c; Pearson and Ramirez, 1990). Mechanoreceptors are also abundant in the tegula, a small but richly innervated sclerite at the wing base (see fig. 2.2A) that is most pronounced in the Diptera, Hymenoptera, Lepidoptera, and Orthoptera (Gettrup and Wilson, 1964; Kien and Altman, 1979; Kutsch et al., 1980). In locusts, sensory activation of the tegula elicits phase shifts in elevator motor neurons and thus influences the timing of the upstroke (Kutsch et al., 1980; Kien and Altman, 1979; Neumann et al., 1982; Pearson and Wolf, 1988; Büschges and Pearson, 1991; Wolf, 1993). Phasic entrainment of the wingbeat frequency may also derive from modulated sensory feedback from the tegula.

Tegulae are highly variable morphologically among insect taxa, but functional correlates of this diversity are unclear. Hindwing tegulae of locusts are more highly innervated than those of the forewings (Büschges et al., 1992), possibly because of the reduced role of the latter wings in aerodynamic force production. The alula, an expanded basal region of the beetle elytron (see Stellwaag, 1914b) may serve a similar function, as may the diverse cuticular and setal coupling mechanisms between fore- and hindwings in various taxa (section 2.2.3). In particular, the frenulum connecting fore- and hindwings in many moths potentially acts as a mechanical sensor (Grodnitsky, pers. comm.). A final form of mechanoreception during flight involves acoustic senses. Chordotonal organs comprise many small vibration-sensitive rodlike bristles and are found in various regions of the insect body. On the wings, these sensory organs may function in the detection of wing bending, but primarily serve as the physiological mechanism underlying function of the tympanum and other auditory organs. Auditory stimuli induce pattern generation in the locust flight motor (Boyan, 1985) and elicit activity patterns similar to ultrasound-induced evasive responses seen elsewhere among the Pterygota (section 7.3.1).

Flies exhibit an extreme modification of wing-base receptors and of the wing itself, miniaturizing the metathoracic wing pair to a pair of gyroscopic halteres (see plate 2G). With the exception of the curious four-winged fossil flies reported by Riek (1977), halteres are a dipteran autapomorphy. Behaviorally, removal of halteres from Diptera is well known to induce functional deficits during flight (e.g., Schelver, 1802; Melin, 1941; Faust, 1952). The haltere base is particularly known for a profusion of campaniform sensillae, consistent with a neurophysiological role in flight stability. During flight, halteres are oscillated at the same frequency as the forewings; this frequency linkage derives from coupled activity in the homologous meso- and metathoracic control muscles (see section 2.1.2). The small size of halteres appears to preclude mechanical oscillation derived aerodynamically from the complex patterns of airflow around the thorax. If oscillating halteres then experience angular accelerations associated with whole-body rotation (and primarily with yaw), fields of campaniform sensillae at the base of the haltere are deformed and transmit afferent information that is proportional to the magnitude of the rotational velocity (Fraenkel and Pringle, 1938; Fraenkel, 1939; Pringle, 1948; Faust, 1952; Schneider, 1953). This sensory transduction of body yaw (and secondarily of pitch and roll) then mediates compensatory wingbeat kinematics. In part, such compensation derives from haltere-derived phase-locking between the wingbeat frequency and the contraction frequency of the synchronous basalar steering muscle in the mesothoracic segment (see Lehmann and Götz, 1996). This neural pathway provides a rapid means of altering activity in the basalar muscle and, by implication, in the transient features of aerodynamic force output. Muscles effecting haltere oscillation are furthermore influenced by descending visual pathways that may indirectly induce the forewing steering motions required for voluntary maneuvers (Chan et al., 1998). Halteres also mediate high-speed head motions in response to externally imposed rotational acceleration of the body (Sandeman, 1980; Hengstenberg, 1988; Nalbach, 1993; Nalbach and Hengstenberg, 1994). As with some optomotor responses, these head motions may in turn stimulate cervical hair fields that initiate compensatory motor output.

Haltere expression and the complexity of associated sensory structures and innervation are phylogenetically diverse and possibly related to flight maneuverability. For example, ancestrally basal dipteran taxa tend to have longer halteres that are more exposed to airflow around the body and that may be less effective gyroscopically (Brauns, 1939; Pringle, 1948). An evolutionary reduction in dipteran abdominal length, with reduced inherent stability in body pitch and yaw, may have paralleled the progressive miniaturization of halteres. To date, however, gyroscopic responsiveness to whole-body rotation has not been systematically correlated with morphological or physiological features of halteres. The profusion of agile nondipteran fliers clearly demonstrates that a high degree of maneuverability does not necessarily require presence of halteres. Many small Coleoptera and Hymenoptera, for example, appear to be as capable of fast-maneuvering flight as their dipteran counterparts.

Additional taxa do, however, exhibit wing miniaturization. In the Strepsiptera, a small order of parasitic insects, the tiny club-shaped forewings of males (females are wingless) are superficially similar to dipteran halteres. Moreover, two fields of campaniform sensillae are present at the base of each forewing (Ulrich, 1930); strepsipteran hindwings appear to function as conventional (unsteady) aerodynamic surfaces. Pix et al. (1993) showed that the forewings of Strepsiptera provide gyroscopic transduction of rotational motion during flight, a function analogous to that of halteres in flies. Moreover, Pix et al. (1993) suggested on anatomical grounds that the reduced elytra of a lymexylid coleopteran (see also Wigglesworth, 1946; Miller, 1971) could act as gyroscopic sensory devices. Certain phengodid and rhiphiphorid beetles also possess shortened clublike elytra that may function gyroscopically (pers. obs.). Comparable wing modifications can be found elsewhere in the pterygote Insecta. For example, the hamulohalteres of male homopteran coccids are greatly miniaturized relative to the aerodynamically active forewings, although the anatomical and physiological studies necessary to establish gyroscopic functions have not been carried out in this taxon. Among the winged insects, however, multiple independent origins of functional halteres seem likely. Some flightless phorid and empidid flies also have miniaturized forewings that appear to be similar to halteres, although their function is unclear.

5.1.2 Central Processing and Motor Output

The peripheral sensors discussed in section 5.1.1 connect both anatomically and physiologically to the central nervous system via descending sensory pathways. For logistical reasons, neurophysiological investigations of both central and peripheral components of the flight rhythm have analyzed tethered and otherwise constrained insects. Much extero- and proprioreceptive feedback, however, derives from wing and body motions that cannot be accurately replicated in tethered insects. Tethered studies are clearly essential for the qualitative identification of sensory pathways, but may be less appropriate for quantitative assessment of integrative sensory feedback and associated motor output during free flight.

Nonetheless, multiple regulatory influences of afferent sensory pathways on flight motor output have been identified. The critical role of such peripheral feedback in modulating the flight rhythm is to accommodate external fluctuations within broader constraints of the mechanically tuned thorax (Delcomyn, 1980). This perspective on sensory modulation suggests that, whereas the flight motor pattern is centrally initiated, the entire nervous system contributes dynamically to the maintenance of a nominal rhythmic equilibrium. Such design permits rapid efferent responses to changes in the sensory environment. To compensate for potential ambiguity in the transduction of external signals, redundancy is provided by a multiplicity of sensory modalities. Wingbeat frequency, for example, can be simultaneously influenced by phasic stimulation of mechanoreceptors along the wings, within the thoracic wall, at the base of the antennae, and within the abdominal cerci. Even proprioceptors internal to flight muscles potentially contribute to motor neuron activation (see Stevenson, 1997). All such afferent sensory input can influence thoracic interneurons that connect to premotor and motor neurons. Whereas the majority of afferent pathways appear to terminate in thoracic ganglia, more direct routes to modifying motor output are also available. For example, optomotor-induced compensatory roll in locusts is effected via sensory inputs to premotor interneurons and to motor neurons, but not to central sites of flight rhythm generation (Reichert and Rowell, 1989). This type of sensory connection provides a more direct means of modifying motor output, and may be a widespread feature of peripheral feedback.

Over longer timescales, hormonal regulation of the flight motor may also be indicated. For example, isolated pterothoracic ganglia in the locust generate rhythmic motor output in the absence of sensory feedback from the wing, but such rhythms deviate quantitatively from the normal motor pattern (see Wilson, 1961; Kammer, 1967, 1985). Application of octopamine and related compounds to isolated ganglia is, however, sufficient to recover quantitatively the basic motor pattern of wing flapping (see Sombati and Hoyle, 1984; Kinnamon et al., 1984; Claassen and Kammer, 1986; Stevenson and Kutsch, 1987; Weisel-Eichler and Libersat, 1996). Octopamine plays a number of diverse roles in flight physiology (Ramirez and Orchard, 1990; Orchard et al., 1993), and fluctuation in hemolymph levels of this molecule may permit both immediate as well as chronic changes in sensitivity of the flight motor to afferent feedback from peripheral receptors.

Further complicating interpretations of sensory feedback are potential interactions between multiple modalities at the level of both cerebral interneurons and the thoracic ganglia. At least in locusts, visual responses seem to be particularly well integrated with concurrent mechanoreceptive cues. For example, the combined anemotactic and visual stimuli associated with flight perturbations elicit responses in cephalic descending neurons that phasically influence motor output and compensatory steering (Camhi and Hinkle, 1974; Reichert et al., 1985; Baader, 1990). Mechanical proprioception by cervical hair fields similarly interacts with visual stimuli at the level of afferent descending neurons (Hensler, 1992a,b). In butterflies, aerodynamic sensory input (antennal deflection) interacts with visual cues to influence wingbeat kinematics (Niehaus, 1981). Input from mechanoreceptors in different regions of the body may also interact and summate at higher levels within the afferent nervous system. Again in locusts, interneuronal integration of feedback from antennae and cephalic setae occurs within the most posterior region of the brain (the tritocerebrum) and is then conveyed to thoracic motor neurons (Bacon and Möhl, 1979, 1983; Bacon and Tyrer, 1979; Möhl and Bacon, 1983). This sensory pathway has been specifically implicated in control of steering during locust flight (Burrows and Pflüger, 1992).

In addition to providing a potential redundancy of information, such interactions among multiple sensory modalities may have important behavioral implications for free-flying insects. Because flight often involves chemoreception and movement within odor plumes (section 5.3.2), odor concentration may exert a neurophysiological influence on motor behavior. In moths, the sex pheromone emitted by females alters motor output of males in plume-tracking behavior, but furthermore enhances neural and behavioral responsiveness underlying the optomotor response (Preiss and Kramer, 1983; Preiss and Futschek, 1985; Olberg and Willis, 1990; Willis and Baker, 1994). Similarly, the quality of a nectar reward influences the kinematic responses of hawkmoths that are tracking moving artificial flowers via visual cues (Farina and Josens, 1994). Multiple interactions between different sensory modalities are thus possible and potentially occur at various levels of processing within the afferent neural system. Precise specification of behavioral and experimental contexts is therefore essential for general significance to be ascribed to experimental manipulations of sensory feedback to the flight motor.

Sensory influences on the flight motor rhythm frequently involve extero- or proprioceptive feedback to compensate for unintentional deviations from a nominal flight equilibrium. Maneuverability, however, requires supplemental mechanisms of flight control. Bypassing, overriding, or desensitization of compensatory feedback loops are required for intentional maneuvers. For example, self-initiated yaw induces contradirectional rotation of the visual field that ordinarily would induce compensatory kinematics to reduce the amount of visual slip. Such optomotor feedback clearly must be overridden during any intentional rotation or translation. As an extreme example, rapid turns punctuate the cruising trajectories of houseflies even during flight within a moving visual environment that accentuates the contradirectional rotation of the visual field (Wagner, 1986c). Further complicating physiological analysis, stimuli that evoke locomotor responses may interact with changes in baseline sensory equilibria initiated by the insect itself (see Rowell, 1988, 1989). Multiple sensory cues thus potentially interact with one another and may further be modulated by the physiological interface with central mechanisms of flight control. Such complexity ultimately lends a greater responsiveness and temporal flexibility to wing and body kinematics during flight.

Aside from the complexities of sensory control, the flight motor of insects is often viewed biomechanically as a fairly regular oscillator that expresses only limited kinematic variability under steady conditions. This perspective, although informative, may actually mislead as to the physical mechanisms underlying the generation of motor patterns. The recent application of nonlinear analysis has revealed a wealth of dynamic complexity in physiological systems once considered to be simple oscillators (e.g., human heartbeat: Babloyantz and Destexhe, 1988; human EEG: Basar, 1990; crustacean stomatogastric system: Mpitsos et al., 1988; see also Olsen and Degn, 1985; Milton et al., 1989). Unpredictable yet deterministically driven motor output can potentially derive from multiple sources of sensory feedback acting on an oscillatory source. Chaotic dynamics characterize many simple mathematical systems that utilize feedback (Mees, 1986). So-called mixed feedback (both positive and negative feedback loops: Mackey and Glass, 1977; Sparrow, 1980; Heiden and Mackey, 1982; Oliveira and Malta, 1987), negative feedback with time delays (Glass et al., 1988), and multiple negative feedback systems (Glass and Malta, 1990) can all result in chaotic output for an appropriate choice of control parameters. The existence of multiple afferent feedback stimuli (both positive and negative in sign) influencing the thoracic ganglia suggests that the insect flight motor is an appropriate neuromuscular system within which to search for chaos. Moreover, many insects (e.g., palatable butterflies; see section 7.3.2) express a high degree of kinematic irregularity, and implications for their neural control of flight may be profound. It would be pleasingly ironic, for example, if the Neotropical butterfly wing flap envisioned to perturb global meteorology (see Lorenz, 1993) were itself to be the consequence of chaotic dynamics.

5.2 MANEUVERABILITY

Maneuverability requires rapid changes in wingbeat kinematics and body orientation to disrupt transiently the nominal equilibrium of steady flight. Flight initiation and termination represent the endpoints of a wide range of unsteady behaviors, and even steady cruising flight derives from continuous corrections to moment-by-moment conditions of instability. Behaviors that delineate the flight envelope include hovering, forward flight, maneuvers, ascent and descent, and even backwards flight. Many insects can transiently fly backwards (e.g., hematophagous and pollinating insects), but only in beetles has this behavior been quantitatively documented (Schneider, 1981c). Sideways flight without roll is also a transient possibility, particularly in Diptera (e.g., Collett and Land, 1975; Nachtigall, 1979b). The use of symmetric wing kinematics in hovering and steady forward flight is fairly well described (section 3.1), but understanding of the asymmetric use of wings and body to generate complex turns and maneuvers is much less complete. Although of direct relevance to fitness in a variety of ecological contexts, the full range of flight behavior has not yet been described, let alone been evaluated biomechanically, for any insect species.

5.2.1 Takeoff and Landing

A wide diversity of environmental factors influences the behavioral decision to take flight, including air temperature, solar radiation, and ambient winds. Startle takeoffs can also be elicited by transient stimuli ranging from predatory attack to abrupt mechanical and visual cues and even chemical odors. Prior to taking off, many insect taxa must unfold their wings. The Coleoptera and the Dermaptera are particularly well known for complex mechanisms of wing unfolding and deployment prior to flight. In some Thysanoptera, the abdomen is used to spread open an array of collapsible wing cilia that increase the effective wing surface area during flight (Ellington, 1980b). Also prior to flight, insect taxa that are facultatively endothermic must engage in preflight warm-up via muscular contractions within the thorax; such warm-up may occur in the absence of obvious wing motions (see Esch and Goller, 1991; Heinrich, 1993). Both wing unfolding and preflight warm-up preclude immediate takeoff in response to adverse stimuli, although selection has presumably been intense during events of predatory attack to depart as quickly as possible. In beetles, however, behavioral reluctance to fly may be partially associated with enhanced cuticular defenses against predation (see section 8.3.2).

During a takeoff, most insects first become airborne by jumping with a rapid extension of their legs. Wing flapping is then stimulated by the tarsal reflex, by which loss of ground contact initiates wing motions (Fraenkel, 1932a; Diakonoff, 1936). Wind-induced flight reflexes (e.g., antennal and setal deformation) may also then contribute to flight initiation. The tarsal reflex is, however, highly useful when studying wing flapping by tethered insects; the legs merely need to be deprived of contact with a substrate. Neurophysiological events associated with the tarsal reflex are not well understood but may be complex. In the cockroach, for example, flight is initiated when sensory inhibition of particular leg receptors is removed, rather than deriving from specific stimulation associated with loss of ground contact (Krämer and Markl, 1978). The tarsal reflex has also been subject to drastic reconfiguration in certain taxa. In aquatic belostomatid hemipterans, removal of tarsal contact initiates not flight but rather swimming reflexes from the legs (Dingle, 1961). For most flying insects, the intensity of the tarsal reflex, the pathways of neural control, and the speed of takeoff are likely to have coevolved for more rapid takeoff.

The neurophysiology of takeoff has been particularly well studied in flies. Startle jumps are initiated by giant interneuronal fibers running from the protocerebrum that activate the thoracic musculature (Levine and Tracey, 1973; Tanouye and Wyman, 1980, Wyman et al., 1984; Bacon and Strausfeld, 1986; Trimarchi and Schneiderman, 1995a,b). Pleurosternal muscles first stiffen the thorax, and action of the synchronous tergotrochanteral and related muscles then initiates both notal deformation and mesothoracic leg extension (Nachtigall and Wilson, 1967; Nachtigall, 1968, 1978; Schouest et al., 1986; Trimarchi and Schneiderman, 1993, 1995c). The absence of pro- or metathoracic leg extension in dipteran jumping supports an evolutionary linkage between leg musculature and the origin of wing flapping (section 6.1.4); wings are present only on the dipteran mesothorax. However, flies missing mesothoracic legs can still jump during takeoff (Nachtigall, 1968; Mulloney, 1969), and, moreover, the tergotrochanteral muscle is absent in some dipteran taxa (King, 1983). Leg extension is therefore not essential for takeoff by flies, and the relative contributions of wings and legs to vertical force production may well vary phylogenetically among fly taxa.

Although most insects use a tarsal reflex to initiate wing flapping, wing motions may also precede actual takeoff in many taxa (e.g., odonates: Pond, 1973; orthopterans: Camhi, 1969; Pond, 1972a,b; Brackenbury, 1990; beetles: Schneider, 1981b; voluntary takeoff of *Drosophila melanogaster*: Trimarchi and Schneiderman, 1995c; surface paddlers and surface skimmers: see section 6.1.4). In a lygaeid hemipteran, tergal muscles terminating in the leg extend the forewings and then initiate the downstroke (Govind and Dandy, 1972). Consequent stretching of the asynchronous dorsolongitudinal muscle may then initiate the flapping rhythm of flight, with leg extension and the take-off jump occurring several wingbeats later. In general, the use of wings in takeoff potentially supplements takeoff momentum provided by leg propulsion (Brackenbury, J., 1991), but close temporal coordination between wing and leg muscles is required, particularly for those muscles with bifunctional roles. In an evolutionary perspective, ancestrally basal taxa might be predicted to exhibit a tighter mechanical and physiological linkage between jumping and flapping than is evident in more derived taxa.

Evolutionary associations between jumping and flight are further evident in the independent origins of saltatorial pterygote taxa, including alticine chrysomelids (flea beetles), various homopterans, and orthopterans (Heinrich, 1993). Specialized jumpers in the Homoptera include the families Cercopidae, Cicadellidae, Fulgoridae, Membracidae, and Psyllidae, all of which use flight in conjunction with a rapid jump to elude predators. The kinematics and mechanisms involved in homopteran jumping have been little studied, although cicadellids tend first to jump and then to extend the wings for flight (Brackenbury, 1990; see also Weber, 1929; Sander, 1957; Dalton, 1975). In all saltatorial taxa, selective forces have likely acted to increase jumping velocities to the minimum values necessary to effect forward flight (see Katz and Gosline, 1993). Jumping to escape predators is also common among many apterygote insects (e.g., Archaeognatha, Collembola and Thysanura), consistent with postulated evolutionary scenarios for winglet evolution in protopterygotes (see chapter 6.1.4). In a flea beetle and probably in other saltatorial taxa, jumping also involves visual targeting of potential landing areas (Brackenbury and Wang, 1995). Choice of destination is likely to yield considerable advantage when escaping from predators, and ancestrally may have characterized jumping by protopterygotes.

In contrast to takeoff, landing requires careful assessment of and reduction in flight speed relative to the immediate environment. The landing reaction of insects typically involves dorsal movement of prothoracic legs and posterior extension of meso- and metathoracic legs.

Accurate landing with near-zero velocity is desirable; undershooting involves additional temporal and energetic expenditure to effect landing, whereas overshooting could result in a possibly deleterious impact. Selection for effective landing mechanisms has therefore probably been substantial. Considerable diversity in the landing styles is also evident. Many beetles, for example, appear to land rather inelegantly. Some butterfly species, by contrast, execute remarkable head-down landings on the vertical trunks of trees. Of possible sensory cues used in landing behavior, only the visual perception of approaching surfaces provides the information necessary to assess body dynamics in rapidly changing circumstances. Extensive work with tethered flies has revealed that the approaching visual stimulus of a landing zone induces appropriate leg extension once a particular sensory threshold is surpassed (e.g., Eckert and Hamdorf, 1980; Borst and Bahde, 1986, 1988a,b; Wittekind, 1988; Borst, 1989). The speed of the landing response typically depends on the expansion velocity and angular subtension of an approaching visual stimulus, but also varies with stimulus size, geometrical composition, and contrast (e.g., Goodman, 1960; Wehrhahn et al., 1981; Borst, 1986). Instantaneous rates of change in the visual field are apparently summed in space and in time, potentially by different subsets of visual receptors (Eckert, 1980; Borst, 1990). Transient or cumulative rates of change in the total visual field can then be compared with preset critical values indicative of imminent collision. Diverse types of moving cues can thus be used to elicit the identical behavioral response of landing.

Algorithmically, one simple and generalizable means of assessing visual flow during target approach is via the tau function (Lee, 1980). This function is given by the ratio of distance from an approaching target to the approach speed, and equals the remaining time-to-contact if approach speed does not change. Empirically, visual control of landing in flies is best described by use of the tau function (Wagner, 1982; Lee et al., 1991). Independent of local visual cues and instantaneous speed, the landing response is initiated once tau exceeds a certain value. Both ground speed and target distance must be evaluated to calculate tau, but instantaneous rates of contrast change, together with optomotor responses, potentially provide this information. Once the landing response is initiated, deceleration will occur rapidly over a period of 100–200 ms until physical contact with a substrate is attained. The precise kinematic means by which such high-speed deceleration is induced are unknown, although changes in wing motions often increase the body angle to yield an increased parasite drag. Landing flies also shift their stroke plane angle anteriorly (Borst and Bahde, 1986), possibly decreasing the downwards pitching moment from the wings and thus rotating the body nose-upward. Landing responses may also vary with geometrical arrangement of the landing zone. An extreme example is the problem of a fly landing upside down on the ceiling of a room. Eyles (1945) concluded that houseflies simply perform a half-roll to effect this maneuver, although a more regular method appears to be dorsal contact of the surface with the forelegs and then full rotation of the body about the axis of contact to land the two additional leg pairs (Hyzer, 1962; Dalton, 1975). Conspicuously absent in all studies of insect landing behavior are experiments using the visual cues, ambient light conditions, and three-dimensional geometries characteristic of natural environments.

5.2.2 Translation and Rotation

Aerial maneuverability encompasses diverse features of flight performance, including linear speeds and accelerations, rates of angular rotation of the body, rates of ascent and descent, and the turning radius necessary to effect a 180° reversal in body orientation. Further examples of kinematic parameters characterizing maneuverability include higher-order derivatives of rotation and translation, the absolute time necessary to reorient and then stop in space, and the ability to implement transient dynamics sequentially in response to changing conditions. Axial and torsional agility are fundamental to all such measures of aerial performance. Axial agility involves the capacity to translate in the forward, lateral, and vertical dimensions, whereas torsional agility indicates rotational capacity about the three mutually orthogonal body axes (i.e., roll, pitch, and yaw; see fig. 1.5). Maneuvers involve production of either simultaneous or temporally decoupled rotational moments together with generation of variable vertical force, thrust, and sideslip. Maneuverability must furthermore be integrated with effective control; rapid maneuvers are advantageous only when desired dynamic states can be captured and held (i.e., handling; see Gal-Or, 1990). Sustained versus transient maxima may also influence the outcome of aerial encounters between volant organisms. Many insect maneuvers involve the complex use of force and moment vectoring to attain aerobatic performance presently unavailable to technology. Analysis of the associated wing and body kinematics is still primarily in the descriptive stage, whereas underlying temporal patterns of force and moment generation remain largely unresolved.

Disruption of a nominal force and moment balance during flight requires modulation of the force balance and/or production of a bilateral asymmetry in forces. Comparing insects (*holoptera*) to birds, Aristotle in *De Incessu Animalium* commented on the morphological absence of a tail in the former taxon. Instead, insects use their wings, bodies, and legs to effect both translational and rotational maneuvers. The kinematic mechanisms associated with variable thrust and vertical force production are discussed in section 3.1.2, whereas the active control of body pitch is treated in section 3.2.1.1. In addition to these mechanisms of flight control, many features of directional change involve rotational roll and yaw. These torques derive from asymmetric force generation by contralateral wings, a possibility enabled by a diversity of rapid and subtle alterations to wingbeat kinematics. Sideslip (i.e., lateral force production) as well as changes in thrust and net vertical force often arise as a consequence of such force asymmetries and torque production. Because the predominant force produced by insects is oriented vertically to offset gravity, even small rolls will yield substantial lateral accelerations as the initially vertical force tilts sideways. Similarly, asymmetry in rotational yaw often induces concomitant roll and sideslip; instantaneous force vectors on the wings vary differentially not only in magnitude but also in direction (e.g., Baker, 1979c; Blondeau, 1981; Alexander, 1986). Different torsional features of maneuverability can thus be linked biomechanically to one another as well as to the axial components of force production.

Both passive and active mechanisms influence the generation of rotational torques. Asymmetric wing motions are the most obvious but not only means of actively producing body roll, pitch, and yaw. Aerodynamic torques acting on the body may also passively act to either accentuate or mitigate body rotations. For any object translating in space, transient deviations from an equilibrium trajectory may either yield further instability via positive mechanical feedback, or they may elicit negative feedback that promotes return to the prior stable condition. For example, as a dragonfly in forward flight yaws slightly, the elongated abdomen transiently experiences a lateral component of airflow in addition to the forward airspeed. As with a wind vane, the ensuing lateral drag then tends to rotate the body about the vertical axis and to realign the body parallel to flow. Such mechanical responsiveness to rapid disruptions of the flight trajectory thus forms the essence of dynamic stability for technological as well as biotic flying machines (Stevens and Lewis, 1992). If stable equilibrium pertains during the flight of insects, restorative aerodynamic torques generated on the bodies may provide for considerable stability in roll, pitch, and yaw. Such passive stability obviates the need for associated sensory transduction of deviations and generation of suitable kinematic responses, but may preclude rapid maneuvers that require positive feedback to promote further instability during flight.

As first discussed by Stellwaag (1916), aerodynamic stability is critically dependent on the position of the center of body mass relative to the location of lift production by the wings. Positive mean positional angles of wings suggest that the center of aerodynamic lift is generally above and forward to the center of body mass located behind the wing base. This mechanical configuration promotes stability in vertical and forward dimensions; thrust production leads the center of mass, and vertical forces act above the point of action for gravitational forces. Deviation of the center of mass from the line of action of horizontal or vertical forces will induce a restorative realignment of the body relative to the point of aerodynamic force production. However, the diversity of unsteady aerodynamic phenomena in wing flapping (section 3.2.2) precludes unequivocal analysis of axial stability based on steady-state mechanisms alone. Sideslip during steady forward flight is generally precluded by bilateral wing motions and by equivalent but offsetting lateral components of aerodynamic force production.

Rotational stability during flight is also facilitated by positive mean positional angles of wings and by the bilateral symmetry of aerodynamic forces generated above the thorax. In steady flow, flapping wings are, on average, canted above the horizontal plane (e.g., a positive dihedral angle) and provide stability in roll and in yaw. As a roll is initiated, for example, the ipsilateral dihedral wing experiences a lateral component of air velocity that increases the angle of attack as well as the absolute airspeed experienced by the wing. The net effect is to increase the forces on the wing that, in turn, act to offset the roll and return to the initial equilibrium condition. As an extreme example, the elytra of beetles are in most taxa are held stationary and dihedrally above the thorax, providing substantial stability in roll (De Souza and Alexander, 1997). Stability in yaw is similarly provided by dihedral wings (Stevens and Lewis, 1992), although unsteady mechanisms may confound such analyses for both roll and yaw stability. Because of the lack of coincidence between centers of aerodynamic lift and body mass, pitch stability in all insects must be actively maintained through nose-downward torque provided by the wings (see section 3.2.1.1).

Additional morphological features of body design may also influence torsional agility, as inertial resistance to acceleration constrains the rapidity of rotational as well as translational motions. In particular, greater moments of inertia about rotational axes reduce accelerational responsiveness to applied aerodynamic torque. The highly elongated odonate body, for example, is difficult to accelerate rotationally in pitch and yaw, but can roll rapidly about its longitudinal axis. Although wings of most insects are small in mass relative to the body and contribute little to rotational moments of inertia, the enlarged wings of certain taxa may hinder rapid rotation. The wings of many butterflies, for example, are characterized by a relatively heavy wing mass positioned substantially distal to the center of body mass. Such positioning will dramatically increase total moments of inertia in yaw and roll; the added mass of air surrounding such large wings (see section 3.2.2.1) will further inhibit rapid body rotations. By contrast, the influence of wing mass on butterfly pitching moments is much smaller given that the anteroposterior location of the center of wing mass is much closer to the center of body mass (Dudley, in prep.). Body moments of inertia in roll and yaw have generally received little attention relative to body responsiveness in pitch (e.g., Ellington, 1984b), but clearly must contribute to torsional agility during maneuvers. Relative dimensions of insect body segments may also influence the ability to carry out particular kinds of translational movements. Sideslip, for example, necessarily induces high parasite drag given that the longitudinal body axis momentarily experiences orthogonal components of airflow. Because the associated drag will then act to slow the laterally translating insect, this and related forces may constrain most such maneuvers to very brief intervals until the body can be rotated via yaw and roll into alignment with the oncoming flow.

Most studies of the wing and body motions used in maneuvers have presented tethered insects with visual stimuli that elicit unbalanced rotational moments and bilateral asymmetry in aerodynamic force production. Because insects do not change their velocity relative to the air or the surroundings in such preparations, the kinematics and mechanics of such responses are most representative of the transient initiation of maneuvers. As a consequence, the characteristics of optomotor feedback and stimulation of sensory pathways supplemental to optomotor input will necessarily be different from those experienced by a free-flying insect. Neither can time-averaged measurements of kinematic asymmetries in tethered insects evaluate potential time dependence in those body rotations and translations that characterize complex maneuvers. Furthermore, equivalent compensatory torques can be generated by different combinations of activity in steering muscles (e.g., Egelhaaf, 1989), suggesting that wingbeat asymmetries are context dependent. Physical constraints on translational and rotational degrees of freedom may also elicit abnormal or escape responses. Some studies have partially addressed the issue of sensory feedback during maneuvers (e.g., Baker, 1979a), but analyses of wing and body motions during free-flying maneuvers are rare.

Nonetheless, extensive work with tethered insects demonstrates that asymmetric wing kinematics are indeed the primary (but not exclusive) active mechanism used by insects to generate torque. Such asymmetries documented to date include differences in the parameters of stroke amplitude (Voss, 1914; Stellwaag, 1916; Hollick, 1940; Hisada et al., 1965; Vogel, 1967a; Götz, 1968; Burton, 1964, 1971; Govind and Burton, 1970; Govind, 1972; Schneider and Krämer, 1974; Cooter, 1979; Götz et al., 1979; Alexander, 1986; Robertson and Reye, 1992; Robertson and Johnson, 1993a; Heide and Götz, 1996; Dawson et al., 1997), stroke plane angle (Stellwaag, 1916; Vogel, 1967a; Govind and Burton, 1970; Kammer, 1971; Dawson et al., 1997), angle of attack of the wings (Dugard, 1967; Faust, 1952; Cooter, 1979; Alexander, 1986; Waldmann and Zarnack, 1988; Zarnack, 1988), rate and timing of wing pronation and supination (Faust, 1952; Baker, 1979b, c; Robertson and Reye, 1992; Dickinson et al., 1993; Dawson et al., 1997), separation between fore-and hindwings (Brodsky and Ivanov, 1975), and phase relationships both between contralateral and between ipsilateral wings (Burton, 1971; Möhl and Zarnack, 1977; Zarnack and Möhl, 1977; Zarnack, 1978b; Cooter, 1979; Thüring, 1986; Schmidt and Zarnack, 1987; Schwenne and Zarnack, 1987; Waldmann and Zarnack, 1988). Bizarre kinematic asymmetries are recorded in some dipteran maneuvers: contralateral wings may beat 90° out of phase, or only one wing may be moving while the other is stationary (see Magnan, 1934; Nachtigall and Wilson, 1967). A great diversity of kinematic mechanisms and wingbeat asymmetries thus potentially comes into play during flight maneuvers.

Much aerodynamic interpretation of such asymmetric wingbeat kinematics has centered on steady-state mechanisms of force production. In yaw, for example, stroke amplitude is typically smaller on the inside of the turn. In turn, the translational velocity of the ipsilateral wing and presumably aerodynamic forces are lowered relative to those of the contralateral wing, yielding an unbalanced rotational torque. Unsteady effects may, however, seriously confound such interpretations, particularly if substantial forces are produced by the wings during rotation at the ends of half-strokes (e.g., Dickinson et al., 1993). Asymmetric wing motions may furthermore disrupt axial as well as rotational equilibrium during flight. Lift on wings is not typically directed vertically, and even slight directional asymmetry in the lift vectors of contralateral wings can substantially alter the magnitude as well as direction of forces acting on the body. Pressure drag alone on wings may underlie some amazingly rapid yaw turns by some taxa. For example, a free-flying dolichopodid fly reversed body orientation by 180° in only 40 ms, apparently positioning one wing orthogonal to oncoming flow and relying on high wing drag to effect rotation in yaw (Land, 1993a). Similar aerodynamic mechanisms probably characterized 180° turns by tsetse flies in 100 ms (see Brady, 1991). No data are available, unfortunately, on the concurrent changes in vertical, lateral, and horizontal forces produced during such remarkable maneuvers.

By contrast, the exemplary study of Wagner (1986a) reconstructed three-dimensional body positions and axial orientations of free-flying houseflies (*Musca domestica*). Directional changes were usually derived from rotations about two body axes simultaneously, but rotations about a single axis were also found. In banked turns, roll angle and sideslip velocity were positively correlated, consistent with a roll-derived lateral component of the otherwise predominantly vertical force vector. Body sideslip was, however, slightly delayed relative to the initiation of roll, demonstrating a substantial inertial lag for body acceleration. In some cases, the conflicting demands of linear acceleration and production of torque may result in trade-offs between axial and torsional agility. For example, *Drosophila* flying in pursuit of conspecifics exhibited an inverse relationship between rotational and translational velocities (Buelthoff et al., 1980). In general, however, the extent of coupling (either obligate or facultative) between rotational and translational components of maneuvers has received little experimental attention, particularly in free-flying insects. Importance should therefore be accorded to evaluation of as many degrees of kinematic freedom during maneuvers as is logistically possible.

Asymmetric wingbeat kinematics that initiate and sustain maneuvers must necessarily derive from bilaterally dissimilar contraction of the thoracic muscles. For example, yaw turns may be associated with reduced muscular activity on the inside of the turn (e.g., Burton, 1964), or with increased firing rates of muscles on the outside of the turn (e.g., Dugard, 1967; Heide, 1975; Baker, 1979b; Zarnack, 1988; Dawson et al., 1997). Bilateral differences in the timing as well as extent of muscle contraction may also be involved. Externally imposed yaw in locusts, for example, invokes a timing shift (i.e., phase advance) in downstroke muscles of the forewing that tends to restore the original body position (Möhl and Zarnack, 1977; Zarnack and Möhl, 1977). Associated phase shifts are much less pronounced in the locust hindwing, possibly because its much larger surface area requires less kinematic compensation to substantially alter aerodynamic forces. In addition to complications deriving from the differential motions of the fore- and hindwings, diverse muscles within a pterothoracic segment may be involved in torsional maneuvers. For example, rolls by tethered locusts can derive from a variety of timing and activity changes in elevator, depressor, and pleuroaxillary muscles (Zarnack and Möhl, 1977; Taylor, 1981b; Elson and Pflüger, 1986; Pflüger and Elson, 1986). Also in locusts, equivalent roll torque can be generated by various combinations of wing asymmetries and, by implication, by different combinations of bilateral asymmetries in the phase and extent of muscle contraction (Thüring, 1986; Schmidt and Zarnack, 1987). Unitary functional explanations of wing motions and patterns of muscle activation associated with a particular maneuver are therefore unlikely to capture the full complexity of kinematic possibilities available to insects.

Not only the wings but also appendicular and axial structures are used for steering by insects. Insects have no counterpart to the avian tail, muscles of which are phasically active and dynamically versatile during wing flapping (e.g., Gatesy and Dial, 1993; Bilo, 1994). However, the abdomen of some insects partially fulfills this functional role. In locusts, lateral abdominal deflection together with lateral motions of meso- and metathoracic legs generate restorative torque during unintentional yaw; this torque derives from the increased pressure drag experienced by the deflected abdomen (Camhi, 1970a). Because drag forces typically increase with the square of airspeed, aerodynamic effects of abdominal deflection are speed dependent and are probably substantial in free-flight maneuvers (Gewecke and Philippen, 1978). In flies, changes in yaw and, to lesser extent, in pitch can also result from abdominal deflection (Götz et al., 1979; Zanker, 1990b). These changes derive primarily from asymmetric aerodynamic torque on the body, whereas concomitant changes in position of the center of body mass are minimal (Zanker, 1988a). Relative to the alteration of wing motions, abdominal deflections provide a less rapid but potentially greater force asymmetry given the large abdominal steering area and associated moment arm. Although yaw torque appears to be the principal mechanical consequence of abdominal deflection, contributions to pitch control (through dorsoventral abdominal movements; see Camhi, 1970b; Brodsky, 1994) and even to roll may be substantial. Electromyographic recordings of tethered insects during simulated maneuvers would further elucidate the relationship between torque generation and activity of abdominal muscles.

In addition to abdominal deflection, asymmetric leg positions can be advantageously used to generate bilateral drag asymmetries and associated torque during maneuvers (e.g., *Drosophila*: Götz et al., 1979; Zanker et al., 1991; locusts: Camhi, 1970a, Arbas, 1986, Lorez, 1995). Whereas production of torque in yaw seems to be the most common aerodynamic use of legs, contributions to roll and pitch generation may also be significant (see Götz et al., 1979). The reduced moment of inertia of legs relative to the large abdominal mass may facilitate a more rapid response by these appendicular structures, although their low cross-sectional area may also limit the magnitude of torque thereby produced. Control of flight stability and generation of maneuvers can thus involve bilaterally asymmetric positioning of multiple control surfaces (wings, abdomen, and legs). The extent to which these functionally independent structures act synergistically during free-flying maneuvers is unknown, although Zanker et al. (1991) determined that asymmetric leg, abdominal, and wing motions were coordinated to produce yaw motions in tethered *Drosophila*. To conclude this discussion of insect maneuverability, the abundance of kinematic data available on torque production by tethered flies and orthopterans serves to highlight the paucity of comparable data for free-flying insects. Anecdotal comparisons of flight maneuverability among insect taxa demand phylogenetic comparisons of axial and torsional agility, together with a systematic assessment of size-dependent aerial performance. As with the power output of muscle (section 4.3.1), thermal sensitivity of maneuverability and underlying sensory responses may also be important in various behavioral and ecological contexts.

5.3 THREE-DIMENSIONAL FLIGHT BEHAVIOR

Most features of insect flight performance involve three-dimensional maneuverability in addition to hovering and straight-line flight. Most importantly, insects must be able to ascend and descend in order to effectively utilize the vertical dimension of space afforded by flight. The kinematic complexity of three-dimensional flight maneuvers is evident in various behavioral contexts ranging from the pursuit of mates to predation. Methodologically, three-dimensional flight paths can be determined using two synchronized cameras in various geometrical orientations (see Van Gheluwe, 1978; Dahmen and Zeil, 1984, 1995; Rayner and Aldridge, 1985; Hardie and Young, 1997). For small insects, a large mirror oriented at 45° to a single camera axis may be sufficient to recover the third spatial dimension (e.g., Okubo et al., 1981; Wehrhahn et al., 1982). Optical range finders coupled with continuous recording of azimuth and elevation can be used to track insects within much larger spatial volumes of interest (see Pennycuick, 1982; Zalucki et al., 1980). From such recordings, the three-dimensional trajectory of a flying animal can be computationally reconstructed (e.g., Tucker, 1991). Particularly with the use of video cameras and infrared illumination for night work, such methods are becoming increasingly sophisticated and field-portable (Riley et al., 1990; Riley, 1993; Ikawa et al., 1994).

5.3.1 Patrolling, Pursuit, and Evasion

Many insects exhibit repetitive flight trajectories that delimit a particular region in space. For example, many odonates, flies, and butterflies vigorously patrol territories and chase away conspecific and heterospecific aerial intruders. Such stereotypical flight patterns usually function in territorial defense and are most generally termed station-holding (Wehner, 1981; Thornhill and Alcock, 1983). Various contexts of sexual selection also involve repetitive flight patterns, most typically in display behaviors during which males advertise to females. Many odonates, for example, use advertisement flights to indicate suitable oviposition sites to females (Thornhill and Alcock, 1983). Similarly, male display in a dolichopodid fly involves flying in repetitive arcs above the female (Land, 1993a). Nest defense by social Hymenoptera often involves station-holding by particular castes in the immediate vicinity of the colony. For example, *Trigona* guard bees hover directly in front of their nest entrance and vigorously repel intruders (Wittman, 1985). Because many Hymenoptera must forage over substantial distances, successful return is possible only with some spatial memory of surrounding landmarks. For this purpose, complex arcing trajectories are used when leaving nests or approaching foraging sites (Zeil, 1993a, b; Lehrer, 1996; Zeil et al., 1996; Voss and Zeil, 1998). Such flights provide the orientational information necessary to visually identify local landmarks and surface patterns that subsequently enable successful return to the nest.

Much territorial behavior in free-flying insects involves pursuit and sometimes capture of conspecific individuals. Chase kinematics and maneuvers have been particularly well examined in dipterans, a taxon for which aerial mating is often of paramount importance (Gruhl, 1924). Chasing behavior comprises three distinct components: visual recognition of and fixation on potential targets, tracking, and the actual chase kinematics. Visual identification of targets requires discrimination of moving conspecifics against an often geometrically complex background. For this purpose, a frequency-dependent and direction-sensitive analysis of the optical flow field is used to locate small objects moving at relatively high velocities (e.g., Virsik and Reichardt, 1974, 1976; Reichardt and Poggio, 1979; Wehrhahn and Hausen, 1980; Kimmerle et al., 1997). Object detection may also be facilitated by neural switching between the panoramic (i.e., global) visual environment and local small-field components of the optomotor response (Egelhaaf, 1989; Kirschfeld, 1994). Once targets are identified, insects rapidly rotate toward the visual stimulus using specific neurophysiological pathways that yield high-frequency torque responses (see Srinivasan and

Bernard, 1977; Reichardt et al., 1983; Egelhaaf, 1987; Egelhaaf et al., 1988; Egelhaaf and Borst, 1993). This pointing of the longitudinal body axis then facilitates axial movements to follow the target object.

Tracking behavior following target identification often occurs prior to territorial chases in flies. Typically one fly follows the other or moves laterally to maintain a constant separation and to stay pointed at the target fly (e.g., Collett and Land, 1975; Brady, 1991; Zeil, 1993d). Motivating factors that underlie a decision to actively chase are not known, although pursuit sequences can also occur immediately upon target sighting. Actual chasing of conspecific and heterospecific intruders involves continuous readjustment of forward, lateral, and rotational velocities. During both tracking and chasing, compensatory optomotor reflexes (section 5.1.1) must be overridden to effect intentional body reorientations and changes in flight velocity. In order to chase and capture, the target's trajectory must be evaluated and the chaser's locomotor output adjusted correspondingly. One possible method is to assess the target's velocity and then to estimate future position, assuming that this vector does not change with time. This predictive strategy for interception involves, in some cases, initial flight away from a moving target if the pursuer's body is not initially oriented at obtuse angles to the target's trajectory (e.g., Collett and Land, 1978).

As an alternative to predictive interception, targets may be continuously tracked and optomotor output correspondingly adjusted in the method known as continuous tracking. Potential kinematic cues used in this behavior include the absolute distance (and its rate of change) to the target, along with vertical and horizontal error angles (and their derivatives) between pursuer and target (see fig. 5.2). For example, airspeeds of chasing male *Musca* flies are proportional to the distance to the target fly (Wehrhahn et al., 1982). Two- and three-dimensional studies of chasing flies have also demonstrated significant correlations between compensatory body rotation and the vertical and horizontal error angles (e.g., Land and Collett, 1974; Wehrhahn, 1979; Wagner, 1986b; Land, 1993b). In such pursuits, both the extent and angular velocity of body rotations change in proportion to the angular differences (and temporal derivatives thereof) between the two flight trajectories, albeit with time lags of 10–30 ms. Such coupling between error angles and body rotations and translations acts to continuously realign the velocity vectors of chaser and target; eventually the target will be overtaken in flight if the chase velocity is sufficiently high (Poggio and Reichardt, 1981; Wehrhahn, 1981). Similar methods of continuous tracking have been described for drone honeybees flying toward a moving queen model (Gries and Koeniger, 1996), and in bibionid flies tracking (but not chasing) territorial intruders (Zeil, 1993d).

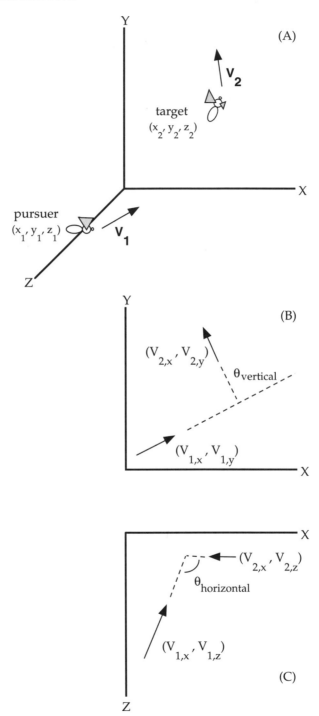

Intermediates to continuous and predictive tracking also exist whereby occasional saccadic readjustments of body orientations punctuate otherwise smooth pursuit trajectories. Continuous tracking is most effective at low translational velocities of chaser and target, whereas saccadic body motions with discontinuous target sampling probably yield better interception for higher target velocities (see Land, 1992, 1993b). Behavioral modulation by insects in response to different target velocities has not been evaluated but may well demonstrate a range of strategies between continuous and discontinuous target tracking. Tracking and interception abilities can also exhibit substantial sexual dimorphism within a species. For example, male flies respond to and track targets more quickly than do females (Srinivasan and Bernard, 1977; Wagner, 1986b), consistent with sexual selection on male performance in the aerial mating systems of many dipterans. Such dimorphism in flight performance is matched by optical and neural differences between the sexes in flies (e.g., Zeil, 1993c). Similar results are likely to apply for all insect taxa in which aerial mate capture by males is important. Such differences may be less pronounced, however, for taxa in which both sexes are also aerial predators (e.g., odonates).

Escape in three dimensions forms the dynamic counterpart to a successful chase, and involves both rapidity and unpredictability of body motions. An important distinction must be made between undirected and directed responses to pursuit (R. B. Srygley, pers. comm.). Examples of the former include changes only in scalar quantities (e.g., changes in airspeed or in the rate of turning), whereas directed responses involve intentional changes in the magnitude as well as direction of kinematic quantities in response to the pursuer's flight trajectory. Successful aerial evasion usually involves flight away from the pursuing insect, and thus precludes visual monitoring of the pursuer's instantaneous flight vector. As a consequence, undirected responses may be more effective in evasion once the escape response has been initiated. For example, male houseflies pursued by conspecifics accelerate and turn more frequently than when in cruising flight (Wagner, 1986b). One universally applicable means of increasing airspeed is to

---

On opposite page:
FIG. 5.2. (A) Relative position (x,y,z) and velocity vector (V) of pursuing and target insects, together with instantaneous vertical (B) and horizontal (C) error angles derived through projection of the corresponding velocity vectors onto the (x,y) and the (x,z) planes, respectively. Note that the longitudinal body axis of a flying insect is not necessarily aligned with its instantaneous velocity vector.

accelerate downwards in an exchange of potential for kinetic energy. The general occurrence of this mechanism is unfortunately not known given the relative paucity of three-dimensional chase trajectories. Neither have chase kinematics during attack by heterospecific aerial predators been investigated for any insect. Because target identification and pursuit occur on the basis of visual cues, the contrast of flying insects against natural backgrounds will also be relevant to successful prey capture and aerial evasion (e.g., Srinivasan and Davey, 1995). Furthermore, flight in nature occurs in the context of ambient light environments that may differ substantially in intensity as well as spectral composition (Endler, 1990). Flight toward darker or visually more complex backgrounds may thus provide further behavioral flexibility to three-dimensional escape strategies.

5.3.2 Orientation in Odor Plumes

Chemotaxis, or movement toward odor sources, is a widespread behavior among both aquatic and terrestrial arthropods. For flying insects, potential targets of chemotaxis include conspecifics advertising sexual receptivity, adult nutritional resources such as rotting fruit, or host plants that serve as sites for oviposition. Selection for efficient methods of source location is particularly intense in the context of long-distance pheromonal advertisement to potential mates, a reproductive behavior common among Lepidoptera (Cardé and Baker, 1984). In such sexual communication systems, individuals that fail to fly to and locate a source of pheromone simply do not mate. Many male moths, for example, fail to locate conspecific females emitting advertisement pheromones, whereas females often mate with the first male to arrive in the immediate vicinity. These empirical observations suggest that selection acts directly on chemotactic responses to increase the rapidity of aerial search behavior. Not only rapid but also energetically efficient search strategies during odor localization may be important. Distances over which male moths follow pheromone plumes, for example, can range up to tens of kilometers. Flight over such distances, as well as maneuvering to localize odor sources over smaller spatial scales, potentially entails substantial searching costs in terms of both energy and time.

Because odors are dispersed by ambient air motions to create a three-dimensional plume, one general approach to finding an odor source is simply to move upwind. Such anemotactic behavior during walking is used by many insects and possibly by apterygotes to locate mates (see Bell et al., 1995), and may have been ancestrally co-opted by winged insects for use in flight. Winds vary substantially in space and time, however, and continuous upwind flight within a plume does not ensure source localization. Transient winds orthogonal to the mean wind vector can displace sections of the plume laterally relative to an overall downwind drift, and plume structures correspondingly tend to be geometrically complex and temporally variable. If flight is sufficiently fast and the target object relatively large and within short range, immediate upwind flight upon entering an odor plume may nonetheless be a viable strategy. Such directed anemotactic responses are characteristic of fast-flying tsetse flies in search of bovine hosts (see Williams, 1994; Brady et al., 1995; Griffiths et al., 1995), as well as by small parasitoid wasps in search of host pupae (Kerguelen and Cardé, 1997). In both cases, the searching insect is initially alerted to host presence by a kairomone plume, chemical odors emanating from a heterospecific host that are not intentionally communicative. Kairomone sources tend to be relatively large and to produce a broad and somewhat diffuse plume (Cardé, 1996). Because of the large associated odor volume, searching insects are less likely to fly out of the plume and an anemotactic strategy for source localization may be most efficient. Visual cues may also serve in close-range source location given that host plants or animals tend to be large relative to the searching insect.

By contrast, pheromone emission usually occurs from the tips of discrete anatomical structures and thus closely resembles a point source of odor. The spatial structure of pheromone plumes correspondingly tends to be more discrete and to exhibit a high degree of sinuosity and lateral displacement (Cardé, 1996). Because upwind flight under such circumstance has a high probability of leaving the plume following initial contact, search strategies supplemental to anemotaxis are required for effective source localization. Typically, anemotaxis within a pheromone plume is coupled with side-to-side casting at progressively greater amplitudes if the plume is subsequently lost (e.g., Kennedy and Marsh, 1974; David et al., 1983; David and Kennedy, 1987; Baker, 1989). Although cross-wind searching for a plume is most efficient when wind direction is relatively invariant (Sabelis and Schippers, 1984), the combination of anemotaxis and casting behavior (i.e., side-to-side movements) provides orientation to source even under turbulent wind conditions. In addition to the context of pheromonal searching, casting behavior also characterizes female moths and fruit flies searching for appropriate oviposition sites by odor (see Willis and Arbas, 1991; Zanen et al., 1994). Dynamics of casting behavior and anemotaxis are not invariant but rather change in response to the geometry of plume structure and local odor concentrations (e.g., Kuenen and Cardé, 1994; Mafra-Neto and Cardé, 1994; Vickers and Baker, 1994). Such indirect sensitivity to local wind conditions and short-term modulation of searching behavior presumably yield an increased over-all search efficiency. Once an odor source has been localized to a particular region (e.g., <1 m distant from the searcher), local concentration gradients are less pronounced and additional sensory cues are used to identify the precise source location. For example, moths tend to reduce flight speeds at higher pheromone concentrations (e.g., Kuenen and Cardé, 1994), suggesting use of supplemental cues such as vision when the source is nearby.

During casting behavior, insects maintain approximately constant groundspeeds when flying in winds of variable speed and direction. For example, moths tracking pheromone plumes in variable-speed winds maintain an approximately constant groundspeed and track angle (the ground track relative to wind direction) by varying both airspeed and flight direction (Marsh et al., 1978; Willis and Cardé, 1990; Willis and Arbas, 1991). Free-flying tsetse flies also maintain constant groundspeeds during natural host-seeking behavior (Griffiths et al., 1995). Because of the potential implications of plume-tracking behavior for reproductive fitness, selection has probably been intense for optomotor mechanisms that adjust airspeed and flight direction in response to varying wind conditions. The temporal scales over which such adjustments occur are also potentially variable (see Willis and Arbas, 1998). Indirect evolutionary effects may derive from the orientational mechanisms used in plume tracking. For example, larger male gypsy moths track plumes at higher air and groundspeeds (Kuenen and Cardé, 1993), suggesting that indirect selection for increased body size might result from the kinematic characteristics of successful searching behavior by males. Experimentally, anemotactic responses in plume tracking can be manipulated using appropriate odor sources and optomotor cues to elicit long-duration laboratory flights from insects (e.g., Miller and Roelofs, 1978; Fadamiro, 1997). Plume structures established under laboratory conditions may, however, differ considerably from outdoor plumes that are subjected to the air turbulence and variable surface contours of natural environments.

5.3.3 Swarming

Swarming and the formation of mass aerial aggregations are common behaviors in a number of winged insect orders. Nuptial flights of termites, Ephemeroptera, Neuroptera, Trichoptera, and many social Hymenoptera are termed noncohesive swarms and serve functions of mating as well as short-range dispersal (Sullivan, 1981; Cooter, 1989). Cohesive swarming flights are associated only with reproductive behavior and are most noted among the Diptera (Downes, 1969; McLachlan and Neems, 1995). This latter category of swarming flight typically involves many males hovering at a visual landmark, and potentially serves as a lek or sexual display formed collectively for the purpose of attracting females. Sex ratios of dipteran swarms tend to be heavily male-biased, with males chasing and mating with female flies that occasionally enter the swarm. Swarming behavior of individual males involves hovering or slow forward flight within the aggregate swarm, and as a consequence can be energetically expensive. Male mosquitoes, for example, expend more than 50% of their available energetic reserves in each nocturnal swarming bout (Yuval et al., 1994). In addition to the costs of steady flight, high energetic expenditure also derives from fast directional changes and continuous relocation within the swarm. Okubo et al. (1977), for example, observed rapid accelerations by swarming midges, documenting avoidance of collision in one instance. Similar short-term variation in flight speed and orientation characterizes individual mosquitoes within a swarm (Gibson, 1985). Although useful for reproductive purposes, swarming may also impose substantial ecological penalties. Large aggregations effectively advertise mating opportunities to conspecifics, but also potentially serve to attract aerial predators (e.g., dragonflies).

In addition to the short-range cohesive and noncohesive swarms of various insect taxa, some Orthoptera and many social Hymenoptera engage in migratory swarming for purposes of long-distance dispersal. Because large numbers of individuals fly in proximity during such behavior, migratory swarming presents distinct problems of coordination such that flying neighbors do not collide. For example, the orientation of individual migratory locusts within swarms may vary somewhat over periods of several seconds (see Waloff, 1972b; Baker et al., 1984), but overall flight directionality and interindividual spacing remain fairly constant over longer time intervals. Swarm cohesiveness in locusts appears to derive from particular mechanisms of collision avoidance that are activated when locusts are in close proximity. Tethered locusts confronted with rapidly approaching targets initiate asymmetric movements of the wings together with abdominal and leg motions that produce evasive yaw (Robertson and Johnson, 1993a). The initiation of this correctional response is visually mediated and is similar to the optomotor mechanisms regulating landing behavior (Robertson and Johnson, 1993b; section 5.2.1). Instead of the aforementioned tau function, however, a more complex mathematical function is calculated neuronally that yields maximum values prior to collision with an approaching object, thereby permitting timely avoidance (Hatsopoulos et al., 1995). An additional mechanism that potentially functions in collision avoidance by locusts is an infrared response mediated neither by visual nor antennal pathways (Robertson et al., 1996). Locust body temperatures are elevated by 5–10°C above ambient air temperature during flight (see Weis-Fogh, 1967b), and proximity to such a radiant heat source could potentially be monitored by infrared sensors that initiate an avoidance response. In support of this hypothesis, the asymmetric wing kinematics associated with such thermal responses are similar to those initiated visually when collision is imminent (Robertson et al., 1996). Multiple sensory pathways are therefore indicated in the maintenance of interindividual spacing and the overall cohesiveness of locust swarms.

5.4 LIMITS TO INSECT FLIGHT PERFORMANCE

Interest in the comparative magnitude of forces produced by flying insects dates to the nineteenth century. For example, Plateau (1865, 1866) fixed small amounts of wax to the thorax and abdomen of diverse insects to determine their maximum load-lifting capacity during free flight. Magnan and Planiol (1933; see also Magnan, 1934) tethered flapping insects to a variable-resistance rotational device in order to compare the maximum forces produced by different insect taxa; odonates were found to produce the greatest force per unit of thoracic muscle mass. Flight muscle mass clearly must influence total force and power production by flying animals, as was first emphasized for insects by Magnan and Perrilliat-Botonet (1932; see also Hartman 1961). More recently, a variety of techniques have been employed to evaluate maximum force and power output of flying insects. Given that insects range over many orders of magnitude in body mass, much of this research has been directed toward understanding allometric constraints on flight performance.

5.4.1 Defining Maximum Performance

Because anthropomorphic conceptions of flight performance potentially incorporate diverse elements of the flight envelope, assessment of maximum performance requires identification of dynamically similar behaviors for which different flying animals can be reliably compared. Examples of such behaviors include maximum flight velocity, maximum linear acceleration, maximum rate of vertical ascent, maximum rotational velocities and accelerations about body axes, and the smallest radius of curvature attainable in three-dimensional maneuvers. Different physical and physiological considerations potentially limit performance in each of these behaviors, and no universal constraint on performance is likely or indeed biologically relevant. Rather, different features of morphological design and flight biomechanics interact in specific physical contexts to impose limits on any given feature of aerial performance.

One ecologically important metric of performance is flight with loads, consisting of either prey, mates, or pollen and nectar loads. Insects are well known to produce forces substantially in excess of body weight, and this capacity is used in a variety of ecological contexts. For example, vespid hymenopterans typically forage for animal prey or nectar and return to the nest with loads ranging from 20 to 70% of body mass (e.g., Herold, 1952; Archer, 1977; Coelho and Hoagland, 1995). Nectar loading in bumblebees can effectively double body mass in some individuals (Heinrich, 1975b). Tsetse flies, which first acquire and then fly with large blood loads, have a substantial ability to transport a transiently increased body mass (Langley, 1970). In an extreme case, ascending flight was recorded in a 42 mg tsetse fly to which had been attached 170 mg of lead ventrally along the abdomen (Hargrove, 1975). Tethered simulations of flight confirm such capacity for excess force production in insects; the vertical force produced by tethered insects, for example, can exceed the body weight by a factor of two (e.g., Schmidt, 1939; Weis-Fogh, 1964a; Lehmann and Dickinson, 1997). A number of morphological and kinematic features might be expected to contribute to maximum takeoff performance (Borin, 1987); wing loading, muscle mass, and maximum stroke amplitude all potentially influence maximum levels of both transient and sustained lift and power production. Allometric variation in maximum acceleration is also likely. The force of muscular contraction is proportional to the muscle cross-sectional area and thus to the square of linear dimensions, whereas the body mass that this force accelerates is generally proportional to the cube of linear dimensions. Whole-body accelerations can only derive from instantaneous muscle force transmitted either directly or indirectly to the surrounding medium, and smaller animals will thus experience higher absolute values of acceleration during locomotion (see Vogel, 1988).

Comparative analyses of maximum aerial performance are, to date, limited to studies of vertical load lifting. By progressively attaching weights of increasing mass to various insects, bats, and birds, Marden (1987) determined the amount of weight that test animals were unable to lift in a vertical takeoff. Maximum load-lifting capacity was then assumed to be intermediate to the load at which no takeoff occurred and the immediately preceding load at which takeoff was successful. Marden (1987) carried out this analysis for taxonomically diverse volant forms ranging over four orders of magnitude in body mass, and concluded that the capacity for vertical force production was linearly proportional to and most strongly predicted by the mass of flight muscle. Morphological parameters such as wing aspect ratio and wing loading were much less significant in predicting the maximum capacity to lift cumulatively applied loads.

Because these results have figured prominently in broader analyses of constraints on animal flight performance (see Ellington, 1991b; Marden, 1994), it is important to evaluate methodological features of these load-lifting experiments that may confound more general biomechanical interpretation. First and foremost, vertical forces during takeoff were not directly measured in these experiments. Takeoff by definition requires a vertical acceleration. Marden (1987) estimated maximum force during takeoff without measuring whole-body accelerations, a quantity that probably exhibits allometric variation. Thus, these load-lifting results do not necessarily address the capacity for maximum acceleration, but rather more narrowly evaluate the ability to lift loads vertically. This restriction renders the results more useful in some ecological contexts but may preclude comparative statements concerning the general capacity for acceleration and evasion, particularly when such motions are not vertically directed. Second, progressive attachment of weight to animals potentially induces fatigue in consecutive bouts of exercise; biomechanical capacity for takeoff is difficult to differentiate from the behavioral motivation to fly. Third, boundary effects potentially influence takeoff aerodynamics (Ellington, 1991b); these and additional biomechanical considerations that bear on energetic estimates during takeoff are discussed in section 5.4.2 (see also Metcalfe and Ure, 1995). Finally, discrete categorization by Marden (1987) of insects as clap-fling or conventional fliers may not accurately characterize the continuum of contralateral wing separation that characterizes flapping flight (section 3.1.2); wing proximity is also likely to be more pronounced under conditions of maximum load and concomitantly increased stroke amplitudes.

Such issues notwithstanding, these empirical results represent an interesting comparative analysis of the capacity of insects to lift loads vertically. Marden (1987) found that, of the morphological parameters examined, relative muscle mass (the ratio of flight muscle mass to body mass) was the strongest statistical predictor of the capacity to generate vertical force supplemental to body mass (see fig. 5.3). Insects can generally lift between 0.5 and 3 times their own body mass. The orders Diptera, Lepidoptera, and Odonata express the highest values of relative muscle mass and correspondingly possess the greatest ability to lift loads. Conversely, the Mantodea, Hemiptera, and Coleoptera have the lowest values of relative muscle mass, reflecting in part the

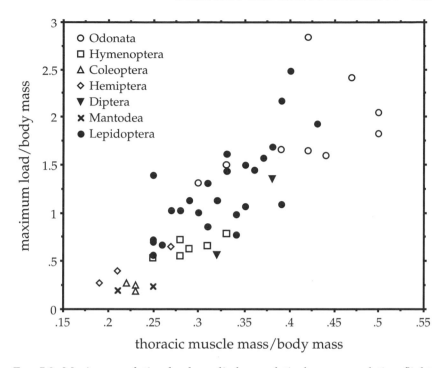

FIG. 5.3. Maximum relative load applied cumulatively versus relative flight muscle mass for fifty insect species from seven orders. Each point represents a species mean derived from data of Marden (1987). Load here refers to the mass lifted that is supplemental to body mass.

greater mass investment in tegminization, sclerotization, and clytrization in these orders. Relative muscle mass is approximately 60% in the Odonata (Marden, 1987) and reaches as values as high as 72% in a chironomid fly (McLachlan and Neems, 1996), consistent with the aerial predation (by odonates) and mating systems in these taxa. Although existing interordinal comparisons are necessarily confounded by phylogenetic relatedness, relative muscle mass clearly exerts an important influence on insect lifting ability. Additional kinematic and morphological factors are also indicated, however, given the substantial variation in load-lifting capacity evident within and among insect orders for any given value of relative muscle mass (see fig. 5.3). Further biomechanical studies that incorporate phylogenetic controls would elaborate this conclusion, as well as providing insight into the remarkable mechanics by which insects can lift many times their own body mass.

Recently, a modified load-lifting method has been used with hummingbirds and euglossine orchid bees that avoids potential complications of cumulatively applied loads. This alternative method requires animals to fly upwards while carrying a thread weighted at fixed intervals with beads of known mass (see Chai et al., 1997). As the animal flies higher, more mass is elevated and must correspondingly be supported by the flapping wings; the animal thus asymptotically approaches maximum lifting capacity in a single continuous vertical motion. As progressively more weight is lifted, both the animal and the string of beads are continuously decelerated. The ever-diminishing kinetic energy of the moving bead string can but minimally offset requirements of vertical weight support from the animal's exertions; at the top of the flight trajectory, the animal is transiently stationary and lifting performance is assumed to be maximal. Cumulative addition of weights is avoided in this approach, and neither the ground effect nor body accelerations are relevant when the animal is motionless and sustaining maximum load at a considerable distance from the takeoff surface. For nine species of euglossine bee, the maximum additional load sustained vertically in this fashion averaged about 95% of body mass (Dillon and Dudley, in prep.). Comparable loads sustained in hovering by different hummingbird species range from 80% to 191% (Chai and Millard, 1997; Chai et al., 1997; see fig. 5.4). Hummingbirds lifting weighted threads also suggest a potential interaction between peak performance and lifting duration: larger hummingbirds lift greater mass over slightly shorter (albeit nonsignificantly different) time intervals (fig. 5.4). In some behavioral contexts (e.g., aerial chases), the ability to sustain maximum effort may be just as important as is the absolute level of locomotor performance. Such trade-offs between capacity and endurance are well described in terrestrial locomotor systems (e.g., the relative roles of aerobic and anaerobic metabolism), but have rarely been evaluated in volant forms. Over longer timescales, the availability of energetic reserves may also influence the outcome of behavioral interactions that elicit peak performance (e.g., Marden and Waage, 1990).

Accelerational escape from both aerial and terrestrial predators represents an important context during which flight muscle mass may significantly influence flight performance and ultimately survivorship. For example, relative muscle mass is significantly higher in palatable relative to unpalatable butterfly species, consistent with the reduced flight speeds and lower evasive capacities characteristic of the latter group (Srygley and Chai, 1990a; Marden and Chai, 1991; see section 7.3.2). For individual animals, both load-carrying of external items as well as internal mass loading (e.g., lipid deposition) increase effec-

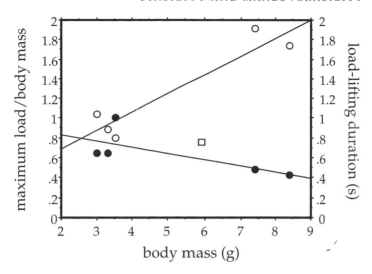

Fɪɢ. 5.4. Maximum relative load (open circles) and duration of maximum effort (closed circles) during asymptotically increasing load-lifting for five hummingbird species (Chai and Millard, 1997; Chai et al., 1997). Maximum relative load is positively correlated with body mass ($r^2 = 0.88$; $P < .02$); duration effort declines slightly but nonsignificantly with body mass ($r^2 = 0.52$; $P > .05$). The open square indicates data for the one hummingbird species tested by Marden (1987) in cumulative loading. Load here refers to the mass lifted that is supplemental to body mass; see text for further details.

tive body mass and yield a de facto reduction in relative muscle mass. In birds, such increases in effective body mass reduce takeoff performance and the ability to escape from predators, as manifested by reduced linear accelerations and rates of ascent (see Hedenström, 1992; Witter and Cuthill, 1993). This effect has been sufficiently pronounced as to have influenced historical patterns of fat loading by great tits in the face of variable predation intensity by sparrowhawks (Gosler et al., 1995). Among insects, such effects may be most pronounced for foraging taxa that carry substantial food volumes (e.g., Wolf and Schmid-Hempel, 1989) and for migrants that must fly with heavy lipid loads. In addition to degrading the ability to accelerate and maneuver, such mass loading may adversely influence flight performance by imposing a nonlinear increase in the power requirements of flight (section 5.4.2).

The rapid maneuvers and accelerations associated with aerial mating may be sensitive to the relative mass of flight muscle. Dragonflies with attached weights exhibit reduced mating success relative to unmanipulated counterparts (Marden, 1989a), and male empidid flies with greater relative muscle mass are more successful in attaining copulation (Marden, 1989b). Behavioral strategies of mate acquisition both within and among butterfly species also exhibit systematic correlation with relative thoracic mass (see Wickman, 1992; Van Dyck et al., 1997). The aerial copulation characteristic of odonates and a variety of other insect taxa imposes even greater demands on weight support by the male. Sexual selection may then favor larger males characterized by a greater absolute mass of flight muscle as well as males with greater relative muscle mass. For example, female scatophagid flies mating with smaller males may experience greater mortality during startle flight of the amplexed pair (Borgia, 1981; see also Davies, 1977), whereas mated caddisfly pairs have a greater tendency to fall into water when males are smaller (Petersson, 1995). The capacity to carry loads also influences selection and aerial transport of prey in some taxa, particularly in predatory insects that must fly with large loads. For example, vespid wasps occasionally fail to take off with anuran larvae removed from their foam nest in predation attempts (Villa et al., 1983). Similarly, yellowjacket wasps reduce the mass of their honeybee prey items prior to flight by removal of body tagmata, thereby mitigating load-lifting constraints and reducing the energetic costs of aerial transport (Coelho and Hoagland, 1995). Sphecid and pompilid Hymenoptera that fly with insect and spider prey, respectively, may experience similar limits to load-carrying performance.

In contrast to the aforementioned results on vertical load-lifting, much less information is available on other kinematic features of maximum performance in insects. As discussed elsewhere (section 3.1.1), paucity of quantitative information precludes comparative analysis of maximum flight velocities and accelerations. The interesting work of May (1991), however, clearly implicates power limits on maximum velocity and acceleration in dragonflies (section 5.4.2). Additional parameters of flight performance relating to maneuverability await similar empirical assessment. Accelerating vertical ascent represents one of the most energetically demanding forms of aerial locomotion, yet comparative data on maximum ascent rates are unavailable. Experimentally, a variety of wind tunnel designs (e.g., Kennedy and Booth, 1963; David and Hardie, 1988) can be used to generate vertical air flows against which small insects must fly to maintain a constant position in a vertically oriented tunnel. By progressively increasing flow velocity, the maximum capacity to fly vertically at a constant velocity can be determined. Aphids and *Drosophila* are particularly good at stationkeeping under such experimental circumstances, although this method should be equally applicable to hummingbirds hovering at a feeder.

5.4.2 Energetic Limits to Performance

The capacity to carry loads, as with as other features of flight performance, can potentially be limited by the amount of power available from the flight muscle. Maximum power output can either be measured directly using work-loop experiments on individual muscles or muscle fibers (section 4.2.3), or can be estimated using either models of aerodynamic power requirements during flight (section 3.3.2) or theoretical models of muscle contraction. In this last approach, the mass-specific power available from muscle (P^*_{muscle}) is assumed to be proportional to muscle strain ε, to the internal myofibrillar stress σ, and to the contraction frequency n (see Pennycuick and Rezende, 1984; Ellington, 1985):

$$P^*_{muscle} = \frac{\sigma \varepsilon n}{\rho}, \tag{5.1}$$

where ρ is the mass density of muscle. Upper limits to contraction frequency and to muscle strain and stress must, in this formulation, ultimately bound muscle power output.

Application of equation (5.1) to synchronous insect flight muscle suggests maximum values for P^*_{muscle} of 80–100 W/kg (Ellington, 1985; Josephson, 1993). Similar calculations for the maximum power of asynchronous flight muscle yield a value of 100–250 W/kg (Ellington, 1985). Note that these values are based on a mass density ρ for striated mammalian muscle (1060 kg/m³) that may overestimate the actual value for insect flight muscle given potentially significant tracheal contributions to total muscle volume. The estimates of Ellington (1985) are generally higher than those obtained in work-loop experiments for both synchronous and asynchronous flight muscles (see table 4.1), but are similar to those obtained from application of aerodynamic models to free-flying insects during maximal performance (see below). Intrinsic to equation (5.1) is an assumption that muscles shorten at constant velocity. Because muscles must accelerate to attain their maximum contraction velocity and then decelerate at the end of force production, the assumption of a constant shortening velocity causes maximum available power to be systematically overestimated (Josephson, 1989). Also, a small but potentially significant amount of work is required to extend muscles after they have contracted, reducing further the net power output over the full contraction cycle. More precise estimates of maximum power output will thus require quantitative specification of contraction kinematics as well as specific knowledge of mass density for the muscle in question.

Allometrically, the flight of large animals may be precluded if the mechanical power required to fly exceeds the power generated by the relevant muscles above some critical body mass. Maximum muscle mass-specific power output is predicted by equation (5.1) to increase monotonically with contraction frequency. Wingbeat frequency varies inversely with body mass in insects (fig. 3.3B), whereas maximum myofibrillar stress and muscle strain should be approximately scale independent (Ellington, 1991b). Mass-specific power output might then be expected to increase at higher wingbeat frequencies and thus at smaller body sizes. For example, many insects (and some bats and birds) are capable of hovering flight, but in general larger animals are less likely to perform this energetically demanding task. Power output asymptotically approaches a limiting value at high contraction frequencies, however, because myofibrils must be replaced morphologically with mitochondria to support increased energy flux as maximum power output rises (see Pennycuick and Rezende, 1984; Ellington, 1985). A similar increase in tracheolar volume may also be necessary to meet an ever-increasing oxygen demand at higher frequencies. Myofibrils may in fact occupy a minority of available volume in some insect flight muscles. Mitochondria, sarcoplasmic reticulum, tracheoles, and hemolymph all contribute (in approximate order of decreasing extent) to the total muscle volume, although allometric and phylogenetic associations of such relative volumetric representation are not known.

In addition to power availability, other physiological factors may act to limit flight performance. For example, the convective supply of metabolites via the hemolymph may be inadequate to support very high rates of muscle metabolism during flight. At the molecular level, muscle contractility ultimately derives from the attachment of and subsequent motion between myosin heads and actin filaments. The maximum number of attachment sites must necessarily limit muscle performance, in that mechanical power output ultimately emerges from the summed action of individual components of contraction. For isolated fibers of insect flight muscle, Pringle and Tregear (1969) calculated that all available myosin units were activated when maximum power output was elicited at the appropriate operating temperatures and frequencies. No such estimates are available for insect flight muscle yielding maximum performance in vivo. If the muscle mass density of myosin can be determined, however, mechanical power output can potentially be calculated from rates of myosin head cycling and the displacement and force generated per cycle (see Hochachka, 1994; Howard, 1997). Such an estimate would provide an independent and informative comparison with other theoretical as well as experimental values for maximum power output.

Empirically, the allometry of maximum mechanical power available from insect flight muscle remains unresolved. One approach to this problem is to estimate individual components of induced, profile, parasite, and inertial power for those flight situations believed to elicit maximum performance. Using the earlier load-lifting data of Marden (1987), Marden (1990) estimated induced power expenditure during presumed hovering flight at maximum load. The Rankine-Froude actuator disk used for this estimate was a circle with radius equal to half the wingspan (the distance between wing tips), rather than the actual area swept out by the beating wings (Ellington, 1991b; Chai et al., 1997; see section 3.2.1.3). Use of the former area for the Rankine-Froude approximation potentially underestimates the induced power requirements of hovering flight. Similarly, spatial and temporal fluctuations in the vortex wake were not incorporated in the induced power estimates of Marden (1990) because the necessary kinematic data were unavailable; such fluctuations can also significantly increase induced power requirements.

Nonetheless, the load-lifting data of Marden (1987) suggest that muscle mass-specific induced power and perhaps maximum power in insects scales with $mass^{0.13}$ (Ellington, 1991b; Marden, 1994). Such an allometry for maximum power presupposes, however, that relative profile power requirements are size independent and constitute an approximately constant 25%–30% of the induced power requirement under conditions of maximum performance (see Ellington, 1991b). Profile power may nonetheless exhibit phylogenetic and allometric variation. For example, the profile power requirement of accelerating odonates is approximately 75% of induced power expenditure (Wakeling and Ellington, 1997c), whereas the comparable value for euglossine bees under submaximal but demanding conditions ranges from 22%–35% (Dudley, 1995b). By contrast, maximum profile power in tethered *Drosophila melanogaster* is an extraordinary 200% of induced power requirements (Lehmann and Dickinson, 1997). Reduced circulatory lift relative to profile drag at low *Re* (section 3.2.1.2) suggests that profile power increasingly dominates flight energetics at smaller body size, a prediction consistent with empirical results for euglossine bees in standard hovering flight (Casey et al., 1992). Neither absolute values nor the allometry of profile power expenditure are adequately resolved for a diversity of insect taxa, but the ratio of profile to induced power does appear to increase at smaller body sizes under conditions of maximum performance.

Also inherent to aforementioned estimates of power during loaded takeoff is the assumption that such load-lifting performance is mechanically equivalent to hovering flight. As with associated vertical force estimates, additional avenues of power expenditure may be indicated. Takeoff by definition requires power to sustain vertical acceleration as well as to increase potential energy of the body mass. For a constant vertical component of velocity V_{vert}, the additional power required to climb represents the gain in potential energy per unit time ($= mgV_{vert}$). For any acceleration a given an instantaneous starting velocity V, further accelerational power ($= maV$) is required in addition to that associated with changes in potential energy. Substantial size dependence in either component of power may potentially alter the power allometries derived by Marden (1990, 1994) and Ellington (1991b). As noted previously, allometric considerations suggest that whole-body accelerations should be inversely proportional to body mass. The magnitude of this effect is unknown for volant organisms during takeoff, but may be substantial across the wide range of body mass evaluated by Marden (1987). Note also that aerodynamic force production resulting in whole-body acceleration is approximately proportional to wing surface area and will scale in proportion only to the square and not the cube of body linear dimensions, again suggesting higher accelerations in smaller animals. Finally, takeoff involves possibly scale-dependent boundary effects that influence airflow and induced power expenditure in hovering (Ellington, 1991b; see section 3.3.3). In sum, these considerations suggest that power estimates from aforementioned load-lifting data be treated cautiously.

Empirical results also suggest that the mechanical power limits postulated from cumulatively applied load-lifting results may underestimate actual available power. For example, values of muscle mass-specific induced power expended by euglossine bees hovering in low-density heliox (21% oxygen/balance helium and nitrogen; see section 8.1.2) and by hummingbirds lifting asymptotically increasing loads exceed the upper 95% confidence limit derived by Marden (1990) for volant animals of equivalent body mass (see Dudley, 1995b; Chai and Millard, 1997; Chai et al., 1997). Such empirical results necessitate a reassessment of induced and total power availability during flight, although few estimates of power output during unequivocal maximum performance are available for free-flying insects (see table 5.1). Both methodological and taxonomic diversity contribute to variability in such estimates of maximum power. If power output of muscles does nonetheless increase at higher contraction frequencies (eq. 5.1), then maximum power of flying insects should follow closely the allometry of wingbeat frequency. Potentially confounding this conclusion would be size dependence of elastic energy storage in the pterothorax as well as allometric change in relative muscle mass across taxa. In general, flight muscle mass among insects tends to follow isometric design

TABLE 5.1

Taxonomic Identity, Body Mass *m*, and Muscle Mass-specific Induced and Total Power Expenditure for Various Free-flying Insects under Aerodynamically Challenging or Limiting Conditions

Order	Genus and Species	*m (mg)*	Induced Power (W/kg Muscle)	Total Power (W/kg Muscle)	Reference
Odonata	*Sympetrum sanguineum*[a]	111.5	86.1	156.2	Wakeling and Ellington (1997c)
Odonata	*Calopteryx splendens*[a]	123.6	82.8	166.1	Wakeling and Ellington (1997c)
Lepidoptera	*Manduca sexta*[b]	1995	93.0	116.6	Willmott and Ellington (1997c)
Hymenoptera	*Xylocopa varipuncta*[c]	910	92.8	110.0	Roberts, Harrison, and Dudley (in prep.)
Hymenoptera	*Euglossa dissimula*[d]	91.0	100.5	131.7	Dudley (1995b)
Hymenoptera	*Euglossa imperialis*[d]	151.1	100.6	127.2	Dudley (1995b)
Hymenoptera	*Eulaema meriana*[d]	819.6	131.8	160.5	Dudley (1995b)
Diptera	*Drosophila melanogaster*[e]	1.0	28	(n/a)	Marden et al. (1997)

[a] Estimates refer to an individual insect in accelerating flight. Wing inertial power is smaller than the aerodynamic power, so that total power expenditure is independent of the extent of elastic energy storage within the pterothorax.

[b] Estimate refers to a single gravid female hovering in what was probably near-maximal performance. Perfect elastic storage of wing inertial energy is assumed.

[c] Maximal performance during hovering failure in low-density heliox. Perfect elastic storage of wing inertial energy is assumed.

[d] Submaximal hovering performance in low-density heliox. Perfect elastic storage of wing inertial energy is assumed.

[e] Induced power calculated as the product of mean body mass for the species and the maximum vertical ascent velocity during forward and upward flight; this approach may underestimate actual expenditure of induced power relative to more detailed formulations (see section 3.3.2).

(e.g., Marden and Chai, 1991), although deviations from isometry may be evident within certain lineages.

As an alternative to interspecific comparisons, the allometry of flight maximum performance can be studied intraspecifically. Natural variance in body size is often substantial, and techniques of allometric engineering can, moreover, be used to manipulate body size both within and among clutches of a given species (see Sinervo et al., 1992). All insects exhibit determinate growth, and adult body size can be systematically varied simply by controlling the amount of food given to larvae. Dwarfed forms of butterflies, for example, have occasionally been described as distinct subspecies even though the adult morphology simply derives from environmental effects of larval food availability. By comparing maximum performance over a range of body sizes within a species, intraspecific allometries can be derived that have no phylogenetic contributions to variance. This approach can be used to evaluate the generality of interspecific allometries, and should permit independent intraspecific tests of biomechanical hypotheses that might be proposed to explain patterns of interspecific variation.

If relative power availability from flight muscle does indeed decline at higher body masses, then power constraints are clearly indicated in the allometry of hovering performance. For example, theoretical models suggest that the mass-specific induced power required to hover should increase with $mass^{0.17}$ given isometric design (see Weis-Fogh, 1973; Ellington, 1984e, 1991; Norberg, 1990). Note that this allometry pertains to hovering with support of body weight support and not necessarily to maximum load-lifting performance. Empirically, mechanical power expenditure of hovering euglossine bees also suggests a slightly positive allometry (see Casey and Ellington, 1989; Casey et al., 1992). Hovering of larger insects thus requires relatively more power, whereas the muscular capacity for power production may concomitantly decline, ultimately limiting performance. A similar restriction may apply to the performance of insects in forward flight. Power expenditure varies with forward airspeed (section 3.3.3), but allometric analysis of flight at physiologically equivalent airspeeds suggests that requisite mass-specific mechanical power will vary as $mass^{0.05-0.19}$ (see Pennycuick, 1968; Rayner, 1979, 1990; Norberg and Rayner, 1987). Again, this positive allometry of the relative power required to fly would combine with any negative allometry of power availability to limit forward flight performance at some critical body mass. Power limits have been specifically implicated for the maximum forward airspeed of dragonflies. May (1991) obtained data on airspeeds and accelerations of free-flying dragonflies, and also measured steady-state drag coefficients on conspecific odonate bodies at the relevant values of Re. The rapid increase in parasite power at high airspeeds (section 3.3.3) was shown to dominate total power at high airspeeds and ultimately to limit performance under various assumptions of power availability. Similarly, the high accelerations exhibited by dragonflies required significant expenditure of power, and may have been limited by the maximum mechanical power available from the flight muscle.

Power availability may also limit flight of insects during external or internal load-carrying (e.g., prey transport and endogenous lipid loading, respectively). Theoretical predictions suggest that the mass-specific power required to fly with added load increases as $mass^{0.5-0.9}$ for invariant wing morphology (Pennycuick, 1975; Norberg and Rayner, 1975; Rayner, 1990; Norberg, 1990, 1995). If prey items or mates are carried externally, additional parasite drag and further expenditure of mechanical power will ensue. Such disproportionate increase in the power required to fly with added mass may have inhibited the evolution of folivory (leaf eating) in flying animals (Dudley and Vermeij, 1992, 1994). The increased digestive machinery necessary to process leaves, together with the low energy density and high mass of the ingested leaves themselves, appear to impose disproportionate energetic demands on volant vertebrates as well as insects. With the sole exception of migratory orthopterans, exclusively folivorous insects tend to be primarily saltatorial or at most only weak fliers.

One possible short-term response to such energetic constraints is simply to increase total flight muscle mass. Although the subject of experimental investigation (Jutsum et al., 1982), muscle hypertrophy in response to training has not yet been demonstrated in insects. Certain behavioral contexts are, however, associated with hypertrophy of flight muscles. For example, dragonflies exhibit a dramatic increase in relative muscle mass prior to establishing territories (Marden, 1989a). Long-distance migrants that engage in premigratory lipid loading would also be particularly likely candidates for such increases in thoracic muscle mass (see section 7.4.3). The monarch butterfly *Danaus plexippus*, for example, increases total body mass via lipid deposition by 20–30% prior to southward migratory flight in the fall (Gibo and McCurdy, 1993). Power requirements of flight must increase substantially under such circumstances, possibly compromising escape responses and energetically limiting flight performance. Although relative muscle mass cannot be determined noninvasively given present techniques, muscle hypertrophy can be deduced indirectly from data on lipid reserves and muscle mass obtained from large-scale sampling among individual insects (e.g., Marden, 1989a).

5.4.3 Aerodynamic Limits to Performance

In contrast to potential energetic limits on flight mechanics, Ellington (1991b) suggested that aerodynamic mechanisms might ultimately constrain aerial performance. Abundant power could potentially be available from the flight muscle, but transduction of this power via wing kinematics to useful aerodynamic work might simply be inadequate to sustain the body weight or to generate thrust. Aerodynamic forces on flapping wings are typically proportional to wing area and will scale in proportion to mass$^{0.67}$ for isometrically designed organisms. Increased demands of weight support must ultimately outstrip force production and limit the upper body size of flying organisms, although this constraint clearly occurs at dimensions well above those of relevance to insects. Instead, aerodynamic mechanisms at the low Re characteristic of insect flight may be inadequate in certain contexts of flight performance. Maximum lifting ability, for example, is defined as the failure of vertical force production to sustain certain loads. Clearly an aerodynamic limit to performance is indicated under such circumstances, but the underlying biomechanical reasons for failure are unclear. The power available from the flight muscle in fact may be greater than that used in maximum load lifting, with the excess only being used in other mechanical circumstances (e.g., fast forward flight). Various features of wingbeat kinematics, unsteady force generation, and muscle energetics may thus interact to constrain flight ability, rendering context dependent any dichotomous distinction between physiological and aerodynamic limits to performance.

The size dependence of maximum force production by flying insects is also difficult to predict given existing uncertainty about those unsteady aerodynamic mechanisms that characterize wing flapping (section 3.2.2). Quasi-steady allometric analysis would nonetheless seem to prelude aerodynamic constraints in hovering flight. The mean lift coefficient required to sustain hovering increases with mass$^{0.33}$ given isometric design (Ellington, 1984f, 1991). The mean Re of the wing chord in hovering flight is proportional to the product of wing area and wingbeat frequency (see Weis-Fogh, 1977; Ellington, 1984f). Given empirically determined interspecific allometries of these two parameters (fig. 3.3B), the Re of wings in hovering flight will increase with mass$^{0.47}$; a similar positive allometry characterizes the Re of insect bodies in forward flight (see also table 1.2). Wings of larger animals thus operate at progressively higher Re, a mechanical condition that facilitates circulation-based lift production (section 3.2.1.2). For example, maximum lift : drag ratios of insect wings in steady-state flow increase at higher Re (see fig. 3.7B). As hovering represents an extreme in terms of lift production, allometric constraints of aerodynamic mechanisms are even less likely for forward flight. The often dramatic lifting performance of insects also indicates substantial reserves of force production. The sensitivity of such performance to changes in size is unclear, although substantial scope exists for aerodynamic modulation under potentially adverse mechanical conditions. For example, euglossine bees in heliox can substantially increase mean lift coefficients of the wings in hovering flight, even though the mean Re of the wing chord decreases by about 40% (Dudley, 1995b; see also Lehmann and Dickinson, 1998).

The flapping of bilaterally paired appendages does, however, impose a geometrical constraint on wing motions that potentially influences performance of all volant animals. The maximum stroke amplitude through which wings can sweep without experiencing contralateral wing interference is on the order of 180° (see fig. 3.2B). The anatomical separation between the wing bases does permit slightly higher stroke amplitudes prior to actual physical contact of the wing tips, although indirect aerodynamic interference between bound vortices on the opposing wings must occur before such contact. In hummingbirds, this feature of body design bounds maximum stroke amplitude and limits the capacity for aerodynamic force production. When hovering in hypodense gas mixtures, hummingbirds literally fall from the air as stroke amplitude approaches 180° (Chai and Dudley, 1995). This geometrical constraint on stroke amplitude also applies during maximum load lifting by hummingbirds (Chai and Millard, 1997; Chai et al., 1997). In both load lifting and flight in hypodense gas mixtures, aerodynamic modulation appears to derive primarily from changes in stroke amplitude. Wingbeat frequency, by contrast, exhibits only negligible changes relative to a substantial increase in stroke amplitude in the face of heightened force and power requirements.

Anatomical determinants of maximum stroke amplitude may similarly limit aerodynamic performance in insects. Resonant characteristics of the pterothoracic apparatus impose a relative constancy on wingbeat frequency (section 3.1.2), whereas modulation of stroke amplitude appears to be the primary means of altering lift and power output. For example, euglossine bees substantially increase stroke amplitude but not wingbeat frequency to generate the additional lift force and power required by flight in hypodense gas mixtures (Dudley, 1995b). However, euglossine bees engaged in maximum load lifting appear to approach limiting stroke amplitudes in the region of 150° (Dillon and Dudley, in prep.). Similarly, mechanical power output in tethered *Drosophila melanogaster* peaks at stroke amplitudes near 180° (Lehmann and Dickinson, 1997, 1998). Maximum stroke amplitude thus appears to impose an insurmountable constraint on lift production for at least two unrelated insect lineages as well as in hummingbirds.

Failure in hovering upon reaching the point of maximum stroke amplitude may coincide with maximum power production by the flight muscle, but such congruence is not a necessary feature of body design. Alternative mechanical contexts may require even greater power from the muscle but reduced lift production by the wings. In forward flight, for example, the wing flapping velocity is supplemented by the airspeed, eliciting a concomitant reduction in the mean lift coefficient required to fly (section 3.2.1.3). The power required to fly may nonetheless exceed that of hovering because of increased parasite drag and and associated power expenditure, particularly at high forward airspeeds (section 3.3.2). No data currently address maximum stroke amplitudes during both forward flight and maximum hovering or load-lifting performance by the same insect taxon. In hummingbirds, however, stroke amplitudes are well below 180° even in fast forward flight (Greenewalt, 1960a). Parasite power is substantially elevated, with power availability and not wingbeat geometry apparently limiting airspeeds of hummingbirds in forward flight (Dudley and Chai, in prep.). A similar relaxation of constraints on wingbeat geometry may occur during transient accelerations using elevated wingbeat frequencies, although the underlying mechanics are much less resolved.

Variable durations of maximum performance further confound comparative analysis of intrinsic constraints on flight. For example, hovering flight of ruby-throated hummingbirds in normoxic (20.9% O_2) but low-density gas mixtures yields an average muscle power output of 133 W/kg over a period of 2–3 seconds (Chai and Dudley, 1995). By contrast, transient lifting performance by this species yields burst power outputs averaging 206 W/kg that are sustained for only one second (see Chai et al., 1997). Quantitative attention must accordingly be paid to the interaction between duration and intensity of maximum flight effort. A variety of physiological, geometrical, and aerodynamic considerations potentially limit insect flight performance over various timescales; different behavioral contexts of maximum effort are not necessarily homologous biomechanically. As a consequence, unequivocal identification of performance limits requires simultaneous determination of the kinematics, aerodynamics, and energetics of flight under well-defined and repeatable conditions of locomotor failure. Given the strong temperature dependence of muscle power output (section 4.3.1), thermal regimes must also be carefully controlled. Future work in this area should precisely specify the behavioral and biomechanical contexts within which performance limits are described.

Finally, existing interspecific analyses of the allometry of power requirements and the availability of muscle power during flight are potentially confounded by nonrandom phylogenetic associations among the taxa under consideration. As an extreme example, the marked physiological differences between synchronous and asynchronous muscle may yield substantially different functions for the size dependence of maximum power output. Only comparative analyses based on reliable phylogenies can remove the effect of shared evolutionary history from size-dependent physiological relationships (see Harvey and Pagel, 1991). The obvious difficulty of this approach lies in obtaining morphological and performance data from a sufficiently high number of species so as to permit meaningful statistical analysis. In the absence of appropriate interspecific sample sizes and resolved phylogenies, however, allometric conclusions concerning power demand and availability can only remain suggestive.

5.5 SUMMARY

Diverse sensory pathways are used by flying insects to monitor and maintain body position in space. Optomotor responses play a particularly important role in stabilizing rotational and translational speeds of the insect with respect to the surrounding environment. Mechanoreceptors associated with the antennae, body hairs, and the abdominal cerci permit continuous sampling of local airflow as induced both by forward motion and by flapping motions of the wings. Mechanoreceptors located within the wings and axillary apparatus continuously monitor cuticular deformations associated with aerodynamic, inertial, and muscularly applied forces. Although rhythmic motor patterns are inherent to interneuronal networks located within thoracic ganglia, peripheral sensory feedback sustains motor pattern generation and provides correctional information in response to short-term disruptions of body orientation and flight direction. The phasic entrainment of multiple mechanoreceptive cues is particularly important for maintaining the wingbeat kinematics exhibited in stable flight.

Maneuverability has been essential for effective exploitation of the aerial domain by winged insects. Takeoff and landing represent the endpoints of a broad spectrum of three-dimensional flight behavior. Often complex disruptions of the nominal force and moment balance of steady flight derive aerodynamically from bilaterally asymmetric wing kinematics, with asymmetric leg or abdominal deflection contributing secondarily to generation of rotational torques. Biomechanical linkage between temporal derivatives of body rotation and translation lies at the core of complex maneuvers. Because many such maneuvers are associated with transient but demanding episodes of natural and sexual selection, the evolutionary pressure for ever-increasing lift and power output has likely been substantial. Depending on context, limits to flight performance may derive from a variety of morphological, biomechanical, and physiological constraints. The concept of a unitary limit to insect flight performance can therefore be misleading. Geometrical constraints to wing motions characterize all flying animals, however, and may be particularly important during maximum hovering and load-lifting performance.

Chapter Six

EVOLUTION OF FLIGHT AND FLIGHTLESSNESS

T HE INCORPORATION OF flight into the locomotor repertoire of insects was clearly a defining event for arthropod if not metazoan evolution. Unfortunately, the evolutionary origins of flight in insects are not well known. Paleontological records of potential transitional forms are absent, and the likely selective forces acting on early winged morphologies can only be surmised, precluding any paleobiological interpretation of this major event in metazoan evolution. Nonetheless, available fossils and biological evidence from extant forms allow some functional hypotheses to be posed and, at least in part, evaluated. Moreover, analysis of ecological factors associated with secondary loss of wings can yield insight into the evolutionary forces that first promoted aerial mobility in insects.

6.1 ORIGIN OF FLIGHT IN HEXAPODS

The origins of flight can be evaluated paleontologically (appearance and diversification in the fossil record), morphologically (anatomical and ontogenetic origins of the protopterygote flight apparatus), and functionally (those selective forces acting on wing evolution relative to the contemporaneous biota and physical environment). Taphonomic considerations relating to the poor fossilization of insects and a global paucity of relevant fossil deposits preclude a highly resolved contextual analysis of wing origins and early diversification of winged insects. However, the evaluation of possible evolutionary scenarios (*sensu* Gans, 1989) can at least delineate likely boundaries for the evolutionary trajectories of flight. Once insects became aerially mobile by the late Devonian or early Carboniferous, extensive diversification yielded, by the early Permian, virtually all the kinematic and morphological variants evident in contemporary pterygotes. General overviews of insect fossil history are provided by Carpenter and Burnham (1985), Wootton (1990), and Brodsky (1994). The insect section of the *Treatise on Invertebrate Paleontology* (Carpenter, 1992) exhaustively documents known insect fossils at the generic level (see also Ross and Jarzembowski, 1993; Labandeira, 1994). The treatment of Handlirsch (1906–1908) provides an interesting classical perspective on insect paleontology.

6.1.1 Fossil Record

One of the most challenging features of insect paleontology is the generally poor preservational state of fossil taxa. Preservational effects are particularly pronounced for the nonmineralized proteinaceous cuticle of arthropods, and the associated fossilization problems and difficulties of morphological interpretation in insect fossils have been widely discussed (e.g., Kukalová-Peck, 1983; Carpenter and Burnham, 1985; Shear and Kukalová-Peck, 1990; Wootton, 1990). Well-preserved insect fossils in amber are unfortunately not known from the Paleozoic. Because of this limited record, the morphology and ecology of extant and phylogenetically basal forms of hexapods have been used to infer the biology of ancestral winged insects. However, such an approach can be profoundly misleading. Although certain extant forms (especially the Paleoptera) are indeed phylogenetically basal, these taxa are highly derived relative to the earliest Paleozoic forms and provide only limited clues for biological reconstruction of ancestral pterygotes. Only additional fossil evidence will resolve this unfortunate state of affairs. In the meantime, reconstructions of protopterygote biology must be viewed with a high degree of skepticism.

The earliest hexapod fossils date from the Lower Devonian, with an apterygote bristletail reported from 390 MYBP (Labandeira et al., 1988; see also Shear et al., 1984; Jeram et al., 1990) and collembolans known from 395 MYBP (Whalley and Jarzembowski, 1981). A diverse nonhexapodan arthropod and vertebrate fauna also characterized the Lower Devonian, including such varied terrestrial predators as scorpions and amphibians (Wood et al., 1985; Shear et al., 1996). Winged insects diverged from apterygotes probably in the Upper Devonian or early Lower Carboniferous, although no record of this event is known from the fossil record. Pterygotes diversified by the Upper Carboniferous into about fifteen described orders (Wootton, 1981a, 1990; Kukalová-Peck, 1991). The earliest winged insect fossils associated with this diversification date from approximately 325 MYBP (Nelson and Tidwell, 1987; Brauckmann, 1991). Thus, the paleontological record has a gap of approximately 55 million years during which no apterygote, pterygote, or transitional forms are known.

The biotic setting for the late Paleozoic radiations of winged insects was distinctly different from the contemporary biosphere (Shear, 1991). Endotherms and angiosperms had not yet evolved. Botanical communities, although primarily composed of vascular plants, were floristically and structurally different from vegetation today. By the late Lower Devonian (~390 MYBP), a terrestrial branching cover of about 1 m in height comprised primarily fernlike lycopsids and psilopsids (Gensel and Andrews, 1984). Evolution of such vascular communities continued through the Middle to Upper Devonian to yield arborescent (treelike) forms of lycopods, sphenopsids, and progymnosperms (Chaloner and Sheerin, 1979; Stewart, 1983; Raven, 1986; Kenrick and Crane, 1997). The diminished stature (< 10 m) and reduced stem densities of such forests suggest that wind speeds were probably higher within vegetational canopies than is the case today (Zalessky, 1949, 1953). Progymnosperm canopies of up to 10 meters in height were likely by the early Carboniferous. By the Upper Carboniferous, some lycopods may have attained heights of 40–50 m (Stewart, 1983). Terrestrial vegetation during the initial period of pterygote evolution thus had a diversity of arborescent forms that potentially provided both geometrical complexity of habitat as well as nutritional resources in the form of photosynthetic and reproductive structures.

Arthropod use of plants for food, shelters, and additional ecological purposes likely began with the initial emergence of plants onto land. Trophic relationships between terrestrial arthropods and vascular plants were established as early as the Upper Silurian (412 MYBP; see Edwards et al., 1995) and Lower Devonian (Banks and Colthart, 1993). Fossil evidence indicates that apterygote insects fed on plants in the Lower Devonian, with these interactions intensifying in the Upper Devonian and through the rest of the late Paleozoic (Scott et al., 1992; Labandeira, 1997). By the Carboniferous, many winged insects had become specialized feeders on stems and plant reproductive structures (Smart and Hughes, 1972; Kevan et al., 1975; Scott and Taylor, 1983; Taylor and Scott, 1983; Scott et al., 1985; Shear and Kukalová-Peck, 1990). Carpenter (1971) indicates that approximately one-half of the described Carboniferous insect fauna had haustellate (sucking) mouthparts used to feed from plant stems, branches, and reproductive structures elevated above the ground. Most insect fossils from mid-Carboniferous ecosystems are known from tropical swamps or adjacent lowlands (Wootton, 1981a) that were likely characterized by elevated air temperatures and high relative humidity. Within these habitats of pterygote diversification, treefalls and associated gaps in the canopy would have generated substantial microclimatic heterogeneity. Reduced humidity, enhanced solar radiation, and increased wind motions are characteristic of such canopy openings. Microclimatic gradients would also have existed across gaps into the closed understory, as well as vertically within vegetational canopies. Early winged insects could thus have utilized diverse structural, thermal, and wind regimes at various spatial scales within their terrestrial habitats.

As with floristic composition and forest structure, the abiotic features of the Paleozoic may have been equally divergent from present-day conditions (Vermeij, 1995). Atmospheric concentrations of carbon dioxide throughout much of the Devonian and Carboniferous were

The possibility that variable oxygen concentrations characterized Paleozoic times, however, suggests a mechanistic hypothesis underlying the evolution of increased body size (Graham et al., 1995). The most immediate physiological effect of an increased oxygen partial pressure is to increase diffusive flux in the tracheal system of insects and many other terrestrial arthropods. As discussed in section 4.1.1, geometrical limits to insect body size imposed by tracheal diffusion will shift upwards as oxygen partial pressure of the atmosphere increases (Graham et al., 1995). This rather obvious possibility has received remarkably little attention in the literature, although Rutten (1966), Schidlowski (1971), Tappan (1974), and Budyko et al. (1987) all briefly suggested that gigantism in the Upper Carboniferous Protodonata would have been associated with a hyperoxic atmosphere (see also Harlé and Harlé, 1911, for a similar density argument). Moreover, a secondary peak of insect gigantism (e.g., among the ephemeropteran family Hexagenitidae; see Carpenter, 1992) appears to occur in the Cretaceous when the atmosphere was similarly hyperoxic.

Because changes in atmospheric oxygen concentration must impinge on all terrestrial organisms, gigantism through mitigation of diffusional constraints might be expected in a diversity of taxa. A compelling feature of the hyperoxic atmosphere model is the widespread taxonomic distribution of arthropod gigantism in the late Paleozoic (Briggs, 1985; Graham et al., 1995). Although hyperoxia does not exclude other potential factors that might select for gigantism (e.g., predator-prey interactions), the causal mechanism of diffusive enhancement would be experienced globally and would be independent of phylogenetic association. Also consistent with the hypothesis of diffusive constraints on oxidative metabolism is the winnowing of giant terrestrial arthropods in parallel with the increasingly hypoxic conditions of the late Permian (Graham et al., 1995; see fig. 6.1A). Most characteristically, the various insect taxa that attained exceptionally large body sizes during the Carboniferous do not persist after the Permian (Carpenter, 1992). Virtually no information is available on the physiological characteristics of extant winged insects under conditions of hyperoxia, although the tracheal system of larval insects exhibits considerable ontogenetic responsiveness to both hypo- and hyperoxia (Loudon, 1988, 1989; Greenberg and Ar, 1996). Accordingly, experimental evaluation of maximum flight performance in adult insects grown up under varying conditions of oxygen availability would be informative. Because aerodynamic force production and the Re during flight are influenced by density and object size, the indirect biomechanical effects of Carboniferous hyperoxia on insect flight performance may also have been significant and warrant experimental investigation (section 6.1.3).

TABLE 6.1
Historical Origination and Extinction of
Pterygote Orders

Geological Period	Ordinal Appearances	Ordinal Extinctions
Quaternary	4	0
Tertiary	4	0
Cretaceous	2	0
Jurassic	1	1
Triassic	4	2
Permian	11	7
Upper Carboniferous	10	0

Source: Data from Carpenter, 1992.

Following the origin and subsequent diversification of the Paleoptera and the exopterygote Neoptera, various neopterous endopterygote lineages appeared by the Upper Carboniferous (~300 MYBP; Labandeira and Phillips, 1996a) and radiated extensively through the
remainder of the Paleozoic. End-Permian insect extinctions were gradual but severe, with eight of twenty-seven orders disappearing altogether and major reductions in at least ten other orders (Sepkoski
and Hulver, 1985; Labandeira and Sepkoski, 1993; see table 6.1). The
remainder of the Phanerozoic (Mesozoic–Recent) has been marked by
further ordinal-level innovation (most notably the Diptera and Hymenoptera in the Triassic, as well as the Lepidoptera in the Jurassic), although the rate of ordinal appearance has declined substantially relative to the remarkable Carboniferous radiations (table 6.1). By contrast,
familial-level rates of origination have been fairly constant since the
mid-Carboniferous (Labandeira and Sepkoski, 1993). Most major insect innovations (e.g., wings, lateral wing flexion, and complete metamorphosis) appeared by the late Permian, with subsequent taxonomic
and morphological diversity then accruing at fairly constant rates to
yield the contemporary biota. Wing acquisition and early evolution of
pterygote body forms occurred in the now distant Paleozoic, rendering evidence on the mechanisms and pathways of wing evolution in
insects difficult to obtain.

6.1.2 Morphological Origin of Wings

Insect wings appear to have evolved only once, and all pterygote insects thus derive from a common wingless ancestor. This assumption
of wing monophyly is widespread among entomologists (e.g., Boudreaux, 1979; Hennig, 1981; Kristensen, 1989, 1991, 1997; Brodsky,

1994; see, however, Lemche, 1940; Manton, 1977; Matsuda, 1981), but no evolutionary analysis, cladistic or otherwise, evaluates this hypothesis directly. A general question relating to wing origins concerns the possible evolution of novel winglike structures, as opposed to substantial modification of preexisting morphological features (Wootton, 1986). A parsimonious perspective on the evolution of insect flight would seem to prefer the latter case. Evolutionary modification of gills, legs, or other such projections to serve in flight would appear to be simpler and more likely to occur than would be the de novo appearance of an aerodynamically effective winglet or wing. Direct proof that evolution acts parsimoniously remains contentious, but this assumption remains a useful working tool when evaluating complex evolutionary scenarios.

Because insect wings are not homologous with limbs as in vertebrate flapping flight, avid attention has been focused on the likely morphological genesis of such novel locomotor structures. The functional analysis of flight by protopterygotes does not necessarily require wing origins to be identified. Nonetheless, ontogenetic and phylogenetic perspectives help to evaluate more accurately those selective forces that may have acted on transitional flying forms. Proposed homologies between the insect wing and polychaete parapodia (Goldschmidt, 1945; Raw, 1956) are unlikely for a variety of reasons (see Snodgrass, 1958). Instead, two hypotheses dating from the nineteenth century dominate contemporary theories of wing origins. Oken (1809–1811) initially suggested that the branchial lamellae or related gill structures of aquatic insect larvae were homologous with adult wings. Larval gills insert laterally on the pleuron and are heavily supplied with veins (tracheation) for the uptake of oxygen. Gegenbaur (1870) and Woodworth (1906) subsequently elaborated on this theory deriving ancestral wings from tracheated pleural extensions. By contrast, Müller (1873, 1875) pointed out that paranotal thoracic lobes in extant Isoptera are not tracheated and could not have originated as gills. In the contrasting paranotal theory of wing evolution, wings derive ancestrally from fixed lateral outgrowths of thoracic and abdominal segments (see Crampton, 1916; Forbes, 1943; Hamilton, 1971; Rasnitsyn, 1981; Quartau, 1985, 1986; Wootton, 1986; Bitsch, 1994). These lateral lobes were initially nonarticulated, in contrast to the initial winglet mobility usually assumed by proponents of the pleural hypothesis.

The pleural and notal theories of wing origins have classically differed by assuming aquatic and terrestrial origins for protopterygotes, respectively. A terrestrial origin for winged insects is supported by most available biological evidence, particularly that relating to the genesis and function of the tracheal system (Little, 1983; Resh and

Solem, 1984; Messner, 1988; Pritchard et al., 1993; see, however, Kuka-lová-Peck, 1978, 1983; Toms, 1984). Use of air-filled tracheae in respiration is a classic adaptation to life on land, and has evolved repeatedly in diverse and unrelated arthropod taxa (Ripper, 1931; Gilyarov, 1970; Boudreaux, 1979; Little, 1983, 1990; Gordon and Olson, 1994). By contrast, aquatic larvae of extant paleopterous orders (Ephemeroptera, Odonata) are often cited in support of pleural origins for wings from gills (e.g., Wigglesworth, 1976; Kukalová-Peck, 1978, 1983). Cladistic analysis for trait directionality of aquatic larvae has not been applied to the Insecta, but secondary acquisition of aquatic habits is the most likely explanation for larvae in such neopterous orders as Megaloptera and Trichoptera, and particularly in the various Coleoptera and Diptera that inhabit freshwater either as larvae or adults (Handlirsch, 1910; Hennig, 1981; Pritchard et al., 1993). Also, the morphological and physiological radiations of tracheal gills within extant Ephemeroptera and Odonata larvae suggest multiple secondary origins and not derivation from a common aquatic ancestor (Hinton, 1968; Pritchard et al., 1993). Such considerations suggest that many insect larvae, although often aquatic, acquired this habit subsequent to the origin of wings. As a consequence, unique homologization of pterothoracic structures with the diversity of tracheal gill forms in extant taxa is unlikely.

Additional phylogenetic perspectives further support this conclusion. Thysanurans are the sister taxon to winged insects (see fig. 1.1) and are primarily if not exclusively terrestrial, as are all other extant noninsect hexapod taxa. Protopterygote terrestriality does not strictly preclude a pleural origin for wings, however; Kingsolver and Koehl (1994) suggested that both traits could have been acquired by hexapods prior to the divergence of the Pterygota from the Thysanura. If terrestriality did indeed characterize protopterygotes, then wing evolution from lobed paranotal extensions would require both subsequent lobe expansion and the development of notal articulation. Acquisition of aquatic larvae must also have occurred early in pterygote evolution, as indicated by the presence of this trait in the phylogenetically basal orders Ephemeroptera and Odonata. In insects generally, larval life histories tend to become more diverse as a greater range of aquatic substrates becomes oxygenated. Particularly intriguing in this regard, enhanced oxygenation of Devonian and Carboniferous freshwaters by atmospheric hyperoxia would have facilitated aquatic invasions by diverse terrestrial animal taxa, including arthropods (Graham et al., 1995). This mechanism for secondary acquisition of aquatic larvae by protopterygotes may have acted in parallel with promotion of flight performance by a hyperdense and hyperoxic atmosphere (fig. 6.1).

However, evidence that larvae of the now extinct Protodonata (ancestral to the Odonata) and of Paleozoic Ephemeroptera were aquatic is in fact equivocal (Wootton, 1988). Similar ambiguity characterizes interpretation of the putatively aquatic nymphs of the order Monura, an ancestral and now extinct apterygote taxon. As with many contemporaneous adult insects, Carboniferous and Permian insect larvae exhibited gigantism relative to extant forms. Increased atmospheric oxygen levels would, of course, have facilitated gigantism in either terrestrial or aquatic larvae of winged insects. Relative to aquatic and neutrally buoyant forms, however, terrestrial insects demonstrate systematic changes in leg size, geometry, and associated musculature to compensate for gravitational loading (Mamayev, 1977). By comparing thoracic and leg dimensions of aquatic pterygotes (and especially of larval Odonata and Ephemeroptera) with those of extant terrestrial apterygotes as well as with fossil larvae of Paleozoic insects, some indication of the likelihood of aquatic life habits in ancestral pterygotes might be obtained. Evaluation of the extant ephemeropteran subimago would be particularly informative in this regard; this developmental instar is transitional between aquatic and terrestrial habits, but is much less flight-capable than the adult imago.

Both larvae and adults of ancestral winged insects probably expressed lateral lobed structures on thoracic and abdominal segments. Although no paleontological evidence directly supports this assertion, prothoracic lobes on winged Carboniferous Paleodictyoptera and Protorthoptera (Carpenter, 1966; Wootton, 1981a), as well as thoracic and abdominal lobes on various Paleozoic insect larvae (Kukalová, 1968; Kukalová-Peck, 1978), provide indirect supporting evidence. In the development of extant pterygotes, homeotic genes typically suppress rather than promote expression of wings on pterothoracic segments, consistent with ancestral presence of serially arranged lobes or protowings (Carroll et al., 1995). Pleural anatomy of the pterygote prothorax is similar to that of the two pterothoracic segments (Snodgrass, 1935), suggesting ancestral similarity among the three thoracic segments with subsequent suppression of lateral lobes from the prothoracic segment. Consistent with this hypothesis, aberrant expression of prothoracic lobes occurs in mutants of some extant winged insects (e.g., Herskowitz, 1949; Ross, 1964). Serially homologous lobes would thus seem to have been present on the three thoracic segments and on many if not all of the abdominal segments of protopterygotes.

Indirect support for the notal origin of wings also comes from the observation that similar extensions are widespread in the nonhexapodan Arthropoda. Lateral lobes are present in trilobites, crustaceans, and myriapods, and in the latter two taxa can be partially articulated (Sharov, 1966; Boudreaux, 1979). Remarkably, hinged and tracheated notal extensions have evolved in a lineage of oribatid mites, but serve no aerodynamic function (Jacot, 1935; Woodring, 1962). Among hexapods, Archaeognatha and Thysanura express lateral flanges on all three thoracic segments, as do various Paleozoic winged insects and some extant nymphs of Ephemeroptera and Plecoptera. Moreover, tracheation of paranotal lobes in extant Thysanura resembles the tracheal patterns of winged insects (see Grassi, 1888; Šulc, 1927). Lateral lobes are initially presumed to be nonarticulated in the paranotal theory of wing origins, in contrast to mobile gills or gill-associated structures of the pleural hypothesis.

However, thoracic deformation by leg musculature can potentially induce notal displacement and motion of paranotal lobes. In extant apterygotes, dorsoventral muscles insert on the notum similar to the action of direct flight muscles in winged insects (see Tiegs, 1955; Matsuda, 1970; Barlet, 1974). Mobility of paranotal lobes might thus emerge indirectly from action of leg musculature acting on the thorax (Becker, 1952; Ewer, 1963; Ewer and Nayler, 1967; see also Weber, 1924a,b). For diverse arthropods, Becker (1958) emphasized the significance of dorsoventral musculature associated with leg motions that also inserts basally on winglike notal projections. At least in locusts, such dorsoventral muscles function both in wing flapping and in walking; bifunctionality in certain leg muscles may be widespread among winged insects (see section 2.1.2). Comparable action of the dorsoventral musculature has not been investigated in extant Thysanura, although indirectly induced movement of lobes in this taxon would suggest a mechanism for acquisition of paranotal mobility in protopterygotes. Intriguingly, thoracic segments of both Thysanura and Archaeognatha possess pleural apodemes, lateral thoracic sutures, and muscles that are probably homologous with the pleural suture and associated musculature of pterygotes (Barlet, 1950, 1988). At least morphologically, many of the elements necessary to effect notal deformation would appear to be present in apterygote insects.

By contrast, the pleural theory of wing origins proposes that wings are homologous with thoracic gills, gill plates, or related pleural structures in ancestrally aquatic hexapods. After approximately a century of disrepute, the pleural theory was revived by Wigglesworth (1973) with a suggestion that wings are homologous with ancestral leg exites, the lobes projecting from individual leg segments in many arthropods. In particular, Wigglesworth (1973) proposed that movable gill plates and ultimately wings derived from thoracic styli (small cuticular projections) in an aquatic apterygote. Thus, such protowings might initially be articulated and serve some respiratory function in water. Styli on leg segments of extant apterygotes were cited by Wigglesworth (1973) in support of this pleural hypothesis, the terrestrial habits of these taxa notwithstanding. For example, Archaeognatha possess usually one or two styli on each coxal leg segment, as well as on most abdominal segments; styli are also present on pregenital abdominal segments of Thysanura (Paclt, 1956; Wygodzinsky, 1987a,b). Additional morphological possibilities include evolutionary transformation of spiracular flaps or of gill covers into wings (Bocharova-Messner, 1971; Leech and Cady, 1994). For example, tracheated abdominal gill covers are common in extant mayflies and superficially resemble insect wings (Woodworth, 1906). Although possibly ancestral and homologous to wings, gills and gill covers may of course be secondarily derived in those extant forms with aquatic larvae. Venational patterns of tracheal gills, for example, differ substantially from those of wings in extant insects (see Crampton, 1916).

Even more elaborate anatomical scenarios have been constructed to explain the origin of wings or winglike structures from leg-associated regions. Kukalová-Peck (1983) proposed the ancestral presence of an additional leg segment, the epicoxa, which ultimately formed the axillary apparatus at the base of the wing. The wing itself might then be derived from the exite of this conjectural segment (see also Kukalová-Peck, 1992, 1997a, b; Czachorowski, 1993). Alternatively, both the exite and endite (the middle lobe) of this epicoxal segment could have contributed to wing formation (Trueman, 1990). The presence of abundant chemoreceptors on the wings of extant insects is also consistent with leg-associated wing origins (Dickinson, Hannaford, and Palka 1997). Numerous developmental and morphological complications are obviously associated with the appearance of an additional leg segment and subsequent incorporation of that segment or its projecting lobes into the lateral body wall. Various ontogenetic studies do suggest, however, that the notum incorporates pleural components during wing development (see Bocharova-Messner, 1959; 1968; Matsuda, 1981; Rasnitsyn, 1981). Wing primordia in mayflies are initiated in the embryonic pleuron and are subsequently incorporated as paranotal extensions (Becker, 1954, 1956). Similarly, wing primordia in dipterans are located embryologically within developing leg tissue (Anderson, 1963; Wieschaus and Gehring, 1976; Williams and Carroll, 1993). Diverse morphological contributions to the adult wing apparatus are thus likely, rendering a strict dichotomy between pleural and notal wing origins of limited utility unless ontogenetic trajectories are more rigorously described.

Additional neontological evidence suggests, however, involvement of leg-related structures in wing ontogeny. Many genes involved in developmental control are common between insect wings and legs (Campbell et al., 1993; Cohen et al., 1993). Averof and Cohen (1997) specifically demonstrated that two genes involved in the formation of crustacean epipodite (i.e., the exite of the coxal leg segment) are homologous to those involved in pterygote wing development. Furthermore, Averof and Cohen (1997) suggested that this homology of gene action demonstrated ancestrally aquatic habits in insects. Although consistent with the possibility that wings evolved from the exites of aquatic forms, such demonstration of gene homology neither uniquely determines structural homology (see Shubin et al., 1997) nor evaluates the possibility of gene co-option during arthropod evolution, for which purposes a detailed comparative study would be necessary. Numerous developmental genes (e.g., *fng, hedgehog, engrailed*) are expressed in various appendages and in axial tissues of both vertebrates and invertebrates, but such genetic commonality does not by itself demonstrate homology among structures (Gaunt, 1997). Exclusion both of the Apterygota and of Myriapoda (a possible sister taxon of hexapods; see section 1.1.1) from the study of Averof and Cohen (1997) is particularly puzzling in this regard. Phylogenetically rigorous studies of those genes controlling wing expression may, however, shed further light on possible associations between insect wing and arthropod leg development.

Frequently cited in support of pleural wing origins are studies of thoracic interneurons involved in flight rhythmogenesis. Morphological and physiological homologies among certain flight-related ganglial interneurons of Ephemeroptera, Odonata, and Orthoptera are consistent with monophyly of flight in winged insects (see Robertson et al., 1982; Dumont and Robertson, 1986; Robertson and Olberg, 1988). Similarly homologous interneurons arise ontogenetically in the first abdominal segment and then fuse with the metathoracic ganglion in these taxa. The presence of such thoracic and abdominal interneurons may indicate ancestral neural control of articulated wings arranged serially along body segments of protopterygotes. Equally possible, however, is that these interneurons served in neuromuscular processes (e.g., running with associated leg and abdominal motions; Ewer, 1963) that existed prior to protowing oscillation. Interneuronal control of muscles in serially homologous body segments is likely to be an ancestral trait of basal hexapods, if not of arthropods generally (see Kutsch and Breidbach, 1994). In this regard, identification and physiological evaluation of interneuronal homology in the thorax and abdomen of extant apterygotes would be highly informative for comparison with results from winged insects.

A necessary condition of the pleural theory is that paranotal lobes of extant basal hexapods, including the fixed wing buds of mayfly and dragonfly nymphs, represent independent secondary fusion events between articulated protowings and the notum (see Kukalová-Peck, 1978, 1987, 1997). Regardless of ontogenetic origins, protowings on thoracic segments are homologized in the pleural theory with abdominal tracheal gills or gill plates of contemporary mayfly nymphs. Contrary to predictions of the pleural theory, however, tracheal gills in extant Ephemeroptera appear to be notal in embryonic origin, ambiguities of anatomical specification notwithstanding (Dürken, 1907, 1923; see also Matsuda, 1981; Carle, 1982). Similarly, morphological studies of Ephemeroptera indicate that muscles of tracheal gills and of the wings are functionally convergent but are not homologous (see Jaworowski, 1896; Heymons, 1899; Dürken, 1907; Börner, 1908; Kluge, 1989). In support of the pleural theory of wing origins, Kukalová-Peck (1978) has suggested that various Paleozoic insect nymphs possessed hinged wings homologous with ancestrally articulated gills along the abdomen and thorax. Although widely accepted, no direct evidence supports this hypothesis; current reconstructions of the nymphal wing base are largely conjectural and badly warrant empirical validation.

To this end, three-dimensional structure and articulation of the wing base in fossil specimens can potentially be evaluated quantitatively using laser scanning methods. This approach uses reflection of laser light from a substrate to reconstruct three-dimensional surface contours at very small spatial scales. The primary indirect evidence for articulation of Paleozoic nymphal wings is an abundance of isolated nymphal wings at several fossil sites, suggesting detachment of these wings at a basal articulation (Kukalová-Peck, 1978). However, this taphonomic hypothesis implicitly assumes that nonarticulated insect wings or notal expansions would, by contrast, remain attached to the thorax during fossilization. Experimental taphonomy with extant dragonfly or mayfly nymphs could be used to test this possibility. Mobility of Paleozoic nymphal wings may also represent a derived condition characteristic of both paleopterous and neopterous orders. Given both unknown taphonomic effects and uncertainty of trait directionality, the issue of winglet mobility in Paleozic nymphs is best regarded as unresolved (see also Wootton, 1981a).

Although rigorously cladistic analyses of hexapod terrestriality are not available, parsimonious evaluation would suggest terrestrial habits in protopterygotes (Bitsch, 1994). By contrast, the morphological origins of wings remain obscure. Comparative evidence from diverse arthropods demonstrates the possibility (but not likelihood) of winglets and ultimately wings evolving from paranotal extensions. In this case, winglet mobility was not ancestral but rather derived in protopterygotes, possibly originating from the action of dorsoventral leg muscles that indirectly induced paranotal deflection. Alternatively, wings may have originated as ancestrally mobile structures such as gills, gill plates, leg exites, or leg styli. Styli on the legs of extant apterygotes present an excellent experimental system with which to test associated hypotheses of aerodynamic utility in such structures. Given the preexisting mobility of the leg and related pleural structures, this pathway for acquisition of wings would seem more parsimonious than an ancestrally stationary wing. Such scenarios presuppose, however, evolutionary fusion of such mobile morphological elements with the body wall and ultimately with the notum. Substantial anatomical and developmental challenges must be overcome in any such evolutionary transformation. Unfortunately, no existing morphological, physiological, or phylogenetic data are sufficient to determine unequivocally the ancestral mobility of winglets in protopterygotes.

Given the necessarily tentative nature of any conclusion concerning wing origins, the most productive line of future enquiry would undoubtedly be to evaluate Upper Devonian and Lower Carboniferous fossil deposits for relevant empirical evidence (Quartau, 1986). Such sites, although indeed rare, do not seem to have been systematically evaluated from the perspective of insect evolution. Finally, Kukalová-Peck (1983) argued that derivation of vertebrate wings from limbs must indicate a functional association between insect wings and ancestral leg segments or other features of the pleuron. This suggestion is, however, confounded by phylogenetic associations of exo- and endoskeletons in arthropods and vertebrates, respectively. Exoskeletal projections such as wings can be remotely activated via indirect cuticular deformation, whereas limbs require internal insertion of musculature to effect mobility. A direct comparison of flight mechanics in vertebrates and insects is therefore not germane from an evolutionary perspective.

6.1.3 Aerodynamic Analysis

Independent of morphological origins, insect wings ultimately serve an aerodynamic function. One immediate prediction of the paranotal theory is that serial winglets enhance gliding performance (Packard, 1898), an effect likely advantageous during aerial escape from predators. Laterally arranged appendages such as gills and limb exites would potentially have the same biomechanical effect. Knowledge of body size and likely gliding airspeeds is fundamental to any analysis of protopterygote aerodynamics, but no guidance is provided by the Paleozoic fossil record on this point. Giant winged insects of the Upper Carboniferous, for example, occur much later than the period relevant to early flight evolution. Thoracic fragments of the Devonian archaeognathan apterygote described by Labandeira et al. (1988) suggest a total body length of 1–5 cm. Flower (1964) used a body length of 1 cm for theoretical calculations of gliding performance in ancestral pterygotes. Flight of much smaller insects would have been dominated by ambient air motions, whereas much larger protopterygotes might have had deleteriously high impact velocities if gliding at equilibrium speeds. In general, an ancestral body length of 2–4 cm seems likely for winged insects (Wootton, 1976).

During the early evolution of pterygotes, winglets would necessarily have been small relative to body dimensions, and gliding performance would have been dominated by aerodynamic properties of the body alone. Using a resolved flow analysis for cylinders (see Ellington, 1991a; section 3.2.1.1), Flower (1964) estimated lift and drag forces on cylinders gliding at various angles relative to horizontal. This analysis suggested that gliding could be effective with a cylindrical body form alone, and that protopterygotes could have effectively glided without winglets. For cylinders of varying length : diameter ratios, Ellington (1991a) used the resolved flow method to determine that glide angle decreases at higher Re, and that glide speeds increase with cylinder length. By implication, larger protopterygotes would have glided at higher Re and at lower glide angles. Equilibrium glide speeds predicted by Flower (1964) and Ellington (1991a) ranged from 5 to 15 m/s, values that are relatively high compared to flight speeds of most extant insects (Kingsolver and Koehl, 1994; see section 3.1.1).

Empirical measurements of lift and drag have shown that addition of winglets to protopterygote body forms can substantially enhance steady-state gliding performance. Kingsolver and Koehl (1985) carried out aerodynamic studies on protopterygote models that were 2–10 cm in length. These models had lateral thoracic winglets of variable length but no abdominal winglets; thoracic winglets were uncambered and devoid of venation or other surface detail. Lift and drag measurements were made on these models at airspeeds in the range of 0.1–2.5 m/s, values substantially lower than the equilibrium gliding speeds predicted by Flower (1964) and Ellington (1991a). The steady-state aerodynamic results of Kingsolver and Koehl (1985) demonstrate that gliding performance increases with winglet size, but that short thoracic winglets (< 20% body length) improve the lift : drag ratio only in relatively large protopterygotes. Somewhat surprisingly, short winglets have little effect on drag during parachuting with the longitudinal body axis oriented orthogonally to relative airflow (see also Ellington, 1991a; Kingsolver and Koehl, 1994). In general, winglets or other lateral projections may be presumed to have enhanced the gliding performance of protopterygotes, although the relative magnitude of this effect may be Re dependent.

Steady-state aerodynamic analyses of protopterygotes necessarily assume invariant gliding trajectories. By contrast, insects that jump and subsequently glide will show an initial descent and rapid speed increase under gravity prior to establishment of an equilibrium glide. If predatory escape initially motivated evolution of flight (see below), then selection may not have acted on gliding per se but rather on the initial aerodynamics of takeoff (Edwards, 1992). The lifting characteristics of winglets in such accelerating flows are unexplored. Alternatively, winglets may have increased total drag during parachuting escape and only secondarily obtained lifting characteristics once suitably enlarged (Ellington, 1991a). The increased air densities and higher Re ensuing from elevated Paleozoic oxygen concentrations (fig. 6.1B) would also have enhanced lifting performance of protopterygote winglets. Even very small paranotal extensions could potentially have conveyed selective aerodynamic advantage during takeoff and/or subsequent gliding. In fact, low aspect ratio wings may yield optimal aerodynamic performance for gliders operating at the Re relevant to pterygote evolution (see Ennos, 1989a). The present-day restriction of wing flapping to adult insects is likely a derived state, whereas selection for jump-initiated flight in juvenile forms may have been ancestrally sufficient to foster the evolution of wings. The aerodynamic effects of ontogenetic variation in body Re and winglet expression could be substantial and represent an unexplored domain for experimental studies of flight evolution. The effects of low-amplitude winglet oscillation on aerodynamic force production are similarly unknown.

Existing theoretical and experimental results may also underestimate the aerodynamic capabilities of ancestral winged insects. Extant apterygotes, for example, are not cylindrical but rather are flattened if not partially cambered dorsoventrally. The external anatomy of apterygotes might therefore be used to model more realistically the body dimensions of protopterygotes. A diversity of morphological features potentially contributes to protopterygote aerodynamics and deserves functional investigation (table 6.2). Winglets, however variable in shape, camber, and relative area, can be evaluated in different serial arrangements along thoracic as well abdominal segments. Alternatively, the aerodynamics of extant apterygotes can be studied empirically. Lateral lobing of Thysanura, for example, can be augmented or

TABLE 6.2
Morphological Features of Protopterygote Design That Potentially Influence
Flight Performance

Body:	Length
	Width/taper
	Relative dimensions of head/thorax/abdomen
	Extent of dorsal convexity/ventral concavity
Winglets:	Number
	Location (thorax and abdomen)
	Area
	Planform
	Camber
	Elevation above/below horizontal
	Rotation about longitudinal axis of winglet
	Flexibility
Abdominal cerci:	Length
	Diameter
	Orientation

reduced to assess effects on both steady gliding performance and take-off jumps. One useful experimental approach would be to evaluate such aerodynamics in gas mixtures corresponding physically to those suggested by a late Paleozoic oxygen pulse (fig. 6.1); both density and Re changes associated with variable atmospheric composition are potentially relevant to flight performance. Through systematic variation of winglet and body configurations, aerodynamic characteristics can be determined for a range of protopterygote morphologies. Such an exercise, in addition to addressing the biomechanical feasibility of specific evolutionary scenarios, would also permit quantitative evaluation of relevant fossil material.

In addition to the production of aerodynamic forces, maintenance of stability is a fundamental requirement for the evolution of flight (Maynard Smith, 1952). If protopterygotes expressed winglets from abdominal as well as thoracic segments, then a nose-down rotational torque would arise from lift forces acting on abdominal winglets and the associated moment arm about the center of body mass. Stability in pitch would then be particularly sensitive to the magnitude and location of such lift forces (Ellington, 1991a). This prediction is consistent with the empirical results of Wootton and Ellington (1991), who evaluated the gliding performance of protopterygote models endowed with both thoracic and abdominal winglets. Steady glides could be obtained from such models, but stability in pitch was critically dependent on the orientation of the abdominal winglets. As noted by Wigglesworth (1963) and Flower (1964), elongated abdominal cerci or filaments could also serve to maintain stability in pitch as well as in yaw. Starting from a condition of stable gliding, transient pitch or yaw would generate dorsoventral and lateral forces, respectively, on such structures. These forces would then act through the elongated moment arm to create torques that would restore the initial body orientation relative to flow. Stability in both yaw and roll may also have derived from positive dihedral angles of winglets along the body (see section 5.2.2). Empirical observations on gliding insect models confirm that roll was probably not a major problem for protopterygotes (see Ellington, 1991a; Wootton and Ellington, 1991).

A final consideration in flight stability concerns effects of winglet mobility. As suggested by various authors (e.g., Haupt, 1941; Wigglesworth, 1963; Flower, 1964; Smart, 1971), winglet motions could be used to control body orientation during flight and while landing. In extant saltatorial taxa, for example, body stability in pitch may be enhanced through bilaterally symmetric wing extension (Brackenbury and Wang, 1995). Asymmetric movements could also have been used to compensate for inadvertent body rotations, providing a motivation for enhanced winglet mobility and ultimately serving as a pathway for acquisition of symmetric wing flapping.

6.1.4 Selective Forces and Scenarios

Diverse hypotheses in the context of both natural and sexual selection have been put forward to explain initial evolution of winglets or wings (e.g., Wigglesworth and contributors, 1963). Although conceptually divergent, these various hypotheses are not mutually exclusive, and none preclude aerodynamic utility of even small protowinglets. Rarely are selective disadvantages of winglets considered, although costs of construction and increased apparency to predators are not negligible concerns. One possibility is that winglike appendages were used in epigamic (courtship) displays for sexual advertisement (Alexander and Brown, 1963; Alexander, 1964), or served as nuptial gifts for potential mates. In extant Thysanura, for example, courtship is close-range and based on a variety of male-initiated stimuli. Protowinglet motions of ancestral pterygotes may additionally have been used in displays and contests among competing males. The present-day restriction of wings to reproductive adult insects might suggest an initial role of protowings in sexual selection, but no evidence is currently available that addresses this hypothesis. Wigglesworth (1963) suggested that wings evolved to facilitate low terminal velocities during aerial dispersal by ambient winds. This possibility is unlikely given that the enhancement of such dispersal usually requires plumose drag-creating structures and not lift-generating wings (Norberg, 1972a; Wootton, 1986; Ellington, 1991a). Also, present-day entrainment of insects by winds and subsequent dispersal typically occurs at body sizes substantially smaller than the postulated sizes of protopterygotes (see section 7.4). Such small body sizes are almost certainly derived in insects, and in most cases are associated with the acquisition of asynchronous flight muscle (section 4.2.2). The enhancement of wind-borne dispersal thus seems an unlikely motivating force behind the initial evolution of insect wings.

Protowings could potentially have functioned in the thermal biology of ancestral pterygotes. Lateral surface areas on the thorax and abdomen could trap air between the insect and substrate, or act as heat exchangers between the insect and the surrounding air (Whalley, 1979). Douglas (1981) similarly proposed that thoracic lobing could advantageously increase body temperature by absorbing solar radiation and also by shielding the body against convective heat loss. Only secondarily would the winglets acquire useful aerodynamic characteristics. Such thermoregulatory hypotheses assume selective advantages to heat gain but exclude possibly deleterious effects of overheating. An increase in body temperature will also increase total metabolic expenditure in ectothermic insects. Kingsolver and Koehl (1985) used the same protopterygote models as in the aforementioned aerodynamic studies to show that short thoracic winglets could increase, via radiative heat gain and conduction to the body, the thoracic temperature relative to the surrounding air. Such short winglets, however, displayed only minimal contributions to the aerodynamic performance of the body in steady-state flow. By contrast, larger winglets would have enhanced aerodynamic performance of protopterygotes while contributing relatively less to the elevation of thoracic temperature. Kingsolver and Koehl (1985) proposed a gradual transition from thermal to aerodynamic roles in the evolution of longer winglets, thereby suggesting continuous advantages during the transformation from winglet to an actual wing with favorable lift characteristics. This functional transition was possibly mediated by isometric size change in protopterygotes themselves—larger insects would accrue progressively greater aerodynamic but lesser thermoregulatory advantages from winglets of the same shape and relative size.

The importance of winglets in altering protopterygote body temperatures, however, is not well established relative to alternative mechanisms of thermoregulation. Heinrich (1993) noted that even large insects can quickly warm up in sunshine because of relatively high ratios of surface area to volume. Small postural changes can rapidly alter convective heat flux, solar load, and the thermal equilibrium of insects. Potential mobility of winglets could also behaviorally influence rates of convective heat loss from the body. Contemporary associations between flight and thermoregulation in winged insects (see section 4.3) are also highly derived and are not necessarily germane to the issue of initial winglet evolution (Heinrich, 1993). More directly relevant would be the demonstration of selective advantages to increased heating rates or higher body temperatures in contemporary apterygotes or in secondarily flightless pterygotes. For example, abdominal temperatures may potentially influence the jumping performance of apterygotes that use abdominal flexion to initiate a jump (see below). Interestingly, Kingsolver and Koehl (1985, 1994) found that abdominal winglets had little influence on the equilibrium thoracic temperature, although effects on equilibrium abdominal temperature were not investigated. Finally, future modeling of heat exchange in protopterygotes can explicitly incorporate changes in the specific heat and thermal conductivity of air as suggested by a late Paleozoic oxygen pulse (Graham et al., 1995). The increased air densities predicted for Devonian and Lower Carboniferous atmospheres (see fig. 6.1B), for example, would tend to promote heat loss from protopterygotes through increased thermal conductivity of air. Ambient air temperatures may also have declined steadily through the Devonian and Carboniferous (Berner, 1994), rendering morphological and behavioral adaptations for heat gain and thermoregulation potentially more advantageous.

These aerodynamic and thermoregulatory hypotheses for winglet functions presuppose terrestrial origins for ancestral insects. By contrast, most pleural hypotheses of wing origins require initial evolution in aquatic contexts and presuppose initially hydrodynamic functions for what ultimately became aerodynamic structures. Bradley (1942) and Mamayev (1977) also suggested a possibly amphibiotic character for protopterygote life in Carboniferous coal swamps. In such contexts, ancestrally mobile winglets could have functioned in air, in water, or at the air-water interface. Such a transition might have been mediated by an ontogenetic switch in habitats—aquatic larval forms could emerge for subsequent terrestrial molts to the adult imago (or to subimagines in transitional forms) that jump in response to predatory attempts (Hutchinson, 1993). This possibility would advantageously combine aerodynamic components of a terrestrial startle response with preexisting articulated larval structures, or with their homologues in the adults. Two observations, however, suggest that an evolutionary transition from initial flapping in water to use in air would have been biomechanically improbable. First, water and air differ approximately by a factor of eight hundred in density, and the *Re* of reciprocating appendages varies correspondingly. Aquatic gills or appendages may move more quickly if projecting into air, but the associated mechanics of muscular contraction are fundamentally mismatched. The very low oscillatory frequencies imposed by the high density of water, for example, would be of limited utility in air. Effective design for protowinglets moving in water thus suggests reduced locomotor effectiveness in air, and the possible functionality of intermediate designs is unclear.

Second and perhaps more important, generation of hydrodynamic drag as posited by aquatic theories of wing origin is fundamentally different from aerodynamic lift production at the *Re* typically assumed for protopterygotes. Use of appendages in drag-based locomotion requires flattened surfaces operating orthogonally to flow (as with gill plates of extant mayfly larvae); such structures are characteristic of many extant larval as well as adult aquatic insects. Airfoils, on the other hand, are most effective with cambered profiles that act at low angles of attack (section 3.2.1). At much lower *Re*, winglets of very small insects might function equally well in air and water, but miniature protopterygotes seem unlikely for reasons previously discussed. Thus, respiratory structures moving in water would be dynamically distinct from flapping winglets operating effectively in air; the nature of any transitional form has, appropriately, yet to be articulated in any general theory of aquatic protopterygotes. Forces of surface tension alone would present a formidable physical barrier to partial body emergence as well as to projection and oscillation of flattened respiratory structures.

A related scenario for the evolution of flight suggests that ancestrally aquatic protopterygotes could have used winglets in air either to drift passively or to skim actively along water surfaces (Marden and Kramer, 1994, 1995; see also Thomas and Norberg, 1996; Samways, 1996). Surface skimming by some extant plecopterans provides an interesting example of this locomotor behavior (Marden and Kramer, 1994, 1995). Because weight support is provided by surface tension, such surface skimming could represent a biomechanically intermediate stage in the transition from aquatic locomotion to free aerial flight. Neither surface skimming nor surface drifting require generation of the vertical forces that predominate in free flight (section 3.2.1), although such forces will be at least partially mitigated by the ground effect provided by proximity of the wings to the water's surface (see section 3.3.1). Using the example of an extant plecopteran genus, Kramer and Marden (1997) suggested that the vertical forces necessary for flight could be generated if upwards body flexion enabled an approximately horizontal stroke plane for wing motions. However, the occurrence of such force generation in present-day forms with flight-capable wings and axillary articulation does not address biomechanical feasibility for relevant protopterygote kinematics and morphology. Similarly, the plecopterans described by Marden and Kramer (1994, 1995) can surface skim with low or artificially reduced wing areas, but must raise and lower their wings to effect both drifting and surface skimming. Such a well-developed wing articulation that serves in flight is unlikely to have been found in protopterygotes, and the relevance of surface skimming and drifting to the pterothoracic mechanisms of ancestral insects remains to be demonstrated. Furthermore, the evolution of free-powered flight from surface skimming or drifting forms presupposes an earlier successful transition across the air-water interface. As discussed previously, the dual functionality of winglets in such a biomechanically discontinuous selection regime is unclear.

Phylogenetic analysis of surface skimming and drifting abilities is presently not available for basal insect orders, but this behavior is probably a derived rather than retained ancestral trait of winged insects (Will, 1995). The low environmental temperatures at which many adult Plecoptera are active, for example, suggests that surface skimming in some taxa is a derived locomotor mechanism associated with low muscle temperatures and correspondingly reduced power output. For example, nearly half of the surface-skimming individuals examined by Marden and Kramer (1994) were also capable of horizontal or ascending flight at air temperatures of 22°C. Interestingly, other examples of surface skimming can also be found in the present-day entomofauna. Among paleopterous insects, some mayflies skim over water following emergence, and certain coenagrionid dragonflies exhibit passive upstream drift using ambient winds (Samways, 1994). Adults of the chironomid fly genus *Clunio* utilize wing flapping to skim along water surfaces in a manner similar to that of aforementioned plecopterans (Oka, 1930; Tokunaga, 1932; see also Hashimoto, 1962, 1976). This behavior is clearly a derived trait for dipterans, as must be underwater wing flapping by some hymenopteran parasitoids (Lubbock, 1863; Matheson and Crosby, 1912) and surface paddling with the wings by certain trichopterans from Lake Baikal (Martynov, 1929, 1935; Kozhov, 1963) and by the chironomid *Pontomyia*. More generally, these examples of surface skimming, paddling, and underwater wing flapping by extant insects illustrate the need for detailed phylogenetic analysis if ancestral character states are to be surmised from contemporary locomotor behaviors (see also Ruffieux et al., 1998). Relative to early insects, all extant forms (from either paleopterous or neopterous orders) must be viewed as highly derived, particularly in light of the unrepresented transitional forms in the Upper Devonian and Lower Carboniferous.

Returning to terrestrial origins of flight, another classic explanation for the evolution of wings has been escape from predators on land. For example, predatory pressure imposed by arachnids could potentially select in hexapods for a variety of defensive adaptations, including jumping and subsequent gliding escape (Bristowe, 1958). Such escape behavior requires sensory perception of an approaching threat as well as neuromuscular activation of leg motions and/or flight. Extensive neurobiological studies support the ancestral presence of dedicated neural pathways underlying pterygote escape behavior. In extant Thysanura (Edwards and Reddy, 1986) and Blattaria (Ritzmann and Fourtner, 1980; Ritzmann, 1984), the displacement of mechanosensitive hairs on abdominal cerci evokes action potentials in giant interneurons that then initiate jumping or running responses. This startle response in ancestral apterygote insects was subsequently co-opted during pterygote evolution to stimulate jumping, wing flapping, and even evasive flight once airborne (see Edwards, 1985, 1992, 1997; Edwards and Palka, 1991; Ganihar et al., 1994; Kutsch and Breidbach, 1994; Libersat, 1994). Such neural associations between cursorial escape, jumping, and flight are clearly consistent with the hypothesis that terrestrial predation motivated jumping escapes and the initial evolution of flight in hexapods. Intriguingly, the tarsal reflex is apparently an ancestral pterygote trait (see section 5.2.1), further strengthening the neural linkage between jumping and wing flapping. Similar startle responses are widespread among animals (Eaton, 1984), and this pathway for the acquisition of flight may be more general. For example, many models of vertebrate flight evolution propose jumping to become airborne prior to the appearance of gliding and ultimately of wing flapping (see Norberg, 1990; Bennett, 1997).

If either paranotal lobes or pleural structures augmented the effectiveness of jumping escapes in insects, then enhanced aerodynamic performance would be strongly selected for in the context of predation. In contrast to theories of wing evolution that require aquatic protopterygotes and a discontinuous selection regime across the air-water interface, escape jumping in terrestrial protopterygotes would from the outset favor gradual and continuous improvement in lift-generating and ultimately flapping winglets. Ancestrally, the potential of selection to yield sophisticated jumping performance is exemplified by the hexapod taxon Collembola, otherwise known as springtails. Collembolans have a highly modified abdominal structure, the furcula, that deflects downwards and enables jumps with considerable horizontal displacement (see Brackenbury and Hunt, 1993). Both morphological and behavioral protoadaptations for jump-mediated glides are also evident among apterygote hexapods, the terrestrial sister taxon of the winged insects. As discussed in section 6.1.2, thoracic paranotal lobes as well as styli on the legs and abdominal segments of apterygotes could potentially serve to generate lift and to facilitate aerial escape. Jumping behavior is well developed in some extant apterygotes; the Archaeognatha (bristletails) can jump distances up to twenty body lengths. Mechanically, such jumps are initiated by dorsal flexion of the abdomen, followed by rapid ventral depression of the abdomen against a substrate (Evans, 1975). Contraction of dorsoventral oblique muscles induces potentially advantageous thoracic curvature during the jump, whereas postural adjustments of the body near the end of the jump are used to decrease the rate of descent. Jumping by protopterygotes may similarly have been characterized by active control of an otherwise ballistic trajectory, together with oscillation of winglets or styli through contraction of the preexisting dorsoventral segmental musculature.

We must also consider the historical context of early pterygote evolution that led to the intense predatory pressure by both invertebrates and vertebrates. A diversity of nonhexapodan insectivorous arthropods, particularly arachnids, was found in Devonian and Carboniferous terrestrial ecosystems (Rolfe, 1980, 1985; Shear and Kukalová-Peck, 1990). As with insects, Carboniferous hyperoxia facilitated gigantism in other arthropods by relaxing constraints on body size (Graham et al., 1995). For example, the aquatic eurypterids were typically 10–20 cm in body length, but many species attained lengths of 100 cm, with individual specimens of *Pterygotus* reaching at least 180 cm (Størmer et al., 1955). Similar giants can be found among terrestrial arthropleurids, diplopods, scorpions, and spiders (Kraus, 1974; Hünicken, 1980; Briggs, 1985; Shear and Kukalová-Peck, 1990). Late Paleozoic tetrapods were also large relative to extant forms. The giant Carboniferous amphibians were primarily insectivores and carnivores, as were many contemporaneous reptiles (Carroll, 1988; Behrensmeyer et al., 1992). An indirect consequence of hyperoxia-mediated gigantism in Carboniferous arthropods may have been the expansion of trophic resources for both vertebrate and invertebrate predators. This expansion in turn may have promoted coevolution between the late Paleozoic insectivores and their prey (Shear and Kukalová-Peck, 1990), particularly as gigantism can function defensively against predation (see Vermeij, 1987).

Although no existing data evaluate predation rates on late Paleozoic terrestrial arthropods, Shear and Kukalová-Peck (1990) noted various morphological characteristics among the early Upper Carboniferous entomofauna that are consistent with the hypothesis of predatory defense. One family of the extinct order Protorthoptera, for example, possessed well-developed prothoracic spines (Burnham, 1983). Extended abdominal cerci characterized many Carboniferous insect nymphs as well as adults of some orders (e.g., Paleodictyoptera), and probably functioned to detect approaching predators. The abdominal cerci as well as caudal filaments of these winged insects morphologically resemble antennae, and possibly served to deflect predatory attack away from the head. For example, caudal filaments and abdominal cerci are fragile in extant Thysanura, particularly at the base, and readily regenerate if broken (Kränzler and Larink, 1980). Such elongated posterior extensions may thus be functionally equivalent to the hindwing projections of lycaenid butterflies that resemble antennae and that misdirect predators (see Robbins, 1981). Comparable strategies using tail distraction and autotomy to confound predators occur in many lizards (see Arnold, 1984). Cryptility may also have been an important feature of protopterygote design, with both dorsoventral flattening and lateral lobing contributing to a reduction of shadows and thus visibility of the insect. Similarly, wing patterning that possibly served in disruptive coloration or mimicry was characteristic of the Carboniferous Paleodictyoptera (Carpenter, 1971). Also consistent with high rates of predation in late Paleozoic ecosystems are various antipredatory structures in contemporaneous terrestrial millipedes (Shear and Kukalová-Peck, 1990). Jumping in response to touch has even been recorded in several extant millipede species (Evans and Blower, 1973; see also Birket-Smith, 1984), although conglobation (rolling up into a ball) is the usual mode of defense.

Jumping hypotheses for the origin of flight in insects frequently presume that such jumps were directed vertically (e.g., Zeuner, 1940; Snodgrass, 1958). This perception undoubtedly arises from the anthropomorphic experience of moving against gravity, but is not universally true. For small arthropods within structurally complex environments such as vegetation, jumping to escape predators may occur at variable directions relative to gravity. The increasing arborescence and geometrical complexity of vegetational stands through the Devonian and into the Carboniferous would have provided a three-dimensional substrate suitable for a diversity of escape strategies, but particularly for jumping escapes and ultimately the evolution of flight in protopterygotes (Kevan et al., 1975). Lateral jumps from branches or leaves, and particularly jumps directed downward but with a horizontal component of velocity, would have enabled access to a diversity of potential refuges both within the vegetational canopy as well as on the ground. Simultaneous gravitational acceleration would also have yielded the increased translational velocities necessary for substantial buildup of lift on thoracic and abdominal projections. Greater horizontal displacement may also have yielded ecological advantages that ultimately selected for increased aerodynamic performance of protopterygotes. For example, access to suitable food, microhabitats, and oviposition sites for females would all be facilitated by an increased capacity for dispersal. Although such behaviors cannot be evaluated in protopterygotes, one might deduce their importance relative to escape from predators from the behavioral contexts of jumping by present-day apterygote insects.

Based on neural, morphological and biomechanical evidence, one can assume that the aerodynamic enhancement of jumping trajectories was probably an important avenue for wing evolution in protopterygotes. The direct link between aerial escape and Darwinian fitness suggests an important role for predator-initiated jumps in protopterygotes. Continuous improvement in jump performance and distance, together with active aerodynamic control of of the trajectory, could potentially select for improved winglets and ultimately for flapping wings. Falsification of this scenario, as of all historical hypotheses, is necessarily a challenging prospect, but substantial indirect support could be derived from manipulations with extant apterygotes. The experimental augmentation of lateral lobing, when coupled with the exposure of apterygotes to actual predators (e.g., centipedes, arachnids), might revealingly simulate late Paleozoic predation events. Such data would help to interpret the relevance of initial takeoff mechanics and associated aerodynamics for successful escape. High rates of failed predatory attack would also suggest strong selection for the evolution of jump-mediated escape, and might be discernible from the fossil record in the form of sublethal damage to wings, bodies, and appendages (see Vermeij, 1987). The complete absence of fossil material from the relevant period of early pterygote evolution unfortunately precludes such an analysis at present.

6.1.5 Evolution of Wingbeat Kinematics

Reconstructions of ancestral wingbeat kinematics are speculative but, depending on various morphological assumptions, can still be informative. If wing articulation is ancestral in pterygotes, as suggested by pleural theories of wing origins, then oscillatory kinematics would have preceded the capacity to fly. If wing articulation is derived, then gliding likely preceded the evolution of large-amplitude wing flapping

(Brodsky, 1994). Incipient flapping motion of paranotal lobes or leg-derived aerodynamic structures could have been induced, however, through the contraction of leg musculature. As with pleural hypotheses, this latter possibility would suggest winglet oscillation from the very outset of pterygote evolution. Cuticular elasticity that restored actively induced thoracic deformations may also have contributed to winglet motions. The loss of abdominal and prothoracic winglets probably occurred in parallel with increased winglet area on the meso- and metathoracic segments; stability in pitch was possibly retained through elongation of abdominal cerci and caudal filaments. At this early stage, pterygotes probably had two homonomous pairs of short wings that oscillated dorsoventrally (Brodsky, 1994). If wings were originally flapped through the action of leg musculature, then adjacent wing pairs probably operated out of phase, like the motion of adjacent ipsilateral legs in terrestrial hexapods. As with wingbeat frequency, stroke amplitudes of equally sized meso- and metathoracic wing pairs were probably comparable. Although such wings could not initially have been rotated about the longitudinal wing axis, inertial and aerodynamic forces would likely have induced limited rotational torques on the wings (see section 2.2.2).

Flapping of even small wings might have contributed to vertical force production in early pterygotes. Kinematic asymmetry between half-strokes is necessary for generation of net vertical forces in forward flight, and potentially derives from variable orientations and/or durations of the down- and upstroke (section 3.2.1.3). Brodsky (1994) assumed a high stroke plane angle, an equivalent half-stroke duration, and a horizontal body axis to model steady-state force production through incipient flapping motions. These assumptions rendered down- and upstroke motions dynamically equivalent, and Brodsky (1994) concluded that no net vertical force could be generated by the flapping wings. However, ancestral insect jumpers and gliders did not necessarily move in the air with the longitudinal body axis parallel to airflow. More likely was some positive body angle relative to flow as well as variable duration of half-strokes, resulting in different angles of attack and kinematic asymmetry between half-strokes. Because flow over low aspect ratio wings is influenced by the presence of the body as well as of adjacent winglets, however, steady-state aerodynamic analysis is inappropriate for the evaluation of incipient wing flapping. Similar calculations of flapping dynamics have been applied to the evolution of bird flight (e.g., Norberg, 1985), but uncertainty in many of the underlying assumptions suggests that empirical studies may be more relevant. For example, effects of body proximity and low aspect ratio wings on incipient flapping dynamics could be investigated experimentally through use of oscillating winglets attached to realistic apterygote models.

Wing-base reduction probably occurred early in pterygote evolution, facilitating active muscular control of wing orientation (see section 2.1.2). Similarly, chordwise asymmetry in wing venation (section 2.2.1) probably appeared as wings became progressively adapted for greater lift production and enhanced resistance to torsion (Haupt, 1941; Rohdendorf, 1943). Increased wing length and area would similarly have required greater stiffness at the wing base to prevent bending. Such morphological increases in wing surface area or advantageous changes in wing shape and camber would have yielded greater aerodynamic forces during gliding and low-amplitude wing flapping. Alternatively, activational changes in the wingbeat frequency and/or stroke amplitude could have increased relative air velocities of the wings and, assuming steady-state aerodynamic mechanisms, would have yielded disproportionately increased lift. For example, Voss (1913) hypothesized a general phylogenetic increase among winged insects in both wingbeat frequency and flapping amplitude. Evolutionary increases in either wing size or in kinematic activation would also have heightened effectiveness in torque generation when the wings were deployed asymmetrically. As wing lift increased, so too would profile and induced drag, requiring additional power expenditure and further investment in thoracic musculature. Wingbeat kinematics and wing geometry are biomechanically coupled in that an increased wing moment of inertia will reduce the flapping frequency if wing motions are influenced by mechanical resonance in the thorax. This decline then generates kinematic scope for a subsequent increase in wingbeat frequency that yields even greater forces of wing flapping. Isometric size increase causing a reduction in wingbeat frequency would have produced similar results. Such mechanical linkage and possibly positive feedback between changes in morphological and activational variables may thus have been an important feature in the evolution of flight kinematics.

Assuming that early pterygotes expressed winglike structures on all three thoracic segments, lift production may have been deleteriously influenced by vortex interaction between adjacent segments (Grodnitsky, 1995). In forward flight, spanwise vorticity is shed continuously from each wing. The associated vortex is of opposite sense to the circulation developed on more posterior wings, and thus interferes with the production of lift. Evolutionary modifications that have partially mitigated this interaction are loss of the prothoracic wings together with morphological reduction of one of the two remaining wing pairs (antero- or posteromotorism), or with physical coupling of the meso- and metathoracic wings (Grodnitsky, 1995). Insects that retain four uncoupled wings tend to operate wing pairs out of phase, thus reducing the impact of potentially deleterious interactions between adjacent wings. The phylogenetic distribution of these alternative strategies to reduce vortex interaction suggests multiple independent origins of reduction in the relative size of one wing pair (see fig. 2.4). Reduction of wing area is of course constrained by the overall demands of force production, and an increased wingbeat frequency may have been one response to such a constraint. Differentiation of pterothoracic segments as well as absolute frequency increases may thus have been established early in pterygote evolution. Consistent with this hypothesis, anteromotorism is characteristic of the most basal pterygote lineage (Ephemeroptera) and is the the most parsimonious reconstruction for the flight apparatus of ancestral winged insects (fig. 2.4).

An impressive feature of the Upper Carboniferous entomofauna is the apparently well-developed flight capacity in most orders. Wings were large relative to horizontally projected body area in most major orders, an observation possibly indicating that increases in wing area preceded the elaboration of wingbeat kinematics in these lineages. For example, low-amplitude flapping, minimal wing rotation, and only limited aerial maneuverability have been hypothesized for various Carboniferous Paleodictyoptera (see Kukalová-Peck, 1983; Brodsky, 1994). Based on wing and body morphology, however, the contemporaneous Protodonata were clearly comparable to extant Odonata in terms of aerial performance, but were perhaps more predisposed to gliding (see Brauckmann and Zessin, 1989; Wootton et al., 1998). Given that air density and Re of the wings and body during flight may have differed in the late Paleozoic (section 6.1.1), however, definitive aerodynamic assessments based on poorly preserved pterothoracic morphologies are precluded. Physiologically, enhanced Paleozoic oxygen levels would also have promoted the high levels of oxidative metabolism necessary to power active flapping flight. Protective modification of the forewings appeared early in pterygote evolution, with the extinct order Protelytroptera exhibiting fully sclerotized and seemingly armored forewings by the mid-Permian. Coleoptera with fully elytrized forewings appeared by the mid- to late Permian, consistent with substantial predatory pressures throughout the late Paleozoic. In general, the well-developed flight apparatus of the Carboniferous and Permian fauna indicates that these insects can contribute little toward our understanding of the initial evolution of wing morphology and wingbeat kinematics. Only relevant fossil specimens and empirical analysis of wing and body aerodynamics for plausible morphologies will address directly the biomechanical origins of flapping flight.

6.2 Evolutionary Consequences of Flight and Flightlessness

Powered flapping flight evolved independently four times during the Phanerozoic, once in the insects and three times among vertebrates. This biomechanical enhancement of mobility more generally suggests selective advantages to flight in diverse ecological situations. Most importantly, flight enables often geometrically complex vegetation to be more effectively used as habitat and resource by volant animals. Relatively high wind speeds further promote dispersal of winged insects to distant ecological resources. In pterygotes as well as in volant vertebrate taxa, flight may have been initially derived from jumps within vegetation. Longer life spans of flying animals relative to their nonvolant counterparts further suggest that predation is a major factor underlying the evolution of aerial locomotion.

6.2.1 Three-Dimensional Mobility

The impact of aerial mobility on insect ecology is most evident in the domain of insect-plant interactions. Noninsect hexapods (e.g., Collembola) that feed on living plant tissue are rare, whereas approximately 85% of extant insect species are phytophagous at some stage in their life cycle (Strong et al., 1984). Most such phytophagy is confined to nonvolant insect larvae, but adult females of these taxa rely on flight to locate larval resources (e.g., the appropriate species of host plant) and to oviposit accordingly. Also, plant reproductive structures as well as apical meristems (growing regions of plant somatic tissue that are often eaten by larvae) are generally located at vegetational extremities. Access to these structures requires either lengthy gressorial excursion or, much more immediately, flight. The fossil record confirms feeding on plants by insects in the Upper Carboniferous and early Permian (e.g., Labandeira and Phillips, 1996a,b; Rasnitsyn and Krassilov, 1996). The geometric heterogeneity of plant structures is also favorable to flight-capable taxa. Because leaf surface areas are typically much greater than the horizontal area over which branches project, flight enables rapid access to a dramatically increased number of habitats and refuges. The branching habit and fractal character of vegetation (Williamson and Lawton, 1991) also enhance the geometrical complexity of habitats that can potentially be used by volant animals.

Along with aerial mobility comes coexistence with ambient winds. Winds provides substantial opportunity for rapid dispersal, but may also constrain mobility if goal-directed motion is desired (section 7.4).

For most contemporary insects, typical wind speeds exceed maximum flight speeds and interfere with many intentional movements that are not directed downwind. This result is primarily a consequence of the small average body sizes (< 1 cm) that characterize the extant insect fauna (Dudley, 1994). Airspeeds would have been substantially higher for ancestral winged insects in the 1–10 cm range of body lengths, enhancing the possibilities for goal-oriented flight in addition to downwind dispersal. Colonization of novel environments can clearly be enhanced via wind-mediated displacement, although such behavior still requires substantial energetic expenditure in that insects must remain aloft in moving airstreams to effect dispersal (section 7.4.2).

A fundamental advantage of flight, however, is that the cost of transport (the energy required to move a unit of body mass per unit of distance) is substantially lower for flying than for running by animals of equal body mass (Schmidt-Nielsen, 1972). This effect arises primarily from the much higher locomotor speeds associated with flight (Schmidt-Nielsen, 1984; Full, 1997), although increased use of muscles for whole-body deceleration during terrestrial locomotion may also be important (Tucker, 1975). Many insects, moreover, can move in two or three different ways. Winged insects retain their legs for walking, whereas adult insects from diverse taxa can also swim (e.g., various Hemiptera and Coleoptera). Similar locomotor diversity is found in many bats and birds, but the use of forelimbs as wings necessarily restricts the range of biomechanical possibilities available to volant vertebrates. By contrast, the morphological novelty of wings has enabled insects to supplement dramatically their locomotor range while simultaneously retaining the ancestral gressorial and cursorial modes of hexapods.

6.2.2 Predation Risk and Longevity

Predator-induced startle and associated jumping responses have been advanced as general protoadaptations for animal flight (section 6.1.4). If such behaviors are to become more refined, then aerial escape and ultimately flight must convey selective advantage. The effectiveness of flight should be most evident in comparisons of survivorship between volant and nonvolant forms under natural conditions, assuming that most mortality is imposed by predation. No intraspecific data are available for insects that evaluate the effects of flight performance on natural survivorship, although indirect data on survivorship of volant vertebrates under field conditions are suggestive. For example, volant endotherms have mortality rates significantly lower than those of nonvolant endotherms of equivalent body mass (Pomeroy, 1990; see also

Tuttle and Stevenson, 1982; Rose, 1991; Holmes and Austad, 1995). Increased survivorship in birds appears to derive directly from greater success at evading predation attempts (Pomeroy, 1990). In natural populations of insects, intraspecific variation in flight capacity among morphs (e.g., Fairbairn and Desranleau, 1987) is potentially correlated with differences in survivorship. Analysis within populations of sublethal damage to the wings and body may indicate, in absence of direct observational data, whether such survivorship differences arise from flight-mediated escapes of variable effectiveness (see Vermeij, 1982, 1987).

Whereas flight likely enhances escape from terrestrial predators, aerial locomotion also exposes volant animals to predation by other fliers (see section 7.3). In this regard, the extent of adult insect mortality imposed by terrestrial relative to aerial predators is taxon-specific and highly variable. Depending on the kinds of predators present in any particular environment, initial escape from a terrestrial predation attempt may also increase the likelihood of attack in the air. Pearson (1989), for example, found that tiger beetles fleeing from lizards became subject to aerial attack by robber flies. Similarly, orthopterans that leap away from army ant swarms are then more subject to attacks by parasitoid flies. Such secondary effects of escape from predators are probably the exception rather than the rule. The number of volant insect species that are themselves insectivorous is certainly small relative to all insect species, although the influence of insect parasitoids as well as of insectivorous bats and birds can hardly be discounted. A final point concerns evolutionary interactions between predation, sexual selection, and maneuverability during flight. Many aspects of mating behavior and mate selection in insects involve flight (see section 5.3), and sexual selection for enhanced flight maneuverability may also yield increased survivorship during both terrestrial and aerial predation attempts. Such positive reinforcement can quickly escalate the evolutionary dynamics of overall selection on flight capacity and maximum performance (section 8.2.2).

6.2.3 Taxonomic Distribution of Insect Flightlessness

As evidence that flight is not universally advantageous, evolutionary loss of wings and of flight capacity is widespread among the pterygote Insecta (fig. 6.2). Approximately 5% of the extant fauna is flightless (Roff, 1994b; see also Marinelli, 1929; La Greca, 1954; Roff, 1990; Wagner and Liebherr, 1992). The Coleoptera provide the greatest absolute number of flightless species, but in relative terms only few beetles exhibit wing reduction or loss. Other orders uniformly exhibit total wing

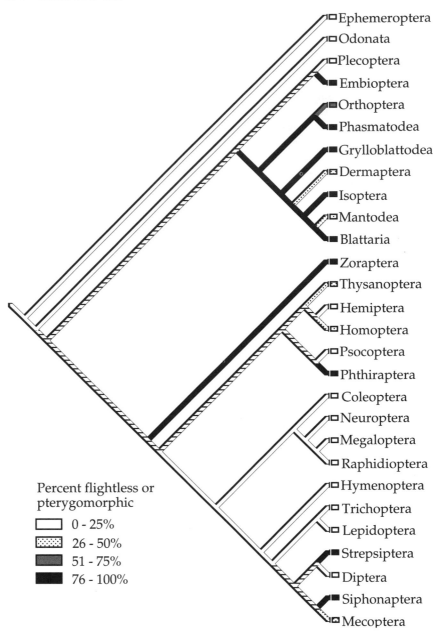

FIG. 6.2. Extent of flightlessness mapped onto the ordinal-level insect phylogeny. Data are taken from Roff (1994); percentage of flightless species here includes those that are pterygomorphic or that exhibit sex-limited flightlessness.

loss, as in the ectoparasitic orders Phthiraptera and Siphonaptera. Large groups of flightless insects can also be found within the Diptera (e.g., the ectoparasitic Braulidae and Hippoboscoidea), Hemiptera, Homoptera, and Lepidoptera, along with some smaller orders (Heppner, 1991; Sattler, 1991). Various hymenopteran taxa (e.g., female Mutillidae, some Ichneumonidae and Chalcidae) are also flightless. In sharp contrast to the monophyletic origin of flight, flightlessness has evolved many independent times in insects, and sometimes multiple times within individual families (Roff, 1994b).

More specifically, flightlessness subsumes a diversity of morphological syndromes, including macroptery (fully winged or alate forms), brachyptery (flightless forms with reduced wing size and/or flight musculature), microptery (as brachyptery, but with very small wings), and aptery (absence of wings). Considerable variation also exists within these broad categories. Many beetles, for example, retain wings but vary intraspecifically in the extent of flight musculature (e.g., Jackson, 1952, 1956; Smith, 1964). Alternatively, flight muscle can be retained but wings are reduced (e.g., Darlington, 1943; Thomas, 1953; Tietze, 1963a, b). Flightlessness may also be reflected by a reduced activity in glycolytic enzymes and in the capacity for mobilization of the metabolic fuels necessary for flight (e.g., Mitchell et al., 1977; Ziegler et al., 1988). Atrophy in anatomical, physiological, and biochemical components of the flight apparatus potentially contributes to secondary flightlessness. The neural regulation of flight does, however, appear to be a highly conservative trait in this regard; flightless stick insects (Order Phasmatodea) nonetheless express electromyographic patterns typical of flight (Kutsch and Kittmann, 1991).

Flightlessness may be obligate within any given species, or it may exhibit intraspecific variation. In this latter case, the term pterygomorphism refers to variable wing expression in response to changing environmental circumstances. Wing expression in pterygomorphic taxa is under endocrinological control and is responsive to such proximate cues as photoperiod, ambient temperature, and population density (Zera and Tobe, 1990; Wagner and Liebherr, 1992). Migratory polymorphism in adult desert locusts, with the solitary phase expressing smaller wings and reduced activity relative to the volant and swarming gregarious phase, is perhaps the most famous example of phenotypic plasticity in insect flight performance. In addition to polymorphisms in wing expression, dealation (the casting off or shedding of wings) is also widespread among at least five insect orders (see Roff, 1989). Ants and termites, for example, are typically wingless except for the reproductive alates, which lose their wings following nuptial flights. Partial shedding of wings occurs following eclosion in some

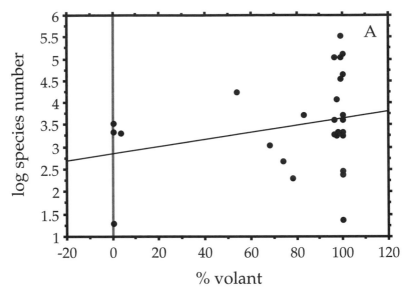

FIG. 6.3A. Log species numbers versus the percentage of volant species (excluding species that are pterygomorphic or that exhibit sex-limited flightlessness) within each order. Ordinal species numbers are from table 1.1; flightlessness data are from Roff (1994). The linear regression is not significant ($r^2 = 0.26$, $P = .18$).

phorid flies (Disney, 1994), and female streblid dipterans similarly shed wings when colonizing their chiropteran hosts. All known cases of dealation occur among taxa that use flight only in the initial searching stages of otherwise essentially sedentary lives.

The widespread taxonomic occurrence of pterygomorphism and secondary flightlessness provides an interesting opportunity to evaluate implications of reduced flight capacity for ordinal-level insect diversification. Among insect orders, the number of insect species shows no correlation with the percentage of volant forms (fig. 6.3A). Because of phylogenetic associations of flightlessness, however, effects of relatedness among insect orders must be incorporated into any such analysis. For example, flightlessness is absent in paleopterous orders and is most pronounced among the neopterous exopterygotes (Roff, 1990; see fig. 6.2). To control statistically for such phylogenetic effects, correlations among either discrete or continuous characters can be assessed using the comparative method known as paired contrasts (Harvey and Pagel, 1991; Purvis and Rambaut, 1995). In this method, character states or values for ancestral phylogenetic nodes are reconstructed using trait data from the extant terminal taxa. For each node within the phylogeny, the associated sister taxa then represent two (or

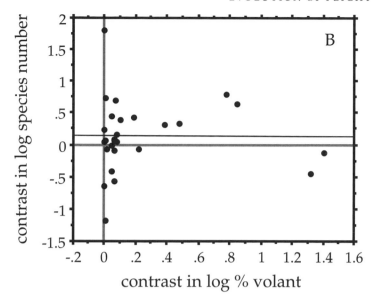

FIG. 6.3B. Independent contrasts between ordinal species numbers and the percentage of volant species, controlling for phylogenetic relatedness using the ordinal-level tree topology of figure 1.1 (see Felsenstein, 1985; Purvis and Rambaut, 1995). Variables were log-transformed to reduce high variance. No significant correlation characterizes the relationship between contrasts in species number and contrasts in the percentage of volant species (r^2 = 0.0001, P = .95). Acquisition of flightlessness has not historically reduced species diversity at the ordinal level.

potentially more) independent lineages that can be contrasted in terms of character states or values. By then examining these statistically independent contrasts for all nodes of the phylogeny, correlations among characters can be derived that are not confounded by phylogenetic association.

In application of the paired contrast method to an ordinal-level insect phylogeny, no correlation is evident between species diversity and the relative extent of flightlessness or pterygomorphism (fig. 6.3B). Insect orders do vary dramatically in the relative extent of flightlessness (fig. 6.3A), and additional factors such as mode of metamorphosis and type of flight muscle (see section 7.1) likely influence species richness. Alternatively, flightlessness may occur following adaptive radiation into particular habitats, in effect historically decoupling diversification from flight capacity. Use of independent contrasts analysis at higher levels of phylogenetic resolution might resolve this issue, particularly given the multiple independent evolution of flightlessness within major orders (e.g., Coleoptera).

6.2.4 Ecological Correlates of Flightlessness

A classic correlate of insect flightlessness is environmental stability (Wagner and Liebherr, 1992). Unfortunately, this concept is difficult to define rigorously at the spatial and temporal scales appropriate to insects. Controlling for taxonomic but not for phylogenetic associations, Roff (1994a) concluded that insect flightlessness occurred more often in permanent woodlands than in more ephemeral habitats. The concept of woodland subsumes a large number of diverse microclimates and niches (e.g., gaps, canopy, understory, soil) for which environmental stability is difficult to define rigorously. In a more restricted ecological context, however, temporal persistence of habitats is correlated with an increased frequency of brachyptery in delphacid homopterans (see Denno et al., 1991; Denno, 1994; Denno et al., 1996). Additional environmental correlates of flightlessness include residence at high elevations or latitudes, residence in low-statured vegetation, subterranean habits, association with marine habitats, and inquilinism (nest dwelling), particularly within nests of social insects (see Dewitz, 1902; Mani, 1968; Byers, 1969; Waloff, 1983; Kavanaugh, 1985; Denno, 1994). Weissflog et al. (1995), for example, documented extraordinary wing and leg atrophy in female phorid flies that reside within colonies of driver ants. Flightlessness also tends to be more pronounced in gregarious parthenogenetic species (e.g., some aphids), among outbreak species (e.g., spring-feeding forest Lepidoptera; Hunter, 1995), and in females (especially of the Diptera and Lepidoptera) whenever the character is sexually dimorphic (Roff, 1990).

One common ecological theme to these diverse manifestations of insect flightlessness is a reduced need for locomotor mobility. The costs of construction and maintenance of the flight apparatus, although difficult to measure, are certainly not negligible. Associated energetic resources may potentially be directed elsewhere in the body if habitat stability does not require flight, and/or if predatory pressure is relaxed such that flight is no longer a major component of the antipredatory repertoire. The most obvious evolutionary response would be an increased investment in reproductive capacity, particularly among females. For example, brachypterous morphs of dimorphic species tend to have higher fecundity (Roff, 1986; Roff and Fairbairn, 1991; Dixon et al., 1993); these morphs must have much-reduced dispersal abilities. Reduced expression of wings under such circumstances may be paralleled by muscle degradation and reduced muscle respiration rates that permit additional energetic allocation to reproductive efforts (e.g., Zera et al., 1997). The enhanced fecundity associated with somatic reallocation of energetic resources will directly select for phenotypic plasticity in wing expression, and can ultimately result in obligate flightlessness given the appropriate environmental conditions. Theoretical modeling also suggests that brachypterous genotypes can spread within macropterous populations given spatially heterogeneous environments (Roff, 1994a).

Additional ecological contexts may also select indirectly for flightlessness. For example, experiments with laboratory populations of *Drosophila melanogaster* have demonstrated increased survivorship of brachypterous forms under windy conditions (L'Héritier et al., 1937). In some limited circumstances, flightlessness may yield certain advantages in sexual selection. Crespi (1988), for example, showed that winglessness in male thrips is correlated with an enhanced ability to fight conspecifics. A final consequence of brachyptery is a reduced ability to flee from predators. Given potential trade-offs between flight muscle and other somatic structures in pterygomorphic species, the ability to fly can clearly disappear if not actively sustained by selection. For example, various avian taxa on isolated islands are well known to have evolved flightlessness in the absence of predators (Carlquist, 1965, 1974; Feduccia, 1996). Reduced resource availability on oceanic islands also correlates with reduced pectoral mass and lower metabolic rates in birds, although no comparable data are available for insects. Nor has predation on flightless insects been systematically evaluated within the associated habitats of mountaintops, high latitudes, and permanent woodlands, although increased survivorship and/or use of antipredatory strategies besides flight may be characteristic of such locations.

Enhanced jumping performance may be one of these latter strategies. For example, flightless Thysanoptera tend to have enlarged hindlegs and possibly an increased jumping ability (Becker, 1958), as also exhibited by some eupelmid Hymenoptera (Gibson, 1986). Although insects of oceanic islands are no more prone to evolve flightlessness than are other insects (Roff, 1990), greater jumping ability has been noted in certain flightless island moths (see Carlquist, 1965). In subAntarctic and Antarctic regions, however, this jumping ability is absent among flightless Diptera and Lepidoptera, apparently because of high winds (P. Convey, pers. comm.). Defense against predation in flightless insects may also derive from the morphological retention of elytrized forewings. In some dimorphic beetles, for example, the micropterous morph loses the hindwings but retains the elytra (Lindroth, 1946). Relative to the macropterous morph of a hemipteran species, the micropterous morph expresses a thickened and more coriaceous region of the forewing that potentially functions in defense (see Zimmerman, 1948; see also Banerjee, 1988a). An additional behavioral response that likely inhibits predation is thanatosis, or the feigning of death. Beetles exhibit a reduced tendency to fly from predators and often use thanatosis as a defensive mechanism (Bleich, 1928). These examples suggest substantial evolutionary interaction between the risk of predation, expression of brachyptery, and evolution of alternative defensive mechanisms. Systematic data on predation rates and the relative use of different antipredatory mechanisms by pterygomorphic species would address this hypothesis directly. Comparisons of males and females in taxa with sexually dimorphic wing expression (e.g., Thomas, 1952) would be particularly informative. Existing associations between flightlessness and jumping are consistent, however, with a major role for predation in motivating not only initial wing evolution but also subsequent loss of flight by insects.

6.3 SUMMARY

Conclusive interpretations of insect flight evolution cannot be achieved due to a meager paleontological record. Much neontological and paleobiological evidence, however, is consistent with the evolution of wings from either leg-derived projections or from the paranotal lobes of terrestrial apterygote insects. Functionally, such lobes may have been selected for enhanced aerodynamic performance and even for their flapping ability from the outset of wing evolution. Additional possibilities such as use in thermoregulation or a role in epigamic display would not preclude lift production on what ultimately became a flying apparatus. Jumping via a startle reflex in response to predation may have been a general selective force that promoted the initial acquisition of flight by all volant animals. In ancestral pterygotes, aerial escape would have been made easier by the lifting characteristics of both winglets and the body; unsteady aerodynamic performance during a jump may have been of greater relevance to survivorship than steady-state glides. Saltatorial apterygote insects have the most plausible morphology with which to model such flight performance experimentally.

Increased survivorship during attempted predation is a powerful evolutionary motivation for enhanced aerial mobility. Reduced mortality rates in volant vertebrates are consistent with a pervasive role of predation in flight evolution. For winged insects, additional advantages of flight include more efficient use and exploitation of terrestrial vegetation together with long-distance dispersal facilitated by ambient air motions. Whereas flapping flight evolved only once in insects, pterygomorphism and secondary flightlessness have independently evolved numerous times. Habitat persistence and a correspondingly reduced need for mobility appear to be major factors underlying reduction in flight capacity. Equally important, however, may be correlated reductions in the risk of predation for habitats characterized by relative environmental stability. In either case, phenotypic plasticity of the flight apparatus permits a fine-tuning of locomotor needs in response to changing environmental circumstances.

Chapter Seven

FLIGHT AND INSECT DIVERSIFICATION

IN ADDITION to their walking and running abilities, winged insects also use flight in a variety of behavioral and ecological contexts (Haskell, 1966). To what extent has pterygote diversification been facilitated by the ability to fly? This important question can be answered only within a phylogenetic context, but major features of the contemporary insect biota have clearly been accompanied by flight-related adaptations. Miniaturization is a defining feature of contemporary insect diversity, deriving in part from the evolutionary acquisition of asynchronous flight muscle and high wingbeat frequencies. The small body size of most winged insects facilitates aeolian (wind-borne) dispersal; evolution of pollinator and parasitoid lifestyles may also be associated in many lineages with size reduction and the ability to hover. For a variety of physiological and phylogenetic reasons, the overlap in body size between extant insects and volant vertebrates is small. Miniaturized pterygotes thus represent a biomechanically distinctive and uniquely diverse form among all animals.

7.1 MINIATURIZATION

An important feature of the terrestrial metazoan biota is the extent to which small animals represent the overwhelming majority of species. Although small body size is characteristic of many invertebrate and vertebrate taxa (May, 1978, 1988), miniaturization is particularly evident among hexapods and largely defines the diversity of winged insects. Their reduced body size is a derived condition given the postulated sizes of protopterygotes (2–4 cm; section 6.1.3), together with the ancestral condition of synchronous flight muscle and low wingbeat frequencies (section 4.2.2). Insect body size thus has had substantial historical change, ranging from small ancestral pterygotes, insect giants of the late Paleozoic, and the present-day predominance of miniaturized forms. No detailed information is available on insect size distributions through geological time, but evidence from the contemporary fauna strongly suggests that miniaturization has been a major feature of pterygote diversification (Chetverikov, 1920). High wingbeat frequencies, novel wing morphologies, and distinctive aerodynamic mechanisms are all associated with these reduced body sizes.

7.1.1 Average Insect Body Size and Phylogenetic Associations

The absence of species-level descriptions for the worldwide insect fauna precludes any detailed analysis of body size distribution. Most insects are small, however, by anthropomorphic standards. Beetles, as the largest order of winged insects (table 1.1), exemplify the small body sizes of most extant pterygotes. In general, mean adult beetle body length lies between 4 and 5 mm (May, 1978; Crowson, 1981). Many corylophid and sphaeriid beetles are miniaturized to body lengths near 0.5 mm, whereas the smallest described adult coleopterans are nanoselliine ptiliids, whose body lengths are on the order of 0.3–0.4 mm (see Dybas, 1966, 1990; Sörensson, 1997). Like beetles, many dipterans and hymenopterans have tiny bodies. Small body size is particularly characteristic of parasitoid and hyperparasitoid taxa. Parasitoids are found primarily among flies (15,000 dipteran parasitoids, mostly within the Bombyliidae and Tachinidae) and in the hymenopteran division Parasitica. This latter category contains about 45 families and 50,000 species of obligate parasitoids (Godfray, 1994). The smallest known adult insects (body length 0.21 mm) are parasitoid mymarid hymenopterans that usually fly, but that also apparently swim with their wings to find and oviposit on host eggs deposited in water (see Matheson and Crosby, 1912; Hagen, 1996). For dipteran and hymenopteran parasitoids as well as for beetles, worldwide species numbers are probably highly underestimated (Noyes, 1978; Crosskey, 1980; Erwin, 1982; Stork, 1988; LaSalle and Gauld, 1991). The taxonomic preponderance of small insects is evidenced by their individual abundances within natural ecosystems. For example, small Coleoptera, Diptera, Hemiptera, and Hymenoptera dominate the insect fauna of a temperate grassland (Siemann et al., 1996). In the tropics, the greatest numbers of flying insects both within forests and along forest edges belong to the orders Coleoptera, Diptera, and Hymenoptera (see Janzen, 1973; Fowler et al., 1993); winged termites are also abundant during nuptial flights following heavy rainfall.

The taxonomic and numerical preponderance of small insects may have been, in part, facilitated by the acquisition of asynchronous flight muscle (Dudley, 1991b). Flight of miniaturized insects requires wingbeat frequencies in excess of 100 Hz (see fig. 3.3B; section 4.2.2). Correspondingly high muscle contraction frequencies can be attained only by the asynchronous flight muscle found in the major endopterygote orders Coleoptera, Diptera, and Hymenoptera (with the possible exception of some Symphyta; see section 4.2.2), as well as among the exopterygote hemipterans, thysanopterans, and some psocopterans and homopterans (see figs. 4.2, 4.3B). In contrast to the monophyletic traits classically associated with insect diversification (i.e., origin of

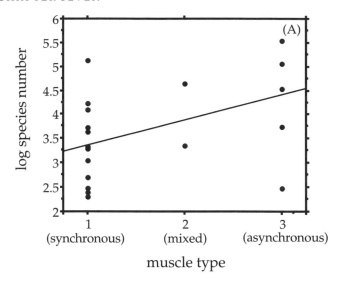

Fig. 7.1A. Log species numbers versus muscle type among insect orders ($r^2 = 0.24$, $P = .015$). Mixed muscle type refers to orders (Homoptera, Psocoptera) within which both muscle types are represented. The Hymenoptera are assumed to be exclusively asynchronous (see section 4.2.2); statistical significance for this and the following independent contrasts analysis is, however, independent of this assumption.

wings, lateral wing flexion, complete metamorphosis; see section 1.1.2), the multiple independent origins of asynchronous flight muscle permit statistical analysis of diversification within a phylogenetic context. Treating insect orders as independent data points, total species count is positively correlated with the presence of asynchronous flight muscle (fig. 7.1A). When paraphyletic homopterans (fig. 4.3A) are included in analyses that control for phylogenetic relatedness, one can correlate the acquisition of asynchronous flight muscle among the pterygotes with enhanced species richness (fig. 7.1B). High wingbeat frequencies and the associated ability to miniaturize have thus been key innovations facilitating adaptive radiation among the winged insects. Higher wingbeat frequencies are furthermore associated with greater force production that in turn enables a reduction in wing area relative to body mass (Dudley, 1991b). This effect is most pronounced in the conversion of one wing pair to elytra (Coleoptera), hemelytra (Hemiptera), or halteres (Diptera), but is characteristic of all taxa with asynchronous flight muscles. Such use of wings in nonaerodynamic and particularly in protective roles has likely played a major role in ordinal-level insect diversification (section 8.3.2).

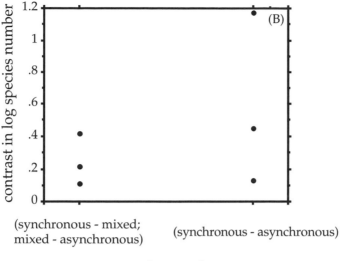

contrast in muscle type

FIG. 7.1B. Independent contrasts between log-transformed ordinal species numbers and muscle type controlling for phylogenetic relatedness. An independent contrast method was implemented using the computer program CAIC (Purvis and Rambaut, 1995); species numbers were log-transformed to reduce high variance. The phylogeny followed the ordinal-level topology of figure 1.1 but with inclusion of a contemporary phylogeny for hemipteran and homopteran lineages (fig. 4.3A). In an independent contrasts test with categorical variables (e.g., muscle type), the null hypothesis is that contrasts in species number are zero on average (Purvis and Rambaut, 1995). Presence of asynchronous muscle either partially or uniformly within a taxon significantly increases the number of species by an average factor of 2.6 (1-tailed sign test, $N = 6$, $P = .016$).

Although body size distributions cannot be similarly analyzed among insect lineages because relevant data are not available for the worldwide fauna, miniaturization has clearly been a general pattern for many pterygote lineages. As mentioned previously, an abundance of miniaturized forms is particularly evident within the orders Coleoptera, Diptera, and Hymenoptera. Based on the ordinal-level designation of muscle types, more than 75% of described insect species have asynchronous flight muscle (Dudley, 1991b). Furthermore, this fraction is a conservative estimate for the relative number of species with this trait. For reasons of sampling bias, smaller insects tend to be less well known; the described beetle, parasitoid, and hyperparasitoid faunas of the tropics are famously incomplete. By contrast, flight muscle in the Lepidoptera, the only other species-rich insect order (see table 1.1) is synchronous. Compared to the other large endopterygote orders, butterflies and moths are fairly well known globally because of the typically much larger body sizes in this taxon. The Microlepidoptera represent most species diversity within the order, but even these small lepidopterans are still substantially larger than the miniaturized pterygotes with asynchronous flight muscle. Several lineages of Lepidoptera may, however, possess asynchronous flight muscle (see section 4.2.2). If it is indeed present, phylogenetic analysis of distribution of this muscle type within the Lepidoptera would provide an independent test of the miniaturization and diversification hypotheses.

What is the lower limit to the body size of winged insects? As body size decreases and the ratio of surface area to body volume increases, the mass of the exoskeleton must increase relative to total body mass unless mean exoskeletal thickness decreases at the same time. If the exoskeleton functions to avoid bending that is imposed either externally by gravity or internally from muscles, then a lower limit on exoskeletal thickness may exist to accommodate structural design. For example, various aquatic crustaceans are much smaller than miniaturized insects, but do not require exoskeletal support against gravity because of their approximately neutral buoyancy. At the other extreme of large body size, endoskeletons become more advantageous than exoskeletons in resisting impact damage, which may in part explain the structural preponderance of the endoskeleton among vertebrates (Currey, 1967). For winged insects, additional factors that can potentially limit miniaturization range from limiting thoracic deformation at high operating frequencies to inordinately high mechanical and metabolic power expenditure at very low body masses. The reduced thoracic volume of smaller insects probably precludes any constraints of oxygen availability, however. Mechanical power production could be increased by reduced tracheolar investment and higher myofibrillar and mitochondrial densities (Rensch, 1948), although ultimately an upper limit on power availability could still apply. Mechanisms of lift production may also be limiting at the low *Re* characteristic of miniaturized insects (see below). A paucity of data on flight energetics and the aerodynamics of small pterygotes still precludes a definitive analysis of the limits to miniaturization.

7.1.2 Morphological and Biomechanical Correlates

In a general review of animal miniaturization, Hanken and Wake (1993) noted that novel morphologies as well as homoplasy (character convergence among unrelated taxa) are common correlates of body size reduction. For miniaturized pterygotes in diverse phylogenetic lineages, wing venation tends to be reduced and consolidated such that the longitudinal wingspan functions as a unitary structural element. Bounded cuticular surface area is necessarily reduced in such a design, but the aerodynamically effective area of the wing is increased by numerous hairlike setae that project peripherally around the wing margin. Setal fringing of the wings is particularly well known from the Thysanoptera, small Psocoptera and Zoraptera, male coccid homopterans, diverse beetle families (and particularly in the families Clambidae, Pselaphidae, Ptiliidae, and Sphaeriidae), various hymenopterans (e.g., mymarids and trichogrammids; see Danforth, 1989), small flies (see Vogel, 1967b), and numerous small trichopterans and moths (see Sakamaki and Akimoto, 1997). Such homoplasy in structural design ultimately derives from functional convergence in response to common mechanical demands (Hanken and Wake, 1993). For example, aquatic crustaceans have fringed locomotor and feeding appendages that are anatomically similar to the wings of small insects and that operate at broadly comparable *Re* (see Cheer and Koehl, 1987; Vogel, 1994). The underlying fluid mechanics of fringed structures oscillating at low *Re* thus appear to have produced convergent appendage design from phylogenetically unrelated arthropod lineages.

The universal kinematic correlate of pterygote miniaturization is a higher wingbeat frequency enabled by the evolution of asynchronous flight muscle. One direct consequence of higher flapping frequencies is that rotational velocities during pronation and supination are also increased, yielding higher rates of wing twisting and bending. Reduced venation and extensive fringing may be aerodynamically advantageous under these circumstances, although patterns of three-dimensional deformation during flapping flight have never been evaluated for such wings. Reduced wing venation likely increases torsional flexibility (Horridge, 1956), whereas setal fringing increases the functional surface area of the wings. For example, close setal spacing on wings of the tiny wasp *Encarsia formosa* inhibits airflow between adjacent setae and effectively converts the wing fringing into an aerodynamically continuous membrane (Ellington, 1975). Flexibility and deformation are probably greater for surfaces constructed of nearly contiguous setae than for wings having a continuous cuticular membrane. The steady-state modeling of Cheer and Koehl (1987) further indicates that intersetal spacing as well as setal diameter can strongly influence patterns of flow through bristled appendages at low *Re*. One biomechanical possibility is that wing torsion combines with patterns of setal deformation and passively induced changes in intersetal spacing to render the wings more responsive to the aerodynamic and inertial forces of flapping.

Estimates of Re for the wings of miniaturized insects range from 10 to 500 (e.g., Horridge, 1956; Pringle, 1957; Bennett, 1973; Weis-Fogh, 1972, 1973; Ellington, 1975); values of Re for individual wing setae are much lower (see table 1.2). Allometric variation in wing dimensions and flapping frequencies also suggests that smaller insects operate at progressively lower Re during hovering flight (see section 5.4.3). Under such conditions, generation of wing circulation and vortex-based lift is increasingly impeded by the frictional forces of fluid viscosity. Setal fringing of wings in very small insects may thus indicate alternative mechanisms of lift production (Thompson, 1917; Horridge, 1956). Vortex generation characterizes wing flapping by *Drosophila melanogaster* operating at Re near 100 (Dickinson and Götz, 1996), but much smaller insects are likely to have greater difficulty using this mechanism to produce lift. For example, steady-state lift on airfoils operating at $Re < 10$ is only linearly proportional to flow velocity, and lift : drag ratios fall below unity (Thom and Swart, 1940). Circulation-based lift is clearly much more costly if not disadvantageous under such circumstances. However, the clap-fling mechanism of lift generation is known empirically to be effective at Re of 32, and likely operates independently of viscous effects at even lower Re (sec. 3.2.2.3). Lower size limits to this mechanism of lift production warrant empirical investigation, particularly for the fringed and presumably more flexible wings typical of the smallest insects.

As an alternative to lift production on wings, miniaturized insects could potentially row through the air by relying on differential drag production between half-strokes to offset the body weight and to generate thrust. Changes in wing orientation, effective surface area, and/ or relative velocity between the down- and upstroke could all contribute to stroke asymmetry and to net force production during flapping. Wing setae could also deform to produce propulsive waves along the wing (Kuethe, 1975). As pointed out by Bennett (1973), wing surface area in minute insects is essentially invariant unless the setae fold together during one half-stroke (as characterizes locomotion of some aquatic beetles; Nachtigall, 1974b). At least for the small wings of Thysanoptera, neither setal folding nor spanwise propulsive waves are evident during oscillation at the approximate wingbeat frequency (Ellington, 1980b). Velocity differences between half-strokes may, however, be effective in generating net forces during flight, as demonstrated empirically for an individual model seta oscillating at Re of 0.2 and 100 (Bennett, 1973). Such a mechanism would necessarily rely on substantial kinematic asymmetry between half-strokes, a possibility not systematically evaluated among small winged insects, although probably absent in the wasp *Encarsia formosa* (Weis-Fogh, 1973). A final aerodynamic consideration concerns the interaction between flow around the wings and the physical boundary formed by the insect's body. Such boundary effects are likely to be much more pronounced at the low *Re* typical of miniaturized insects (see Loudon et al., 1994).

Energetically, the relative costs of three-dimensional aerial mobility may in fact be reduced for small insects. Because the wing advance ratio in flight increases approximately in proportion to $mass^{0.20}$ (section 3.1.2), smaller insects fly at progressively lower advance ratios and at reduced absolute flight speeds. Mass-specific power expenditure in forward flight increases with $mass^{0.05-0.19}$ (section 5.4.2), suggesting reduced relative costs of horizontal movement to miniaturized forms. Similarly, mass-specific induced power expenditure in hovering flight scales approximately with $mass^{0.17}$ (see section 5.4.2). Hovering is thus energetically more feasible for smaller animals, which fact in part links the profusion of small dipterans, hymenopterans, and coleopterans to the slow forward flight and hovering required for much of the pollination, mating, and oviposition behavior in these taxa. Also, the generation of vertical forces supplemental to body weight (i.e., vertical acceleration) as well as ascent at constant velocity will require additional expenditure of power that likely exhibits an allometry similar to that of induced power. Miniaturization can thus be favored energetically if vertical ascent is a major component of flight behavior. Finally, mechanical analogs of miniaturized flying insects have been constructed in the form of flapping microrobots and microair vehicles, technologies still early in development (e.g., Suzuki et al., 1994).

7.2 POLLINATION

With the exception of the boreal coniferous forests, global vegetation is dominated by angiosperms. The very use of the word "flora" to denote all plant species within a particular region indicates the extent to which flowers have largely supplanted ancestral modes of reproduction in vascular plants. Angiosperm diversification in the early Cretaceous (~140 MYBP) initiated novel features of terrestrial ecology that now largely define the physical structure of ecosystems worldwide, and moreover that promoted innovation in animal behaviors such as frugivory and pollination. Coevolutionary interaction between angiosperms and winged insects likely began at the very outset of flower evolution, facilitating adaptive radiations in both taxa that continue to this day.

7.2.1 Paleobiological Background

Indirect evidence supports an active role of insects in plant pollination as early as the Carboniferous (Taylor and Millay, 1979; Shear, 1991). Although specific mutualistic interactions have not been demonstrated paleontologically, feeding by insects on plant reproductive structures and inadvertent transfer of pollen were probably widespread in the late Paleozoic and throughout the Mesozoic. By the late Jurassic, incipient pollen feeding and pollination by insects of flowering plants had likely evolved from preexisting pterygote associations with pteridophytes and gymnosperms (see Crepet, 1983; Crepet and Friis, 1987; Crepet et al., 1991; Pellmyr, 1992; Crane et al., 1995; Labandeira and Phillips, 1996b; Ren, 1998). Subsequent angiosperm diversification and spread within the terrestrial flora has been substantially influenced by dedicated insect pollinators. Because animal behaviors are particularly difficult to surmise from the fossil record, little is known about pollination modes of early angiosperms. Pollinators were probably generalists in terms of flower selection if pollen was, as seems likely, the initial nutritive reward. Floral fidelity together with morphological and behavioral specialization by pollinators were then rapidly promoted by the evolution of energy-rich nectar rewards. Remarkable obligate relationships have evolved between particular plant taxa and various lineages of pollinating insects. Among the most renowned of these associations is the tight species specificity and coevolution between orchids and bees (Dressler, 1982; Roubik, 1989). The thysanopteran pollinators of Indo-Malayan Dipterocarpaceae (Appanah and Chan, 1981) and the agaonid wasps that pollinate figs (Janzen, 1979; Wiebes, 1979) represent equally interesting but much less studied coevolved systems.

In spite of such important contemporary associations, the historical contributions of insect pollination to angiosperm taxonomic diversification remain unclear. Nor are possible reciprocal effects on insect speciation well understood. With the exception of the Lepidoptera, all major pollinating insect orders were present if not well diversified prior to the origin of angiosperms. Historical rates of angiosperm family origination have been fairly constant since the Cretaceous and are independent of major insect radiations over the same time period (Labandeira and Sepkoski, 1993). Also, species richness of contemporary angiosperm families does not appear to be associated with biotic mode of pollination (Ricklefs and Renner, 1994). Such large-scale taxonomic analyses may, however, be of insufficient resolution as to discern coevolutionary interactions between particular plant and insect lineages. Most importantly, angiosperm diversification may have been promoted by the potent combination of insect pollination with vertebrate dispersal of seeds and fruit over long distances (Regal, 1977). Only through statistical comparison of well-resolved phylogenies for both insects and angiosperms can the issue of reciprocal enhancement of species diversity be addressed definitively.

7.2.2 Flight Performance and Pollinator Effectiveness

Numerous biotic as well as abiotic factors influence the efficiency of pollen transfer between flowers (see Barth, 1991; Proctor et al., 1996). For flying insects to evolve into dedicated pollinators, the net costs of locating flowers must ultimately be less than the overall benefit obtained from feeding on floral nectar and possibly acquiring additional substances (e.g., pollen, chemical fragrances). In turn, floral strategies of attraction and reward have evolved to correspond more closely to the energetic characteristics and dispersal abilities of associated pollinators (Heinrich and Raven, 1972). Flowers pollinated by vertebrates, for example, are often physically larger and offer much greater nectar volumes than flowers pollinated by generally much smaller insects. Effective attraction of pollinators requires the production of nectar of sufficient quantity and energetic quality (e.g., carbohydrate concentration), but the overall energy balance of plants must obviously limit the extent of such investment. On ecological timescales, behavioral strategies of individual pollinators are also influenced by the quantity and quality of nectar obtained from flowers (Heinrich, 1975a, 1983). Many flowers attract a large taxonomic diversity of pollinators, rendering tight coevolutionary relationships with particular insect taxa unlikely. At the other extreme along the spectrum of mutualistic specificity, many orchid species have dispensed altogether with nectar rewards. Instead, morphological similarity of floral structures to female wasps elicits copulatory behavior from conspecific male wasps, ultimately yielding pollination among individual orchids of the same species.

One of the most striking examples of the interaction between pollinator energetics and floral rewards involves a comparison of pollination syndromes across altitudinal gradients. At higher elevations, reduced air density and lower air temperatures substantially increase the metabolic costs of facultative endothermy and of flight (see Heinrich and Raven, 1972; Dudley and Chai, 1996). The reduced muscle power output at lower operating temperatures (section 4.3.1) appears to combine with higher energetic demands of flight to impose body size restrictions on insect flight performance. Among pollinators, larger and more strongly endothermic insects (e.g., bumblebees) as well as birds typically predominate at higher elevations. Hummingbirds, for example, gradually supplant insect pollinators along altitudinal transects in the montane Neotropics. Pollination-specific responses of flowers to alpine climates are less clear, in part because of covariance in plant photosynthetic capacity and net production with light availability and air temperature. Intraspecific variation in floral nectar rewards across altitudinal transects (e.g., Cruden et al., 1983) may nonetheless reflect coevolutionary interaction with animal pollinators in response to common physical factors of the environment.

Contemporary insect pollinators are dominated by representatives of the Coleoptera, Diptera, Hymenoptera (in particular, the bee superfamily Apoidea), Lepidoptera, and Thysanoptera (see Kevan and Baker, 1983; Proctor et al., 1996). With the exception of the Lepidoptera, insects from these orders possess asynchronous flight muscle and are often miniaturized forms that hover at flowers either prior to or during pollination. Hovering permits an increased rate of flower visitation (e.g., Heinrich, 1983) and probably enhances the ability prior to pollination to inspect flowers for presence of predators (e.g., crab spiders in the family Thomisidae). Other than hawkmoths (Sphingidae) and some butterflies (e.g., heliconiine and ithomiine nymphalids), few insects with synchronous muscle engage in sustained hovering during pollination. Among pollinating vertebrates, only hummingbirds and some bats (e.g., glossophagine phyllostomids; see section 7.5.2) can hover to attain the rapid rates of flower visitation and pollen transfer characteristic of many insect pollinators. Pollination specificity correspondingly appears to be more diffuse in vertebrates, with tight coevolutionary mutualisms described primarily from the hummingbirds (e.g., Grant and Grant, 1968).

In addition to enhancing the ability to hover, body miniaturization in pollinators may also increase the efficiency of pollen transport over long distances. Small insects can act as wind-dispersed pollen vectors that become aerially entrained by and displaced in the direction of prevailing winds (see section 7.4). For example, minute fig-pollinating wasps disperse up to 14 km during a lifespan of only several days (Nason et al., 1998). Much if not all of such dispersal is probably downwind given the slow flight of such small insects relative to ambient wind speeds (section 7.4.2). Relative to the comparable aerial displacement of pollen grains from wind-pollinated plant taxa, however, miniaturized insects can also express behavioral specificity of flower choice upon arriving in the vicinity of nectar-bearing plants. Pollination systems involving miniaturized insects thus combine the advantages of wind-assisted dispersal with specialized sensory and perhaps morphological adaptations that may heighten mutualistic benefits for both plant and insect. For larger insects, a more active control of flight directionality is enabled by their generally higher flight speeds (section 3.1.1). Such insects may correspondingly be more specialized to pollinate widely separated individual plants or clumps of plants. For example, tropical tree species are taxonomically diverse but typically occur at low adult densities. Bees that fly long distances and follow regular daily routes (i.e., traplining) are the major pollinators of such plants (see Janzen, 1971; Frankie, 1975, 1976; Dressler, 1982; Bawa et al., 1985; Roubik, 1989). Evolutionary associations between pollinator body sizes and plant pollination strategies are not well studied, but larger insects clearly have the potential to act as more intentional pollen vectors over greater distances.

7.3 PREDATION

Predation-based selection for enhanced flight performance has presumably been intense during the evolution of winged insects. Some insect taxa capture prey in the air, whereas a far greater number are preyed upon by both insect and vertebrate predators. The substantial ecological impact of avian and chiropteran predators on insect populations worldwide demonstrates ongoing selection for enhanced mechanisms of predator detection, deterrence, and evasion. These forces of natural selection interact synergistically with the demands of sexual selection in many taxa to accelerate the evolution of aerial maneuverability (section 8.2.3).

7.3.1 Aerial Predation on Insects

Few insect taxa contain aerial predators, but those insects that do capture prey in the air are taxonomically diverse. The Odonata are perhaps the best studied of such groups, with the three extant odonate suborders relying primarily on flying insects but also capturing prey at rest on foliage. Dragonflies also occasionally forage in swarms (Corbet, 1962). One interesting evolutionary offshoot within the odonates is the clade of helicopter damselflies (Pseudostigmatidae), an exclusively New World taxon noted for large body size (wing lengths up to 88 mm) along with very low wingbeat frequencies (Rüppell and Fincke, 1989; Fincke, 1992; see plate 2A). These predators capture orb-weaving spiders by hovering in front of the web, adeptly flying forwards, grabbing the prey, and then rapidly reversing direction to avoid entanglement. As with the Odonata generally, such precise control of flight is testimony to sophisticated flight performance in what is a phylogenetically basal order.

In addition to the Odonata, several families of Diptera are formidable aerial predators. The Asilidae, or robber flies, visually monitor insects flying in the immediate vicinity of their perched position. Chasing ensues if potential prey are judged suitable victims. Aerial predatory behavior characterizes flies of the family Empididae; males often present captured prey items as nuptial gifts to conspecific females. Among other aerial predators of their fellow Insecta are the sphecid Hymenoptera (e.g., beewolves: philanthiine sphecids; cicada killers: nyssoniine sphecids), the bittacid Mecoptera, and various Neuroptera (Ascalaphidae, Nemopteridae, and some Myrmeleontidae). Life as an aerial predator involves substantial risks as well as potential gain. For example, increased exposure both to other insect predators as well as to volant vertebrates may require the evolution of supplemental escape tactics or other mechanisms of predator deterrence (e.g., aposematism and unpalatability; see section 7.3.2). Consistent with this hypothesis, dragonflies are famously difficult to catch, whereas the more weakly flying damselflies (Odonata: Zygoptera) tend to be less conspicuous and to forage close to vegetation. Similarly, many asilid flies morphologically resemble stinging hymenopterans, whereas helicopter damselflies may actually be unpalatable to potential predators (O. M. Fincke, pers. comm.).

Terrestrial insects are often the prey of vertebrates (Rohdendorf, 1970b), whereas insects that feed on vertebrates are typically aquatic or ectoparasitic. Birds are the predominant diurnal predators of insects, and a majority of bird families include representatives that feed regularly or even exclusively on insects (Morse, 1975). Although many families of birds typically catch insects in the air during sallying behavior (e.g., the generalized insectivorous families Muscicapidae and Tyrannidae), most insectivorous birds search for insects on vegetational substrates in gleaning or related foraging behavior (Morse, 1975; Remsen and Robinson, 1990). Insects are typically at rest prior to such capture attempts, although escape from foliage- and twig-gleaning birds often involves evasive flight and subsequent aerial pursuit by the insectivore in question. Some avian taxa use wing and body motions intentionally in a flush-pursue behavior that elicits flight responses from otherwise hidden insects (see Remsen and Robinson, 1990). In addition to the various taxa that forage on substrate for insects, the bird families Apodidae (swifts) and Hirundinidae (swallows) feed primarily on the diverse insect microfauna dispersed upwards by ambient winds (section 7.4.1). Hummingbirds (Trochilidae) supplement their primary nectarivorous habits through aerial capture of insects, mostly small dipterans and hymenopterans (see Stiles, 1995). Ecologically, the impact of such generalist avian insectivores is undoubtedly substantial. The avian fauna of Panama, for example, includes one hundred species of tyrannid flycatchers alone, whereas the Tyrannidae form the largest New World bird family (see Ridgeley and Gwynne, 1989).

In addition to such generalist feeders on insects, specialized avian insectivores include the indigenous Neotropical families Bucconidae (puffbirds), Momotidae (motmots), and Galbulidae (jacamars), as well as the Old World Coraciidae (rollers), Dricruridae (drongos), Eurylaimidae (broadbills), and Meropidae (bee eaters). These birds typically sally forth from perches to capture flying insects and exhibit a high degree of maneuverability during sometimes extended aerial chases. Morphological specializations of these insectivores include elongated and highly mobile tails, together with often long, narrow beaks that facilitate prey capture and subsequent manipulation (see plate 4). Choice of prey items is often taxonomically restricted and is characterized by a refined discriminatory sense of insect suitability based on the behavioral and visual characteristics of potential prey (e.g., Chai, 1986, 1988). Such dedicated foraging behavior illustrates substantial selection for improved antipredatory characteristics in target prey items, and particularly in relatively large insects such as lepidopterans and orthopterans.

Before the emergence of echolocating microchiropteran bats by the early Tertiary (50 MYBP; see Jepsen, 1970), flying insects were able to exploit the night skies with but limited exposure to nocturnal insectivores. Two extant avian orders (Caprimulgiformes and Strigiformes) contain nocturnally foraging species that feed primarily on free-flying insects, although their ecological impact is likely to be much less than that of bats (Morse, 1975). Throughout the Tertiary and into recent times, the taxonomically diverse array of microchiropterans (with about 850 described species) has likely imposed substantial predatory pressure on nocturnal insect populations. Two historical outcomes of such predation have been behavioral shifts to diurnality (e.g., day-flying moths in the families Sesiidae, Uraniidae, and Zygaenidae) as well as the evolution of ultrasound detectors and bat-evading mechanisms (see Miller, 1983; May, 1991). Ultrasound-induced maneuvers are well-known within the Insecta, classically demonstrated among moths (e.g., Treat, 1955; Roeder, 1962, 1967), but also described in lacewings (Miller and Olsen, 1979; Olsen and Miller, 1979; Madsen and Miller, 1987), crickets (Moiseff et al., 1978; May et al., 1988), mantids (Yager et al., 1990; Yager and May, 1990), locusts (Robert, 1989), and beetles (Spangler, 1988; Hoy et al., 1989; Yager and Spangler, 1997).

The multiple independent origins of such evasive behaviors indirectly demonstrate the strong selective forces that have historically acted on winged insects to avoid aerial predation.

Ultrasound-induced evasion by insects typically involves wing, abdominal, and leg motions similar to those used during maneuvers (section 5.2.2). For example, locusts reacting to simulated bat echolocation respond with abdominal deflection and asymmetries in forewing kinematics (Robert and Rowell, 1992a; Dawson et al., 1997). In mantids, evasive responses involve turning and powered dives; these responses have been shown in field experiments to dramatically enhance survivorship during attack by bats (Yager et al., 1990). One extreme response to ultrasonic stimulation involves rapid positioning of hindlegs into the path of the beating wings (e.g., May and Hoy, 1990). This behavior, as with the rapid wing flip of lacewings (Miller and Olsen, 1979), likely yields a highly unpredictable aerodynamic response that interferes with interception strategies used by oncoming predators. Tiger beetles similarly move the elytra into the stroke plane of the wings, disrupting the flight trajectory and moreover producing ultrasonic clicks that potentially confuse echolocating bats (Yager and Spangler, 1997). Other insect taxa such as arctiid tiger moths possess dedicated sound-producing structures (microtymbals) that generate ultrasound in response to echolocation stimuli. Such countersignals may confound predation attempts, and/or may communicate distastefulness in an effort to dissuade potential bat predators. Ever-escalating coevolutionary interaction between bats and insects has thus led to the appearance of diverse evasive and deterrent mechanisms among insects as well as potentially more sophisticated methods of prey detection and discrimination by bats. For example, the reflection characteristics of ultrasonic echoes from certain insect taxa may enable bats to effect body size determination and perhaps taxon-specific identification (Schnitzler et al., 1983; Kober and Schnitzler, 1990). The extant richness of bat echolocation strategies, exhibiting variable pulse structures and differing patterns of frequency and amplitude modulation, further demonstrates the capacity of coevolutionary interaction between predator and prey to generate morphological, physiological, and behavioral diversity among all participants.

Many reptiles, amphibians, and arachnids are formidable predators of winged insects whenever these are at rest, recently landed, or merely walking about. Some pterygote taxa (e.g., reduviid assassin bugs) similarly capture and feed on other insects. The ecological impact of such predation on insect populations is not well documented, but selection for enhanced jumping and takeoff performance is one likely result over an evolutionary timescale. Many spider taxa also construct webs to capture flying insects. Although visually impressive, the ecological and evolutionary impacts of web capture on flying insects may be small relative to the effects of volant vertebrates (see Wise, 1993). For example, evolution of Batesian and Müllerian mimicry in butterflies, hymenopterans, and other insect taxa (section 7.3.2) appears to have been driven primarily in response to avian predation. Some insects do, however, exhibit specific behavioral responses to the presence of spider webs. Flying dipterans often visually detect these structures prior to contact and maneuver rapidly to avoid entanglement (Craig, 1986). Similarly, the slow oscillatory flight and extended legs typical of many phylogenetically basal flies may, in part, serve to detect webs prior to body entanglement (W. G. Eberhard, pers. comm.).

Predators often impose sublethal damage on prey during unsuccessful capture attempts (Vermeij, 1982, 1987). Predator-induced wing damage is widespread in many insects, particularly in butterflies, and can result in both symmetric and asymmetric reduction of wing area (e.g., Carpenter, 1937). Experiments that systematically evaluate insect flight abilities following wing area reduction have never been carried out, perhaps for fear of experimenter ridicule. Only one study has examined the consequences of symmetric reduction in wing area. Airspeeds of a pierid butterfly increased as the hindwings were gradually reduced in size, whereas the same manipulation on the metathoracic wings of a beetle resulted in a decreased airspeed (Demoll, 1918). A related manipulation in hymenopterans and moths involves mechanical decoupling of fore- and hindwings through extirpation of the hamuli or frenulum, respectively. Insects subject to such manipulations retain the ability to fly (Rabaud, 1933), but potential impairment of flight performance is unclear. In molting hummingbirds, a reduction in wing area through feather loss is correlated with a dramatic reduction in body mass (easing lift requirements) and a decline in hovering performance (Chai, 1997). For insects as well as volant vertebrates, the loss of wing area clearly mandates compensatory force production, but neither kinematic changes nor possible impairment of flight or escape abilities have been systematically investigated.

Of equal interest are possible compensatory responses to bilateral asymmetries in wing size and shape. For reasons either of failed predation or structural fatigue, asymmetric wing damage can be widespread in nature and may impose substantial energetic and aerodynamic costs. Slower flight speeds or reduced maneuverability under such circumstances may increase vulnerability to predation. For example, houseflies captured by barn swallows exhibit greater wing asymmetry relative to their uncaptured counterparts (Møller, 1996; see also

Møller and Swaddle, 1997), suggesting impairment of flight performance. Insects with such wing damage or asymmetry are, however, clearly capable of straight as well as turning flight. Locusts with one of four wings shortened or altogether removed can nonetheless compensate and sustain forward flight (e.g., Kamada and Kinosita, 1947; Scherenstein, 1968; Wilson, 1968; Simmons, 1977b; Baker and Cooter, 1979a). Similarly, free-flying *Drosophila* with asymmetric wing damage can achieve straight flight trajectories as well as maneuvers (Wolf and Heisenberg, 1990). These examples of compensation convincingly demonstrate neuromuscular flexibility of the flight motor system in response to substantial morphological deficits.

7.3.2 Predation, Erratic Flight, and Mimicry

Apparently as an evolutionary response to aerial predation, many insects fly rapidly, erratically, and unpredictably; other insect taxa fly slowly and regularly. Butterflies are particularly well known for erratic and irregular flight paths. Some moth taxa also exhibit highly variable flight paths (e.g., Callahan, 1965). When flying in extended trajectories, various Orthoptera and Homoptera (e.g., Cicadidae) also fly with seemingly unpredictable changes in direction. Why fly irregularly if the goal of flight is rectilinear translation in space? One classical explanation of such versatile protean displays is that these flight patterns function as warning signals indicating difficulty of capture and associated prey unprofitability (see Humphries and Driver, 1970; Driver and Humphries, 1988). This hypothesis has never been tested experimentally, but is consistent with observations of high rates of vertebrate aerial predation on insects and presumably with associated selection for enhanced defensive responses. Many insects also fly erratically when either threatened or actually attacked by a potential predator. When free-flying palatable butterflies perceive a predatory attack, both the complexity and possibly the speed of the already erratic three-dimensional flight trajectory transiently increase (Young, 1971; Nikolaev, 1974; Srygley, Oliveira, and Dudley, pers. obs.). Marden and Chai (1991) suggested that butterflies typically escape predation attempts through a net upwards movement, although such a reaction is not generally characteristic of this taxon. For example, only a minority of butterfly species in a lowland Ecuadorian rain forest fly upwards in response to attempted capture (DeVries, pers. comm.), and this behavior is similarly rare among Malaysian and Papuan butterflies (pers. obs.). The three-dimensional escape responses of flying insects have never been evaluated quantitatively, however, and together with chase strategies of aerial predators represent a fruitful area for future investigation.

The production of continuously erratic flight paths can involve substantial energetic expenditure. Variable accelerations, repeated changes of direction, and vertical displacement of the body may all dramatically elevate the costs of flight relative to direct translation of the body over an equivalent horizontal distance (see Dudley, 1991a; May, 1991). Such increased costs of flight probably act as a constraint on the evolution of continuously erratic locomotion, and may in part explain the restriction of such flight behavior to a very limited subset of winged insects. Two-dimensional analyses of erratically flying butterflies reveal substantial vertical deviations in the flight path relative to insect body dimensions (Chai and Srygley, 1990; Dudley, 1991a), but the three-dimensional geometry of such flight paths remains to be evaluated. Mathematical parameters useful for describing such flight paths include turning rate (absolute change in angular orientation per unit time), sinuosity (the absolute change in angular orientation per unit distance), and local curvature. The phylogenetic origins of irregular flight paths among butterflies are unclear given the absence of such quantitative information. The substantial impact of avian predation on large and conspicuous diurnal insect taxa suggests, however, that this deterrent mechanism and the evolution of unpalatability (see below) are fundamental adaptations of butterflies and, more generally, of diurnal Lepidoptera.

In contrast to the irregular flight paths of many palatable butterflies, a large number of unpalatable butterfly taxa (e.g., ithomiine, danaine, and many heliconiine nymphalids) tend to fly slowly and regularly. Rather than relying on erratic flight paths that indicate a difficulty to capture them, such butterfly taxa are chemically defended against predators and advertise their unpalatability via aposematic warning coloration (e.g., flashy red, yellow, and black wing patterns). Bates (1862) and Wallace (1865) first noted that palatable butterflies fly substantially faster than unpalatable butterflies, an observation confirmed empirically for butterflies flying in insectaries (Chai, 1986, 1988, 1990; Chai and Srygley, 1990) and during free flight (Srygley and Dudley, 1993; Dudley and Srygley, 1994; see also DeVries, 1987). Unpalatability in butterflies is a trait difficult to confirm without extensive feeding trials and/or chemical assays. In landmark studies, Chai (1986, 1988, 1996) evaluated feeding responses of a specialized avian insectivore, the rufous-tailed jacamar (see plate 4). The palatability of Neotropical butterflies to this bird species can be dichotomously characterized, as can be morphological and behavioral features (Chai, 1986, 1988). Palatable butterflies are generally erratic fliers, whereas aposematic and unpalatable butterflies tend to fly with more regular flight paths. Palatable butterflies also tend to be brightly colored on the dorsal wing surface. Ventral wing surfaces, however, have highly cryptic patterns, rendering the butterfly difficult to detect when wings are folded together dorsally at rest.

Additional morphological features differentiate palatable and unpalatable butterflies and are used by jacamars to assess prey suitability (Chai, 1986, 1988). In palatable butterflies, the center of body mass tends to lie more anteriorly, consistent with increased pitching responsiveness of the body and thus with enhanced maneuverability (Srygley and Dudley, 1993; Srygley and Kingsolver, 1998). By contrast, unpalatable butterflies are characterized by a relative reduction in thoracic mass and correspondingly greater mass allocation to relatively longer bodies, consistent with a reduced need for evasiveness and rapid flight (see Chai and Srygley, 1990; Srygley and Chai, 1990a; Marden and Chai, 1991; Srygley and Kingsolver, 1998). Muscle fiber diameters in unpalatable danaine (milkweed) butterfly species also tend to be relatively greater than in palatable butterflies (George and Bhakthan, 1960). A number of morphological and physiological characteristics are thus correlated with the evolution of unpalatability in butterflies. The reduced demand for rapid flight from predators also appears to relax thermoregulatory demands; unpalatable butterflies tend to fly in cooler microhabitats and exhibit reduced thoracic temperatures (Srygley and Chai, 1990b). Such ecophysiological correlates of unpalatability suggest complex evolutionary responses to the often intense predation pressure imposed by visually oriented diurnal insectivores.

A famous and visually compelling phenomenon in tropical biology is mimicry among butterflies (Bates, 1862; Wallace, 1865, 1867). Batesian mimicry involves convergence between an unpalatable model species and one or more palatable mimic species. Because the unpalatable model is avoided by predators through learned aversive responses, mimics obtain a selective advantage through morphological and, in some cases, behavioral resemblance to rarely attacked prey. In a Müllerian mimicry complex, by contrast, two or more unpalatable species resemble one another and mutually reinforce aversive deterrence of would-be predators. Batesian and Müllerian butterfly mimics often exhibit similar wingbeat kinematics and flight styles as well as remarkable convergence in color patterns, wing shape, and relative body dimensions. For example, morphological, behavioral, and even thermal characteristics of certain Batesian mimics from a Neotropical butterfly fauna resemble those of their unpalatable models (Chai, 1986; Srygley and Chai, 1990b). Morphological convergence may include not only similarity in wing color and pattern, but also convergence in such flight-related parameters as wing mass and position of the center of body mass (Srygley, 1994a, 1999). Batesian mimics of unpalatable butterflies do retain relatively higher thoracic muscle masses that presumably enhance escape options if they are attacked by either naive or particularly perceptive predators (Srygley and Chai, 1990a).

Batesian mimicry in butterflies is often sexually dimorphic, with only the female of a particular species resembling an unpalatable model (Wallace, 1865). Because of the increased demands of reproduction, females of palatable butterfly species tend to have relatively larger abdomens and a reduced investment in thoracic muscle mass (Srygley and Chai, 1990a; Marden and Chai, 1991). As a consequence, female butterflies may have reduced maneuverability relative to males, and additionally may be more vulnerable to predators during extended searches for oviposition sites. Empirically, females of palatable butterflies do appear to be more subject to avian attack than are males (e.g., Ohsaki, 1995). The restriction of Batesian mimicry to females is thus consistent with an enhanced need for such antipredatory measures as protective resemblance. Sex-limited mimicry may in fact be most pronounced in those taxa most subject, for ecological reasons, to attack by visually oriented predators. A related prediction is that the relative morphological and kinematic similarities of mimicry complexes are proportional to the intensity of historical predation on a species that may still be evident today in some habitats.

Biomechanical convergence in flight styles among different species was termed "locomotor mimicry" by Srygley (1994a). In addition to the Batesian resemblance of unpalatable models, locomotor mimicry also encompasses potential kinematic similarities among Müllerian mimics as well as among palatable butterflies that fly irregularly in a variety of displays. Despite Brower's (1995) objections, such behavioral convergence is an extension of unpalatability into the more general area of prey unprofitability and the associated deterrent effects on predator behavior (see Baker and Parker, 1979). An excellent example of locomotor mimicry between two erratically flying and palatable butterfly species (*sensu* Chai, 1986) is provided by the heliconiine nymphalid *Dryas iulia* and the nymphaline nymphalid *Marpesia petreus* (Srygley, 1994a). Wing coloration is roughly similar between the two species as well as in the related species *Marpesia berania*. The daggerwing genus *Marpesia*, however, differs substantially in wing outline and expresses a hindwing extension that at close range immediately identifies the genus (plate 5). From distances greater than several meters, however, the wing and body motions of *Marpesia petreus* in migratory flight are remarkably similar to those of the heliconiine butterfly, even to experienced observers (Srygley, Oliveira, and Dudley, pers. obs.). All three species are essentially sympatric and widely distributed throughout the Neotropics (DeVries, 1987), so that exposure during flight to a common set of predators is likely. The advertisement of unprofitability through an erratic flight trajectory is thus a highly viable hypothesis for the striking convergence in wingbeat kinematics and flight trajectory among these species. Most importantly, behavioral similarity in flight seems to be more accentuated than the convergent wing coloration so paradigmatic of classical mimicry theory.

Such similarity in flight styles between two unrelated butterfly species nicely illustrates the evolutionary potential of aerial predation to select for locomotor mimicry that is underpinned by both kinematic and morphological resemblance. Similar results pertain to the convergence of flight characteristics within Müllerian mimicry complexes. For example, Srygley (1999) demonstrated locomotor and morphological convergence within two mimetic groups of unpalatable *Heliconius* butterflies; phylogenetic contributions to differences in flight kinematics and anatomy were of secondary importance. As with morphological characters, locomotor convergence may thus occur within the context of either Batesian or Müllerian mimicry. Further analyses of mimetic similarity among butterflies as well as within other insect taxa (e.g., Beebe and Kenedy, 1957; Linsley et al., 1961) are clearly called for, preferably with the statistical incorporation of phylogenetic hypotheses and with a quantitative assessment of both locomotor similarity and resemblance in color pattern. Remarkable interordinal mimicry complexes that include phylogenetically unrelated insect taxa deserve comparable evaluation (e.g., Marshall and Poulton, 1902; Carpenter and Ford, 1933; Fraser, 1936; Silberglied and Eisner, 1969; Hespenheide, 1973; Opler, 1981).

7.4 Long-Range Dispersal and Migration

The flight of many insects is often characterized by either inadvertent or intentional dispersal. As with many organisms, insects can be physically swept upwards and transported by storms and other large-scale atmospheric phenomena. Drifting with ambient winds (aeolian dispersal) is another mode of transport available to many small insects. One specific form of dispersal is through migration, uni- or bidirectional motions enabled by specific locomotor behaviors and morphologies (Dingle, 1989, 1996). The study of insect dispersal and migration involves diverse biotic and abiotic factors, including endocrinological regulation of behavior (Pener, 1985), genetic polymorphism for dispersal-related morphology (Harrison, 1980; Zera and Denno, 1997), lability of the flight muscle (Johnson, 1976), meteorological influences (Pedgley, 1982; Drake and Farrow, 1988, 1989; Rainey, 1989), and the energetics of flight (Pennycuick, 1969; Rayner, 1990; Dudley, 1995a).

Classic books by Williams (1958) and Johnson (1969) also provide strategic overviews of insect migratory behavior. Insect dispersal and migration can be studied at the level of proximate mechanisms that influence the ability and motivation to migrate, or at ultimate levels of causation that address evolutionary origins and maintenance of such behavior (Dingle, 1996). From an aerodynamic perspective, long-range displacement of flying insects can be viewed as a mechanistic phenomenon influenced primarily by the biomechanical and physiological features of flight. Also clear, however, is the fact that insect dispersal is often strongly size dependent, and that the associated energetics of flight can be a major selective factor underlying the evolution of migratory behavior.

7.4.1 Diversity of Insect Migrants

Continuous aerial entrainment by winds combines with the large individual numbers of insects worldwide (section 1.1.2) to result in a substantial but transient population of insects moving within the troposphere (0–12 km from the earth's surface). This fauna within the vertical air column is sometimes called the aerial plankton, and is studied primarily by deploying nets from airplanes (see Felt, 1928; Glick, 1939; Johnson, 1969). Intercontinental movements of insects as well as continuous fallout of the aerial fauna on distant mountaintops (see Mani, 1962) are perhaps the most dramatic consequences of such long-range dispersal. Although macropterous insects of all sizes may engage in aerial dispersal and migration, smaller insects tend to dominate the aerial fauna (Gislén, 1948; Johnson, 1969), partly because smaller species are more abundant in the insect fauna as a whole (section 7.1.1). However, smaller insects also tend to be relatively more abundant at higher altitudes within the air column (Glick, 1939; Johnson, 1957, 1969; Hespenheide, 1977). Small body size facilitates both incidental and intentional dispersal by wind, an effect that arises from the low airspeeds of most insects relative to ambient air motions (section 7.4.2). Such opportunities for both short- and long-range displacement have combined with the inherent capacity of flight to yield opportunities for colonization that are perhaps unsurpassed by other terrestrial animals.

Dispersal of insects by ambient winds is impressive from taxonomic as well as geographical perspectives. Representatives from all major and many minor insect orders have been recorded from aerial samples taken over land (Johnson, 1969). Equally striking instances of long-distance dispersal among most taxa have been documented from aerial samples obtained over open seas (Bowden and Johnson, 1976).

Virtually all such insects found aloft are winged adults, although larvae from both hemi- and holometabolous orders are also occasionally present. Representatives from most insect orders also engage in specific migratory displacement, although these behaviors are most pronounced in the Lepidoptera and Orthoptera (see Fraenkel, 1932b; Johnson, 1969). Remarkable simultaneous directional migrations by representatives of five or more insect orders have also been described at various tropical locales (e.g., Beebe, 1949). Such temporal and geographical convergence in migratory behavior among unrelated taxa is strongly suggestive of common underlying causes that promote the evolution of long-range intentional displacement.

7.4.2 Active and Passive Mechanisms

Motion of the atmosphere can be substantial near the earth's surface, and many features of insect dispersal and migration are influenced by ambient winds. Taylor (1958) introduced the concept of the flight boundary layer to studies of insect dispersal. This boundary layer extends from the surface of the earth vertically to some height above which ambient wind speeds exceed insect flight speeds. Within the flight boundary layer, insect airspeeds are greater than typical wind speeds, thereby enabling flight directionality to be maintained actively through wing flapping. Above the flight boundary layer, however, flight directionality is primarily if not exclusively determined by the prevailing winds. Because wind speed increases exponentially with increased distance from the earth's surface (Geiger, 1965), even fast-flying insects must at some height eventually cross the flight boundary layer. Motion of the insect relative to the earth's surface will then be downwind, although active flapping flight is still necessary to maintain vertical position in the moving air. Observation of downwind flight does not by itself unequivocally confirm flight above the boundary layer for any particular taxon, but simply indicates that ambient air motions are contributing to an unknown extent to the insect's motion. Flight boundary layers also vary dynamically because of spatial and temporal heterogeneity in ambient winds. This effect combines with the general absence of accurate airspeed data (section 3.1.1) to make the depth of the flight boundary layer difficult to specify for any particular insect (Dudley, 1995a).

The ability to decouple the flight trajectory from ambient winds depends on the relative magnitude of insect airspeeds and wind speeds. Allometric study of the former quantity suggests that body size exerts a major influence on the evolution of migratory behavior. In particular, most insects have body lengths of less than 1 cm and attain maxi-

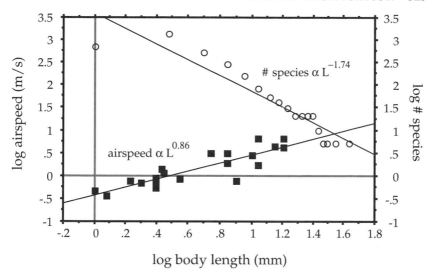

FIG. 7.2. Comparative allometries of insect flight speed and body size distribution (modified from Dudley, 1994). Most insect species fly at airspeeds well below 5 m/s. Least-square regressions as follows: log airspeed = 0.86 (log body length) −.396, r^2 = 0.79, P < .001; log species number = −1.74 (log body length) + 3.6, r^2 = 0.86, P < .001.

mal airspeeds that only rarely exceed ambient air motions (see Dudley, 1994). Systematic comparison of insect size distributions, the allometry of insect airspeeds, and typical values for wind speed demonstrates the generality of this observation. May (1978) compiled a body-length distribution for all species of British Coleoptera. Unfortunately, no other such compilation is presently available to evaluate possible geographic or taxonomic variance in the ensuing distribution, but the allometry of species number is likely to be similar for most other insect faunas. Data relating insect airspeed to body length were given by Johnson (1969) for a variety of insect orders; these data do not necessarily refer to maximal attainable airspeed, but rather to some chosen airspeed in a laboratory context. Allometries of the species distribution of May (1978) and the flight speed data of Johnson (1969) can then be compared to assess the fraction of the insect fauna likely to exceed any particular airspeed (see fig. 7.2).

This analysis suggests that only about 40% of insect species are likely to fly at airspeeds greater than 1 m/s, dropping to 22% of the fauna exceeding 2 m/s and only 8% capable of surpassing 5 m/s (Dudley, 1994). By contrast, a wind of 1 m/s is barely discernible to humans, and is a typical diurnal value within many tropical forests.

However, wind speeds are much higher above the tropical forest canopy, in the range of 5–10 m/s (e.g., Haddow and Corbet, 1961; Allen et al., 1972; Thompson and Pinker, 1975; Aoki et al., 1978; Windsor, 1990). Typical diurnal winds in tropical savannah and above temperate-zone vegetation can be even stronger (see Rumney, 1968). Although aforementioned distributions for insect body size and flight speed are only approximate, a substantial fraction of all insect species flies at airspeeds that rarely exceed ambient air motions. The majority of the world's insect fauna must then always fly within 1–2 m from the ground or within vegetational canopies if intentionally directed flight is to be attained. A more detailed analysis of insect flight performance (particularly for taxa with body mass below 10 mg) would quantitatively elaborate this result, but the qualitative conclusion is likely to be unchanged. Most insects will be able to disperse or migrate only via wind-induced displacement. In the evolution of migratory habits, the magnitude of insect flight speed relative to wind intensity has therefore likely played a major role.

Insect displacement by winds was once viewed as an entirely passive physical process to which behavioral decision making made no contribution. Kennedy (1951), however, suggested that a rigorous dichotomy between active and passive mechanisms of dispersal was inaccurate. Even for those insects whose flight trajectories are dominated by ambient air motion, the actual decision to take off and initiate aerial dispersal is subject to various behavioral and physiological influences (e.g., Dixon and Mercer, 1983; Washburn and Washburn, 1984). In aphids, intentional takeoff is matched by active aerial searching for a suitable host plant; horizontal displacements are facilitated but not necessarily dominated by ambient winds (Loxdale et al., 1993). Similar behavioral control characterizes active downwards flight by aphids once the final destination has been identified (Thomas et al., 1977). Active regulation of position via controlled ascent and descent may also be used to maintain constant altitude in vertically moving air volumes. Although many small insects and even some larger ones use thermal updrafts to become airborne and to stay aloft (Johnson, 1969), insects must hover in the absence of sustaining vertical air motions if their horizontal displacement is to be sustained. The energetic cost of hovering is comparable to or greater than that of slow forward flight (section 4.1.1), so that even passive drifting with winds involves considerable energetic expenditure.

Wind-mediated displacement above the flight boundary layer is widespread among insects, but many insects also actively regulate directionality within the boundary layer. Wind-drifting taxa are often small and have asynchronous flight muscle (e.g., Diptera, Homoptera), whereas renowned lepidopteran and orthopteran migrants tend to be much larger and possess synchronous flight muscle (section 7.1.1). Larger insects have greater control of the flight trajectory because of typically higher airspeeds (section 3.1.1), but the difficulties of decoupling airspeed from ground speed for flying insects render measurements of flight orientation relative to prevailing winds difficult to obtain (Johnson, 1969). Limited data do suggest that dispersing insects of various sizes often fly upwind within their flight boundary layer, thus establishing directionality independent of ambient air motions. For example, Lutz (1927) found that the majority of insects trapped nocturnally near the ground (primarily dipterans) flew against prevailing winds. Kennedy (1939) found a similar tendency of mosquitoes to fly directly into airflow under laboratory conditions. Foraging honeybees increase airspeeds when flying in a headwind and similarly decrease airspeed in tailwinds (Park, 1923; von Frisch and Lindauer, 1955; see also Heran and Lindauer, 1963). Maintenance of a constant ground speed via optomotor responses (section 5.1.1) is one obvious mechanism by which insects could indirectly assess and compensate for the effects of adverse winds under such conditions. An additional cue that might be used by flying insects to determine wind direction is the potentially variable orientation of wind gusts over small spatial scales, although data in support of this hypothesis remain murky (see Riley and Reynolds, 1986).

Migratory butterflies are particularly well known for powered directional flight within the boundary layer (see Williams, 1930; Baker, 1978; Walker, 1985). Such behavior requires sophisticated means of identifying and maintaining a constant flight orientation when flying in winds of variable speed and direction. For example, Neotropical butterfly migrants adjust airspeeds and flight directions in order to compensate for drift induced by adverse winds (Srygley et al., 1996; see also Walker and Riordan, 1981). Behavioral manipulations reveal use of a time-compensated sun compass in these migrants that helps to establish the preferred geographical direction of flight (Oliveira et al., 1998). Similar orientational mechanisms likely characterize upwind migrations of various Diptera, Hymenoptera, and Odonata (see Williams, 1958; Mikkola, 1986). Migratory flight within the boundary layer, when coupled with rapid directional adjustment of the flight trajectory, is of clear advantage in identifying and moving to nectar-bearing flowers or other ecological resources. Such a vision-based strategy would clearly be less effective at night, as would be any optomotor-based assessment of ground speed (e.g., Riley et al., 1988). Diurnal insect migrants thus appear to follow one of two general methods correlated with body size: large insects utilizing powered flight within the boundary layer, or convective dispersal of small insects flying above their corresponding boundary layer. Unfortunately, a detailed survey of insect migration habits relative to body size and flight speed has never been carried out. Some large insects clearly migrate using wind-assisted horizontal displacement. The monarch butterfly (*Danaus plexippus*) flies primarily within the flight boundary layer, but occasionally soars with ambient winds to effect long-distance migration across the North American continent (Gibo and Pallett, 1979; Gibo, 1981, 1986). Other long-range diurnal migrants (e.g., swarming locusts) similarly fly with the wind when outside of their flight boundary layer (see Johnson, 1969; Brown, 1970).

Additional ecological factors potentially influence insect dispersal strategies. Diurnal flight exposes insect migrants to attack by numerous avian insectivores, and may additionally impose significant thermal stress, particularly in xeric and tropical regions. Flying at night reduces the problem of heat overload (an effect also obtained at higher altitudes), but may substitute some highly effective chiropteran predators for their diurnal avian counterparts. Because nocturnal air motions are generally reduced in magnitude, the height of the flight boundary layer is correspondingly increased. However, nocturnal winds can still be high relative to the flight speeds of small insects. Relative to diurnally active insects, Taylor (1974) and Taylor et al. (1979) found that fewer small insect taxa crossed the nocturnal boundary layer into higher levels of rapidly moving air. Similarly, downwind nocturnal migration is most common among larger insects (see Riley and Reynolds, 1986; Drake and Farrow, 1988; Burt and Pedgley, 1997). The difficulty of measuring air- and windspeeds simultaneously for high-flying nocturnal migrants precludes any definitive statement concerning the relative contribution of prevailing winds to long-distance displacement in such taxa.

7.4.3 Migrant Physiology and Aerodynamics

Energetic costs of flight must influence migration strategies both for wind-assisted dispersal as well as for directed flight within the boundary layer. Small insects being convectively dispersed must nonetheless maintain a forward airspeed and offset their body weight, or alternatively must reduce their flight velocity so as to hover within the moving air volume. If the goal of dispersal is to maximize the horizontal distance traveled, then flight at the minimum power speed would maximize time aloft and thus the extent of wind-assisted displacement. By contrast, migratory flight within the boundary layer would likely occur at the maximum range speed that optimizes horizontal coverage (see section 3.3.3). No data are presently available that assess insect airspeeds and rates of energetic expenditure during convective dispersal. Moreover, aerodynamic models that predict optimal airspeeds are seriously compromised if energy stores can be replenished during migration (see Pennycuick, 1989; Walsberg, 1990). For boundary layer migrants in particular, the fixed availability of endogenous reserves may be substantially augmented through nectar feeding or related strategies during long-distance flights.

For example, the diurnal Neotropical moth *Urania fulgens* engages in long-distance migratory flights that may cover several thousand kilometers through Latin America. Comparison of estimated metabolic rates with endogenous lipids suggests that these moths can fly only 110 km or less using stored reserves (DeVries and Dudley, 1990; Dudley and DeVries, 1990). Extensive nectar feeding en route is therefore essential to obtain the energy for the long-distance flights characteristic of this and probably many other diurnal lepidopteran migrants. Choice of airspeeds during continuous powered flight within the boundary layer may correspondingly vary with instantaneous rates of energy uptake as well as with net rates of energetic expenditure. A complete description of migratory energetics must thus contain not only information on energetic costs of flight, but also of energetic uptake prior to and following migration. For example, total energy uptake during the course of a migratory period can exceed premigratory hypertrophy in stored reserves (e.g., Brower, 1985); the relative extent and predictability of postmigratory nutritional resources may similarly influence strategies of migration. Furthermore, ecological costs of flight and feeding (e.g., exposure to predators) may vary dramatically along a migratory pathway. Comprehensive data on rates of energetic uptake and expenditure are difficult if not impossible to obtain for individual migrating insects. Comparison of feeding rates and flight energetics among different individuals might, however, be logistically tractable for many diurnally migrating butterflies and moths.

In many insect migrants, a flight-oogenesis syndrome has been classically described whereby migration typically occurs prereproductively (Johnson, 1969). The flight apparatus must necessarily be well developed for migration, but the ovaries and other elements of the reproductive tract in migrant females typically remain immature. Why might dispersal occur before reproduction in migratory phenotypes? The usual ecological explanation (e.g., Dingle, 1972) is that fecundity is higher when dispersal occurs immediately prior to reproduction, a hypothesis for which there exists some empirical support. For example, fecundity of insects in tethered flight is often inversely proportional to flight duration, although long-duration flight in insect migrants does not necessarily reduce total fecundity (see Roff, 1977; Dingle, 1985; Rankin et al., 1986; Gunn et al., 1989). Moreover, both flight activity and oogenesis immediately following dispersal may in some insects be stimulated by the same endocrinological pathway (Rankin and Burchsted, 1992). The discrete categorization suggested by the flight-oogenesis syndrome may thus be inappropriate from an integrative physiological perspective. Oviposition and successful larval development are clearly likely to be more successful if adult reproductives disperse from deteriorating to more appropriate habitat. This observation does not by itself, however, preclude flight by female insects with mature reproductive capacity.

In contrast to hypotheses based on total fecundity, specific biomechanical and physiological constraints on reproductive development have been less well investigated in migratory insects. The biosynthesis of energetic substrates used in flight, together with construction and physiological maintenance of flight muscle, can potentially reduce allocation to reproductive structures (see Zera et al., 1994, 1997). Lability of the flight apparatus is in fact a general feature of many migratory insects, and is particularly evident in polymorphic species that exhibit trade-offs between flight muscle and ovarian mass (see Zera and Denno, 1997; Tanaka and Suzuki, 1998). Because lipid is the optimal storage medium for long-distance migrants (section 4.1.2) but can also be a major egg constituent, flight metabolism and reproductive allocation may compete directly for biochemical substrate. Migratory polymorphisms in wing morphology are similarly evident at the level of energetic reserves. For example, solitary phase locusts express essentially none of the lipids necessary for long-distance flight that are present in the gregarious migratory morphs (Chino et al., 1992). Energy availability and differential allocation of resources within the body of insects thus underpin, at least partially, prereproductive migratory behavior in many taxa.

Because wing dimensions do not change with either increased lipid reserves or egg loads, one additional consequence of lipid loading and mass hypertrophy prior to migratory flight is an inordinate increase in the mechanical and metabolic power required to fly (see section 5.4.2). Increased body mass through allocation to reproductive structures will thus have a disproportionately negative effect on migrant flight energetics. Given that the energetic costs of flight are so high (section 4.1.1), mass allocation to reproductive structures and particularly to oogenesis can be advantageously postponed until migratory activity has terminated. One possible exception to this conclusion, however, may be found in boundary-layer migrants that oviposit along the length of a migratory pathway (e.g., many butterflies). In addition to influencing energetic expenditure, substantial lipid reserves and/or egg loads may impair other features of flight performance. Lipid reserves of migrant insects are typically 10–20% of total body mass (e.g., DeVries and Dudley, 1990), probably reducing maneuverability and evasive capacity in response to predatory attack. The biomechanical means of compensation for lipid-derived mass increase have not been evaluated in migrant insects, although migrating individuals often exhibit significant differences in wing length and area relative to nonmigratory conspecifics (see Angelo and Slansky, 1984).

What then is an optimal energetic strategy for lipid loading prior to migration? Premigratory lipid accumulation will obviously increase the maximum range attainable during migration, but the magnitude of this effect decreases exponentially with added lipid mass (see Rayner, 1990). Increases in body volume through lipid loading or oogenesis will also increase body drag somewhat, correspondingly increasing the parasite power and total power required to fly (Pennycuick, 1975). Gradual use of lipid reserves through the course of a migration reduces total body mass. In theory, this reduction should in turn lower the maximum range speed (which varies in proportion to mass$^{0.5}$; see Pennycuick, 1978). A similar reduction in flight speed should characterize migrants maximizing time spent aloft. Insects at the beginning of long-distance flights would thus be predicted to be flying fastest, and then to reduce gradually their airspeed as the migration proceeds. In a comparison of different migrating individuals within a species, insects with higher lipid load relative to body mass should be faster fliers. Detailed surveys of variation in airspeeds during migration would be necessary to test these theoretical predictions, but no such data are available for any insect species. Nor do aforementioned aerodynamic models incorporate potential confounding effects of variable energy uptake during the course of a migratory period.

7.5 COMPARISON OF INSECT AND VERTEBRATE FLIGHT

Within the extant fauna, flapping flight is confined to bats, birds, and insects. Phylogenetic associations of body design fundamentally distinguish the winged insects from volant vertebrates; virtually disjunct distributions of body size also render the kinematics and aerodynamics of insect flight distinct from those of bats and birds. With the impressive exception of hummingbirds and some nectar-feeding bats, physiological and biomechanical features of flight in winged insects are found nowhere else in the animal kingdom.

7.5.1 Mechanics of Flight

Allometric considerations dominate any mechanical comparison of insect and vertebrate flight. Volant vertebrates are, on average, three to five orders of magnitude heavier than the typical adult pterygote. Some insect taxa (e.g., various acridid orthopterans and scarabaeid coleopterans) do venture into a body mass domain of 10–50 g that is more typical of flying vertebrates, but such overlap in size is the rare exception, given average body dimensions of the extant insect fauna (section 7.1.1). Two immediate consequences of differing body size distributions between insects and volant vertebrates are substantially different flight speeds and wingbeat frequencies. Flight speeds of birds and bats are typically much higher than those of insects (see Rayner, 1987; Norberg and Rayner, 1987; section 3.1.1). With the exception of the hummingbirds, wingbeat frequencies of insects dramatically exceed those of volant vertebrates (see fig. 3.3B). Given these major differences in body size, forward airspeed, and wingbeat frequency, the mean values of *Re* for the wings and bodies of flying insects are typically two to four orders of magnitude lower than comparable values for flying vertebrates. The dependence of body lift and drag on *Re* is substantial and indicates that insects experience relatively higher body drag than do bats and birds (see section 3.2.1.1). Differences in lift production between winged insects and volant vertebrates will similarly be influenced by widely divergent *Re* for flapping wings, although the *Re* dependence of unsteady lift mechanisms remains to be determined experimentally (see sections 3.2.1.2 and 3.2.2). These latter mechanisms are, however, likely to be much more pronounced in miniaturized pterygotes than among bats and birds.

Few comparative data are available on differences in maneuverability among volant animal taxa. Marden and Chai (1991) suggested that maximum accelerations of butterflies are comparable to those of their avian predators, but this conclusion was based on vertical load-lifting data that are not necessarily relevant to translational accelerations during forward flight (see section 5.4.1). Only limited data have been obtained on the maximum accelerations of flying insects (section 3.1.1). Nor have other kinematic features of flight maneuverability been systematically compared across volant animal taxa. Given isometric body design, accelerations are likely greater at lower body masses (section 5.4.1). Insects thus fly more slowly than volant vertebrates, but may also be capable of greater linear as well as rotational accelerations. Unfortunately, the relative contributions of these and related features of axial and torsional agility to evasive capacity are unknown, and furthermore are likely to be highly dependent on the behavioral context of the interaction. For example, greater accelerational ability may be of little use to a flying insect if the primary escape response consists of undirected and possibly unpredictable disruption of its flight trajectory. Given the substantial impact of aerial predation on winged insects (section 7.3.1), three-dimensional studies of attack sequences would usefully determine those kinematic features that underlie both insect evasiveness as well as predatory strategies during attempted capture.

As with all arthropods, insects have exoskeletons, articulated joints, and cuticular projections (including the wings) that extend from body segments. In contrast to vertebrate wings that are flapped by the action of intrinsic musculature, wing motions of insects must be induced by the thoracic deformation and indirect transmission of forces through the wing base (section 2.1). Such mechanical activation is likely enabled by the elastic properties of the cuticular exoskeleton, thus restricting its presence among volant taxa to the winged insects. Very small insects have fringed wings that likely serve aerodynamic roles particular to wing flapping at low Re (see section 7.1.2). Larger insects, by contrast, fly using cambered and continuous wing surfaces that superficially resemble the wings of bats and birds. Longitudinal flexion of the insect wing is, however, severely restricted relative to that of the longitudinally articulated vertebrate wing. Therefore, dramatic wing bending during the upstroke and associated reduction in effective surface area are not kinematically feasible for most insect wings. This morphological restriction precludes insects from employing the aerodynamically variable upstroke typical of vertebrates flying within particular airspeed ranges (Norberg, 1990). Capacities for deformation in wing contour and for wing rotation about the longitudinal axis at the ends of half-strokes are, however, much greater in insects than in bats and birds. For example, substantial camber reversal between the down- and upstroke is facilitated by the flexibility of the wing cuticle and by particular patterns of wing venation (section 2.2.2). Such features of morphological design can promote dynamic equivalence between the half-strokes and ultimately facilitate the evolution of hovering flight.

Additional phylogenetic associations of body design differentiate physiological features of flight performance in insects and volant vertebrates. The tracheal system of insects, apart from limited ventilation in fairly large taxa, typically requires only costs of construction and maintenance to effect exchange of respiratory gases (see section 4.1.1). The preclusion of large body size via diffusive constraints within the tracheal system likely contributes to the minimal overlap in body size between present-day insects and flying vertebrates. By contrast, pulmonary respiration in bats and birds necessitates active ventilation and concomitant energetic expenditure that is exacerbated by the metabolic demands of flight. Convective transport of oxygen by the vertebrate respiratory and cardiac systems imposes potentially limiting nondiffusive steps on metabolic performance. The ability of the heart to pump oxygen-bearing blood may in fact limit muscle power output in flying vertebrates (see Bishop and Butler, 1995; Bishop, 1997). Respiratory constraints on flight performance, if applicable to insects, clearly must refer to the tracheal system alone given its unitary role in supplying oxygen to the flight muscle. Finally, the facultative endothermy characteristic of many flying insects (section 4.3) contrasts sharply with the continuous endothermy and associated energetic expenditure of extant vertebrate fliers. In part, heat derived from flight metabolism may partially offset the thermal demands of endothermy in bats and birds (e.g., Chai et al., 1998). In large flying insects, however, thermoregulation emerges as an indirect consequence of muscular contraction and imposes no net energetic cost per se (see section 4.3). For most insects, sustained endothermy during flight would be energetically impossible given the increased surface area : volume ratio definitionally characteristic of small body size.

7.5.2 Hummingbirds as Insect Analogs

Among volant vertebrates, hummingbirds (Trochilidae) represent the only example of evolutionary convergence on winged insects. Over 330 species of hummingbird are found exclusively in the Neotropics, with a small number of species seasonally migrating into North America. Species richness of the Trochilidae is matched by a remarkable diversity in bill, wing, and tail shape (see Greenewalt, 1960a). The flight abilities of hummingbirds are renowned and range from trans-Caribbean migration to impressive chases, display dives, and maneuvers. Most importantly, the nectarivorous lifestyle of hummingbirds is been paralleled by the evolution of high wingbeat frequencies and a spectacular ability to hover that is unparalleled among vertebrates. The highest recorded wingbeat frequency among trochilids is approximately 80 Hz (Greenewalt, 1962), representing the fastest known reciprocating vertebrate appendage. As with insects, an inverse relationship between wing length or body mass and wingbeat frequency (see fig. 3.3B) is suggestive of tightly tuned mechanical resonance within the hummingbird flight apparatus.

Additional morphological and kinematic features of hummingbirds are suggestive of pterygote design. Fusion of radial wing bones has eliminated the possibility of longitudinal wing flexion during the upstroke, a condition found neither in chiropterans nor in other avian lineages. Instead, the hummingbird wing functions at an approximately constant surface area during both half-strokes, as is typical of wing flapping in insects. Similarly, hovering hummingbirds exhibit camber reversal and substantial wing rotation about the longitudinal axis at the ends of half-strokes. Hovering flight is also characterized by nearly sinusoidal wing oscillation and approximately equal force production by the down- and upstroke, patterns otherwise only known from the insects (Weis-Fogh, 1972). Somewhat surprisingly, the kinematics of forward flight in hummingbirds have never been systematically analyzed. Modulation of stroke amplitude, however, does appear to be one important means of airspeed regulation (see Greenewalt, 1960a). Wing circulation is probably reversed between half-strokes, yielding temporal patterns of force production and a concertina-type vortex wake very similar to those of flying insects (Pennycuick, 1988; Rayner, 1995; see section 3.2.2.5). Major differences in flight performance between hummingbirds and insects necessarily derive from the typically greater size of the former taxon, in addition to the presence of a tail that likely assists maneuvers.

Flight metabolism in hummingbirds is also convergent on that of insects. Among vertebrates, trochilids exhibit the highest known rates of mass-specific oxygen uptake and represent an important system for evaluating limits to animal locomotor performance (see Suarez et al., 1991; Suarez, 1992; Hochachka, 1994; Chai and Dudley, 1995; Dudley and Chai, 1996). For example, hummingbird flight muscles have little or no anaerobic capacity, as is otherwise found in volant vertebrates. High rates of oxygen flux in hummingbird flight muscle are matched by mitochondrial densities that approach 40% of total muscle volume, and by cristal surface densities within mitochondria that are near a theoretical maximum (see Srere, 1985; Mathieu-Costello et al., 1992). Flight metabolism of hummingbirds is therefore highly similar to that of winged insects. Digestive, pulmonary, and circulatory systems of hummingbirds are similarly up-regulated to sustain the elevated metabolic rates of flight (Suarez et al., 1991). As with bumblebees in forward flight (section 4.1.1), hummingbirds exhibit nearly constant rates of oxygen consumption when hovering and in forward flight up to speeds of 7 m/s (Berger, 1985; see fig. 4.1). This finding also suggests that mechanical power expenditure and, by implication, flight aerodynamics of hummingbirds are similar to those of insects with fairly high wingbeat frequencies.

Behaviorally, hummingbirds can hover at nectar-bearing flowers for extended periods of time (e.g., up to 30 s). No other vertebrate taxon is capable of such sustained hovering. Glossophagine phyllostomid bats are, however, effective pollinators of flowers in the Neotropics, hovering for up to 4 seconds through the generation of vertically oriented lift in both the down- and the upstroke (see von Helverson, 1993; Winter et al., 1998). Among insects, hummingbirds are most comparable to hawkmoths; some diurnal sphingids feeding from flowers can in fact be mistaken for hovering trochilids. At equivalent body masses, hummingbirds and hawkmoths in hovering flight exhibit similar rates of oxygen consumption, a result consistent with common aerodynamic solutions to the problem of aerial nectarivory (Bartholomew, 1987). Biomechanical comparison of these two taxa, particularly in the context of hovering, might thus reveal fundamental physical factors underlying animal flight performance. Comparative metabolic analysis at levels of maximum performance could alternatively demonstrate fundamental differences between pulmonary and tracheal respiratory systems in limiting aerobic capacity. Also, hummingbirds are most common both intra- and interspecifically at midmontane elevations in the Neotropics. Compensatory responses to hypoxic and hypodense conditions during flight are easily studied in the laboratory and can demonstrate physiological as well as aerodynamic limits to performance (see Chai and Dudley, 1996; Dudley and Chai, 1996; Dudley, 1998). Such analyses, when repeated among different hummingbird species and combined with estimates of hummingbird phylogenetic relatedness (Bleiweiss et al., 1997), will ultimately elucidate evolutionary trajectories in what is a functionally unique and exciting vertebrate lineage.

7.6 SUMMARY

Evolution of small body size has occurred repeatedly among winged insects, yielding a profusion of diverse miniaturized forms that represent the majority of insect and even metazoan taxa. Asynchronous flight muscle and high wingbeat frequencies are the key features underlying dramatic reduction in insect body size. Aerodynamic studies have rarely investigated the taxonomically diverse and numerically abundant miniaturized insects for which highly viscous flows are relevant. Nor are unsteady mechanisms of force production well described for wing flapping at low Re. Anthropomorphic bias and technical limitations have obviously focused attention on large insects, but a broad biomechanical understanding of pterygote diversity will only come from the kinematic and aerodynamic study of smaller species. Various consequences of body size reduction among winged insects include a facilitated ability to hover, heightened dispersal by ambient winds, and possibly enhanced coevolution of insect pollinators with angio-sperms. Reduced flight speed at small body size does, however, make it more difficult for an insect to direct its aerial movement. Much of the extant insect fauna flies ceaselessly yet futilely against prevailing winds that alone determine the direction of flight.

Because of their reduced flight speeds and high individual numbers, insects are particularly susceptible to predation by volant vertebrates. Diverse radiations of insectivorous bats and birds worldwide indicate substantial predatory pressure on nocturnal and diurnal flying insects, respectively. In turn, selection for heightened defensive strategies by insects has been intense and has likely resulted in coevolution between aerial predators and their prey. For example, many nocturnally flying insects listen to bat echolocation calls and correspondingly engage in evasive action, in turn promoting further improvement in chiropteran strategies of echolocation and prey interception. Among diurnal in-sects, defensive mechanisms directed toward vertebrate predators in-clude erratic flight styles, mimicry, and unpalatability. Miniaturization in numerous pterygote lineages, together with diffusive constraints on maximum insect body size, has resulted in essentially nonoverlapping distributions of body size between extant insects and volant verte-brates. Alone among flying vertebrates, hummingbirds exhibit mor-phological, aerodynamic, and behavioral features convergent on pter-ygote design.

Chapter Eight

FUTURE DIRECTIONS IN INSECT

FLIGHT BIOMECHANICS

MORE SO THAN ANY OTHER taxonomically based discipline, insect biology is an ever-expanding realm of investigation. The publication *Entomology Abstracts*, for example, cites about one thousand new papers appearing each month in the general area of entomology. Much of this material relates solely to basic descriptions of insect diversity, be it taxonomic, morphological, or physiological; approximately seven thousand new insect species are described annually (Lane, 1992). Insects have served historically as key taxa with which to derive fundamental concepts in ecology, evolution, and behavior. However, kinematic and aerodynamic diversity within the winged Insecta has only partially been described at ordinal levels of comparison, and intraordinal patterns of biomechanical diversification remain to be investigated. Combined analyses of physical flight mechanisms, locomotor performance in nature, and phylogenetic associations bring the promise of an elaborate and sophisticated functional understanding of insect diversity.

8.1 Aerodynamic Mechanisms

Unsteady mechanisms are prominent features of force production through wing flapping (section 3.2.2), and an important goal of future studies will be visualization, quantitative description, and fundamental physical explanation of unsteady aerodynamic flows. Many such studies will involve mechanical analyses of oscillating winglike structures without explicit appplication to specific life forms. Although it is essential from a physical perspective, such an approach may be biologically incomplete in that morphological and kinematic similarities to actual insects may be substantially diminished. Key features of wing aerodynamics that are unique to animal flight may also be overlooked. Within an explicitly biological context, the use of physically variable gas media provides a potent experimental tool for exploring the range of kinematic and aerodynamic mechanisms available to free-flying insects.

8.1.1 Unsteady Airflows and Vortex Mechanisms

As with vertebrate flight, studies of vortex wakes generated by free-flying insects may elucidate patterns of unsteady force production and expenditure of mechanical power (section 3.2.2). The quantitative description of three-dimensional flow fields is complicated and involves such modern techniques as particle image- and particle tracking velocimetry (PIV and PTV, respectively). These methods visually record the motion of small particles entrained within flow fields around object and reconstruct their velocities accordingly. The more direct approach of laser Doppler velocimetry (LDV) measures velocity components directly from the optical phase shifts of light reflected from moving tracer particles. Such methods are most readily applied to either flapping physical models or restrained insects (e.g., Ellington et al., 1996; Willmott et al., 1996); implementation with free-flying insects will be methodologically challenging. At present, the quantitative analysis of flow fields requires a high degree of optical resolution and tightly constrained experimental geometries (e.g., Stamhuis and Videler, 1995). Hummingbirds at feeders are, however, remarkably stable and exhibit highly predictable wing motions. Butterflies with low wingbeat frequencies may similarly be appropriate candidates for free-flight visualization studies. Future progress in instrumentation should also render quantitative methods of flow field analysis more widely applicable to specific contexts of insect flight than is currently the case.

Although technological applications of insect flapping flight are at present remote, considerable scope exists for the use of physical models to explore the aerodynamic consequences of variation in wing kinematic parameters. Construction of flapping devices that mimic wing motions in two or more degrees of freedom can potentially demonstrate the structure of ensuing flow fields (e.g., van den Berg and Ellington, 1996a, b) as well as the magnitude of instantaneous forces generated by the moving wings. Kinematic parameters can be subtly yet systematically varied through digital manipulation of the microactuators that drive such flapping devices. Similar measurements implemented with invariant kinematics and wing morphology but within physically variable gas media (see below) can be used to evaluate Re dependence of force production. Biomechanical consequences of wing shape and flexibility can also be investigated through use of flapping physical models of variable geometries and structural composition.

8.1.2 Physically Variable Gas Media

Empirical studies of animal flight performance have generally been carried out within the contemporary atmosphere, although Chadwick

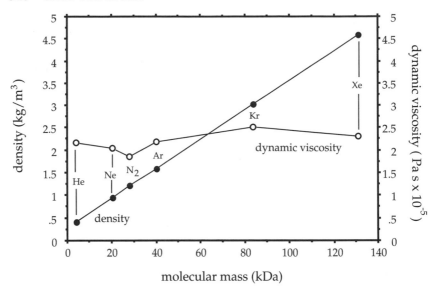

FIG. 8.1. Variation in gas density and dynamic viscosity as a function of the major component's molecular mass in two-component mixtures of 21% oxygen with either diatomic nitrogen or one of the noble gases (helium, neon, argon, krypton, and xenon; use of radon is precluded logistically because of rapid radioactive decay). Density and viscosity values refer to standard temperature and pressure. Density changes following noble gas substitution for nitrogen are substantial whereas variation in dynamic viscosity is much smaller.

and Williams (1949; see also Chadwick, 1951; Sotavalta, 1952b) utilized heliox (20.9% oxygen/79.1% helium) to investigate density dependence of wingbeat frequency in tethered insects. The outer electron shell of helium is full (as with all noble gases), rendering the molecule biologically inert in most circumstances (see Schreiner, 1966, 1968; Hamilton et al., 1970). Replacement of atmospheric nitrogen by helium and other noble gases affords a 10-fold variation in air density with oxygen concentration unchanged, whereas concomitant changes in dynamic viscosity are much smaller (Dudley and Chai, 1996; see fig. 8.1). Use of such novel aerodynamic media thus permits the non-invasive manipulation of animal flight mechanics. Methods for calculating the physical properties of variable composition gas mixtures can be found in Reid et al. (1987).

In addition to altering flight aerodynamics, experimental use of such gas mixtures may influence the respiratory and thermal physiology of flying animals. Gas diffusion coefficients vary in inverse (but nonlinear) proportion to gas density, whereas thermal conductivity increases in direct proportion to density (see Paganelli et al., 1975; Reid et al.,

1987). The use of low-density heliox to study insect flight, for example, will induce concomitant increases in tracheal diffusion of both oxygen and carbon dioxide. The mechanical work required for tracheal ventilation will also decrease at lower air densities. Aerodynamic phenomena per se are not influenced by such physiological effects, but muscle performance and aerobic metabolism underlying flight in physically variable gas mixtures potentially reflect these associated changes in thermal and respiratory physiology. The use of three- and four-component gas mixtures does, however, permit construction of variable density media with oxygen diffusivity and thermal conductivity similar to those of normal unmanipulated air (Dudley and Chai, 1996). Changes in thermal conductivity may similarly be offset by parallel changes in air temperature. Various inert synthetic molecules (e.g., halocarbons) further expand the range of physical gas properties available for application to experimental flight studies.

Flight of animals in variable density mixtures enables a wide range of kinematic and aerodynamic modulation to be expressed and quantified within laboratory contexts. For example, hummingbirds and euglossine bees hovering freely in low density but normoxic mixtures have demonstrated exceptionally high levels of aerodynamic force production and mechanical power output (see section 5.4.2). Available power reserves as well as fundamental limits to aerial performance can be demonstrated in such flights. By contrast, intrusive experimental manipulations such as tethering or attachment of weights may confound the physiological capacity for flight with the behavioral motivation to fly or to engage in wing flapping under potentially abnormal experimental circumstances. An additional noninvasive manipulation comparable to the use of physically variable gas mixtures involves either offsetting or augmenting gravitational acceleration so as to alter the effective body mass of flying animals. Apart from limited work on the flight of *Manduca sexta* in microgravity (see section 3.2.1.1), variable gravity vectors have not been used in the study of insect flight biomechanics. In addition to the study of short-term kinematic and energetic responses to variable acceleration vectors, chronic ontogenetic exposure to hypo- and hypergravity might demonstrate phenotypic plasticity in those pterothoracic structures required for a dramatically altered force balance during flight.

8.2 INSECT FLIGHT BIOMECHANICS IN NATURE

Most studies of insect flight kinematics and aerodynamics have been carried out under laboratory conditions. However, the full utility of biomechanical analysis will only be realized in situations relevant to natural field conditions. Opportune advances in instrumentation clearly are facilitating the description and analysis of complex behavioral and physiological phenomena. It is equally clear that the preceding statement has always been true throughout the twentieth century given the historical rate of technological change. Nonetheless, ample opportunity now exists to describe diverse features of insect flight performance in various natural contexts. Because of ongoing and accelerating anthropogenic destruction, the most pressing need for such studies resides within tropical ecosystems.

8.2.1 Ecophysiology of Flight

Given the overall significance of flight for pterygote ecology, field investigations can supply an informed perspective on those physical as well as biomechanical factors that influence expression of flight-related behavioral and morphological traits. One of the most interesting ecological contexts for insect flight occurs across altitudinal gradients; both air density and oxygen partial pressure drop by about 50% as elevation increases from sea level up to 5000 m. Such changes in the physical properties of air, together with the adiabatic decrease in air temperature at higher elevations, may significantly influence the flight-related morphology, physiology, behavior, and altitudinal distribution of insects and volant vertebrates (see Mani, 1962, 1968; Dudley and Chai, 1996). Most importantly, the experimental decoupling of air density, oxygen partial pressure, and air temperature in laboratory contexts permits physiological and aerodynamic consequences of high-altitude flight to be systematically analyzed.

To date, altitudinal variation in insect flight performance has not been demonstrated, although suitable candidates for investigation readily suggest themselves. For example, mosquitoes are reported to be capable of flight at normoxic air densities only 20% of normobaric density (Galun and Fraenkel, 1961); those factors presently restricting malarial mosquitoes to elevations below 2000 m are unknown. Bumblebees are well-known montane and cold-climate specialists that exhibit remarkable species diversity across altitudinal gradients (e.g., Williams, 1991). One bumblebee species has even been recorded at altitudes as high as 5600 m near Qomolangma Feng, a.k.a. Mount Everest (Williams, 1985). In spite of the major pollinating role played by bumblebees in alpine ecosystems, patterns of intra- and interspecific morphological variation across altitudinal gradients have never been investigated in *Bombus* nor in any other insect taxon. The remarkable taxonomic diversity and high degree of sympatry among bumblebee species in the mountains of central Asia (e.g., Tian Shan, Kunlun Shan, western Sichuan; see Williams, 1994) present an excellent opportunity to carry out such comparisons.

One immediate biomechanical prediction is that relative wing length should increase at higher altitudes to offset the increased induced power associated with lower air densities (eq. 3.7). In *Drosophila robusta*, absolute wing length increases at greater elevations whereas thoracic length remains constant (Stalker and Carson, 1948; see also Hepburn et al., 1998). Interspecific comparisons of sympatric bird taxa similarly demonstrate longer wings at higher altitudes (e.g., Hamilton, 1961; Mayr, 1963). Biomechanical correlates of altitudinal density reduction have been most convincingly shown in hummingbirds. At greater elevations, hummingbirds exhibit interspecifically both longer wings and lower values for wing loading that presumably have the effect of reducing induced power expenditure (see Feinsinger et al., 1979; Epting, 1980). Under laboratory conditions, metabolic and mechanical power expenditure by hovering hummingbirds increases at reduced total pressure (Berger, 1974a,b) and under hypodense but normoxic conditions (Chai and Dudley, 1995). Maximum flight performance may be similarly influenced at higher altitudes, although oxygen availability may also limit metabolic capacity in hovering flight. Under field conditions, relatively simple load-lifting methods (see Chai et al., 1997) can easily be applied to obtain extensive comparative data on the maximum load lifting performance of both insects and hummingbirds

Additional ecological aspects of insect flight performance relate to the temporal and spatial characteristics of flight activity. Transient, daily, and lifetime flight distances of volant insects are largely unknown. Hocking (1953) reviewed a large literature relating to net flight displacements over the lifetime primarily of hematophagous Diptera; values ranged widely from about 0.5 km to as much as 320 km. Released *Drosophila pseudoobscura* exhibit net daily displacements of about 260 m, with net horizontal translation of about 700 m over the insect's life (Powell and Dobzhansky, 1976). The well-established allometry of flight speeds in insects (section 3.1.1) does not necessarily indicate greater net displacement through flight. For example, large periodical cicadas (*Magicicada* spp.) typically move less than one kilometer during their lifetime (see White et al., 1983). On shorter timescales (e.g., several days), dipterans typically engage in net displacements on the order of kilometers (e.g., Lindquist et al., 1951; Schoof and Silverly, 1954; MacLeod and Donnelly, 1963; Coyne et al., 1982; see also Johnson, 1969). By contrast, parasitic oestrid flies potentially fly hundreds of kilometers in a single day, probably in order to locate their nocturnally mobile ungulate hosts (see Nilssen and Anderson,

1995). Systematic relationships between net and total displacement over various timescales remain to be elucidated, but will certainly be influenced by body mass and by phylogenetic association of the taxon in question.

Temporal partitioning of insect flight activity in natural environments is equally ill-resolved. With the exception of migrating or dispersing forms, most insects seem to spend the majority of the day at rest. Tsetse flies, for example, fly only about 5–30 min/day (see Brady, 1972; Bursell and Taylor, 1980). In general, distributions of flight times tend to be positively skewed, with most flights being of short duration and the rarer longer flights forming an extended distributional tail (see Davis, 1980). Total flight time relative to other behaviors can sometimes be determined simply by visual tracking of insects within subdivided and marked three-dimensional habitats (e.g., Aluja et al., 1989; Casas and Aluja, 1997). Unfortunately, most flight activities involve rapid three-dimensional movements through vegetation that obscures continuous visual monitoring. Increased resolution of radiotelemetric systems may, however, facilitate detailed field analyses of individual flight activity, cumulative displacement, and mean ground speed over different timescales. For example, radar tracking devices have recently been developed that can record the location and two-dimensional flight trajectories of insects within a radius of several hundred meters; the electronic tags attached to the target insect weigh less than 5 mg (see Riley et al., 1996; Roland et al., 1996; Riley et al., 1998). Remote telemetry of physiological data is also technologically feasible; Kutsch et al. (1993) used a 0.42 g radiotelemetric device to transmit muscle action potentials from free-flying locusts (see also Swihart and Baust, 1965). Most such methods require fairly large insects so as to avoid mass-associated perturbation of natural flight performance. Progressive electronic miniaturization does, however, bode well for future application to smaller taxa.

Insect associations with particular microclimates similarly influence the spatial and temporal partitioning of flight behavior. Anthropogenic fragmentation of habitats may select for intraspecifically variable flight tendencies among different habitat patches (see Dempster, 1991; Taylor and Merriam, 1995). Within forests and vegetational stands generally, substantial vertical gradients are found in such physical factors as relative humidity, solar radiation, air temperature, and wind speed (see Rosenberg et al., 1983; Parker, 1995). Partly as a consequence, insects flying within vegetation tend to exhibit nonrandom patterns of vertical distribution. For example, most insects flying in tropical rain forests are found in the canopy where sunshine is most intense (see Sutton and Hudson, 1980; Rees, 1983; Sutton, 1983). Similarly, insect species numbers are much higher along edges of tropical forests than within the closed understory (e.g., Janzen, 1973; Fowler et al., 1993). Filtering of particular wavelengths by foliage may also alter the spectral composition of light across vertical gradients, thus altering the apparent color of light reflected from objects. Perhaps partly for this reason, butterflies in some tropical forests exhibit vertical stratification by color pattern (see Burd, 1994). Environmental heterogeneity is also evident on even smaller spatial scales. For example, smaller pollinating insects tend to be more active within sunflecks that mitigate thermoregulatory demands of flight (see Herrera, 1997). Influences of microclimatic variability on the frequency and duration of flight may well be most pronounced in miniaturized insects given their physiological sensitivity to local environmental conditions.

The duration and speed of flight directly influence associated rates of energetic expenditure (section 4.1.1). Few studies address the metabolic costs of flight under natural conditions, although washout of radioactive cesium has been used to estimate field energetics of tsetse flies (see Taylor, 1978). The doubly labeled water technique (using isotopes of both hydrogen and oxygen) has great potential in this regard given its proven utility with free-ranging vertebrates. This method relies on dilution within the body of an injected volume of doubly labeled water; subsequently quantified washout from the body of both water and carbon dioxide enables estimates of metabolic rate if the respiratory quotient is also known. For bumblebees tethered on a rotational flight mill, such isotopically based estimates of flight metabolism correlate well with simultaneous measurements of the rate of carbon dioxide production (Wolf et al., 1996). Field metabolic rates of bumblebees, and probably of other hymenopterans, can thus be reliably measured over variable time intervals provided that individual insects can be injected with doubly labeled water and later recaptured for analysis. Particularly when combined with simultaneous radiotelemetric monitoring of insect position, this technique will provide a powerful interpretive framework for studies of natural flight physiology.

8.2.2 Predation, Sexual Selection, and Maneuverability

Maneuverability and the regulation of aerodynamic forces are dominant requirements for effective flight behavior (chapter 5). General features of those kinematic asymmetries that contribute to body rotations have been broadly delineated in insects (section 5.2.2), but biomechanical understanding of three-dimensional flight trajectories and maneuvers is much less complete. Relevant insect systems, however, exist with which to implement both descriptive as well as experimental studies of free-flight maneuverability. For example, hawkmoths hovering at flowers oscillate forwards and backwards in response to appropriate optomotor cues (Farina et al., 1995). Euglossine bees hovering at chemical baits that mimic natural orchid fragrances are similarly capable of tracking three-dimensional oscillations of the scent source (pers. obs.; see plate 6). Study of such controlled flight performance offers an excellent opportunity to describe those wing and body kinematics underlying visually induced (and experimentally manipulable) translations and rotations. If such tracking behavior can be evaluated in the jet of a wind tunnel, then maneuverability during forward flight can be similarly assessed.

Actual use of the intrinsic capacity for maneuverability remains largely undescribed for naturally occurring behaviors. Most such events will occur in contexts either of mate selection or interaction between predator and prey. Documentation of the rates of successful predation or mating will be an informative component of such studies. If selection has historically been strong for increased flight maneuverability, then high rates of successful evasion may be evident in encounters between flying insects and their predators. On the other hand, low rates of capture characterize male flies pursuing females (Wagner, 1986b), suggesting further evolutionary scope for enhanced capture methods by males. Because of intrinsically higher accelerations (section 5.4.1), smaller insects may well be more successful in those mating situations requiring a high degree of aerial agility (see Ghiselin, 1974; McLachlan, 1986b). For example, relative success in mate capture is inversely proportional to body size in certain odonate taxa (Convey, 1989; Banks and Thompson, 1985) and in chironomid flies (McLachlan and Allen, 1987). Such an association between body size and mating success may select for smaller males, although additional features of lifetime reproductive effort (e.g., longevity) contribute to an overall selective advantage for larger individuals (see Neems et al., 1990).

Moreover, forces of intra- and intersexual selection can act synergistically on neural and biomechanical underpinnings of maneuverability, thereby promoting rapid evolution in such traits. For example, hovering displays of male stenogastrine vespid wasps are directed toward females but may also be used in territorial interactions among males, sometimes escalating into chases and even physical impact (Turillazzi, 1983). Even stronger selection may occur if sexually selected traits are also of advantage in the context of natural selection. Dragonflies defend territories and chase out male conspecifics (intrasexual selection), pursue females (intersexual selection), capture prey items in the air (natural selection for aerial attack), and evade both aerial and terrestrial predators via flight (natural selection for escape mechanisms). Enhanced maneuverability is clearly advantageous in each behavioral context. Similarly, Moore (1990) has documented intra- and intersexual selection acting on morphological characters of territorial dragonflies. Territorial success increases monotonically with increasing body length, whereas success in mating and fertilization peaks at intermediate body lengths. Contributions of aerial maneuverability to these different components of fitness are unknown in this system.

Pterygote evolution has probably been accelerated through multiple modes of selection acting on flight performance. Other interesting trade-offs are possible in sexually selected systems. In stag beetles, for example, intra- and interspecific comparisons suggest that larger mandibles used in male-male combat are produced at the expense of wing surface area and possibly of flight muscle (Kawano, 1997). Flight performance may thus be adversely influenced by sexual selection acting on secondary male characters. A general hypothesis would suggest that natural and sexual selection have worked in concert to accelerate the evolution of maneuverability and maximum flight performance among insects. The Odonata may well be the most appropriate taxon with which to test this hypothesis phylogenetically.

8.3 EXPLORING INSECT DIVERSITY

Primarily because the necessary data are unavailable even for such basic morphological traits as body mass and wing area, a comprehensive functional analysis of insect flight diversity would be premature. Ordinal-level comparisons of flight-related morphology have been made throughout this book, but familial, generic, and interspecific analyses are not at present a meaningful possibility. However, modern phylogenetic methods coupled with quantitative study of key biomechanical characters will be powerful tools in future analyses of pterygote diversification. Ideally, such studies will include the functional explanation of taxonomic diversity within the Coleoptera, the most species-rich insect order.

8.3.1 Flight Biomechanics in Phylogenetic Contexts

The advent of comparative methods as applied to modern phylogenies has substantially enhanced the capacity to test adaptive hypotheses in biology (Brooks and McLennan, 1991; Harvey and Pagel, 1991). Functional morphology is particularly well suited to such an approach, not merely for distributional analyses of particular adaptations, but also for tests that associate particular characters with clade diversification. Asynchronous flight muscle is one such trait (section 7.1), but numerous others might be located within the arena of flight biomechanics. For example, tegminization and elytrization have historically yielded major differences in wing and body morphology among insect orders, and may have indirectly facilitated adaptive radiation within the Coleoptera (see below). Unique and surprising biomechanical attributes will also emerge from more detailed surveys of insect flight mechanics. The kinematics and aerodynamics of miniaturized insects are areas that have received little experimental attention (section 7.1.2) but hold considerable promise for demonstration of novel flight mechanisms. Closer examination of seemingly aberrant or divergent taxa will also be a powerful method for discovering evolutionary innovations. For example, asynchronous muscle has not yet been described from the Lepidoptera, although its presence (if not that of transitional muscle types) is certainly possible (section 4.2.2). Many small moths in particular would seem to require fairly high wingbeat frequencies given their small body size, reduced wing area, and fringed wing morphology convergent on that of miniaturized asynchronous taxa.

Knowledge of trait distribution is the foundation of comparative analysis, but the most detailed descriptions of free-flight kinematics are restricted to neopterous insects (and particularly the Diptera and Hymenoptera). Smaller orders are much less well known. The Hemiptera and the paraphyletic taxon Homoptera have been but minimally sampled, and nothing appears to be known about the flight mechanics of termites or zorapterans. Even within major orders, only a few taxa are likely to have been studied, and much broader surveys are clearly called for. Acquisition of kinematic data may well be facilitated by modern noninvasive optical methods that can precisely determine three-dimensional orientation and the contour of moving wings (e.g., Zeng et al., 1996; Zeng, Matsumoto et al., 1996; Zeng, Matsumoto, and Kawachi, 1996a–e). Tethered simulations of flight will be useful in this regard to delimit the range of kinematic performance potentially available for free flight performance. Attention must equally be paid to the behavioral and ecophysiological context of free-flying insects if wingbeat kinematics are to be meaningfully compared among taxa. Paucity of data at present precludes all but anecdotal comparison of maneuverability among winged insects, although the increased wingbeat frequencies associated with asynchronous muscle may well facilitate aerodynamic responsiveness during flight.

8.3.2 Biodiversity and the Coleoptera

Beetle speciation numerically defines insect if not metazoan diversity (table 1.1), and the order Coleoptera represents a distinct morphological universe unto itself. The flight of beetles is, by contrast, remarkably unstudied. Detailed anatomical treatments of the beetle thorax (e.g., Stellwaag, 1914a; Rüschkamp, 1927; Matsuda, 1970) have no parallel literature on the functional morphology of flight, even though the complexity of thoracic musculature and wing articulation in Coleoptera rivals that of the Diptera and Hymenoptera (see Pringle, 1957; Matsuda, 1970). Both venational patterns and wing-folding mechanisms are highly diverse among beetle taxa (Schneider, 1975b,1978a; Hammond, 1979; Kukalová-Peck and Lawrence, 1993; Brackenbury, 1994b). One general feature of beetles is a forward shift of the metathoracic wing base, a change that likely occurred in parallel with elytral disassociation from aerodynamic force production, but also with acquisition of stabilizing functions by the elytra in some cases (section 2.2.3). Schneider (1978b, 1987) provided interpretations of thoracic muscle action in beetles that readily lend themselves to future electromyographic evaluation. Virtually no experimental work has been carried out on coleopteran kinematics and aerodynamics, although Termier et al. (1971) measured forces on tethered beetles and found interesting phasic differences relative to similar recordings on insects from other orders (see section 3.2.2). Grodnitsky and Morozov (1995) visualized vortex wakes in tethered beetles, whereas wing kinematics of various free-flying beetles were described qualitatively by Brackenbury (1994b). Apart from this latter work, little information is available on the free-flight kinematics of Coleoptera.

General statements as to the reduced flight tendencies of beetles are widespread, but focal studies of flight frequency and duration are unfortunately not available for any coleopteran taxon under natural circumstances. In many beetle taxa, substantial body sclerotization and consequent mass increase may influence flight energetics sufficiently as to alter overall patterns of locomotor activity. Anecdotal reports of limited maneuvering abilities in beetles seem to refer primarily to large taxa, whereas small beetles compose the majority of the extant fauna (section 7.1.1) and seem, at least superficially, to be just as agile and flight-capable as many dipterans and hymenopterans. Small silphids, scarabaeine and cetoniine scarabaeids, and chrysomelids, for example, appear to be highly capable in hovering, forward flight, and various aerial maneuvers. Some large beetles are also quite active if not actually aggressive; predatory tiger beetles (Cicindelidae) are particularly known for quick takeoffs and rapid flight. Flight initiation in beetles may be somewhat slowed in many taxa by the requisite unfolding of the wings from beneath the elytra (Rüschkamp, 1927). Wing unfolding and takeoff seem to be quite rapid in some large beetles as well as in smaller taxa. The relevant kinematics have been obtained neither for the Coleoptera nor for the Dermaptera, which exhibit an even more complex geometrical folding of the hindwings (see Kleinow, 1966).

Any fundamental explanation of biotic diversity must interpret the extremely high number of extant beetle species. The British biologist J.B.S. Haldane once attributed beetle abundance to the inordinate fondness of some divine creator for this particular taxon; beetle diversity remains to this day an unexplained source of embarrassment to biologists. Of biological features contributing to insect diversification generally, complete metamorphosis has often been cited as an important trait given that the four largest insect orders (Coleoptera, Diptera, Hymenoptera, and Lepidoptera) are holometabolous. Among these four orders, three of the four possess asynchronous flight muscle and exhibit a general trend toward body miniaturization. The one exception (Lepidoptera) may in part have acquired high species diversity as a consequence of almost exclusively phytophagous larval diets within the order. Phytophagy and lineage diversification are strongly associated in many insect taxa (e.g., Mitter et al., 1988), and many beetles are also plant eaters in adult as well as larval stages (see Strong et al., 1984). Moreover, historical switches from gymnosperm to angiosperm hosts increased diversification rates among phytophagous beetle clades (Farrell, 1998). The greater species richness of Coleoptera relative to that of the Lepidoptera, however, suggests that phytophagy alone is insufficient to explain the dramatic interordinal differences in species numbers that are evident today.

As a further consideration, species richness among insect orders may in part derive from flight-related characteristics. Among the Diptera, Coleoptera, and Hymenoptera, one evolutionary possibility afforded by acquisition of asynchronous flight muscle and increased wingbeat frequencies is that an entire wing pair may be used for purposes other than aerodynamic force generation (Dudley, 1991b). Dipteran halteres do not have a direct aerodynamic role because of their dramatically reduced size, and beetle elytra generally make only small contributions to total force production (section 3.2.1). Beetles are, however, unique in that the structural defense provided by elytra coexists with holometaboly and body miniaturization, a combination found in no other insect order. Other holometabolous orders either retain both wing pairs in uncoupled formation, mechanically couple the ipsilateral wings, or use only one wing pair aerodynamically. Only the beetles have successfully engaged ipsilateral wings in the functionally independent roles of protection and aerodynamic force production during flapping flight. Tegminization and elytrization do occur in various other insect orders, but with the exception of the Coleoptera are confined to the hemimetabolous exopterygotes (see section 2.2.3).

A complete phylogenetic analysis of beetle species richness relative to that of other orders would be premature given potential contributions of holometaboly, body miniaturization, phytophagy, and additional traits that may facilitate diversification. Only application of modern comparative methods to well-resolved phylogenies will provide convincing statistical support for particular hypotheses of adaptive radiation among insect clades. Nonetheless, features of wing morphology and flight performance specific to beetles may well underlie their remarkable taxonomic and morphological diversity. Given the salient role that flight plays in so many features of insect life, it would be surprising if the combined effects of miniaturization and wing elytrization have not historically contributed to generation of diversity in this magnificent order.

GLOSSARY

Accessory pulsatile organs — small muscular structures that, in the thorax, pump hemolymph through the wing veins

Actuator disk — area across which the wings apply downwards momentum to the surrounding air

Advance ratio — ratio of the translational velocity of the body to the mean flapping velocity of the wings

Aeolian — transported by wind

Afferent (cf. *efferent*) — towards a central region, particularly in reference to impulse conduction from a sensory receptor to the central nervous system

Alary — of or pertaining to the wings

Alary buds — regions of larval tissue from which adult wings arise

Alate — a winged form or adult

Allopatric — disjunct in geographical distribution

Alula — an expanded axillary region of the beetle elytron

Amphibiotic — characterized by larval development in water and adult life on land

Amplexiform — coupling of wings effected by substantial overlap between fore- and hindwings

Amplexus — coupling between male and female during mating behavior

Anal vein — the most posterior longitudinal wing vein

Ancestral (cf. *derived*) — inherited from an earlier form

Anemotaxis — movement in response to air motion or wind

Angle of attack — angular orientation of a wing or wing chord relative to oncoming airflow

Annulate — formed of ringlike segments

Anteromotorism (adj. *anteromotoric*; cf. *posteromotorism*) — primary use of the forewings to generate aerodynamic forces

Apical — pertaining to the tip or apex (as of a wing)

Apodeme — internal projection of the exoskeleton to which muscles are attached

Aposematic — possessing warning coloration suggestive of unpalatability

Appendicular (cf. *axial*) — of or pertaining to the appendages

Aptery — winglessness

Apterygote — ancestrally wingless insect

Aspect ratio — ratio of wing length to the mean wing chord

Asynchronous muscle (cf. *synchronous muscle*) — type of insect flight muscle that exhibits repetitive stretch-activated contraction in response to activation by a single nervous impulse

Autapomorphy (cf. *synapomorphy*) — a derived character unique to a particular taxon

Autotomy — self-amputation of a part or appendage

Axial (cf. *appendicular*) — of or pertaining to the longitudinal body axis

Axillary — of or pertaining to the wing articulation

Axillary apparatus — articulation between the wing and thorax

Axillary sclerites — major sclerites of the wing articulation

Basal — phylogenetically ancestral; pertaining to the base or point of attachment

Bimotorism (adj. *bimotoric*) — equivalent use of fore- and hindwings in aerodynamic force generation

Biramous (cf. *uniramous*) — with two branches, as of an appendage

Body angle — orientation of the longitudinal body axis relative to horizontal

Boundary layer — gradient of reduced flow speeds between free stream flow and the surface of an object

Brachyptery — syndrome of shortened or reduced wings

Camber — condition of having a slightly arched or convex surface

Campaniform sensillae — dome-shaped mechanoreceptors in the cuticle

Canard — a small stabilizing surface located forward of the wings

Casting behavior — zigzag motions of male insects flying upwind within pheromone plumes

Caudal filament — threadlike projection from the terminal abdominal segment

Cercus (pl. *cerci*) — usually paired sensory appendages on either the terminal or the penultimate abdominal segment

Character state — condition of a particular character

Chemotaxis — movement in response to chemical stimuli

Circulation — rotational intensity of a vortex; movement of hemolymph

Claval furrow — line of flexion between the cubitus and anal veins

Clavus — region of wing posterior to the claval furrow

Contradirectional (cf. *syndirectional*) — in or towards the opposite direction

Contralateral (cf. *ipsilateral*) — of or pertaining to the opposite side

Coriaceous — leathery

Corpora cardiaca — neurosecretory organs behind the brain

Costa — the most anterior longitudinal wing vein

Coxa — basal leg segment

Cubitus — longitudinal vein posterior to the medial vein

Cursorial (cf. *gressorial*) — adapted for running

Cuticle — extracellular secretion of epidermal cells that constitutes the arthropod exoskeleton

Dealation — shedding of wings

Derived (cf. *ancestral*) — modified or specialized relative to the ancestral condition

Dihedral — upwards inclination of wings relative to the horizontal plane

Diptery (cf. *tetraptery*) — condition of having two wings

Distal (cf. *proximal*) — away from the base or point of origin of a structure

Downstroke (cf. *upstroke*) — dorsoventral motion of the wing in the stroke plane

Drag (cf. *lift*) — resistive component of aerodynamic force parallel to relative air motion

Eclosion — emergence of an adult insect from either its final larval instar (hemimetabolous taxa) or from the pupa (holometabolous taxa)

Efferent (cf. *afferent*) — away from a central region, particularly in reference to impulse conduction from the central nervous system to effectors such as muscles

Elevational angle — angle of inclination of the wing relative to the stroke plane

Elytrization — evolutionary conversion of the forewings into elytra

Elytron (pl. *elytra*) — the thickened and sclerotized forewing of beetles

Endite (cf. *exite*) — lobe projecting from the middle of an arthropod leg segment

Endopterygote (cf. *exopterygote*) — insect with wings that develop internally and that undergoes complete metamorphosis

Endothermic — characterized by internal (endogenous) production of heat and concomitant elevation of body temperature

Epigamic — used in courtship to attract the opposite sex

Epimeron (cf. *episternum*) — posterior region of the thoracic pleuron

Episternum (cf. *epimeron*) — anterior region of the thoracic pleuron

Exite (cf. *endite*) — lobe projecting from the outer border of an arthropod leg segment

Exopterygote (cf. *endopterygote*) — insect with wing buds that develop externally

Exteroception (cf. *proprioception*) — sensory transduction of stimuli external to the organism

Flexion — the act of bending

Folivory — ingestion of leaves

Frenulum (pl. *frenula*) — hairs projecting from the hindwing of many Lepidoptera that couple the fore- and hindwings

Furca (pl. *furcae*) — apodeme at the base of thoracic segments upon which certain wing depressor muscles terminate

Geometrical angle of attack — orientation of a wing or wing chord relative to the stroke plane

Gressorial (cf. *cursorial*) — adapted for walking

Halteres — miniaturized hindwings of Diptera

Hamuli — hooks along hindwing of Hymenoptera that couple the fore- and hindwings

Hamulohalteres — reduced hindwings of male Coccoidea

Hematophagy — ingestion of blood

Hemelytron — thickened forewing of Hemiptera

Hemimetaboly (cf. *holometaboly*) — incomplete metamorphosis

Hemolymph — blood of insects

Heteronomy (cf. *homonomy*) — condition of morphological and functional dissimilarity among segments or associated structures

Hexapod — arthropod with six legs

Holometaboly (cf. *hemimetaboly*) — complete metamorphosis by means of a pupal stage

Homeotic gene — regulatory gene determining segment and appendage identity

Homology — condition of equivalent phylogenetic origin

Homonomy (cf. *heteronomy*) — condition of morphological and functional similarity among body segments or associated structures (e.g., wings)

Homoplasy — character convergence among unrelated taxa

Humeral plate — preaxillary sclerite at the base of the costal vein

Hyperoxia — condition of increased oxygen partial pressure relative to normoxia

Hypoxia — condition of decreased oxygen partial pressure relative to normoxia

Inquilinism — habit of living within the nest of a different species

Interordinal (cf. *intraordinal*) — among different taxonomic orders

Intraordinal (cf. *interordinal*) — within a taxonomic order

Ipsilateral (cf. *contralateral*) — of or pertaining to the same side

Johnston's organ — mechanoreceptive sensory structure at the antennal base

Jugal fold — line of wing flexion behind the final anal vein

Jugum — basal region of wing posterior to the jugal fold

Kairomone — chemical signal transferred interspecifically that benefits only the receiver

Kelvin's theorem — law of conservation for circulation

Leading edge (cf. *trailing edge*) — the forward edge of a moving wing; the anterior wing margin

Lift (cf. *drag*) — component of aerodynamic force perpendicular to relative air motion

Lumen — cavity of a tube

Macroptery — syndrome of fully sized wings

Mechanoreception — sensory transduction of deformation or bending

Medial vein — longitudinal vein posterior to the radius

Median flexion line — line of bending anterior to the median vein

Mesothorax — the middle thoracic segment

Metathorax — the most posterior thoracic segment

Microptery — syndrome of very small wings

Moment arm — length across which an applied force acts to yield torque

Moment balance — equilibrium between moments or torques of opposite sense

Monophyletic (cf. *paraphyletic, polyphyletic*) — derived from a single ancestral taxon

Neoptera (adj. *neopterous*; cf. *Paleoptera*) — taxonomic grouping of insects characterized by their ability to flex the wings backward along the abdomen

Normobaria — pressure of one standard atmosphere (760 mm Hg)

Normoxia — oxygen partial pressure of ~159 mm Hg (i.e., 20.9% O_2 at standard temperature and pressure)

Notum (pl. *nota*; adj. *notal*; cf. *sternum*) — dorsal sclerite of a thoracic segment

Ocelli — simple eyes located dorsally on the insect head

Ommatidium (pl. *ommatidia*) — discrete functional unit of the insect compound eye

Optical flow field — relative motion of the surrounding visual environment

Optomotor response — visually-induced compensatory wing or body movements

Ordinal — of or pertaining to an insect order

Paleoptera (adj. *paleopterous*; cf. *Neoptera*) — taxonomic grouping of insects characterized by the absence of backward wing flexion, of extant insects comprising the Ephemeroptera and Odonata

Paranotum (pl. *paranota*) — lateral extension of the notum

Paraphyletic (cf. *monophyletic, polyphyletic*) — pertaining to a taxonomic group that does not include all taxa with common ancestors

Phasic (cf. *tonic*) — temporally variable or oscillatory

Pheromone — chemical signal directed toward a conspecific

Phragma (pl. *phragmata*) — internal cuticular partition

Phytophagy — feeding on plants

Pilosity — covering of hair

Pitch (cf. *roll, yaw*) — body rotation about a transverse axis (i.e., about any body axis perpendicular to the sagittal plane)

Planform — the two-dimensional projection of wing area

Pleural suture — external groove of the pleuron separating the episternum from the epimeron

Pleuron (pl. *pleura*) — lateral sclerite of a thoracic segment

Plume — odor trail, particularly of pheromones

Plumose — featherlike

Polyphyletic (cf. *monophyletic, paraphyletic*) — pertaining to a taxonomic category with members derived from multiple ancestors

Positional angle — angular location of the longitudinal wing axis within the stroke plane

Posteromotorism (adj. *posteromotoric*; cf. *anteromotorism*) — primary use of the hindwings to generate aerodynamic forces

Pressure drag (cf. *skin friction*) — drag force associated with the pressure gradient that characterizes the flow field around an object

Pronation (cf. *supination*) — rotation of the wing such that the leading edge moves nose-downwards

Proprioception (cf. *exteroception*) — sensory transduction of stimuli originating within the organism, particularly with reference to position and movement of the body in space

Protean display — erratic and unpredictable movement

Prothorax — the most anterior thoracic segment

Protopterygote — phylogenetically ancestral winged insect

Proximal (cf. *distal*) — near to the base or point of origin of a structure

Pteralia — preaxillary and axillary sclerites of the wing articulation

Pterostigma (pl. *pterostigmata*) — pigmented wing region near the leading edge of a wing

Pterothorax — the wing-bearing segments of the thorax

Pterygomorphism — phenotypic plasticity or variable forms of wing expression

Pterygote — winged insect

Radius — longitudinal vein posterior to and branching behind the subcosta

Recirculation bubble — local region of continuously recirculating airflow, either at the leading edge or within a venational trough

Remigium — wing region anterior to the claval furrow

Resilin — rubberlike protein at the wing base of many insects

Reynolds number — the ratio of inertial to viscous forces acting on an object moving within a fluid

Rhythmogenesis — generation of the neural pattern of muscle contraction during flight

Roll (cf. *pitch, yaw*) — body rotation about the longitudinal axis

Saccadic — discontinuous, jumplike

Saltatorial — pertaining to or adapted for jumping

Sarcomere — contractile subunit of a muscle fiber

Scalariform — ladderlike

Sclerite — discrete cuticular plate of the body wall

Sclerotization — hardening of the cuticle

Scutellum (pl. *scutella*; cf. *scutum*) — posterior region of the notum

Scutum (cf. *scutellum*) — anterior region of the notum

Seta (pl. *setae*) — bristlelike projection from the cuticle

Sideslip — lateral motion of the body

Simple harmonic motion — sinusoidal motion

Skin friction (cf. *pressure drag*) — viscous drag in the boundary layer of an object moving in a fluid

Spiracle — external opening of the tracheal system

Station-holding — stereotypical patrolling of a particular air volume

Steady-state flow — airflow over an object at fixed relative speed and orientation

Sternum (cf. *notum*) — ventral sclerite of a thoracic segment

Stroke plane — approximate plane delineated by the down- and upstrokes

Stroke plane angle — orientation of the stroke plane relative to horizontal

Stylus (pl. *styli*) — a small anatomical projection from the cuticle

Subcosta — longitudinal wing vein immediately posterior to the costa

Supination (cf. *pronation*) — rotation of the wing such that leading edge moves nose-upwards

Sympatric (cf. *allopatric*) — overlapping in geographical distribution

Synapomorphy (cf. *autapomorphy*) — a derived character shared by multiple taxa

Synchronous muscle (cf. *asynchronous muscle*) — muscle type characterized by a single contraction in response to activation by one nervous impulse

Syndirectional (cf. *contradirectional*) — in or towards the same direction

Tagma (pl. *tagmata*) — group of body segments forming a functional unit

Tagmosis — organization of the body into distinct regions or tagmata

Taphonomy — study of the processes of fossilization

Tarsal reflex — induction of wing flapping through removal of leg contact with ground

Tau function — the ratio of distance from an approaching target to the approach speed

Tegmen (pl. *tegmina*) — the thickened forewing of Blattaria, Dermaptera, Mantodea, Orthoptera, and Phasmatodea

Tegminization — evolutionary conversion of the forewings into tegmina

Tegula (pl. *tegulae*) — preaxillary sclerite anterior to base of costa

Tergum — dorsal sclerite of any arthropod body segment

Tetraptery (cf. *diptery*) — condition of having four wings

Thanatosis — feigning of death

Tonic (cf. *phasic*) — constant or changing only slowly

Tracheal system — insect respiratory system, comprising a network of internal air tubes

Track angle — angle between wind direction and the track of body displacement over the ground

Trailing edge (cf. *leading edge*) — the rear edge of a moving wing; the posterior wing margin

Trichia (pl. *trichiae*) — hair or hairlike projection from the cuticle

Trochanter — leg segment immediately distal to the coxa

Tymbal — pertaining to the abdominal sound-producing organ of some Homoptera

Uniramous (cf. *biramous*) — unbranched, as of an appendage

Upstroke (cf. *downstroke*) — ventrodorsal motion of the wing in the stroke plane

Vagility — ability to move or migrate

Vannus — expanded basal region of the hindwing in some taxa

Visual slip — movement of the visual field

Volant — capable of flight

Vortex — rotating flow field

Wagner tension field — mechanical stress in a membrane evident as folds that are parallel to the major axis of imposed tension

Wing articulation — junction between wing and thorax

Wingbeat frequency — frequency of oscillation of the wings in the stroke plane

Wing chord — line segment connecting the leading and trailing wing edges that, in planform, is perpendicular to the longitudinal wing span

Yaw (cf. *pitch, roll*) — body rotation about a sagittal axis (i.e., about any axis perpendicular to both longitudinal and transverse axes)

REFERENCES

HISTORICAL NOTE: Chabrier (1822) published the first book-length treatment of insect flight and related morphology based on his series of memoirs published from 1820 to 1822 (see also Chabrier, 1829). The scientific study of insect flight biomechanics may properly be said to have begun with the work of Marey in the mid-nineteenth century. Utilizing the newly developed technique of cinematography, Marey carried out extensive studies of animal physiology, terrestrial locomotion, and animal flight, including that of insects (see bibliography in Braun, 1992). Pettigrew (1872, 1873) wrote extensive comparative surveys of animal flight. Various experimental contributions and overviews of insect flight mechanics through the end of the nineteenth and into the early part of the twentieth century were provided by Erhard (1913), Voss (1913, 1914), Demoll (1918), and Oehmichen (1920). Paralleling these studies, progress in descriptive anatomy enabled Snodgrass (1909, 1927, 1929, 1935) and Ritter (1911) to evaluate insect flight from the perspective of thoracic morphology.

The second major book on insect flight was written by Magnan (1934), who described wing kinematics and manipulations of flight in various insects, and for the first time implemented flow visualization of wing flapping and analysis of the limits to insect flight performance. Wartime circumstances did not impede physical analyses of flapping flight by von Holst and Küchemann (1941) and Küchemann and von Holst (1941); von Holst (1943) also constructed flapping devices modeled on insect wing motions. At the same time, Melin (1941) published a biologically oriented survey of insect flight while Sotavalta (1941) began his epic series of publications on the wingbeat frequency of insects. The science of aeronautics as applied to insect flight was termed "entomoptetics" by Jordanoglou (1949a); this curious neologism appears not to have been used since first publication. Jordanoglou (1949b, c) also constructed and evaluated a technological analogue of the flying insect termed the "entomopter" (see also Kahn, 1950). In the same year, a fundamentally different muscle type termed "asynchronous muscle" was described by Pringle (1949). The midcentury review of Hocking (1953) emphasized insect flight speeds, maximum range, and the limits to flight performance.

Later in the 1950s, a pioneering analysis of the mechanics of locust flight was published by Torkel Weis-Fogh and Martin Jensen. Utilizing locusts in tethered simulations of free flight, wingbeat kinematics (Weis-Fogh, 1956a), aerodynamics (Jensen, 1956), mechanisms of sensory control (Weis-Fogh, 1956b), and cuticular properties (Jensen and Weis-Fogh, 1962) were systematically investigated. Pringle (1957) wrote the first modern treatise on the biomechanics of insect flight, emphasizing aerodynamics, muscle physiology, and sensory systems. Subsequent technological and conceptual advances in functional morphology, neurobiology, and unsteady aerodynamics have contributed exponentially to the flood of contemporary research on flight biomechanics that we are privileged to experience today. Because the pathbreaking work of

Marey emphasized photographic analyses of flying insects, it is also appropriate to mention here the pictorial collections of Dalton (1975, 1977, 1982; see also Kuribayashi, 1981), as well as the book by Brackenbury (1992).

Abbott, R.H. (1973). The effects of fibre length and calcium ion concentration on the dynamic response of glycerol extracted insect fibrillar muscle. *J. Physiol.* **231**:195–208.

Abbott, R.H., and Steiger, G.J. (1977). Temperature and amplitude dependence of tension transients in glycerinated skeletal and insect fibrillar muscle. *J. Physiol.* **266**:13–42.

Ahmad, A. (1984). A comparative study on flight surface and aerodynamic parameters of insects, birds and bats. *Ind. J. Exp. Biol.* **22**:270–278.

Ahmadi, A.R., and Widnall, S.E. (1986). Energetics and optimum motion of oscillating lifting surfaces of finite span. *J. Fluid Mech.* **162**:261–282.

Aidley, D.J. (1985). Muscular contraction. In *Comprehensive Insect Physiology, Biochemistry, and Pharmacology*, vol. 5 (ed. G.A. Kerkut and L.I. Gilbert), pp. 407–437. Oxford: Pergamon Press.

Aidley, D.J. (1989). *The Physiology of Excitable Cells.* 3d ed. Cambridge, U.K.: Cambridge University Press.

Albano, E., and Rodden, W.P. (1969). A doublet-lattice method for calculating lift distributions on oscillating surfaces in subsonic flows. *AIAA J.* **7**:279–285.

Alcock, J., Gwynne, D.T., and Dadour, I.R. (1989). Acoustic signaling, territoriality, and mating in whistling moths, *Hecatesia thyridion* (Agaristidae). *J. Insect Behav.* **2**:27–37.

Alexander, D.E. (1984). Unusual phase relationships between the forewings and hindwings in flying dragonflies. *J. Exp. Biol.* **109**:379–383.

Alexander, D.E. (1986). Wind tunnel studies of turns by flying dragonflies. *J. Exp. Biol.* **122**:81–98.

Alexander, R.D., (1964). The evolution of mating behaviour in arthropods. *Symp. R. Ent. Soc. Lond.* **2**:78–94.

Alexander, R.D., and Brown, W.L. (1963). Mating behavior and the origin of insect wings. *Occas. Papers Mus. Zool. Univ. Mich.* **628**:1–19.

Alexander, R.McN. (1988). *Elastic Mechanisms in Animal Movement.* Cambridge, U.K.: Cambridge University Press.

Alexander, R.McN. (1997). Optimum muscle design for oscillatory movements. *J. theor. Biol.* **184**:253–259.

Alexander, R.McN., and Bennet-Clark, H.C. (1977). Storage of elastic strain energy in muscle and other tissues. *Nature* **265**:114–117.

Allen, L.H., Lemon, E., and Muller, L. (1972). Environment of a Costa Rican forest. *Ecology* **53**:102–111.

Altman, J. (1983). Sensory inputs and the generation of the locust flight motor pattern: from the past to the future. In *BIONA-report 2* (ed. W. Nachtigall), pp. 127–136. Stuttgart: Gustav Fischer.

Altman, J.S. (1975). Changes in the flight motor pattern during the development of the Australian plague locust, *Chortoicetes terminifera. J. Comp. Physiol.* A **97**:127–142.

Altman, J.S., and Tyrer, N.M. (1974). Insect flight as a system for the study of the development of neuronal connections. In *Experimental Analysis of Insect Behaviour* (ed. L.B. Browne), pp. 159–179. Heidelberg: Springer-Verlag.

Aluja, M., Prokopy, R.J., Elkinton, J.S., and Laurence, F. (1989). A novel approach for tracking and quantifying the movement patterns of insects in three dimensions under semi-natural conditions. *Environ. Ent.* **18**:1–7.

Amos, W.B. and Miller, P.L. (1965). The supply of oxygen to the active flight muscle of *Petrognatha gigas* (F.) (Cerambycidae). *Entomologist* **98**:88–94.

Anderson, D.T. (1963). The embryology of *Dacus tryoni*. 2. Development of imaginal discs in the embryo. *J. Embryol. Exp. Morph.* **11**:339–351.

Anderson, S.O., and Weis-Fogh, T. (1964). Resilin. A rubberlike protein in arthropod cuticle. *Adv. Insect Physiol.* **2**:1–65.

Andersson, M. (1994). *Sexual Selection*. Princeton, N.J.: Princeton University Press.

Angelo, M.J., and Slansky, F. (1984). Body building by insects: Trade-offs in resource allocation with particular reference to migratory species. *Fla. Ent.* **67**:22–41.

Aoki, M., Yabuki, K., and Koyama, H. (1978). Micrometeorology of Pasoh Forest. *Malays. Nat. J.* **30**:149–159.

Appanah, S., and Chan, H.T. (1981). Thrips: pollinators of some dipterocarps. *Malays. For.* **44**:234–252.

Arbas, E.A. (1986). Control of hindlimb posture by wind-sensitive hairs and antennae during locust flight. *J. Comp. Physiol.* A **159**:849–857.

Archer, M.E. (1977). The weights of forager loads of *Paravespula vulgaris* (Linn.) (Hymenoptera: Vespidae) and the relationship of load weight to forager size. *Ins. Soc.* **24**:95–102.

Archer, R.D., Sapuppo, J., and Betteridge, D.S. (1979). Propulsion characteristics of flapping wings. *Aero. J.* **83**:355–371.

Arnett, R.H. (1985). *American Insects: A Handbook of the Insects of America North of Mexico*. New York: Van Nostrand Reinhold.

Arnold, E.N. (1984). Evolutionary aspects of tail shedding in lizards and their relatives. *J. Nat. Hist.* **18**:127–169.

Arnold, J.W. (1964). Blood circulation in insect wings. *Mem. Ent. Soc. Can.* **38**:3–60.

Ashworth, J., and Luttges, M. (1986). Comparisons in three-dimensionality in the unsteady flows elicited by straight and swept wings. *AIAA Paper 86-2280*.

Ashworth, J., Mouch, T., and Luttges, M. (1988). Application of forced unsteady aerodynamics to a forward swept wing X-29 model. *AIAA Paper 88-0563*.

Atkins, D.M. (1958). Observations on the flight, wing movements and wing structure of male *Priacma serrata* (Lec.) (Coleoptera: Cupedidae). *Can. Ent.* **90**:339–347.

Atkins, M.D. (1960). A study of the flight of the Douglas-Fir Beetle, *Dendroctonus pseudotsugae* Hopk. (Coleoptera: Scolytidae) II. Flight movements. *Can. Ent.* **92**:941–954.

Attila, U. (1947). Betrachtung des Flügelschlages bei Insekten an Hand eines physikalischen Modells. *Acta Ent. Fenn.* **5**:1–9.

Auber, J. (1963). Ultrastructure de la junction myo-épidermique chez les Diptères. *J. Microscopie* **2**:325–336.

Auber, J. (1967). Distribution of the two kinds of myofilaments in insect muscles. *Amer. Zool.* **7**:451–456.

Auber, M.J. (1967). Particularités ultrastructurales des myofibrilles des muscles du vol chez des Lépidoptères. *C.R. Acad. Sci. Paris* D **264**:621–624.

Auerswald, L., and Gäde, G. (1995). Energy substrates for flight in the blister beetle *Decapotoma lunata* (Meloidae). *J. Exp. Biol.* **198**:1423–1431.

Autrum, H. (1949). Neue Versuche zur optischen Auflösungsvermögen fliegender Insecten. *Experientia* **5**:271–277.

Autrum, H. (1950). Die Belichtungspotentiale und das Sehen der Insecten (Untersuchungen an *Calliphora* und *Dixippus*). *Z. vergl. Physiol.* **32**:176–227.

Autrum, H. (1952). Über zeitliches Auflösungsvermögen und Primärvorgängen im Insektenauge. *Naturwissenschaften* **39**:290–297.

Autrum, H. (1958). Electrophysiological analysis of the visual system in insects. *Expl. Cell Res. Suppl.* **5**:426–439.

Autrum, H., and Gallwitz, U. (1951). Zur Analyse der Belichtungspotentiale des Insektenauges. *Z. vergl. Physiol.* **33**:407–435.

Autrum, H., and Stoecker, M. (1950). Die Verschmelzungsfrequenzen des Bienenauges. *Z. Naturforsch.* **5B**:38–43.

Averof, M., and Akam, M. (1995). Insect-crustacean relationships: insights from comparative developmental and molecular studies. *Phil. Trans. R. Soc. Lond.* B **347**:293–303.

Averof, M., and Cohen, S.M. (1997). Evolutionary origin of insect wings from ancestral gills. *Nature* **385**:627–630.

Azuma, A. (1992). *The Biokinetics of Flying and Swimming*. Tokyo: Springer-Verlag.

Azuma, A., and Kawachi, K. (1979). Local momentum theory and its application to the rotary wing. *J. Aircraft* **16**:6–14.

Azuma, A., and Watanabe, T. (1988). Flight performance of a dragonfly. *J. Exp. Biol.* **137**:221–252.

Azuma, A., Nasu, K., and Hayashi, T. (1983). An extension of the local momentum theory to rotors operating in a twisted flow field. *Vertica* **7**:45–59.

Azuma, A., Azuma, S., Watanabe, I., and Furuta, T. (1985). Flight mechanics of a dragonfly. *J. Exp. Biol.* **116**:79–107.

Baader, A. (1990). The posture of the abdomen during locust flight: regulation by steering and ventilatory interneurones. *J. Exp. Biol.* **151**:109–131.

Baader, A., Schäfer, M., and Rowell, C.H.F. (1992). The perception of the visual flow field by flying locusts: a behavioural and neuronal analysis. *J. Exp. Biol.* **165**:137–160.

Babloyantz, A., and Destexhe, A. (1988). Is the normal heart a periodic oscillator? *Biol. Cybern.* **58**:203–211.

Bacon, J., and Möhl, B. (1979). Activity of an identified wind interneurone in a flying locust. *Nature* **278**:638–640.

Bacon, J., and Möhl, B. (1983). The tritocerebral commissure giant (TCG) wind-sensitive interneurone in the locust. I. Its activity in straight flight. *J. Comp. Physiol.* A **150**:439–452.

Bacon, J., and Tyrer, M. (1979). Wind interneurone input to flight motor neurones in the locust, *Schistocerca gregaria*. *Naturwissenschaften* **66**:116–117.

Bacon, J.P., and Strausfeld, N.J. (1986). The dipteran 'Giant fibre' pathway: neurons and signals. *J. Comp. Physiol.* A **158**:529–548.

Bailey, W.J. (1982). Resonant wing systems in the Australian whistling moth *Hecatesia* (Agaristidae, Lepidoptera). *Nature* **272**:444–446.

Baker, G.T. (1976). Insect flight muscle: maturation and senescence. *Gerontology* **22**:334–361.

Baker, P.S. (1979a). Flying locust visual responses in a radial wind tunnel. *J. Comp. Physiol.* A **131**:39–47.

Baker, P.S. (1979b). The role of forewing muscles in the control of direction in flying locusts. *J. Comp. Physiol.* A **131**:59–66.

Baker, P.S. (1979c). The wing movements of flying locusts during steering behaviour. *J. Comp. Physiol.* A **131**:49–58.

Baker, P.S., and Cooter, R.J. (1979a). The natural flight of the migratory locust, *Locusta migratoria* L. I. Wing movements. *J. Comp. Physiol.* A **131**:79–87.

Baker, P.S., and Cooter, R.J. (1979b). The natural flight of the migratory locust, *Locusta migratoria* L. II. Gliding. *J. Comp. Physiol.* A **131**:89–94.

Baker, P.S., Gewecke, M., and Cooter, R.J. (1981). The natural flight of the migratory locust, *Locusta migratoria* L. III. Wing-beat frequency, flight speed and attitude. *J. Comp. Physiol.* A **141**:233–237.

Baker, P.S., Gewecke, M., and Cooter, R.J. (1984). Flight orientation of swarming *Locusta migratoria*. *Physiol. Ent.* **9**:247–252.

Baker, R.R. (1978). *The Evolutionary Ecology of Animal Migration*. New York: Holmes and Meier.

Baker, R.R., and Parker, G.A. (1979). The evolution of bird coloration. *Phil. Trans. R. Soc. Lond.* B **287**:63–130.

Baker, T.C. (1989). Pheromones and flight behavior. In *Insect Flight* (ed. G.J. Goldsworthy and C.H. Wheeler), pp. 231–255. Boca Raton, Fla.: CRC Press.

Balmford, A., Thomas, A.L.R., and Jones, I.L. (1993). Aerodynamics and the evolution of bird tails. *Nature* **361**:628–631.

Banerjee, S. (1988a). Organisation of wing cuticle in *Locusta migratoria* Linnaeus, *Tropidacris cristata* Linnaeus and *Romalea microptera* Beauvais (Orthoptera: Acrididae). *Int. J. Insect Morphol. and Embryol.* **17**:313–326.

Banerjee, S. (1988b). The functional significance of wing architecture in acridid Orthoptera. *J. Zool., Lond.* **215**:249–267.

Banks, H.P., and Colthart, B.J. (1993). Plant-animal-fungal interactions in Early Devonian trimerophytes from Gaspé, Canada. *Amer. J. Bot.* **80**:992–1001.

Banks, M.J., and Thompson, D.J. (1985). Lifetime mating success in the damselfly *Coenagrion puella*. *Anim. Behav.* **33**:1175–1183.

Barber, S.B., and Pringle, J.W.S. (1966). Functional aspects of flight in belostomatid bugs (Heteroptera). *Proc. R. Soc. Lond.* B **164**:21–39.

Baret, P., Fouarge, A., Bullens, P., and Lints, F. (1994). Life-span of *Drosophila melanogaster* in highly oxygenated atmospheres. *Mech. Ageing Dev.* **76**:25–31.

Barlet, J. (1950). La question des pièces pleurales du thorax des Machilides. *Bull. Ann. Soc. R. Ent. Belg.* **86**:179–190.

Barlet, J. (1974). La musculature thoracique d'*Oncojapyx basilewskyi* Pagè (Aptérygotes Diplures). *Bull. Ann. Soc. R. Ent. Belg.* **110**:91–141.

Barlet, J. (1988). Considérations sur le squelette thoracique des insectes Aptérygotes. *Bull. Ann. Soc. R. Ent. Belg.* **124**:171–187.

Barth, F.G. (1991). *Insects and Flowers: The Biology of a Partnership.* Princeton, N.J.: Princeton University Press.

Bartholomew, G.A. (1987). Interspecific comparison as a tool for ecological physiologists. In *New Directions in Ecological Physiology* (ed. M.E. Feder, A.F. Bennett, W.W. Burggren, and R.B. Huey), pp. 11–35. Cambridge, U.K.: Cambridge University Press.

Bartholomew, G.A., and Barnhart, M.C. (1984). Tracheal gases, respiratory gas exchange, body temperature and flight in some tropical cicadas. *J. Exp. Biol.* **111**:131–144.

Bartholomew, G.A., and Casey, T.M. (1977). Endothermy during terrestrial activity in large beetles. *Science* **195**:882–883.

Bartholomew, G.A., and Casey, T.M. (1978). Oxygen consumption of moths during rest, pre-flight warm-up, and flight in relation to body size and wing morphology. *J. Exp. Biol.* **76**:11–25.

Bartholomew, G.A., and Heinrich, B. (1973). A field study of flight temperature in moths in relation to body weight and wing loading. *J. Exp. Biol.* **58**:123–135.

Bartholomew, G.A., and Heinrich, B. (1978). Endothermy in African dung beetles during flight, ball making, and ball rolling. *J. Exp. Biol.* **73**:65–83.

Basar, E. (ed.) (1990). *Chaos in Brain Function.* Berlin: Springer-Verlag.

Basibuyuk, H.H., and Quicke, D.L.J. (1997). Hamuli in the Hymenoptera (Insecta) and their phylogenetic implications. *J. Nat. Hist.* **31**:1563–1585.

Bastian, J. (1972). Neuro-muscular mechanisms controlling a flight maneuver in the honeybee. *J. Comp. Physiol.* **77**:126–140.

Batchelor, G.K. (1967). *An Introduction to Fluid Dynamics.* Cambridge, U.K.: Cambridge University Press.

Bates, H.W. (1862). Contributions to an insect fauna of the Amazon valley. Lepidoptera: Heliconidae. *Trans. Linn. Soc. Lond.* **23**:495–566.

Bawa, K.S., Bullock, S.H., Perry, D.R., Coville, R.E., and Grayum, M.H. (1985). Reproductive biology of tropical lowland rain forest trees. II. Pollination systems. *Amer. J. Bot.* **72**:346–356.

Bechert, D.W., Hoppe, G., and Reif, W.-E. (1985). On the drag reduction of the shark skin. *AIAA Paper No. 85–0546.*

Becker, E.G. (1952). The problem of the origin and development of the wing in insects. Chapter 1. Precursors of the insect wing. *Vestn. Mosk. Univ. Ser. Phys. Math. Nat. Sci.* **9**:59–68. (In Russian.)

Becker, E.G. (1954). The problem of the origin and development of the wing in insects. Chapter 2. The structure, mechanics and origin of the flying apparatus of mayflies (Ephemeroptera). *Vestn. Mosk. Univ. Ser. Fiz.-Mat. Estestv. Nauk* **5**:119–129. (In Russian.)

Becker, E.G. (1956). The problem of the origin and development of the wing in insects. Chapter 3. The mesothorax of mayflies (Ephemeroptera) and the evolution of the insect wing apparatus. *Vestn. Mosk. Univ. Ser. Fiz.-Mat. Estestv. Nauk* **6**:105–110. (In Russian.)

Becker, E.G. (1958). The problem of the origin and development of the wing in insects. Chapter 5. A contribution to the knowledge of ontogeny and phylogeny of the organs of flight in Orthoptera. *Ent. Rev., Wash.* **37**:671–677.

Beebe, W. (1949). Insect migration at Rancho Grande in north-central Venezuela. General account. *Zoologica* **34**:107–110.

Beebe, W., and Kenedy, R. (1957). Habits, palatability and mimicry in thirteen ctenuchid moth species from Trinidad, B.W.I. *Zoologica* **42**:147–157.

Beenakkers, A.M.T., Van der Horst, D.J., and Van Marrewijk, W.J.A. (1981). Role of lipids in energy metabolism. In *Energy Metabolism in Insects* (ed. R.G.H. Downer), pp. 53–100. New York: Plenum Press.

Beenakkers, A.M.T., Van der Horst, D.J., and Van Marrewijk, W.J.A. (1984). Insect flight muscle metabolism. *Insect Biochem.* **14**:243–260.

Beenakkers, A.M.T., Van der Horst, D.J., and Van Marrewijk, W.J.A. (1985). Biochemical processes directed to flight muscle metabolism. In: *Comprehensive Insect Physiology, Biochemistry, and Pharmacology*, vol. 10 (*Biochemistry*) (ed. G.A. Kerkut and L.I. Gilbert), pp. 451–486. Oxford: Pergamon.

Behrensmeyer, A.K., Damuth, J.D., DiMichele, W.A., Sues, H.-D., and Wing, S.L. (1992). *Terrestrial Ecosystems through Time: Evolutionary Paleoecology of Terrestrial Plants and Animals*. Chicago: University of Chicago Press.

Bell, J., Kipp, L.R., and Collins, R.D. (1995). The role of chemo-orientation in search behavior. In *Chemical Ecology of Insects 2* (ed. R.T. Cardé and W.J. Bell), pp. 105–152. New York: Chapman and Hall.

Belotserkovsky, S.M., Gulyayev, V.V., and Nisht, M.I. (1974). On the study of flight of insects and birds. *Dokl. Akad. Nauk SSSR* **219**:567–570. (In Russian.)

Belton, P. (1986). Sounds of insects in flight. In *Insect Flight: Migration and Dispersal* (ed. W. Danthanarayana), pp. 60–70. Berlin: Springer-Verlag.

Bennet-Clark, H.C., and Ewing, A. (1968). The wing mechanisms involved in the courtship of *Drosophila*. *J. Exp. Biol.* **49**:117–128.

Bennett, L. (1966). Insect aerodynamics: vertical sustaining force in near-hovering flight. *Science* **152**:1263–1266.

Bennett, L. (1970). Insect flight: lift and rate of change of incidence. *Science* **167**:177–179.

Bennett, L. (1973). Effectiveness and flight of small insects. *Ann. Ent. Soc. Amer.* **66**:1187–1190.

Bennett, L. (1975). Insect aerodynamics near hovering. In *Swimming and Flying in Nature*, vol. 2 (ed. T.Y.-T. Wu, C.J. Brokaw, and C. Brennen), pp. 815–828. New York: Plenum Press.

Bennett, L. (1976). Induced airflow created by large hovering beetles. *Ann. Ent. Soc. Amer.* **69**:985–990.

Bennett, L. (1977). Clap and fling aerodynamics—an experimental evaluation. *J. Exp. Biol.* **69**:261–272.

Bennett, S.C. (1997). The arboreal leaping theory of the origin of pterosaur flight. *Hist. Biol.* **12**:265–290.

Benson, W.W., Haddad, C.F.B., and Zikán, M. (1989). Territorial behavior and dominance in some heliconiine butterflies (Nymphalidae). *J. Lep. Soc.* **43**:33–49.

Bentley, D.R., and Hoy, R.R. (1970). Postembryonic development of adult motor patterns in crickets: a neural analysis. *Science* **170**:1409–1411.

Berger, M. (1974a). Energiewechsel von Kolibris beim Schwirrflug unter Höhenbedingungen. *J. Ornithol.* **115**:273–288.

Berger, M. (1974b). Oxygen consumption and power of hovering hummingbirds at varying barometric and oxygen pressures. *Naturwissenschaften* **61**:407.

Berger, M. (1985). Sauerstoffverbrauch von Kolibris (*Colibri coruscans* and *C. thalassinus*) beim Horizontalflug. In *BIONA-report 3* (ed. W. Nachtigall), pp. 307–314. Stuttgart: Gustav Fischer.

Berner, R.A. (1990). Atmospheric carbon dioxide levels over Phanerozoic time. *Science* **249**:1382–1385.

Berner, R.A. (1991). A model for atmospheric CO_2 over Phanerozoic time. *Amer. J. Sci.* **291**:339–376.

Berner, R.A. (1994). GEOCARB II: a revised model of atmospheric CO_2 over Phanerozoic time. *Amer. J. Sci.* **294**:56–91.

Berner, R.A. (1998). The carbon cycle and CO_2 over Phanerozoic time: the role of land plants. *Phil. Trans. R. Soc. Lond.* B **353**:75–82.

Berner, R.A., and Canfield, D.E. (1989). A new model for atmospheric oxygen over Phanerozoic time. *Amer. J. Sci.* **289**:333–361.

Bertsch, A. (1984). Foraging in male bumblebees (*Bombus lucorum* L.): maximizing energy or minimizing water load? *Oecologia* **62**:325–336.

Betteridge, D.S., and Archer, R.D. (1974). A study of the mechanics of flapping wings. *Aero. Quart.* **25**:129–142.

Betts, C.R. (1986a). The comparative morphology of the wings and axillae of selected Heteroptera. *J. Zool., Lond. (B)* **1**:255–282.

Betts, C.R. (1986b). Functioning of the wings and axillary sclerites of Heteroptera during flight. *J. Zool., Lond. (B)* **1**:283–301.

Betts, C.R., and Wootton, R.J. (1988). Wing shape and flight behaviour in butterflies (Lepidoptera: Papilionoidea and Hesperioidea): a preliminary analysis. *J. Exp. Biol.* **138**:271–288.

Bilo, D. (1994). Course control during flight. In *Perception and Motor Control in Birds* (ed. M.N.O. Davies and P.R. Green), pp. 227–247. Berlin: Springer-Verlag.

Bilo, D., Lauck, A., and Nachtigall, W. (1985). Measurement of linear body accelerations and calculation of the instantaneous aerodynamic lift and thrust in a pigeon flying in a wind tunnel. In *BIONA-report 3* (ed. W. Nachtigall), pp. 87–108. Stuttgart: Gustav Fischer.

Bilo, D., Lauck, A., Wedekind, F., Rothe, H.-J., and Nachtigall, W. (1982). Linear accelerations of a pigeon flying in a wind tunnel. *Naturwissenschaften* **69**:345–346.

Bilton, D. (1994). The flight apparatus and flying ability of *Hydroporus glabriusculus* (Coleoptera, Dytiscidae), with a brief review of structural modifications in flightless beetles. *Ent. Tidskr.* **115**:23–32.

Birket-Smith, S.J.R. (1984). *Prolegs, Legs and Wings of Insects*. Copenhagen: Scandinavian Science Press.

Bishop, C.M. (1997). Heart mass and the maximum cardiac output of birds and mammals: implications for estimating the maximum aerobic power input of flying animals. *Phil. Trans. R. Soc. Lond.* B **352**:447–456.

Bishop, C.M., and Butler, P.J. (1995). Physiological modelling of oxygen consumption in birds during flight. *J. Exp. Biol.* **198**:2153–2163.

Bitsch, J. (1994). The morphological groundplan of Hexapoda: critical review of recent concepts. *Ann. Soc. Ent. Fr. (n.s)* **30**:103–129.

Blackburn, T.M., and Gaston, K.J. (1994). Animal body size distributions: patterns, mechanisms and implications. *Trends Ecol. Evol.* **9**:471–474.

Bleich, O.E. (1928). Thanatose und Hypnose bei Coleopteren. *Z. wiss. Biol.* (A) **10**:1–61.

Bleiweiss, R., Kirsch, J.A.W., and Matheus, J.C. (1997). DNA hybridization evidence for the principal lineages of hummingbirds (Aves: Trochilidae). *Molec. Biol. Evol.* **14**:325–343.

Blondeau, J. (1981). Aerodynamic capabilities of flies, as revealed by a new technique. *J. Exp. Biol.* **92**:155–163.

Blondeau, J., and Heisenberg, M. (1982). The three-dimensional optomotor torque system of *Drosophila melanogaster*. Studies on wildtype and the mutant *optomotor-blind*[H31]. *J. Comp. Physiol.* A **145**:321–329.

Bocharova-Messner, O.M. (1959). Development of wings in the early postembryonic stages of developing dragonflies (Class Odonata). *Trud. Inst. Morf. Zhiv. Im. Severtsova* **27**:187–200. (In Russian.)

Bocharova-Messner, O.M. (1968). The principles of ontogenesis of the pterothorax in Polyneoptera, with reference to the problem of origin and evolution of the insect flight apparatus. In *Problems of Insect Functional Morphology and Embryology* (ed. D.M. Fedotov), pp. 3–26. Moscow: Nauka. (In Russian.)

Bocharova-Messner, O.M. (1971). On the origin of flight apparatus of insects. In *Proceedings of the XIIIth International Congress of Entomology*, vol. 1 (Moscow, 1968), p. 232. Leningrad: Nauka.

Bocharova-Messner, O.M., and Aksyuk, T.S. (1981). Tunnel formation by the wings of the diurnal butterfly during flight (Lepidoptera, Rhopalocera). *Dokl. Akad. Nauk SSSR* **260**:1490–1493. (In Russian.)

Boettiger, E.G. (1960). Insect flight muscles and their basic physiology. *Ann. Rev. Ent.* **5**:1–16.

Boettiger, E.G., and Furshpan, E. (1952). The mechanics of flight movements in Diptera. *Biol. Bull.* **102**:200–211.

Boore, J.L., Collins, T.M., Stanton, D., Daehler, L.L., and Brown, W.M. (1995). Deducing the pattern of arthropod phylogeny from mitochondrial DNA rearrangements. *Nature* **376**:163–165.

Boore, J.L., Lavrov, D.V., and Brown, W.M. (1998). Gene translocation links insects and crustaceans. *Nature* **392**:667–668.

Borgia, G. (1981). Mate selection in the fly *Scatophaga stercoraria*: female choice in a male-controlled system. *Anim. Behav.* **29**:71–80.

Borin, A.A. (1987). On takeoff capacity of flying animals. *Dokl. Akad. Nauk SSSR* **293**:1256–1258. (In Russian.)

Börner, C. (1908). Die Tracheenkiemen der Ephemeriden. *Zool. Anz.* **33**:806–823.

Borst, A. (1986). Time course of the houseflies' landing response. *Biol. Cybern.* **54**:379–383.

Borst, A. (1989). Temporal processing of excitatory and inhibitory motion stimuli in the fly's landing system. *Naturwissenschaften* **76**:531–534.

Borst, A. (1990). How do flies land? *Bioscience* **40**:292–299.

Borst, A., and Bahde, S. (1986). What kind of movement detector is triggering the landing response of the housefly? *Biol. Cybern.* **55**:59–69.

Borst, A., and Bahde, S. (1988a). Visual information processing in the fly's landing system. *J. Comp. Physiol.* A **163**:167–173.

Borst, A., and Bahde, S. (1988b). Spatio-temporal integration of motion. A simple strategy for safe landing in flies. *Naturwissenschaften* **75**:265–267.

Boudreaux, B.H. (1979). *Arthropod Phylogeny with Special Reference to Insects.* New York: John Wiley and Sons.

Bourgogne, J. (1951). Ordre des Lépidopteres. In *Traité de Zoologie, Anatomie, Systématique, Biologie. Insectes Supérieurs et Hémiptéroides*, vol. 10 (ed. P. Grassé), pp. 174–448. Paris: Masson et Cie.

Bowden, J., and Johnson, C.G. (1976). Migrating and other terrestrial insects at sea. In *Marine Insects* (ed. L. Cheng), pp. 97–117. Amsterdam: North Holland.

Boyan, G.S. (1985). Auditory input to the flight system of the locust. *J. Comp. Physiol.* A **156**:79–91.

Brackenbury, J. (1990). Wing movements in the bush-cricket *Tettigonia viridissima* and the mantis *Ameles spallanziana* during natural leaping. *J. Zool., Lond.* **220**:593–602.

Brackenbury, J. (1991). Wing kinematics during natural leaping in the mantids *Mantis religiosa* and *Iris oratoria*. *J. Zool., Lond.* **223**:341–356.

Brackenbury, J., and Hunt, H. (1993). Jumping in springtails: mechanism and dynamics. *J. Zool., Lond.* **229**:217–236.

Brackenbury, J.H. (1991). Kinematics of take-off and climbing flight in butterflies. *J. Zool., Lond.* **224**:251–270.

Brackenbury, J.H. (1992). *Insects in Flight.* London: Blandford.

Brackenbury, J.H. (1994a). Hymenopteran wing kinematics: a qualitative study. *J. Zool., Lond.* **233**:523–540.

Brackenbury, J.H. (1994b). Wing folding and free-flight kinematics in Coleoptera (Insecta): a comparative study. *J. Zool., Lond.* **232**:253–283.

Brackenbury, J.H., and Wang, R. (1995). Ballistics and visual targeting in flea-beetles (Alticinae). *J. Exp. Biol.* **198**:1931–1942.

Bradley, J.C. (1942). The origin and significance of metamorphosis and wings among insects. In *Proceedings of the Eighth American Science Congress*, pp. 303–309. Washington, D.C.: Department of State.

Brady, J. (1972). Spontaneous, circadian components of tsetse fly activity. *J. Insect Physiol.* **18**:471–484.

Brady, J. (1991). Flying mate detection and chasing by tsetse flies (*Glossina*). *Physiol. Ent.* **16**:153–161.

Brady, J., Griffiths, N., and Paynter, Q. (1995). Wind speed effects on odour source location by tsetse flies (*Glossina*). *Physiol. Ent.* **20**:293–302.

Brauckmann, C. (1991). Arachniden und Insekten aus dem Namurium von

Hagen-Vorhalle (Ober-Karbon; West-Deutschland). *Veröffentl. Fuhlrott-Museum* **1**:1–215.

Brauckmann, C., and Zessin, W. (1989). Neue Meganeuridae aus dem Namurium von Hagen-Vorhalle (BRD) und die Phylogenie der Meganisoptera. *Dtsch. Ent. Z., N.F.* **36**:177–215.

Braun, M. (1992). *Picturing Time: The Work of Etienne-Jules Marey (1830–1904).* Chicago: University of Chicago Press.

Brauns, A. (1939). Morphologische und physiologische Untersuchungen zum Halterenproblem unter besonderer Berücksichtigung brachypterer Arten. *Zool. Jb. Abt. Allg. Zool. Physiol.* **59**:245–390.

Breidbach, O., and Kutsch, W. (1990). Structural homology of identified motoneurones in larval and adult stages of hemi- and holometabolous insects. *J. Comp. Neurol.* **297**:392–409.

Briggs, D.E.G. (1985). Gigantism in Palaeozoic arthropods. *Special Papers in Paleontology*, no. 33, p. 157.

Bristowe, W.S. (1958). *The World of Spiders.* London: Collins.

Brodsky, A.K. (1970). On the role of wing pleating in insects. *Zh. Evol. Biokhim. Fiziol.* **6**:470–471. (In Russian.)

Brodsky, A.K. (1981). Aerodynamic peculiarities of insect flight. IV. Data on the flight of the mayfly *Ephemera vulgata* L. *Vestn. Leningr. Univ. Biol.* **15**:12–18. (In Russian.)

Brodsky, A.K. (1982). Evolution of the wing apparatus in stoneflies (Plecoptera). Part IV. Wing kinematics and general conclusion. *Ent. Obozr.* **61**:491–500. (In Russian.)

Brodsky, A.K. (1985a). Kinematics of insect wings in horizontally maintained flight (a comparative study). *Ent. Rev., Wash.* **64**(3):56–73.

Brodsky, A.K. (1985b). Some new principles of insect flight. *Dokl. Akad. Nauk SSSR* **283**:1491–1495. (In Russian.)

Brodsky, A.K. (1986a). A new approach to the study of flapping flight in insects. *Obsch. Ent.* **68**:4–7. (In Russian.)

Brodsky, A.K. (1986b). Flight of insects with high wingbeat frequency. *Ent. Obozr.* **65**:269–279. (In Russian.)

Brodsky, A.K. (1991). Vortex formation in the tethered flight of the peacock butterfly *Inachis io* L. (Lepidoptera, Nymphalidae) and some aspects of insect flight evolution. *J. Exp. Biol.* **161**:77–95.

Brodsky, A.K. (1994). *The Evolution of Insect Flight.* Oxford: Oxford University Press.

Brodsky, A.K., and Grodnitsky, D.L. (1985). Tethered flight aerodynamics of the European skipper *Thymelicus lineola* Ochs. (Lepidoptera, Hesperiidae). *Ent. Obozr.* **64**:484–492. (In Russian.)

Brodsky, A.K., and Ivanov, V.P. (1975). Aerodynamic peculiarities of insect flight. III. Flow around the wings of the mayfly *Ephemera vulgata* L. (Ephemeroptera). *Vestn. Leningr. Univ. Biol.* **3**:7–10. (In Russian.)

Brodsky, A.K., and Ivanov, V.D. (1983a). Functional assessment of wing structure in insects. *Ent. Rev., Wash.* **62**:35–52.

Brodsky, A.K., and Ivanov, V.D. (1983b). Visualization of the air flow around a flying insect. *Dokl. Acad. Nauk SSSR* **271**:742–745. (In Russian.)

Brodsky, A.K., and Ivanov, V.D. (1984). The role of vortices in insect flight. *Zool. Zh.* **63**:197–208. (In Russian.)

Brodsky, A.K., and Ivanov, V.D. (1985). Traces of flying insects. *Priroda* **10**:74–79. (In Russian.)

Brodsky, A.K., and Vorobjov, N.N. (1990). Gliding of butterflies and role of the wing scale cover in their flight. *Ent. Obozr.* **69**:241–256. (In Russian.)

Brooks, D.R., and McLennan, D.A. (1991). *Phylogeny, Ecology, and Behavior: A Research Program in Comparative Biology.* Chicago: University of Chicago Press.

Brower, A.V.Z. (1995). Locomotor mimicry in butterflies? A critical review of the evidence. *Phil. Trans. R. Soc. Lond.* B **347**:413–425.

Brower, L.P. (1985). New perspectives on the migration biology of the monarch butterfly, *Danaus plexippus* L. *Contrib. Mar. Sci.* **27**:748–785.

Brown, E.S. (1970). Nocturnal insect flight direction in relation to the wind. *Proc. R. ent. Soc. Lond. (A)* **45**:39–43.

Buchmann, S.L. (1983). Buzz pollination in angiosperms. In *Handbook of Experimental Pollination Ecology* (ed. C.E. Jones and R.J. Little), pp. 73–113. New York: Van Nostrand Reinhold.

Buchthal, F., and Weis-Fogh, T. (1956). Contribution of the sarcolemma to the force exerted by resting muscle of insects. *Acta Physiol. Scand.* **35**:345–364.

Buchthal, F., Weis-Fogh, T., and Rosenfalck, P. (1957). Twitch contractions of isolated flight muscle of locusts. *Acta Physiol. Scand.* **39**:246–276.

Buckholz, R.H. (1981). Measurements of unsteady periodic forces generated by the blowfly flying in a wind tunnel. *J. Exp. Biol.* **90**:163–173.

Buckholz, R.H. (1986). The functional role of wing corrugations in living systems. *ASME J. Fluids Eng.* **108**:93–97.

Budyko, M.I., Ronov, A.B., and Yanshin, A.L. (1987). *History of the Earth's Atmosphere.* Berlin: Springer-Verlag.

Buelthoff, H., Poggio, T., and Wehrhahn, C. (1980). 3-D analysis of the flight trajectories of flies (*Drosophila melanogaster*). *Z. Naturforsch.* **35C**:811–815.

Bull, M.L. (1910). Sur les inclinaisons du voile de l'aile de l'insecte pendant le vol. *C.R. Acad. Sci., Paris* **150**:129–131.

Bullard, B. (1983). Contractile proteins of insect flight muscle. *Trends Biochem. Sci.* **8** (Feb.):68–70.

Bullard, B., Leonard, K., Larkins, A., Butcher, G., Karlik, C., and Fyrberg, E. (1988). Troponin of asynchronous flight muscle. *J. Molec. Biol.* **204**:621–637.

Burd, M. (1994). Butterfly wing colour patterns and flying heights in the seasonally wet forest of Barro Colorado Island, Panama. *J. Trop. Ecol.* **10**:601–610.

Burkhardt, D. (1977). On the vision of insects. *J. Comp. Physiol.* A **120**:33–50.

Burkhardt, D., and Schneider, G. (1957). Die Antennen von *Calliphora* als Anzeiger der Fluggeschwindigkeit. *Z. Naturforsch.* **12B**:139–143.

Burkhardt, G., and Wegener, G. (1994). Glycogen phosphorylase from flight muscle of the hawk moth, *Manduca sexta*: purification and properties of three interconvertible forms and the effect of flight on their interconversion. *J. Comp. Physiol.* B **164**:261–271.

Burnham, L. (1983). Studies on Upper Carboniferous insects: 1. The Geraridae (Order Protorthoptera). *Psyche* **90**:1–57.

Burrows, M. (1973). The role of delayed excitation in the co-ordination of some metathoracic flight motoneurons of a locust. *J. Comp. Physiol.* **83**:135–164.

Burrows, M. (1975a). Monosynaptic connexions between wing stretch receptors and flight motor neurons of the locust. *J. Exp. Biol.* **62**:189–219.

Burrows, M. (1975b). Co-ordinating interneurones of the locust which convey two patterns of motor commands: their connexions with flight motoneurones. *J. Exp. Biol.* **63**:713–733.

Burrows, M. (1977). Flight mechanisms of the locust. In *Identified Neurons and Behavior of Arthropods* (ed. G. Hoyle), pp. 339–356. New York: Plenum Press.

Burrows, M. (1996). *The Neurobiology of an Insect Brain*. Oxford: Oxford University Press.

Burrows, M., and Pflüger, H.J. (1992). Output connections of a wind sensitive interneurone with motor neurones innervating flight steering muscles in the locust. *J. Comp. Physiol.* A **171**:437–446.

Bursell, E. (1981). The role of proline in energy metabolism. In *Energy Metabolism in Insects* (ed. R.G.H. Downer), pp. 135–154. New York: Plenum Press.

Bursell, E., and Taylor, P. (1980). An energy budget for *Glossina* (Diptera: Glossinidae). *Bull. Ent. Res.* **70**:187–196.

Burt, P.J.A., and Pedgley, D.E. (1997). Nocturnal insect migration: effects of local winds. *Adv. Ecol. Res.* **27**:61–92.

Burton, A.J. (1964). Nervous control of flight orientation in a beetle. *Nature* **204**:1333.

Burton, A.J. (1971). Directional change in a flying beetle. *J. Exp. Biol.* **54**:575–585.

Burton, A.J., and Sandeman, D.C. (1961). The lift provided by the elytra of the Rhinoceros Beetle, *Oryctes boas* Fabr. *S. Afr. J. Sci.* **57**:107–109.

Büschges, A., and Pearson, K.G. (1991). Adaptive modifications in the flight system of the locust after the removal of wing proprioceptors. *J. Exp. Biol.* **157**:313–333.

Büschges, A., Ramirez, J.M., Driesang, R., and Pearson, K.G. (1992). Connections of the forewing tegulae in the locust flight system and their modification following partial deafferentation. *J. Neurobiol.* **23**:31–43.

Bushnell, D.M., and Moore, K.J. (1991). Drag reduction in nature. *Ann. Rev. Fluid Mech.* **23**:65–79.

Byers, G.W. (1969). Evolution of wing reduction in crane flies (Diptera: Tipulidae). *Evolution* **23**:346–354.

Byrne, D.N., Buchmann, S.L., and Spangler, H.G. (1988). Relationship between wing loading, wingbeat frequency and body mass in homopterous insects. *J. Exp. Biol.* **135**:9–23.

Callahan, P.S. (1965). A photoelectric-photographic analysis of flight behaviour in the Corn Earworm, *Heliothis zea*, and other moths. *Ann. Ent. Soc. Amer.* **58**:159–169.

Camhi, J.M. (1969). Locust wind receptors. III. Contribution to flight initiation and lift control. *J. Exp. Biol.* **50**:363–373.

Camhi, J.M. (1970a). Yaw-correcting postural changes in locusts. *J. Exp. Biol.* **52**:519–531.

Camhi, J.M. (1970b). Sensory control of abdomen posture in flying locusts. *J. Exp. Biol.* **52**:533–537.

Camhi, J.M., and Hinkle, M. (1974). Response modification by the central flight oscillator of locusts. *J. Exp. Biol.* **60**:477–492.

Camhi, J.M., Sumbre, G., and Wendler, G. (1995). Wing-beat coupling between flying locust pairs: preferred phase and lift enhancement. *J. Exp. Biol.* **198**: 1051–1063.

Campbell, B.C., Steffen-Campbell, J.D., and Gill, R.J. (1994). Evolutionary origin of whiteflies (Hemiptera: Sternorrhyncha: Aleyrodidae) inferred from 18S rDNA sequences. *Insect Molec. Biol.* **3**:73–88.

Campbell, B.C., Steffen-Campbell, J.D., Sorensen, J.T., and Gill, R.J. (1995). Paraphyly of Homoptera and Auchenorrhyncha inferred from 18S rDNA nucleotide sequences. *Syst. Ent.* **20**:175–194.

Campbell, G., Weaver, T., and Tomlinson, A. (1993). Axis specification in the developing *Drosophila* appendage: the role of *wingless, decapentaplegic,* and the homeobox gene *aristaless. Cell* **74**:1113–1123.

Candia Carnevali, M.D., and Reger, J.F. (1982). Slow-acting flight muscles of saturniid moths. *J. Ultrastruct. Res.* **79**:241–249.

Candy, D.J. (1978). The regulation of locust flight muscle metabolism by octopamine and other compounds. *Insect Biochem.* **8**:177–182.

Candy, D.J. (1989). Utilization of fuels by the flight muscles. In *Insect Flight* (ed. G.J. Goldsworthy and C.H. Wheeler), pp. 305–319. Boca Raton, Fla.: CRC Press.

Cardé, R.T. (1996). Odour plumes and odour-mediated flight in insects. In *Olfaction in Mosquito-Host Interactions* (ed. G.R. Bock and G. Cardew), pp. 54–70. Chichester, U.K.: John Wiley and Sons.

Cardé, R.T., and Baker, T.C. (1984). Sexual communication with pheromones. In *Chemical Ecology of Insects* (ed. W.J. Bell and R.T. Cardé), pp. 356–383. London: Chapman and Hall.

Carle, F.L. (1982). Thoughts on the origin of insect flight. *Ent. News* **93**:159–172.

Carlquist, S. (1965). *Island Life: A Natural History of the Islands of the World.* Garden City, N.Y.: Natural History Press.

Carlquist, S. (1974). *Island Biology.* New York: Columbia University Press.

Carmean, D., and Crespi, B.J. (1995). Do long branches attract flies? *Nature* **373**:666.

Carpenter, F.M. (1939). The Lower Permian insects of Kansas. Part 8. Additional Megasecoptera, Protodonata, Odonata, Homoptera, Psocoptera, Protelytroptera, Plecoptera and Protoperlaria. *Proc. Am. Acad. Arts Sci.* **73**:29–70.

Carpenter, F.M. (1953). The geological history and evolution of insects. *Amer. Scient.* **41**:256–270.

Carpenter, F.M. (1963). Studies on Carboniferous insects from Commentry, France. Part IV. The genus *Diaphonoptera* and the order Diaphanopterodea. *Psyche* **70**:120–128.

Carpenter, F.M. (1966). The Lower Permian insects of Kansas. Part 11. The orders Protorthoptera and Orthoptera. *Psyche* **73**:46–88.

Carpenter, F.M. (1971). Adaptations among Paleozoic insects. *Proc. N. Amer. Paleont. Conv.* **1969**:1236–1251.

Carpenter, F.M. (1976). Geological history and evolution of the insects. *Proc. 15th In. Congr. Ent. (Wash.)* **1**:63–70.

Carpenter, F.M. (1992). *Treatise on Invertebrate Paleontology. Part R, Arthropoda 4, Volumes 3 and 4 (Hexapoda).* Lawrence: University of Kansas Press.

Carpenter, F.M., and Burnham, L. (1985). The geological record of insects. *Ann. Rev. Earth Planet. Sci.* **13**:297–314.

Carpenter, G.D.H. (1937). Further evidence that Birds do attack and eat Butterflies. *Proc. Zool. Soc. Lond.* **12**:223–247.

Carpenter, G.D.H., and Ford, E.B. (1933). *Mimicry.* London: Methuen.

Carroll, R.L. (1988). *Vertebrate Paleontology and Evolution.* New York.: W. H. Freeman.

Carroll, S.B. (1995). Homeotic genes and the evolution of arthropods and chordates. *Science* **376**:479–485.

Carroll, S.B., Weatherbee, S.D., and Langeland, J.A. (1995). Homeotic genes and the regulation and evolution of insect wing number. *Nature* **375**:58–61.

Cartar, R.V. (1992). Morphological senescence and longevity: an experiment relating wing wear and life span in foraging wild bumble bees. *J. Anim. Ecol.* **61**:225–231.

Casas, J., and Aluja, M. (1997). The geometry of search movements in insects in plant canopies. *Behav. Ecol.* **8**:37–45.

Casey, T.M. (1976). Flight energetics of sphinx moths: power input during hovering flight. *J. Exp. Biol.* **64**:529–543.

Casey, T.M. (1980). Flight energetics and heat exchange of gypsy moths in relation to air temperature. *J. Exp. Biol.* **64**:529–543.

Casey, T.M. (1981a). A comparison of mechanical and energetic estimates of flight cost for hovering sphinx moths. *J. Exp. Biol.* **91**:117–129.

Casey, T.M. (1981b). Energetics and thermoregulation of *Malacosoma americanum* (Lepidoptera: Lasiocampidae) during hovering flight. *Physiol. Zool.* **54**:362–371.

Casey, T.M. (1981c). Insect flight energetics. In *Locomotion and Energetics in Arthropods* (ed. C.F. Herreid and C.R. Fourtner), pp. 419–452. New York: Plenum Press.

Casey, T.M. (1988). Thermoregulation and heat exchange. *Adv. Insect Physiol.* **20**:119–146.

Casey, T.M. (1989). Oxygen consumption during flight. In *Insect Flight* (ed. G.J. Goldsworthy and C.H. Wheeler), pp. 257–272. Boca Raton, Fla.: CRC Press.

Casey, T.M., and Ellington, C.P. (1989). Energetics of insect flight. In *Energy Transformation in Cells and Organisms* (ed. W. Wieser and E. Gnaiger), pp. 200–210. Stuttgart: Georg Thieme.

Casey, T.M., Ellington, C.P., and Gabriel, J.M. (1992). Allometric scaling of muscle performance and metabolism: insects. In *Hypoxia and Mountain Medicine* (ed. J.R. Sutton, G. Coates, and C.S. Houston), pp. 152–162. Oxford: Pergamon Press.

Casey, T.M., May, M.L., and Morgan, K.R. (1985). Flight energetics of euglossine bees in relation to morphology and wing stroke frequency. *J. Exp. Biol.* **116**:271–289.

Chabrier, J. (1822). *Essai sur le Vol des Insectes et Observations sur Quelques Parties de la Mécanique des Mouvemens Progressifs de l'Homme et des Animaux Vertébrés*. Paris: A. Belin.

Chabrier, J. (1829). Explication du vol des Oiseaux et des Insectes. *Ann. Sci. Nat. Zool.* **16**:499–508.

Chadwick, L.E. (1940). The wing motion of the dragonfly. *Bull. Brooklyn ent. Soc.* **35**:109–112.

Chadwick, L.E. (1951). Stroke amplitude as a function of air density in the flight of *Drosophila*. *Biol. Bull.* **100**:15–27.

Chadwick, L.E., and Gilmour, D. (1940). Respiration during flight in *Drosophila repleta* Wollaston: the oxygen consumption considered in relation to the wing-rate. *Physiol. Zool.* **13**:398–410.

Chadwick, L.E., and Williams, C.M. (1949). The effects of atmospheric pressure and composition on the flight of *Drosophila*. *Biol. Bull.* **97**:115–137.

Chai, P. (1986). Field observations and feeding experiments on the responses of rufous-tailed jacamars (*Galbula ruficauda*) to free-flying butterflies in a tropical rainforest. *Biol. J. Linn. Soc.* **29**:161–189.

Chai, P. (1988). Wing coloration of free-flying Neotropical butterflies as a signal learned by a specialized avian predator. *Biotropica* **20**:20–30.

Chai, P. (1990). Relationships between visual characteristics of rainforest butterflies and responses of a specialized insectivorous bird. In *Adaptive Coloration in Invertebrates* (ed. M. Wicksten), pp. 31–60. Galveston: Sea Grant College Program, Texas A & M University.

Chai, P. (1996). Butterfly visual characteristics and ontogeny of responses to butterflies by a specialized tropical bird. *Biol. J. Linn. Soc.* **58**:37–67.

Chai, P. (1997). Hummingbird hovering energetics during moult of primary flight feathers. *J. Exp. Biol.* **200**:1527–1536.

Chai, P., and Dudley, R. (1995). Limits to vertebrate locomotor energetics suggested by hummingbirds hovering in heliox. *Nature* **377**:722–725.

Chai, P., and Dudley, R. (1996). Limits to flight energetics of hummingbirds hovering in hypodense and hypoxic gas mixtures. *J. Exp. Biol.* **199**:2285–2295.

Chai, P., and Millard, P. (1997). Flight and size constraints: hovering performance of large hummingbirds under maximal loading. *J. Exp. Biol.* **200**:2757–2763.

Chai, P., and Srygley, R.B. (1990). Predation and the flight, morphology, and temperature of Neotropical rain-forest butterflies. *Amer. Nat.* **135**:748–765.

Chai, P., Chang, A.C., and Dudley, R. (1998). Flight thermogenesis and energy conservation in hovering hummingbirds. *J. Exp. Biol.* **201**:963–968.

Chai, P., Chen, J.S.C., and Dudley, R. (1997). Transient hovering performance of hummingbirds under conditions of maximal loading. *J. Exp. Biol.* **200**:921–929.

Chai, P., Harrykissoon, R., and Dudley, R. (1996). Hummingbird hovering performance in hyperoxic heliox: effects of body mass and sex. *J. Exp. Biol.* **199**:2745–2755.

Chaloner, W.G., and Sheerin, A. (1979). Devonian macrofloras. *Spec. Pap. Palaeontol.* 23:145–161.

Chaloner, W.G., Scott, A.C., and Stephenson, J. (1991). Fossil evidence for plant-arthropod interactions in the Palaeozoic and Mesozoic. *Phil. Trans. R. Soc. Lond.* B **333**:177–186.

Chalwatzis, N., Hauf, J., van de Peer, Y., Kinzelbach, R., and Zimmermann, F.K. (1996). 18S ribosomal RNA genes of insects: primary structure of the genes and molecular phylogeny of the Holometabola. *Ann. Ent. Soc. Amer.* **89**:788–803.

Chan, W.P., and Dickinson, M.H. (1996). *In vivo* length oscillations of indirect flight muscles in the fruit fly *Drosophila virilis J. Exp. Biol.* **199**:2767–2774.

Chan, W.P., Prete, F., and Dickinson, M.H. (1998). Visual input to the efferent control system of a fly's "gyroscope." *Science* **280**:289–292.

Chance, M.A.C. (1975). Air flow and the flight of a noctuid moth. In *Swimming and Flying in Nature*, vol. 2 (ed. T.Y.-T. Wu, C. J. Brokaw, and C. Brennen), pp. 829–842. New York: Plenum Press.

Cheng, L. (ed.) (1976). *Marine Insects.* Amsterdam: North-Holland.

Chappell, M.A. (1982). Temperature regulation of carpenter bees (*Xylocopa californica*) foraging in the Colorado desert of southern California. *Physiol. Zool.* **55**:267–280.

Chappell, M.A. (1984). Temperature regulation and energetics of the solitary bee *Centris pallida* during foraging and intermale mate competition. *Physiol. Zool.* **57**:215–225.

Cheer, A.Y.L., and Koehl, M.A.R. (1987). Paddles and rakes: fluid flow through bristled appendages of small organisms. *J. theor. Biol.* **129**:17–39.

Chen, S., and Young, B. (1943). Vision and flight: an experimental study on the cicada *Cryptotympana pustulata. Sinensia* **14**:55–60.

Cheng, H.K., and Murillo, L.E. (1984). Lunate tail swimming propulsion as a problem of curved lifting line in unsteady flow. Part 1. Asymptotic theory. *J. Fluid Mech.* **143**:327–350.

Chetverikov, S.S. (1920). The fundamental factor of insect evolution. *Ann. Rep. Smithsonian Inst. 1918*, pp. 441–449.

Childress, S. (1981). *Mechanics of Swimming and Flying.* Cambridge, U.K.: Cambridge University Press.

Chino, H. (1985). Lipid transport: biochemistry of hemolymph lipophorin. In *Comprehensive Insect Physiology, Biochemistry, and Pharmacology*, vol. 10 (ed. G.A. Kerkut and L.I. Gilbert), pp. 115–35. Oxford: Pergamon Press.

Chino, H., Lum, P.Y., Nagao, E., and Hiraoka, T. (1992). The molecular and metabolic essentials for long-distance flight in insects. *J. Comp. Physiol.* B **162**:101–106.

Choe, J. (1992). *Zoraptera* of Panama with a review of the morphology, systematics, and biology of the order. In *Insects of Panama and Mesoamerica: Selected Studies* (ed. D. Quintero and A. Aiello), pp. 249–256. Oxford: Oxford University Press.

Chopra, M.G. (1974). Hydromechanics of lunate-tail swimming propulsion. *J. Fluid Mech.* **64**:375–391.

Chopra, M.G. (1976). Large amplitude lunate-tail theory of fish locomotion. *J. Fluid Mech.* **74**:161–182.

Chopra, M.G., and Kambe, T. (1977). Hydromechanics of lunate-tail swimming propulsion. Part 2. *J. Fluid Mech.* **79**:49–69.

Chow, C.-Y., and Huang, M.-K. (1982). The initial lift and drag of an impulsively started airfoil of finite thickness. *J. Fluid Mech.* **118**:393–409.

Church, N.S. (1960a). Heat loss and the body temperatures of flying insects. I. Heat loss from evaporation of water from the body. *J. Exp. Biol.* **37**:171–185.

Church, N.S. (1960b). Heat loss and the body temperatures of flying insects. II. Heat conduction within the body and its loss by radiation and convection. *J. Exp. Biol.* **37**:186–212.

Claassen, D.E., and Kammer, A.E. (1986). Effects of octopamine, dopamine, and serotonin on production of flight motor output by thoracic ganglia of *Manduca sexta. J. Neurobiol.* **17**:1–14.

Clark, H.W. (1940). The adult musculature of the anisopterous dragonfly thorax (Odonata, Anisoptera). *J. Morphol.* **67**:523–565.

Cloupeau, M., Devillers, J.F., and Devezeaux, D. (1979). Direct measurements of instantaneous lift in desert locust; comparison with Jensen's experiments on detached wings. *J. Exp. Biol.* **80**:1–15.

Cockbain, A.J. (1961). Water relationships of *Aphis fabae* Scop. during tethered flight. *J. Exp. Biol.* **38**:175–180.

Coelho, J.R. (1991). The effect of thorax temperature on force production during tethered flight in honeybee (*Apis mellifera*) drones, workers, and queens. *Physiol. Zool.* **64**:823–835.

Coelho, J.R., and Hoagland, J. (1995). Load-lifting capacities of three species of yellowjackets (*Vespula*) foraging on honey-bee corpses. *Funct. Ecol.* **9**:171–174.

Coelho, J.R., and Mitton, J.B. (1988). Oxygen consumption during hovering is associated with genetic variation of enzymes in honey-bees. *Funct. Ecol.* **2**:141–146.

Coggshall, J.C. (1972). The landing response and visual processing in the milkweed bug, *Oncopeltus fasciatus. J. Exp. Biol.* **57**:401–413.

Cohen, B., Simcox, A.A., and Cohen, S.M. (1993). Allocation of the thoracic imaginal primordia in the *Drosophila* embryo. *Development* **117**:597–608.

Collatz, K.-G., and Wilps, H. (1986). Aging of flight mechanism. In *Insect Aging* (ed. K.-G. Collatz and R.S. Sohal), pp. 55–72. Berlin: Springer-Verlag.

Collett, T.S. (1980a). Angular tracking and the optomotor response. An analysis of visual reflex interaction in a hoverfly. *J. Comp. Physiol.* A **140**:145–158.

Collett, T.S. (1980b). Some operating rules for the optomotor system of a hoverfly during voluntary flight. *J. Comp. Physiol.* A **138**:271–282.

Collett, T.S., and King, A.J. (1975). Vision during flight. In *The Compound Eye and Vision of Insects* (ed. G.A. Horridge), pp. 437–466. Oxford: Clarendon Press.

Collett, T.S., and Land, M.F. (1975). Visual control of flight behaviour in the hoverfly, *Syritta pipiens* L. *J. Comp. Physiol.* A **99**:1–66.

Collett, T.S., and Land, M.F. (1978). How hoverflies compute interception courses. *J. Comp. Physiol.* A **125**:191–204.

Comstock, J.H. (1918). *The Wings of Insects.* Ithaca, N.Y.: Comstock.

Convey, P. (1989). Influences on the choice between territorial and satellite behaviour in male *Libellula quadrimaculata* Linn. (Odonata: Libellulidae). *Anim. Behav.* **109**:125–141.

Cooper, P.D., Schaffer, W.M., and Buchmann, S.L. (1985). Temperature regulation of honey bees (*Apis mellifera*) foraging in the Sonoran desert. *J. Exp. Biol.* **114**:1–15.

Cooter, R.J. (1973). Flight and landing posture in hoppers of *Schistocerca gregaria* (Forsk.). *Acrida* **2**:307–317.

Cooter, R.J. (1979). Visually induced yaw movements in the flying locust, *Schistocerca gregaria* (Forsk.). *J. Comp. Physiol.* A **131**:67–78.

Cooter, R.J. (1989). Swarm flight behavior in flies and locusts. In *Insect Flight* (ed. G.J. Goldsworthy and C.H. Wheeler), pp. 165–203. Boca Raton, Fla.: CRC Press.

Cooter, R.J., and Baker, P.S. (1977). Weis-Fogh clap and fling mechanism in *Locusta*. *Nature* **269**:53–54.

Corbet, P.S. (1962). *A Biology of Dragonflies.* London: Witherby.

Coyne, J.A., Boussy, I.A., Prout, T., Bryant, S.H., Jones, J.S., and Moore, J.A. (1982). Long-distance migration of *Drosophila*. *Amer. Nat.* **119**:589–595.

Crabtree, B., and Newsholme, E.A. (1972). The activities of lipases and carnitine palmitoyltransferase in muscles from vertebrates and invertebrates. *Biochem. J.* **130**:697–705.

Crabtree, B., and Newsholme, E.A. (1975). Comparative aspects of fuel utilization and metabolism by muscle. In *Insect Muscle* (ed. P.N.R. Usherwood), pp. 405–500. London: Academic Press.

Craig, C.L. (1986). Orb-web visibility: the influence of insect flight behavior and visual physiology on the evolution of web designs within the Araneoidea. *Anim. Behav.* **34**:54–68.

Crampton, G. (1916). The phylogenetic origin and the nature of the wings of insects according to the paranotal theory. *J. N.Y. Ent. Soc.* **24**:1–39.

Crampton, G.C. (1924). The phylogeny and classification of insects. *Pomona J. Ent. Zool.* **16**:33–47.

Crane, P.R., Friis, E.M., and Pedersen, K.R. (1995). The origin and early diversification of angiosperms. *Nature* **374**:27–33.

Crepet, W.L. (1983). The role of insect pollination in the evolution of the angiosperms. In *Pollination Biology* (ed. L. Real), pp. 29–50. Orlando, Fla.: Academic Press.

Crepet, W.L., and Friis, E.M. (1987). The evolution of insect pollination in angiosperms. In *The Origins of Angiosperms and Their Biological Consequences* (ed. E.M. Friis, W.G. Chaloner, and P.R. Crane), pp. 181–201. Cambridge, U.K.: Cambridge University Press.

Crepet, W.L., Friis, E.M., and Nixon, K.C. (1991). Fossil evidence for the evolution of biotic pollination. *Phil. Trans. R. Soc. Lond.* B **333**:187–195.

Crespi, B.J. (1988). Adaptation, compromise, and constraint: the development, morphometrics, and behavioral basis of a fighter-flier polymorphism in male *Hoplothrips karnyi*. *Behav. Ecol. Sociobiol.* **23**:93–104.

Crosskey, R.W. (1980). Family Tachinidae. In *Catalogue of the Diptera of the Afrotropical Region* (ed. R.W. Crosskey), pp. 822–882. London: British Museum (Natural History).

Crowson, R.A. (1981). *The Biology of the Coleoptera*. London: Academic Press.

Cruden, R.W., Hermann, S.M., and Peterson, S. (1983). Patterns of nectar production and plant-pollinator coevolution. In *The Biology of Nectaries* (ed. B. Bentley and T. Elias), pp. 80–125. New York: Columbia University Press.

CSIRO (1991). *The Insects of Australia*. 2d ed. Canberra: CSIRO.

Cullen, M.J. (1974). The distribution of asynchronous muscle in insects with special reference to the Hemiptera: an electron microscope study. *J. Ent.* 49A:17–41.

Currey, J.D. (1967). The failure of exoskeletons and endoskeletons. *J. Morphol.* 123:1–16.

Curtsinger, J.W., and Laurie-Ahlberg, C.C. (1981). Genetic variability of flight metabolism in *Drosophila melanogaster*. I. Characterization of power output during tethered flight. *Genetics* 98:549–564.

Czachorowski, S. (1993). How and from what did insect wings originate? *Przegl. Zool.* 37:207–218. (In Polish.)

D'Abrera, B. (1986). *Sphingidae Mundi—Hawkmoths of the World*. Faringdon, U.K.: E. W. Classey.

Dahmen, H.-J., and Zeil, J. (1984). Recording and reconstructing three-dimensional trajectories: a versatile method for the field biologist. *Proc. R. Soc. Lond.* B 222:107–113.

Dalton, S. (1975). *Borne on the Wind*. New York: Reader's Digest Press.

Dalton, S. (1977). *The Miracle of Flight*. New York: McGraw-Hill.

Dalton, S. (1982). *Caught in Motion: High-speed Nature Photography*. New York: Van Nostrand Reinhold.

Daly, H.V. (1963). Close-packed and fibrillar muscles of the Hymenoptera. *Ann. Ent. Soc. Amer.* 56:295–306.

D'Andrea, M., and Carfì, S. (1988). Spines on the wing veins in Odonata. 1. Zygoptera. *Odonatologica* 17:313–335.

D'Andrea, M., and Carfì, S. (1989). Spines on the wing veins in Odonata. 2. Anisozygoptera and Anisoptera. *Odonatologica* 18:147–178.

Danforth, B. (1989). The evolution of hymenopteran wings: the importance of size. *J. Zool., Lond.* 218:247–276.

Danforth, B.N., and Michener, C.D. (1988). Wing folding in the Hymenoptera. *Ann. Ent. Soc. Amer.* 81:342–349.

Danzer, A. (1956). Der Flugapparat der Dipteren als Resonanzsystem. *Z. vergl. Physiol.* 38:259–283.

Darlington, P.J. (1943). Carabidae of mountains and islands: data on the evolution of isolated faunas and on atrophy of wings. *Ecol. Monogr.* 13:37–61.

Darwin, F.W., and Pringle, J.W.S. (1959). The physiology of insect fibrillar muscle. I. Anatomy and innervation of the basalar muscle of lamellicorn beetles. *Proc. R. Soc. Lond.* B 151:194–203.

Dathe, H.H., Oehme, H., and Kitzing, H. (1984). Zur Konfiguration des Hubstrahles rüttelnder Vögel. *Zool. Jb. Abt. Allg. Zool. Physiol.* 88:387–403.

David, C.T. (1978). The relationship between body angle and flight speed in free-flying *Drosophila*. *Physiol. Ent.* 3:191–195.

David, C.T. (1979a). Height control by free-flying *Drosophila*. *Physiol. Ent.* 4:209–216.

David, C.T. (1979b). Optomotor control of speed and height by free-flying *Drosophila*. *J. Exp. Biol.* **82**:389–392.

David, C.T. (1982). Compensation for height in the control of groundspeed by *Drosophila* in a new, 'barber's pole' wind tunnel. *J. Comp. Physiol.* A **147**:485–493.

David, C.T. (1984). The dynamics of height stabilization in *Drosophila*. *Physiol. Ent.* **9**:377–386.

David, C.T. (1985). Visual control of the partition of flight force between lift and thrust in free-flying *Drosophila*. *Nature* **313**:48–50.

David, C.T. (1986). Mechanisms of directional flight in wind. In *Mechanisms in Insect Olfaction* (ed. T.L. Payne, M.C. Birch, and C.E.J. Kennedy), pp. 49–57. Oxford: Clarendon Press.

David, C.T., and Hardie, J. (1988). The visual responses of free-flying summer and autumn forms of the black bean aphid, *Aphis fabae*, in an automated flight chamber. *Physiol. Ent.* **13**:277–284.

David, C.T., and Kennedy, J.S. (1987). The steering of zigzagging flight by male gypsy moths. *Naturwissenschaften* **74**:194–196.

David, C.T., Kennedy, J.S., and Ludlow, A.R. (1983). Finding of a sex pheromone source by gypsy moths released in the field. *Nature* **303**:804–806.

Davies, I. (1974). The effect of age and diet on the ultrastructure of hymenopteran flight muscle. *Exp. Geront.* **9**:215–219.

Davies, N.B. (1977). Prey selection and social behaviour in wagtails (Aves: Motacillidae). *J. Anim. Ecol.* **46**:37–57.

Davis, M.A. (1980). Why are most insects short fliers? *Evol. Theory* **5**:103–111.

Davis, R.A., and Fraenkel, G. (1940). The oxygen consumption of flies during flight. *J. Exp. Biol.* **17**:402–407.

Dawson, J.W., Dawson-Scully, K., Robert, D., and Robertson, R.M. (1997). Forewing asymmetries during auditory avoidance in flying locusts. *J. Exp. Biol.* **200**:2323–2335.

DeLaurier, J.D. (1993a). An aerodynamic model for flapping-wing flight. *Aero. J.* **97** (April):125–130.

DeLaurier, J.D. (1993b). The development of an efficient ornithopter wing. *Aero. J.* **97** (May):153–162.

Delcomyn, F. (1980). Neural basis of rhythmic behavior in animals. *Science* **210**:492–498.

Demoll, R. (1918). *Der Flug der Insekten und der Vögel*. Jena: Gustav Fischer.

Dempster, J.P. (1991). Fragmentation, isolation, and mobility of insect populations. In *The Conservation of Insects and Their Habitats* (ed. N.M. Collins and J.A. Thomas), pp. 143–153. London: Academic Press.

Denno, R.F. (1994). The evolution of dispersal polymorphisms in insects: the influence of habitats, host plants and mates. *Res. Pop. Ecol.* **36**:127–135.

Denno, R.F., Roderick, G.K., Olmstead, K.L., and Döbel, H.G. (1991). Density-related migration in planthoppers (Homoptera: Delphacidae): the role of habitat persistence. *Amer. Nat.* **138**:1513–1541.

Denno, R.F., Roderick, G.K., Peterson, M.A., Huberty, A.F., Döbel, H.G., Eubanks, M.D., Losey, J.E., and Langellotto, G.A. (1996). Habitat persistence underlies intraspecific variation in the dispersal strategies of planthoppers. *Ecol. Monogr.* **66**:389–408.

Denny, M.W. (1993). *Air and Water: The Biology and Physics of Life's Media.* Princeton, N.J.: Princeton University Press.

De Souza, M.M., and Alexander, D.E. (1997). Passive aerodynamic stabilization by beetle elytra (wing covers). *Physiol. Ent.* **22**:109–115.

Desutter-Grandcolas, L. (1995). Functional forewing morphology and stridulation in crickets (Orthoptera, Grylloidea). *J. Zool., Lond.* **236**:243–252.

DeVries, P.J. (1987). *The Butterflies of Costa Rica.* Princeton, N.J.: Princeton University Press.

DeVries, P.J., and Dudley, R. (1990). Morphometrics, airspeeds, thermoregulation and lipid reserves of migrating *Urania fulgens* (Uraniidae) moths in natural free flight. *Physiol. Zool.* **63**:235–251.

Dewitz, J. (1902). Der Apterismus bei Insekten, seine künstliche Erzeugung und seine physiologische Erklärung. *Arch. Anat. Physiol. Abt. Physiol.* **1902**: 61–67.

Dewitz, J. (1920). Über die Entstehung rudimentärer Organe bei den Tieren. 4. Die Beeinflussung der Flügelbildung bei Insekten durch Kälte und Blasäuregase. *Zool. Jb. Abt. Allg. Zool. Physiol.* **37**:305–312.

Diakonoff, A. (1936). Contributions to the knowledge of the fly reflexes and the static sense in *Periplaneta americana* L. *Arch. Neerl. Physiol.* **21**:117–128.

Dial, K.P., Biewener, A.A., Tobalske, B.W., and Warrick, D.R. (1997). Mechanical power output of bird flight. *Nature* **390**:67–70.

Dickinson, M.H. (1990a). Comparison of encoding properties of campaniform sensilla on the fly wing. *J. Exp. Biol.* **151**:245–261.

Dickinson, M.H. (1990b). Linear and nonlinear encoding properties of an identified mechanoreceptor on the fly wing measured with mechanical noise stimuli. *J. Exp. Biol.* **151**:219–244.

Dickinson, M.H. (1992). Directional sensitivity and mechanical coupling dynamics of campaniform sensilla during chord-wise deformations of the fly wing. *J. Exp. Biol.* **169**:221–233.

Dickinson, M.H. (1994). The effects of wing rotation on unsteady aerodynamic performance at low Reynolds numbers. *J. Exp. Biol.* **192**:179–206.

Dickinson, M.H. (1996). Unsteady mechanisms of force generation in aquatic and aerial locomotion. *Amer. Zool.* **36**:537–554.

Dickinson, M., and Götz, K. (1993). Unsteady aerodynamic performance of model wings at low Reynolds numbers. *J. Exp. Biol.* **174**:45–64.

Dickinson, M., and Götz, K. (1996). The wake dynamics and flight forces of the fruit fly *Drosophila melanogaster*. *J. Exp. Biol.* **199**:2085–2104.

Dickinson, M.H., and Lighton, J.R.B. (1995). Muscle efficiency and elastic storage in the flight motor of *Drosophila*. *Science* **268**:87–89.

Dickinson, M.H., and Tu, M.S. (1997). The function of dipteran flight muscle. *Comp. Biochem. Physiol.* **116**A:223–238.

Dickinson, M.H., Hannaford, S., and Palka, J. (1997). The evolution of insect wings and their sensory apparatus. *Brain Behav. Evol.* **50**:13–24.

Dickinson, M.H., Lehmann, F.-O., and Chan, W.P. (1998). The control of mechanical power in insect flight. *Amer. Zool.* **38**:718-728.

Dickinson, M.H., Lehmann, F.-O., and Götz, K.G. (1993). The active control of wing rotation by *Drosophila*. *J. Exp. Biol.* **182**:173–189.

Dickinson, M.H., Hyatt, C.J., Lehmann, F.-O., Moore, J.R., Reedy, M.C., Simcox, A., Tohtong, R., Vigoreaux, J.O., Yamashita, H., and Maughan, D.W. (1997). Phosphorylation-dependent power output of transgenic flies: an integrated study. *Biophys. J.* **73**:3122–3134.

Dingle, H. (1961). Flight and swimming reflexes in giant water bugs. *Biol. Bull.* **121**:117–128.

Dingle, H. (1972). Migration strategies of insects. *Science* **175**:1327–1335.

Dingle, H. (1985). Migration. In *Comprehensive Insect Physiology, Biochemistry, and Pharmacology*, vol. 9, *Behavior* (ed. G.A. Kerkut and L.I. Gilbert), pp. 375–415. Oxford: Pergamon Press.

Dingle, H. (1989). The evolution and significance of migratory flight. In *Insect Flight* (ed. G.J. Goldsworthy and C.H. Wheeler), pp. 99–114. Boca Raton, Fla.: CRC Press.

Dingle, H. (1996). *Migration: The Biology of Life on the Move.* New York: Oxford University Press.

Disney, R.H.L. (1994). *Scuttle Flies: The Phoridae.* London: Chapman and Hall.

Dixon, A.F.G., Horth, S., and Kindlmann, P. (1993). Migration in insects: costs and strategies. *J. Anim. Ecol.* **62**:182–190.

Dixon, A.F.G., and Mercer, D.R. (1983). Flight behaviour in the sycamore aphid: factors affecting take-off. *Ent. Exp. Appl.* **33**:43–49.

Dohle, W. (1988). *Myriapoda and the Ancestry of Insects.* Manchester, U.K.: Manchester Polytechnic.

Dohle, W. (1997). Myriapoda-insect relationships as opposed to an insect-crustacean sister group relationship. In *Arthropod Relationships* (ed. R.A. Fortey and R.H. Thomas), pp. 305–315. London: Chapman and Hall.

Douglas, M.M. (1981). Thermoregulatory significance of thoracic lobes in the evolution of insect wings. *Science* **211**:84–86.

Downer, R.G.H., and Matthews, J.R. (1976). Patterns of lipid distribution and utilisation in insects. *Amer. Zool.* **16**:733–745.

Downes, J.A. (1969). The swarming and mating flight of Diptera. *Ann. Rev. Ent.* **14**:271–293.

Downey, J.C., and Allyn, A.C. (1975). Wing-scale morphology and nomenclature. *Bull. Allyn Mus.* **31**:1–32.

Drake, V.A., and Farrow, R.A. (1988). The influence of atmospheric structure and motions on insect migration. *Ann. Rev. Ent.* **33**:183–210.

Drake, V.A., and Farrow, R.A. (1989). The 'aerial plankton' and atmospheric convergence. *Trends Ecol. Evol.* **4**:381–385.

Dressler, R.L. (1982). Biology of the orchid bees (Euglossini). *Ann. Rev. Ecol. Syst.* **13**:373–394.

Driver, P.M., and Humphries, D.A. (1988). *Protean Behaviour: The Biology of Anarchization.* Oxford: Clarendon Press.

Dudley, R. (1990). Biomechanics of flight in Neotropical butterflies: morphometrics and kinematics. *J. Exp. Biol.* **150**:37–53.

Dudley, R. (1991a). Biomechanics of flight in Neotropical butterflies: aerodynamics and mechanical power requirements. *J. Exp. Biol.* **159**:335–357.

Dudley, R. (1991b). Comparative biomechanics and the evolutionary diversification of flying insect morphology. In *The Unity of Evolutionary Biology* (ed. E.C. Dudley), pp. 503–514. Portland: Dioscorides Press.

Dudley, R. (1992). Aerodynamics of flight. In *Biomechanics (Structures & Systems): A Practical Approach* (ed. A.A. Biewener), pp. 97–121. Oxford: Oxford University Press.

Dudley, R. (1994). Aerodynamics of insect dispersal and the constraint of body size. In *Proceedings of the 13th International Biometeorology Congress*, vol. 3 (ed. A.R. Maarouf, N.N. Barthakur, and W.O. Haufe), part 2, pp. 1035–1041. Downsview, Ontario: Environment Canada.

Dudley, R. (1995a). Aerodynamics, energetics, and reproductive constraints of migratory flight in insects. In *Insect Migration: Tracking Resources through Space and Time* (ed. V.A. Drake and A.G. Gatehouse), pp. 303–319. Cambridge, U.K.: Cambridge University Press.

Dudley, R. (1995b). Extraordinary flight performance of orchid bees (Apidae: Euglossini) hovering in heliox (80% He/20% O_2). *J. Exp. Biol.* **198**:1065–1070.

Dudley, R. (1998). Atmospheric oxygen, giant Paleozoic insects and the evolution of aerial locomotor performance. *J. Exp. Biol.* **201**:1043–1050.

Dudley, R., and Chai, P. (1996). Animal flight mechanics in physically variable gas mixtures. *J. Exp. Biol.* **199**:1881–1885.

Dudley, R., and DeVries, P.J. (1990). Flight physiology of migrating *Urania fulgens* (Uraniidae) moths: kinematics and aerodynamics of natural free flight. *J. Comp. Physiol.* A **167**:145–154.

Dudley, R., and Ellington, C.P. (1990a). Mechanics of forward flight in bumblebees. I. Kinematics and morphology. *J. Exp. Biol.* **148**:19–52.

Dudley, R., and Ellington, C.P. (1990b). Mechanics of forward flight in bumblebees. II. Quasi-steady lift and power requirements. *J. Exp. Biol.* **148**:53–88.

Dudley, R., and Gans, C. (1991). A critique of symmorphosis and optimality models in physiology. *Physiol. Zool.* **64**:627–637.

Dudley, R., and Srygley, R.B. (1994). Flight physiology of Neotropical butterflies: allometry of airspeeds during natural free flight. *J. Exp. Biol.* **191**:125–139.

Dudley, R., and Vermeij, G.J. (1992). Do the power requirements of flapping flight constrain folivory in flying animals? *Funct. Ecol.* **6**:101–104.

Dudley, R., and Vermeij, G.J. (1994). Energetic constraints of folivory: leaf fractionation by frugivorous bats. *Funct. Ecol.* **8**:668.

Duelli, P. (1985). A new functional interpretation of the visual system of male scale insects (Coccida, Homoptera). *Experientia* **41**:1036.

Dugard, J.J. (1967). Directional change in flying locusts. *J. Insect Physiol.* **13**:1055–1063.

Dumont, J.P.C., and Robertson, R.M. (1986). Neuronal circuits: an evolutionary perspective. *Science* **233**:849–853.

Dumortier, B. (1963). Morphology of sound emission apparatus in Arthropods. In *Acoustic Behavior of Animals* (ed. R.-G. Busnel), pp. 277–345. Amsterdam: Elsevier.

Dürken, B. (1907). Die Tracheenkiemenmuskulatur der Ephemeriden unter

Berücksichtigung der Morphologie des Insektenflügels. *Z. wiss. Zool.* **87**:435–550.

Dürken, B. (1923). Die postembryonale Entwicklung der Tracheenkiemen und ihrer Muskulatur bei *Ephemerella ignita*. *Zool. Jb. Abt. Anat. Ontog. Tiere* **44**: 439–614.

D'Urso, V. (1993). The wing coupling apparatus in *Peloridium hammoniorum* Breddin, 1897. *Spixiana* **16**:133–139.

D'Urso, V., and Ippolito, S. (1994). Wing-coupling apparatus of Auchenorrhyncha (Insecta: Homoptera). *Int. J. Insect Morphol. & Embryol.* **23**:211–224.

Duve, H. (1975). Intracellular localization of trehalase in thoracic muscle of the blowfly, *Calliphora erythrocephala*. *Insect Biochem.* **5**:299–311.

Dybas, H.S. (1966). Evidence for parthenogenesis in the featherwing beetles, with a taxonomic review of a new genus and eight new species (Coleoptera: Ptiliidae). *Fieldiana, Zool.* **51**:11–52.

Dybas, H.S. (1990). Insecta: Coleoptera Ptiliidae. In *Soil Biology Guide* (ed. D.L. Dindal), pp. 1093–1112. New York: John Wiley and Sons.

Eaton, R.C. (ed.) (1984). *Neural Mechanisms of Startle Behavior.* New York: Plenum Press.

Eckert, H. (1980). Orientation sensitivity of the visual movement detection system activating the landing response of the blowflies, *Calliphora*, and *Phaenicia*: a behavioural investigation. *Biol. Cybern.* **37**:235–247.

Eckert, H., and Hamdorf, K. (1980). Excitatory and inhibitory response components in the landing response of the blowfly, *Calliphora erythrocephala*. *J. Comp. Physiol.* A **138**:253–264.

Edmunds, G.F., and Traver, J.R. (1954). The flight mechanics and evolution of the wings of Ephemeroptera, with notes on the archetype insect wing. *J. Wash. Acad. Sci.* **44**:390–400.

Edmunds, M. (1974). *Defence in Animals: A Survey of Anti-Predator Defences.* Harlow, U.K.: Longman.

Edwards, D., Selden, P.A., Richardson, J.B., and Axe, L. (1995). Coprolites as evidence for plant-animal interaction in Siluro-Devonian terrestrial ecosystems. *Nature* **377**:329–331.

Edwards, G.A., Ruska, H., and DeHarven, É. (1958). Neuromuscular junctions in flight and tymbal muscles of the cicada. *J. Biophys. Biochem. Cytol.* **4**:251–254.

Edwards, J.S. (1985). Predator evasion and the origin of insect flight: an exercise in evolutionary neuroethology. *Soc. Neurosci. Abstr* **11**:497.

Edwards, J.S. (1992). Giant interneurons and the origin of insect flight. In *Nervous Systems: Principles of Design and Function* (ed. R.N. Singh), pp. 485–495. Bombay: Wiley Eastern.

Edwards, J.S. (1997). The evolution of insect flight: implications for the evolution of the nervous system. *Brain Behav. Evol.* **50**:8–12.

Edwards, J.S., and Palka, J. (1991). Insect neural evolution—a fugue or an opera? *Semin. Neurosci.* **3**:391–398.

Edwards, J.S., and Reddy, G.R. (1986). Mechanosensory appendages in the firebrat (*Thermobia domestica*, Thysanura): a prototype system for terrestrial predator invasion. *J. Comp. Neurol.* **243**:535–546.

Edwards, R.H., and Cheng, H.K. (1982). The separation vortex in the Weis-Fogh circulation-generation mechanism. *J. Fluid Mech.* **120**:463–473.

Egelhaaf, M. (1987). Dynamic properties of two control systems underlying visually guided turning in house-flies. *J. Comp. Physiol.* A **161**:777–783.

Egelhaaf, M. (1989). Visual afferences to flight steering muscles controlling optomotor response of the fly. *J. Comp. Physiol.* A **165**:719–730.

Egelhaaf, M. (1990). Spatial interactions in the fly visual system leading to selectivity for small-field motion. *Naturwissenschaften* **77**:182–185.

Egelhaaf, M., and Borst, A. (1993). Motion computation and visual orientation in flies. *Comp. Biochem. Physiol.* **104**A:659–673.

Egelhaaf, M., Hausen, K., Reichardt, W., and Wehrhahn, C. (1988). Visual course control in flies relies on neuronal computation of object and background motion. *Trends Neurosci.* **11**:351–358.

Eggers, A., Preiss, R., and Gewecke, M. (1991). The optomotor yaw response of the desert locust, *Schistocerca gregaria*. *Physiol. Ent.* **16**:411–418.

Elder, H.Y. (1975). Muscle structure. In *Insect Muscle* (ed. P.N.R. Usherwood), pp. 1–73. London: Academic Press.

Ellington, C.P. (1975). Non-steady-state aerodynamics of the flight of *Encarsia formosa*. In *Swimming and Flying in Nature*, vol. 2 (ed. T.Y.-T. Wu, C.J. Brokaw, and C. Brennen), pp. 783–796. New York: Plenum Press.

Ellington, C.P. (1978). The aerodynamics of normal hovering flight: three approaches. In *Comparative Physiology—Water, Ions and Fluid Mechanics* (ed. K. Schmidt-Nielsen, L. Bolis, and S.H.P. Maddrell), pp. 327–345. Cambridge, U.K.: Cambridge University Press.

Ellington, C.P. (1980a). Vortices and hovering flight. In *Instationäre Effekte an schwingenden Tierflügeln* (ed. W. Nachtigall), pp. 64–101. Wiesbaden, Germany: Franz Steiner.

Ellington, C.P. (1980b). Wing mechanics and take-off preparations of *Thrips* (Thysanoptera). *J. Exp. Biol.* **85**:129–136.

Ellington, C.P. (1984a). The aerodynamics of hovering insect flight. I. The quasi-steady analysis. *Phil. Trans. R. Soc. Lond.* B **305**:1–15.

Ellington, C.P. (1984b). The aerodynamics of hovering insect flight. II. Morphological parameters. *Phil. Trans. R. Soc. Lond.* B **305**:17–40.

Ellington, C.P. (1984c). The aerodynamics of hovering insect flight. III. Kinematics. *Phil. Trans. R. Soc. Lond.* B **305**:41–78.

Ellington, C.P. (1984d). The aerodynamics of hovering insect flight. IV. Aerodynamic mechanisms. *Phil. Trans. R. Soc. Lond.* B **305**:79–113.

Ellington, C.P. (1984e). The aerodynamics of hovering insect flight. V. A vortex theory. *Phil. Trans. R. Soc. Lond.* B **305**:115–144.

Ellington, C.P. (1984f). The aerodynamics of hovering insect flight. VI. Lift and power requirements. *Phil. Trans. R. Soc. Lond.* B **305**:145–181.

Ellington, C.P. (1985). Power and efficiency of insect flight muscle. *J. Exp. Biol.* **115**:293–304.

Ellington, C.P. (1991a). Aerodynamics and the origin of insect flight. *Adv. Insect Physiol.* **23**:171–210.

Ellington, C.P. (1991b). Limitations on animal flight performance. *J. Exp. Biol.* **160**:71–91.

Ellington, C.P. (1995). Unsteady aerodynamics of insect flight. In *Biological Fluid Dynamics* (ed. C.P. Ellington and T.J. Pedley), pp. 109–129. Cambridge, U.K.: Company of Biologists.

Ellington, C.P., Machin, K.E., and Casey, T.M. (1990). Oxygen consumption of bumblebees in forward flight. *Nature* **347**:472–473.

Ellington, C.P., van den Berg, C., Willmott, A.P., and Thomas, A.L.R. (1996). Leading-edge vortices in insect flight. *Nature* **384**:626–630.

Elson, R., and Pflüger, H.-J. (1986). The activity of a steering muscle in flying locusts. *J. Exp. Biol.* **120**:421–441.

Endler, J.A. (1990). On the measurement and classification of colour in studies of animal colour patterns. *Biol. J. Linn. Soc.* **41**:315–352.

Ennos, A.R. (1987). A comparative study of the flight mechanism of Diptera. *J. Exp. Biol.* **127**:355–372.

Ennos, A.R. (1988a). The importance of torsion in the design of insect wings. *J. Exp. Biol.* **140**:137–160.

Ennos, A.R. (1988b). The inertial cause of wing rotation in Diptera. *J. Exp. Biol.* **140**:161–169.

Ennos, A.R. (1989a). The effect of size on the optimal shapes of gliding insects and seeds. *J. Zool., Lond.* **219**:61–69.

Ennos, A.R. (1989b). Inertial and aerodynamic torques on the wings of Diptera in flight. *J. Exp. Biol.* **142**:87–95.

Ennos, A.R. (1989c). The kinematics and aerodynamics of the free flight of some Diptera. *J. Exp. Biol.* **142**:49–85.

Ennos, A.R. (1995). Mechanical behaviour in torsion of insect wings, blades of grass and other cambered structures. *Proc. R. Soc. Lond.* B **259**:15–18.

Ennos, A.R., and Wootton, R.J. (1989). Functional wing morphology and aerodynamics of *Panorpa germanica* (Insecta: Mecoptera). *J. Exp. Biol.* **143**:267–284.

Epting, R.J. (1980). Functional dependence of the power for hovering on wing disc loading in hummingbirds. *Physiol. Zool.* **53**:347–357.

Erhard, E. (1913). Der Flug der Tiere. *Verh. Dtsch. Zool. Gesellsch.* **23**:201–225.

Erwin, T.L. (1982). Tropical forests: Their richness in Coleoptera and other arthropod species. *Coleopt. Bull.* **36**:74–75.

Erwin, T.L. (1988). The tropical forest canopy: the heart of biotic diversity. In *Biodiversity* (ed. E.O. Wilson), pp. 123–129. Washington, D.C.: National Academy Press.

Esch, H., and Goller, F. (1991). Neural control of fibrillar muscles in bees during shivering and flight. *J. Exp. Biol.* **159**:419–431.

Esch, H., Nachtigall, W., and Kogge, S.N. (1975). Correlations between aerodynamic output, electrical activity in the indirect flight muscles and wing positions of bees flying in a servomechanically controlled wind tunnel. *J. Comp. Physiol.* A **100**:147–159.

Esch, H.E., and Burns, J.E. (1996). Distance estimation by foraging honeybees. *J. Exp. Biol.* **199**:155–162.

Ettema, G.J.C., Huijing, P.A., and DeHaan, A. (1992). The potentiating effect of prestretch on the contractile performance of rat gastrocnemius medialis muscle during subsequent shortening and isometric contraction. *J. Exp. Biol.* **165**:121–136.

Evans, M.E.G. (1975). The jump of *Petrobius* (Thysanura, Machilidae). *J. Zool., Lond.* **176**:49–65.

Evans, M.E.G., and Blower, J.G. (1973). A jumping millipede. *Nature* **246**:427–428.

Evans, P.D., and Siegler, M.V.S. (1982). Octopamine mediated relaxation of maintained and catch tension in locust skeletal muscle. *J. Physiol.* **324**:93–112.

Ewer, D.W. (1963). On insect flight. *J. Ent. Soc. S. Afr.* **26**:3–13.

Ewer, D.W., and Nayler, L.S. (1967). The pterothoracic musculature of *Deropeltis erythrocephala*, a cockroach with a wingless female, and the origin of wing movements in insects. *J. Ent. Soc. S. Afr.* **30**:18–33.

Ewing, A.W. (1979). The role of feedback during singing and flight in *Drosophila melanogaster*. *Physiol. Ent.* **4**:329–337.

Eyles, E.D. (1945). "How does a fly land on the ceiling?" *Proc. R. Ent. Soc. Lond. (A)* **20**:14–15.

Fadamiro, H.Y. (1997). Free flight capacity determination in a sustained flight tunnel: effects of age and sexual state on the flight duration of *Prostephanus truncatus*. *Physiol. Ent.* **22**:29–36.

Fairbairn, D., and Desranleau, L. (1987). Flight threshold, wing muscle histolysis, and alary polymorphism: correlated traits for dispersal tendency in the Gerridae. *Ecol. Ent.* **12**:13–24.

Farina, W.M., and Josens, R.B. (1994). Food source profitability modulates compensatory responses to a visual stimulus in the hawk moth *Macroglossum stellatarum*. *Naturwissenschaften* **81**:131–133.

Farina, W.M., Kramer, D., and Varjú, D. (1995). The response of the hovering hawkmoth *Macroglossum stellatarum* to translatory pattern motion. *J. Comp. Physiol. A* **176**:551–562.

Farina, W.M., Varjú, D., and Zhou, Y. (1994). The regulation of distance to dummy flowers during hovering flight in the hawk moth *Macroglossum stellatarum*. *J. Comp. Physiol. A* **174**:239–247.

Farrell, B.D. (1998). "Inordinate fondness" explained: why are there so many beetles? *Science* **281**:555–559.

Faure, J.C. (1932). The phases of locusts in South Africa. *Bull. Ent. Res.* **23**:293–405.

Faust, R. (1952). Untersuchungen zum Halterenproblem. *Zool. Jb. Abt. Allg. Zool. Physiol.* **63**:325–366.

Favier, D., Maresca, C., and Rebont, J. (1982). Dynamic stall due to fluctuations of velocity and incidence. *AIAA J.* **20**:865–871.

Fayyazuddin, A., and Dickinson, M.H. (1996). Haltere afferents provide direct, electrotonic input to a steering motor neuron in the blowfly, *Calliphora*. *J. Neurosci.* **16**:5225–5232.

Feduccia, A. (1996). *The Origin and Evolution of Birds*. New Haven, Conn.: Yale University Press.

Feener, D.H., and Brown, B.V. (1997). Diptera as parasitoids. *Ann. Rev. Ent.* **42**:73–97.

Feinsinger, P., Colwell, R.K., Terborgh, J., and Chaplin, S.B. (1979). Elevation and the morphology, flight energetics, and foraging ecology of tropical hummingbirds. *Amer. Nat.* **113**:481–497.

Felt, E.P. (1928). Dispersal of insects by air currents. *Bull. N.Y. State Mus.* **274**:59–129.

Fincke, O.M. (1992). Behavioural ecology of the Giant Damselflies of Barro Colorado Island, Panama (*Odonata: Zygoptera: Pseudostigmatidae*). In *Insects of Panama and Mesoamerica: Selected Studies* (ed. D. Quintero and A. Aiello), pp. 102–113. Oxford: Oxford University Press.

Finlayson, L.H. (1975). Development and degeneration. In *Insect Muscle* (ed. P.N.R. Usherwood), pp. 75–149. London: Academic Press.

Fittkau, E.J., and Klinge, H. (1973). On biomass and trophic structure of the central Amazonian rain forest ecosystem. *Biotropica* **5**:2–14.

Fitzhugh, G.H., and Marden, J.H. (1997). Maturational changes in troponin T expression, Ca^{2+} sensitivity and twitch contraction kinetics in dragonfly flight muscle. *J. Exp. Biol.* **200**:1473–1482.

Flower, J.W. (1964). On the origin of flight in insects. *J. Insect Physiol.* **10**:81–88.

Forbes, W.T.M. (1943). The origin of wings and venational types in insects. *Amer. Midl. Nat.* **29**:381–405.

Fourtner, C.R., and Randall, J.B. (1982). Studies on cockroach flight: the role of continuous neural activation of non-flight muscles. *J. Exp. Zool.* **221**:143–154.

Fowler, H.G., Silva, C.A., and Venitcinque, E. (1993). Size, taxonomic and biomass distributions of flying insects in Central Amazonia: Forest edge vs. understory. *Rev. Biol. Trop.* **41**:755–760.

Fraenkel, G. (1932a). Untersuchungen über die Koordination von Reflexen und automatisch- nervösen Rhythmen bei Insekten. I. Die Flugreflexe der Insekten und ihre Koordination. *Z. vergl. Physiol.* **16**:371–393.

Fraenkel, G. (1932b). Die Wanderungen der Insekten. *Ergebn. Biol.* **9**:1–238.

Fraenkel, G. (1939). The function of the halteres of flies (Diptera). *Proc. Zool. Soc. Lond.* A **109**:69–78.

Fraenkel, G., and Pringle, J.W.S. (1938). Halteres of flies as gyroscopic organs of equilibrium. *Nature* **141**:919–920.

Francis, R.H., and Cohen, J. (1933). The flow near a wing which starts suddenly from rest and then stalls. *Rep. Memo. Aero. Res. Comm.*, no. 1561.

Frankie, G.W. (1975). Tropical forest phenology and pollinator plant coevolution. In *Coevolution of Plants and Animals* (ed. L.E. Gilbert and P.H. Raven), pp. 192–209. Austin: University of Texas Press.

Frankie, G.W. (1976). Pollination of widely dispersed trees by animals in Central America, with an emphasis on bee pollination systems. In *Variation, Breeding, and Conservation of Tropical Forest Trees* (ed. J. Burley and B. T. Stiles), pp. 151–159. London: Academic Press.

Fraser, F.C. (1936). *The Fauna of British India, Including Ceylon and Burma. Odonata.* Vol. III. London: Taylor and Francis.

Fraser, P.J. (1977). Cercal ablation modifies tethered flight behaviour of cockroach. *Nature* **268**:523–524.

Freymuth, P. (1989). An unsteady model of animal hovering. *Lect. Notes Eng.* **54**:231–245.

Freymuth, P. (1990). Thrust generation by an airfoil in hover modes. *Exp. Fluids* **9**:17–24.

Friedrich, M., and Tautz, D. (1995). Ribosomal DNA phylogeny of the major extant arthropod classes and the evolution of myriapods. *Nature* **376**:165–167.

Friedrich, R.W., Spatz, H.-C., and Bausenwein, B. (1994). Visual control of wing beat frequency in *Drosophila*. *J. Comp. Physiol.* A **175**:587–596.

Fryer, G. (1996). Reflections on arthropod evolution. *Biol. J. Linn. Soc.* **58**:1–55.

Full, R.J. (1997). Invertebrate locomotor systems. In *Handbook of Physiology; Section 13: Comparative Physiology*, vol. 2 (ed. W.H. Dantzler), pp. 853–930. New York: Oxford University Press.

Gal-Or, B. (1990). *Vectored Propulsion, Supermaneuverability and Robot Aircraft.* New York: Springer-Verlag.

Galun, R., and Fraenkel, G. (1961). The effect of low atmospheric pressure on adult *Aedes aegyptii* and on housefly pupae. *J. Insect Physiol.* **7**: 161–176.

Ganihar, D., Libersat, F., Wendler, G., and Camhi, J.M. (1994). Wind-evoked evasive responses in flying cockroaches. *J. Comp. Physiol.* A **175**:49–65.

Gans, C. (1989). Stages in the origin of vertebrates: analysis by means of scenarios. *Biol. Rev.* **64**:221–268.

Gans, C., Dudley, R., Aguilar, N.M. and Graham, J.B. (1999). Late Paleozoic atmospheres and biotic evolution. *Hist. Biol.* **13**:199–219.

Garland, T., and Huey, R.B. (1987). Testing symmorphosis: does structure match functional requirements? *Evolution* **41**:1404–1409.

Garrick, I.E. (1936). Propulsion of a flapping and oscillating airfoil. *NACA Report No. 527*, pp. 419–427.

Gaston, J. (1991). The magnitude of global insect species richness. *Conserv. Biol.* **5**:283–296.

Gatehouse, A.G. (1997). Behavior and ecological genetics of wind-borne migration by insects. *Ann. Rev. Ent.* **42**:475–502.

Gatesy, S.M., and Dial, K.P. (1993). Tail muscle activity patterns in walking and flying pigeons (*Columba livia*). *J. Exp. Biol.* **176**:55–76.

Gauld, I., and Bolton, B. (eds.) (1988). *The Hymenoptera.* Oxford: Oxford University Press.

Gaunt, S.J. (1997). Chick limbs, fly wings and homology at the fringe. *Nature* **386**:324–325.

Gee, C.E., and Robertson, R.M. (1994). Effects of maturation on synaptic potentials in the locust flight system. *J. Comp. Physiol.* A **175**:437–447.

Gee, C.E., and Robertson, R.M. (1996). Recovery of the flight system following ablation of the tegulae in immature adult insects. *J. Exp. Biol.* **199**:1395–1403.

Gegenbaur, C. (1870). *Grundzüge der vergleichenden Anatomie.* Leipzig: Wilhelm Engelmann.

Geiger, R. (1965). *The Climate Near the Ground.* Cambridge, Mass.: Harvard University Press.

Geiger, G., and Poggio, T. (1977). On head and body movements of flying flies. *Biol. Cybern.* **25**:177–180.

Gensel, P.G., and Andrews, H.N. (1984). *Plant Life in the Devonian.* New York: Praeger.

George, J.C., and Bhakthan, N.M.G. (1960). A study on the fibre diameter and certain enzyme concentrations in the flight muscles of some butterflies. *J. Exp. Biol.* **37**:308–315.

George, J.C., Vallyathan, N.V., and Scaria, K.S. (1958). Lipase activity in insect flight muscle. *Experientia* **14**:250–251.

Gettrup, E. (1962). Thoracic proprioceptors in the flight system of locusts. *Nature* **193**:498–499.

Gettrup, E. (1966). Sensory regulation of wing twisting in locusts. *J. Exp. Biol.* **44**:1–16.

Gettrup, E., and Wilson, D.M. (1964). The lift-control reaction of flying locusts. *J. Exp. Biol.* **41**:183–190.

Gewecke, M. (1967). Die Wirkung von Luftströmung auf die Antennen und das Flugverhalten der blauen Schmeißfliege (*Calliphora erythrocephala*). *Z. vergl. Physiol.* **54**:121–164.

Gewecke, M. (1970). Antennae: another wind-sensitive receptor in locusts. *Nature* **225**:1263–1264.

Gewecke, M. (1972). Antennen und Stirn-Scheitelhaare von *Locusta migratoria* L. als Luftströmungs-Sinnesorgane bei der Flugsteuerung. *J. Comp. Physiol.* **80**:57–94.

Gewecke, M. (1975). The influence of the air-current sense organs on the flight behavior of *Locusta migratoria*. *J. Comp. Physiol.* A **103**:79–95.

Gewecke, M., and Heinzel, H.G. (1980). Aerodynamic and mechanical properties of the antennae as air-current sense organs in *Locusta migratoria*. I. Static characteristics. *J. Comp. Physiol.* A **139**:357–366.

Gewecke, M., and Kutsch, W. (1979). Development of flight behaviour in maturing adults of *Locusta migratoria*: I. Flight performance and wing-stroke parameters. *J. Insect Physiol.* **25**:249–253.

Gewecke, M., and Niehaus, M. (1981). Flight and flight control by the antennae in the Small Tortoiseshell (*Aglais urticae* L., Lepidoptera). I. Flight balance experiments. *J. Comp. Physiol.* A **145**:249–256.

Gewecke, M., and Philippen, J. (1978). Control of the horizontal flight-course by air-current sense organs in *Locusta migratoria*. *Physiol. Ent.* **3**:43–52.

Gewecke, M., and Schlegel, P. (1970). Die Schwingungen der Antenne und ihre Bedeutung für die Flugsteuerung bei *Calliphora erythrocephala*. *Z. vergl. Physiol.* **67**:325–362.

Gewecke, M., Heinzel, H.-G., and Philippen, J. (1974). Role of antennae of the dragonfly *Orthetrum cancellatum* in flight control. *Nature* **249**:584–585.

Ghiselin, M.T. (1974). *The Economy of Nature and the Evolution of Sex.* Berkeley: University of California Press.

Gibo, D.L. (1981a). Altitudes attained by migrating monarch butterflies, *Danaus plexippus* (Lepidoptera, Danaidae) as reported by glider pilots. *Can. J. Zool.* **59**:571–572.

Gibo, D.L. (1981b). Some observations on soaring flight in the mourning cloak butterfly (*Nymphalis antiopa* L.) in southern Ontario. *J. N.Y. Ent. Soc.* **89**:98–101.

Gibo, D.L. (1981c). Some observations on slope soaring in *Pantala flavescens* (Odonata: Libellulidae). *J. N.Y. Ent. Soc.* **89**:184–187.

Gibo, D.L. (1986). Flight strategies of migrating monarch butterflies (*Danaus plexippus* L.) in southern Ontario. In *Insect Flight: Dispersal and Migration* (ed. W. Danthanarayana), pp. 172–184. Berlin: Springer-Verlag.

Gibo, D.L., and McCurdy, J.A. (1993). Lipid accumulation by migrating monarch butterflies (*Danaus plexippus* L.). *Can. J. Zool.* **71**:76–82.

Gibo, D.L., and Pallett, M.J. (1979). Soaring flight of monarch butterflies, *Danaus plexippus* (Lepidoptera: Danaidae), during the late summer migration in southern Ontario. *Can. J. Zool.* **57**:1393–1401.

Gibson, G. (1985). Swarming behaviour of the mosquito *Culex pipiens quinquefasciatus*: a quantitative analysis. *Physiol. Ent.* **10**:283–296.

Gibson, G.A.P. (1986). Mesothoracic skeletomusculature and mechanics of flight and jumping in Eupelminae (Hymenoptera, Chalcidoidea: Eupelmidae). *Can. Ent.* **118**:691–719.

Gilbert, C., Gronenberg, W., and Strausfeld, N.J. (1995). Oculomotor control in calliphorid flies: head movements during activation and inhibition of neck motor neurons corroborate neuroanatomical predictions. *J. Comp. Neurol.* **361**:285–297.

Gilbert, F.S. (1984). Thermoregulation and the structure of swarms in *Syrphus ribesii* (Syrphidae). *Oikos* **42**:249–255.

Gilmour, K.M., and Ellington, C.P. (1993a). *In vivo* muscle length changes in bumblebees and the *in vitro* effects on work and power. *J. Exp. Biol.* **183**:101–113.

Gilmour, K.M., and Ellington, C.P. (1993b). Power output of glycerinated bumblebee flight muscle. *J. Exp. Biol.* **183**:77–100.

Gilyarov, M.S. (1970). *Regularities in Arthropod Adaptation to Terrestrial Life.* Moscow: Nauka. (In Russian.)

Gislén, T. (1948). Aerial plankton and its condition of life. *Biol. Rev.* **23**:109–126.

Glaser, A.E., and Vincent, J.F.V. (1979). The autonomous inflation of insect wings. *J. Insect Physiol.* **25**:315–318.

Glass, L., Beuter, A., and Larocque, D. (1988). Time delays, oscillations and chaos in physiological control systems. *Math. Biosci.* **90**:111–125.

Glass, L., and Malta, C.P. (1990). Chaos in multi-looped negative feedback systems. *J. theor. Biol.* **145**:217–223.

Glick, P.A. (1939). The distribution of insects, spiders, and mites in the air. *U.S. Dept. Agric. Tech. Bull.* **673**:1–150.

Godfray, H.C.J. (1994). *Parasitoids: Behavioral and Evolutionary Ecology.* Princeton, N.J.: Princeton University Press.

Goldschmidt, R.B. (1945). The structure of Podoptera, a homoeotic mutant of *Drosophila melanogaster. J. Morphol.* **77**:71–103.

Gomi, T., Okuda, T., and Tanaka, S. (1995). Protein synthesis and degradation in the flight muscles of adult crickets (*Gryllus bimaculatus*). *J. Exp. Biol.* **198**:1071–1077.

Goodman, L.J. (1959). Hair plates on the first cervical sclerites of the Orthoptera. *Nature* **183**:1106–1107.

Goodman, L.J. (1960). The landing responses of insects. I. The landing response of the fly, *Lucilia sericata*, and other Calliphorinae. *J. Exp. Biol.* **37**:854–878.

Goodman, L.J. (1965). The role of certain optomotor reactions in regulating stability in the rolling plane during flight in the desert locust, *Schistocerca gregaria*. *J. Exp. Biol.* **42**:385–407.

Goosey, M.W., and Candy, D.J. (1980). The D-octopamine content of the haemolymph of the locust, *Schistocerca americana gregaria*, and its elevation during flight. *Insect Biochem.* **10**:393–397.

Gordon, M.S., and Olson, E.C. (1995). *Invasions of the Land: The Transitions of Organisms from Aquatic to Terrestrial Life*. New York: Columbia University Press.

Gosler, A.G., Greenwood, J.J.D., and Perrins, C. (1995). Predation risk and the cost of being fat. *Nature* **377**:621–623.

Götz, K.G. (1968). Flight control in *Drosophila* by visual perception of motion. *Kybernetik* **4**:199–208.

Götz, K.G. (1987). Course-control, metabolism and wing interference during ultralong tethered flight in *Drosophila melanogaster*. *J. Exp. Biol.* **128**:35–46.

Götz, K.G. and Wandel, U. (1984a). Optomotor control of the force of flight in *Drosophila* and *Musca*. I. Homology of wingbeat-inhibiting movement detectors. *Biol. Cybern.* **51**:129–134.

Götz, K.G. and Wandel, U. (1984b). Optomotor control of the force of flight in *Drosophila* and *Musca*. II. Covariance of lift and thrust in still air. *Biol. Cybern.* **51**:135–139.

Götz, K.G., Hengstenberg, B., and Biesinger, R. (1979). Optomotor control of wing beat and body posture in *Drosophila*. *Biol. Cybern.* **35**:101–112.

Govind, C.K. (1972). Differential activity in the coxo-subalar muscle during directional flight in the milkweed bug, *Oncopeltus*. *Can. J. Zool.* **50**:901–905.

Govind, C.K., and Burton, A.J. (1970). Flight orientation in a coreid squash bug (Heteroptera). *Can. Ent.* **102**:1002–1007.

Govind, C.K., and Dandy, J.W.T. (1972). Non-fibrillar muscles and the start and cessation of flight in the milkweed bug, *Oncopeltus*. *J. Comp. Physiol.* **77**:398–417.

Grabow, K., and Rüppell, G. (1995). Wing loading in relation to size and flight characteristics of European Odonata. *Odonatologica* **24**:175–186.

Grace, B., and Shipp, J.L. (1988). A laboratory technique for examining the flight activity of insects under controlled environment conditions. *Int. J. Biometeorol.* **32**:65–69.

Graham, J.B., Dudley, R., Aguilar, N., and Gans, C. (1995). Implications of the late Palaeozoic oxygen pulse for physiology and evolution. *Nature* **375**:117–120.

Graham, J.M.R. (1983). The lift on an aerofoil in starting flow. *J. Fluid Mech.* **133**:413–425.

Grant, K.A., and Grant, V. (1968). *Hummingbirds and Their Flowers*. New York: Columbia University Press.

Granzier, H.L.M., and Wang, K. (1993). Interplay between passive tension and strong and weak binding cross-bridges in insect indirect flight muscle. *J. Gen. Physiol.* **101**:235–270.

Gray, J.R., and Robertson, R.M. (1994). Activity of the forewing stretch receptor in immature and mature adult locusts. *J. Comp. Physiol.* A **175**:425–435.

Greenberg, S., and Ar, A. (1996). Effects of chronic hypoxia, normoxia and hyperoxia on larval development in the beetle *Tenebrio molitor. J. Insect Physiol.* **42**:991–996.

Greenewalt, C.H. (1960a). *Hummingbirds.* Garden City, N.Y.: Doubleday.

Greenewalt, C.H. (1960b). The wings of insects and birds as mechanical oscillators. *Proc. Amer. Phil. Soc.* **104**:605–611.

Greenewalt, C.H. (1962). Dimensional relationships for flying animals. *Smithsonian Misc. Collect.* **144**:1–46.

Gressitt, J.L. (1965). Biogeography and ecology of land arthropods of Antarctica. In *Biogeography and Ecology in Antarctica* (ed. J. van Mieghem and P. van Oye), pp. 431–490. The Hague: Dr. W. Junk.

Gressitt, J.L., Samuelson, G.A., and Vitt, D.H. (1968). Moss growing on living Papuan moss-forest weevils. *Nature* **217**:765–767.

Gries, M., and Koeniger, N. (1996). Straight forward to the queen: pursuing honeybee drones (*Apis mellifera* L.) adjust their body axis to the direction of the queen. *J. Comp. Physiol.* A **179**:539–544.

Griffiths, N., Paynter, Q., and Brady, J. (1995). Rates of progress up odour plumes by tsetse flies: a mark-release video study of the timing of odour source location by *Glossina pallidipes. Physiol. Ent.* **200**:100–108.

Grodnitsky, D.L. (1992). Free and tethered flight of butterflies (Papilionida, Papilionoidea). *Zool. Zh.* **71**:21–28. (In Russian.)

Grodnitsky, D.L. (1993). Preliminary data on body movement of freely flying butterflies. *Zool. Zh.* **72**:84–94. (In Russian.)

Grodnitsky, D.L. (1995). Evolution and classification of insect flight kinematics. *Evolution* **49**:1158–1162.

Grodnitsky, D.L. (1999). *Form and Function of Insect Wings: The Evolution of Biological Structures.* Baltimore, MD.: Johns Hopkins University Press.

Grodnitsky, D.L., and Dudley, R. (1996). Vortex visualization during free flight of heliconiine butterflies (Lepidoptera: Nymphalidae). *J. Kans. Ent. Soc.* **69**: 199–203.

Grodnitsky, D.L., and Kozlov, M.V. (1985). Functional morphology of the wing apparatus and flight peculiarities in primitive moths (Lepidoptera: Micropterigidae, Eriocraniidae). *Zool. Zh.* **64**:1661–1671. (In Russian.)

Grodnitsky, D.L., and Kozlov, M.V. (1987). Experiment-dependent kinematics in tethered flight of a codling moth *Laspeyresia pomonella* (Lepidoptera, Tortricidae). *Zool. Zh.* **66**:1314–1320. (In Russian.)

Grodnitsky, D.L., and Kozlov, M.V. (1990). Wing functional morphology in some species of the suborder Papilionina (Lepidoptera). *Vestn. Zool.* **2**:58–64. (In Russian.)

Grodnitsky, D.L., and Kozlov, M.V. (1991). Evolution and functions of wings and their scale covering in butterflies and moths (Insecta: Papilionida= Lepidoptera). *Biol. Zentralbl.* **110**:199–206.

Grodnitsky, D.L., and Morozov, P.P. (1992). Flow visualization experiments on tethered flying green lacewings *Chrysopa dasyptera. J. Exp. Biol.* **169**:143–163.

Grodnitsky, D.L., and Morozov, P.P. (1993). Vortex formation during tethered flight of functionally and morphologically two-winged insects, including evolutionary considerations of insect flight. *J. Exp. Biol.* **182**:11–40.

Grodnitsky, D.L., and Morozov, P.P. (1994). Morphology, flight kinematics, and deformation of the wings in holometabolous insects (Insecta: Oligoneoptera = Scarabaeiformes). *Russ. Ent. J.* **3**:3–32.

Grodnitsky, D.L., and Morozov, P.P. (1995). The vortex wakes of flying beetles. *Zool. Zh.* **74**:66–72. (In Russian.)

Gronenberg, W., and Strausfeld, N.J. (1990). Descending neurons supplying the neck and flight motor of Diptera: physiological and anatomical characteristics. *J. Comp. Neurol.* **302**:973–991.

Gruhl, K. (1924). Paarungsgewohnheiten der Dipteren. *Z. wiss. Zool.* **122**:205–280.

Guermond, J.-L., and Sellier, A. (1991). A unified unsteady lifting-line theory. *J. Fluid Mech.* **229**:427–451.

Gullan, P.J., and Kosztarab, M. (1997). Adaptations in scale insects. *Ann. Rev. Ent.* **42**:23–50.

Gunn, A., Gatehouse, A.G., and Woodrow, K.P. (1989). Trade-off between flight and reproduction in the African armyworm moth, *Spodoptera exempta*. *Physiol. Ent.* **14**:419–427.

Gürgey, E., and Thiele, F. (1991). Numerical simulation of the viscous flow over an oscillating airfoil. In *Experimental and Numerical Flow Visualization* (ed. B. Khalighi, M.J. Braun, and C.J. Freitas), pp. 377–383. New York: ASMA.

Gursul, I., and Ho, C.-M. (1992). High aerodynamic loads on an airfoil submerged in an unsteady stream. *AIAA J.* **30**:1117–1119.

Gustafson, K., and Leben, R. (1991). Computation of dragonfly aerodynamics. *Comp. Phys. Comm.* **65**:121–132.

Gustafson, K., Leben, R., and McArthur, J. (1992). Lift and thrust generation by an airfoil in hover modes. *Comp. Fluid Dyn. J.* **1**:47–57.

Haas, F., and Wootton, R.J. (1996). Two basic mechanisms in insect wing folding. *Proc. R. Soc. Lond.* B **263**:1651–1658.

Haddow, A.J., and Corbet, P.S. (1961). Entomological studies from a high tower in Mpanga Forest, Uganda. II. Observations on certain environmental factors at different levels. *Trans. R. Ent. Soc. Lond.* **113**:257–269.

Hagen, K.S. (1996). Aquatic Hymenoptera. In *An Introduction to the Aquatic Insects of North America*, pp. 474–483. 3d ed. Dubuque, Iowa: Kendall/Hunt.

Hamilton, K.G.A. (1971). The insect wing, Part I. Origin and development of wings from notal lobes. *J. Kans. Ent. Soc.* **44**:421–433.

Hamilton, K.G.A. (1972). The insect wing, Part IV. Venational trends and the phylogeny of the winged orders. *J. Kans. Ent. Soc.* **45**:295–308.

Hamilton, R.W., Cohen, J.D., Doebbler, G.F., Exposito, L.F., King, J.M., Smith, K.H., and Schreiner, H.R. (1970). Biochemical and metabolic effects of a six-month exposure of small animals to a helium-oxygen atmosphere. *Space Life Sci.* **2**:57–99.

Hamilton, T.H. (1961). The adaptive significances of intraspecific trends of variation in wing length and body size among bird species. *Evolution* **15**:180–195.

Hammond, P.M. (1979). Wing-folding mechanisms of beetles, with special reference to investigations of adephagan phylogeny. In *Carabid Beetles: Their*

Evolution, Natural History, and Classification (ed. T.L. Erwin, G.E. Ball, and D.R. Whitehead), pp. 113–180. The Hague: Dr. W. Junk.

Hampson, G.F. (1892). On stridulation in certain Lepidoptera, and on the distortion of the hind wings in the males of certain Ommatophorinae. *Proc. Zool. Soc. Lond.* **1892**:188–193.

Hanauer-Thieser, U., and Nachtigall, W. (1995). Flight of the honey bee VI: energetics of wind tunnel exhaustion flights at defined fuel content, speed adaptation and aerodynamics. *J. Comp. Physiol.* B **165**:471–483.

Handlirsch, A. (1906–1908). *Die fossilen Insekten und die Phylogenie der rezenten Formen. Ein Handbuch für Pälontologen und Zoologen.* Leipzig: Wilhelm Engelmann. (Published in nine parts over three years.)

Handlirsch, A. (1910). Einige interessante Kapitel der Paläo-Entomologie. *Verh. Zool.-Bot. Gesellsch. Wien* **60**:160–185.

Hanegan, J.L., and Heath, J.E. (1970). Temperature dependence of the neural control of the moth flight system. *J. Exp. Biol.* **53**:629–639.

Hanken, J., and Wake, D.B. (1993). Miniaturization of body size: organismal consequences and evolutionary significance. *Ann. Rev. Ecol. Syst.* **24**:501–519.

Hankin, E.H. (1921a). The problem of soaring flight. *Proc. Camb. Phil. Soc.* **20**:222–227.

Hankin, E.H. (1921b). The soaring flight of dragon-flies. *Proc. Camb. Phil. Soc.* **20**:460–465.

Hannes, F. (1926). Bienenflugton und Flügelschlagzahl. *Biol. Zentralbl.* **46**:129–142.

Hanström, B. (1928). *Vergleichende Anatomie des Nervensystems der Wirbellosen Tiere unter Berücksichtigung seiner Funktion.* Berlin: Julius Springer.

Harcombe, E.S., and Wyman, R.J. (1978). The cyclically repetitive firing sequences of identified *Drosophila* flight motoneurons. *J. Comp. Physiol.* A **123**:271–279.

Hardie, J., and Young, S. (1997). Aphid flight-track analysis in three dimensions using video techniques. *Physiol. Ent.* **22**:116–122.

Hardy, A.C., and Milne, P.S. (1938). Studies in the distribution of insects by aerial currents. Experiments in aerial tow-netting from kites. *J. Anim. Ecol.* **7**:199–229.

Hargrove, J.W. (1975). The flight performance of tsetse flies. *J. Insect Physiol.* **21**:1385–1395.

Harlé, É., and Harlé, A. (1911). Le vol de grands reptiles et insectes disparus semble indiquer une pression atmosphérique élevée. *Bull. Soc. Geol. Fr., 4th ser.* **11**:118–121.

Harper, D.G., and Blake, R.W. (1989). A critical analysis of the use of high-speed film to determine maximum accelerations of fish. *J. Exp. Biol.* **142**:465–471.

Harrison, J.F. (1986). Caste-specific changes in honeybee flight capacity. *Physiol. Zool.* **59**:175–187.

Harrison, J.F. (1997). Ventilatory mechanism and control in grasshoppers. *Amer. Zool.* **37**:73–81.

Harrison, J.F., and Hall, H.G. (1993). African-European honeybee hybrids have low nonintermediate metabolic capacities. *Nature* **363**:258–260.

Harrison, J.F., and Lighton, J.R.B. (1998). Oxygen-sensitive flight metabolism in the dragonfly *Erythemis simplicicollis*. *J. Exp. Biol.* **201**:1739–1744.

Harrison, J.F., Nielsen, D.I., and Page, R.E. (1996). Malate dehydrogenase phenotype, temperature and colony effects on flight metabolic rate in the honeybee, *Apis mellifera*. *Funct. Ecol.* **10**:81–88.

Harrison, J.F., Fewell, J.H., Roberts, S.P., and Hall, H.G. (1996). Achievement of thermal stability by varying metabolic heat production in flying honeybees. *Science* **274**:88–90.

Harrison, R.G. (1980). Dispersal polymorphisms in insects. *Ann. Rev. Ecol. Syst.* **11**:95–118.

Hart, M.H. (1978). The evolution of the atmosphere of the earth. *Icarus* **33**:23–39.

Hartman, F.A. (1961). Locomotor mechanisms of birds. *Smithsonian. Misc. Collec.* **143**:1–91.

Harvey, P.H., and Pagel, M.D. (1991). *The Comparative Method in Evolutionary Biology.* Oxford: Oxford University Press.

Hashimoto, H. (1962). Ecological significance of the sexual dimorphism in marine chironomids. *Sci. Rep. Tokyo Kyo. Dai.* B **151**:221–252.

Hashimoto, H. (1976). Non-biting midges of marine habitats (Diptera: Chironomidae). In *Marine Insects* (ed. L. Cheng), pp. 337–414. New York: Elsevier.

Haskell, P.T. (1957). The influence of flight noise on behaviour in the desert locust *Schistorcerca gregaria* (Forsk.). *J. Insect Physiol.* **1**:52–75.

Haskell, P.T. (1959). Function of certain prothoracic hair receptors in the Desert Locust. *Nature* **183**:1107.

Haskell, P.T. (1966). Flight behaviour. In *Insect Behaviour* (ed. P.T. Haskell), pp. 29–58. London: Royal Entomological Society.

Hatsopoulos, N., Gabbiani, F., and Laurent, G. (1995). Elementary computation of object approach by a wide-field visual neuron. *Science* **270**:1000–1003.

Haupt, H. (1941). Die ältesten geflügelten Insekten und ihre Beziehungen zur Fauna der Jetztzeit. *Z. Naturwiss. (Halle)* **94**:60–121.

Hausen, K., and Egelhaaf, M. (1989). Neural mechanisms of visual course control in insects. In *Facets of Vision* (ed. D.G. Stavenga and R.C. Hardie), pp. 391–324. Berlin: Springer-Verlag.

Haussling, H.J. (1979). Boundary-fitted coordinates for accurate numerical solution of multibody flow problems. *J. Comput. Physics.* **30**:107–124.

Hedenström, A. (1992). Flight performance in relation to fuel load in birds. *J. theor. Biol.* **158**:535–537.

Hedenström, A., and Alerstam, T. (1995). Optimal flight speed of birds. *Phil. Trans. R. Soc. Lond.* B **348**:471–487.

Heide, G. (1968). Flugsteuerung durch nicht-fibrilläre Flugmuskeln bei der Schmeißfliege *Calliphora. Z. vergl. Physiol.* **59**:456–460.

Heide, G. (1971a). Die Funktion der nicht-fibrillären Flugmuskeln von *Calliphora*. Teil I. Lage, Insertionsstellen und Innervierungsmuster der Muskeln. *Zool. Jb. Abt. Allg. Zool. Physiol.* **76**:87–98.

Heide, G. (1971b). Die Funktion der nicht-fibrillären Flugmuskeln von *Calli-*

phora. Teil II. Muskuläre Mechanismen der Flugsteuerung und ihre nervöse Kontrolle. *Zool. Jb. Abt. Allg. Zool. Physiol.* **76**:99–137.

Heide, G. (1974). The influence of wingbeat synchronous feedback on the motor output systems in flies. *Z. Naturforsch.* **29C**:739–744.

Heide, G. (1975). Properties of a motor output system involved in the optomotor response of flies. *Biol. Cybern.* **20**:99–112.

Heide, G. (1979). Proprioceptive feedback dominates the central oscillator in the patterning of the flight motoneuron output in *Tipula* (Diptera). *J. Comp. Physiol.* A **134**:177–189.

Heide, G., and Götz, K.G. (1996). Optomotor control of course and altitude in *Drosophila melanogaster* is correlated with distinct activities of at least three pairs of flight steering muscles. *J. Exp. Biol.* **199**:1711–1726.

Heidelbach, J., Dambach, M., and Böhm, H. (1991). Processing wing flick-generated air-vortex signals in the African cave cricket *Phaeophilacris spectrum*. *Naturwissenschaften* **78**:277–278.

Heiden, U., an der, and Mackey, M.C. (1982). The dynamics of production and destruction: analytic insight into complex behaviour. *J. math. Biol.* **8**:345–364.

Heinrich, B. (1971). Temperature regulation of the sphinx moth, *Manduca sexta*. I. Flight energetics and body temperature during free and tethered flight. *J. Exp. Biol.* **54**:141–152.

Heinrich, B. (1972). Patterns of endothermy in bumblebee queens, drones and workers. *J. Comp. Physiol.* **77**:65–79.

Heinrich, B. (1975a). Energetics of pollination. *Ann. Rev. Ecol. Syst.* **6**:139–170.

Heinrich, B. (1975b). Thermoregulation in bumblebees. II. Energetics of warm-up and free flight. *J. Comp. Physiol.* B **96**:155–166.

Heinrich, B. (1977). Why have some animals evolved to regulate a high body temperature? *Amer. Nat.* **111**:623–640.

Heinrich, B. (1979). Keeping a cool head: honeybee thermoregulation. *Science* **205**:1269–1271.

Heinrich, B. (1980). Mechanisms of body-temperature regulation in honeybees, *Apis mellifera*. II. Regulation of thoracic temperature at high air temperatures. *J. Exp. Biol.* **85**:73–97.

Heinrich, B. (1983). Insect foraging energetics. In *Handbook of Experimental Pollination Biology* (ed. C.E. Jones and R.J. Little), pp. 187–214. New York: Scientific and Academic Editions.

Heinrich, B. (1993). *The Hot-blooded Insects: Strategies and Mechanisms of Thermoregulation*. Cambridge, Mass.: Harvard University Press.

Heinrich, B., and Bartholomew, G.A. (1979). Roles of endothermy and size in inter- and intraspecific competition for elephant dung in an African dung beetle. *Physiol. Zool.* **52**:484–496.

Heinrich, B., and Buchmann, S.L. (1986). Thermoregulatory physiology of the carpenter bee, *Xylocopa varipuncta*. *J. Comp. Physiol.* B **156**:557–562.

Heinrich, B., and Mommsen, T.P. (1985). Flight of winter moths near 0 °C. *Science* **228**:177–179.

Heinrich, B., and Raven, P.H. (1972). Energetics and pollination ecology. *Science* **176**:597–602.

Heinzel, H.-G. (1983). Rezeption von Luftströmungen und ihre Bedeutung für den Flug der Wanderheuschrecke. In *BIONA-report 2* (ed. W. Nachtigall), pp. 52–69. Stuttgart: Gustav Fischer.

Heinzel, H.-G., and Dambach, M. (1987). Travelling air vortex rings as potential communication signals in a cricket. *J. Comp. Physiol.* A 160:79–88.

Heinzel, H.-G., and Gewecke, M. (1987). Aerodynamic and mechanical properties of the antennae as air-current sense organs in *Locusta migratoria*. II. Dynamic characteristics. *J. Comp. Physiol.* A 161:671–680.

Heisenberg, M., and Wolf, R. (1984). *Vision in Drosophila*. Berlin: Springer-Verlag.

Hengstenberg, R. (1988). Mechanosensory control of compensatory head roll during flight in the blowfly, *Calliphora erythrocephala* Meig. *J. Comp. Physiol.* A 163:151–165.

Hengstenberg, R., Sandeman, D.C., and Hengstenberg, B. (1986). Compensatory head roll in the blowfly *Calliphora* during flight. *Proc. R. Soc. Lond.* B 227:455–482.

Hennig, W. (1981). *Insect Phylogeny*. Chichester, U.K.: John Wiley and Sons.

Hensler, K. (1992a). Neuronal co-processing of course deviation and head movement in locusts. I. Descending deviation detectors. *J. Comp. Physiol.* A 171:257–271.

Hensler, K. (1992b). Neuronal co-processing of course deviation and head movement in locusts. II. Thoracic interneurons. *J. Comp. Physiol.* A 171:273–284.

Hepburn, H.R., and Ball, A. (1973). On the structure and mechanical properties of beetle shells. *J. Mater. Sci.* 8:618–623.

Hepburn, H.R., Youthed, C., Illgner, P., Radloff, S.E., and Brown, R.E. (1998). Production of aerodynamic power in mountain honeybees (*Apis mellifera*). *Naturwissenschaften* 85:389-390.

Heppner, J.B. (1991). Brachyptery and aptery in Lepidoptera. *Tropical Lepidoptera* 2:11–40.

Heran, H. (1955). Versuche über die Windkompensation der Bienen. *Naturwissenschaften* 42:132–133.

Heran, H. (1956). Ein Beitrag zur Frage nach der Wahrnehmungsgrundlage der Entfernungsweisung der Bienen. *Z. vergl. Physiol.* 38:168–218.

Heran, H. (1959). Wahrnehmung und Regelung der Fluggeigengeschwindigkeit bei *Apis mellifica*. *Z. vergl. Physiol.* 42:103–163.

Heran, H., and Lindauer, M. (1963). Windkompensation und Seitenwindkorrektur der Bienen beim Flug über Wasser. *Z. vergl. Physiol.* 47:39–55.

Herbst, H.G., and Freund, K. (1962). Kinematik der Flügel bei ventilierenden Honigbienen. *Dtsch. Ent. Z.*, N.F. 9:1–29.

Herold, W. (1952). Beobachtungen über die Arbeitsleistung einiger Arbeiter von *Vespa germanica* F.-*Dolichovespula germanica* F. *Biol. Zentralbl.* 71:461–469.

Herrera, C.M. (1997). Thermal biology and foraging responses of insect pollinators to the forest floor irradiance mosaic. *Oikos* 78:601–611.

Herskowitz, I.H. (1949). Hexaptera, a homoeotic mutant in *Drosophila melanogaster*. *Genetics* 34:10–25.

Hespenheide, H. (1977). Dispersion and the size composition of the aerial insect fauna. *Ecol. Ent.* **2**:139–141.

Hespenheide, H.A. (1973). A novel mimicry complex: beetles and flies. *J. Ent. (A)* **48**:49–56.

Heukamp, U. (1984). Sensory regulation of the pleuroalar muscles in the migratory locust. *Naturwissenschaften* **71**:481–482.

Heymons, R. (1899). Der morphologische Bau des Insektenabdomens. Eine kritische Zusammenstellung der wesentlichsten Forschungsergebnisse auf anatomischen und embryologischen Gebiete. *Zool. Centralbl.* **6**:537–556.

Hill, A.V. (1950). The dimensions of animals and their muscular dynamics. *Sci. Progr.* **38**:209–230.

Hillerton, J.E., Reynolds, S.E., and Vincent, J.F.V. (1982). On the indentation hardness of insect cuticle. *J. Exp. Biol.* **96**:45–52.

Hind, A.T., and Gurney, W.S.C. (1997). The metabolic cost of swimming in marine homeotherms. *J. Exp. Biol.* **200**:531–542.

Hinton, H.E. (1948). On the origin and function of the pupal stage. *Trans. R. Ent. Soc. Lond.* **99**:395–409.

Hinton, H.E. (1963). The origin and function of the pupal stage. *Proc. R. Ent. Soc. Lond.* (A) **38**:77–85.

Hinton, H.E. (1968). Spiracular gills. *Adv. Insect Physiol.* **5**:65–161.

Hisada, M., Tamasige, M., and Suzuki, N. (1965). Control of the flight of the dragonfly *Sympetrum darwinianum* Selys. I. Dorsophotic response. *J. Fac. Sci. Hokkaido Univ. VI, Zool.* **15**:568–577.

Hlavac, T.F. (1974). *Merope tuber* (Mecoptera): A wing-body interlocking mechanism. *Psyche* **81**:303–306.

Hochachka, P.W. (1987). Patterns of O_2-dependence of metabolism. *Adv. Exp. Med. Biol.* **222**:143–149.

Hochachka, P.W. (1994). *Muscles as Molecular and Metabolic Machines.* Boca Raton, Fla.: CRC Press.

Hocking, B. (1953). The intrinsic range and speed of flight of insects. *Trans. R. Ent. Soc. Lond.* **104**:223–345.

Hodkinson, I.D., and Casson, D. (1991). A lesser predilection for bugs: Hemiptera (Insecta) diversity in tropical rain forests. *Biol. J. Linn. Soc.* **43**:101–109.

Hoff, W. (1919). Der Flug der Insekten und der Vögel. *Naturwissenschaften* **7**:159–162.

Holland, H.D. (1984). *The Chemical Evolution of the Atmosphere and Oceans.* Princeton, N.J.: Princeton University Press.

Hölldobler, B., and Wilson, E.O. (1990). *The Ants.* Cambridge, Mass.: Harvard University Press.

Hollick, F.S.J. (1940). The flight of the dipterous fly *Muscina stabulans* Fallén. *Phil. Trans. R. Soc. Lond.* B **230**:357–390.

Holmes, D.J., and Austad, S.N. (1995). The evolution of avian senescence patterns: implications for understanding primary aging processes. *Amer. Zool.* **35**:307–317.

Homberg, U. (1994). Flight-correlated activity changes in neurons of the lateral accessory lobes in the brain of the locust *Schistocerca gregaria*. *J. Comp. Physiol.* A **175**:597–610.

Homsher, E., and Kean, C.J. (1978). Skeletal muscle energetics and metabolism. *Ann. Rev. Physiol.* **138**:301–318.

Horowits, R., Kempner, E.S., Bisher, M.E., and Podolsky, R.J. (1986). A physiological role for titin and nebulin in skeletal muscle. *Nature* **323**:160–164.

Horridge, G.A. (1956). The flight of very small insects. *Nature* **178**:1334–1335.

Horridge, G.A. (1965). Arthropoda: details of the groups. In *Structure and Function in the Nervous System of Invertebrates*, Vol. 2 (ed. T.H. Bullock and G.A. Horridge), pp. 1165–1270. San Francisco: W.H. Freeman.

Horridge, G.A., and Marcelja, L. (1992). On the existence of 'fast' and 'slow' directionally sensitive motion detector neurons in insects. *Proc. R. Soc. Lond.* B **248**:247–254.

Horsmann, U., Heinzel, H.-G., and Wendler, G. (1983). The phasic influence of self-generated air current modulations on the locust flight motor. *J. Comp. Physiol.* A **150**:427–438.

Hough-Goldstein, J.A., Geiger, J., Chang, D., and Saylor, W. (1993). Palatability and toxicity of the Colorado Potato Beetle (Coleoptera: Chrysomelidae) to domestic chickens. *Ann. Ent. Soc. Amer.* **86**:158–164.

Houghton, G. (1964). Generalized hovering-flight correlation for insects. *Nature* **204**:666–668.

Howard, J. (1997). Molecular motors: structural adaptations to cellular functions. *Nature* **389**:561–567.

Hoy, R., Nolen, T., and Brodfuehrer, P. (1989). The neuroethology of acoustic startle and escape in flying insects. *J. Exp. Biol.* **146**:287–306.

Hoyle, G. (1983). *Muscles and Their Neural Control.* New York: John Wiley and Sons.

Hughes, A.L., and Hughes, M.K. (1995). Small genomes for better flyers. *Nature* **377**:391.

Humphries, D.A., and Driver, P.M. (1970). Protean defence by prey animals. *Oecologia* **5**:285–302.

Hünicken, M.A. (1980). A giant fossil spider (*Megarachne servinei*) from Bajo de Veliz, Upper Carboniferous, Argentina. *Bol. Acad. Nac. Cienc. (Cordoba)* **53**: 317–328.

Hunter, A.F. (1995). The ecology and evolution of reduced wings in forest macrolepidoptera. *Evol. Ecol.* **9**:275–287.

Hutchinson, G.E. (1993). *A Treatise on Limnology. Volume IV. The Zoobenthos* (ed. Y.H. Edmondson). New York: John Wiley and Sons.

Huxley, J., and Barnard, P.C. (1988). Wing-scales of *Pseudoleptocerus chirindensis* Kimmins (Trichoptera: Leptoceridae). *Zool. J. Linn. Soc.* **92**:285–312.

Hyzer, W.G. (1962). Flight behavior of a fly alighting on a ceiling. *Science* **137**:609–610.

Ikawa, T., Okabe, H., Mori, T., Urabe, K., and Ikeshoji, T. (1994). A method for reconstructing three-dimensional positions of swarming mosquitoes. *J. Insect Behav.* **7**:237–248.

Ikeda, K., and Boettiger, E.G. (1965a). Studies on the flight mechanism of insects—II. The innervation and electrical activity of the fibrillar muscles of the bumble bee, *Bombus. J. Insect Physiol.* **11**:779–789.

Ikeda, K., and Boettiger, E.G. (1965b). Studies on the flight mechanism of insects—III. The innervation and electrical activity of the basalar fibrillar flight muscle of the beetle, *Oryctes rhinoceros. J. Insect Physiol.* **11**:791–802.

Ilius, M., Wolf, R., and Heisenberg, M. (1994). The central complex of *Drosophila melanogaster* is involved in flight control: studies on mutants and mosaics of the gene *ellipsoid body open. J. Neurogen.* **9**:189–206.

Ivanov, V.D. (1985). A comparative analysis of wing kinematics in caddisflies (Insecta: Trichoptera). *Ent. Obozr.* **64**:273–284. (In Russian.)

Ivanov, V.D. (1990). A comparative analysis of flight aerodynamics of caddisflies (Insecta: Trichoptera). *Zool. Zh.* **69**:46–60. (In Russian.)

Izumi, K., and Kuwahara, K. (1983). Unsteady flow field, lift and drag measurements of impulsively started elliptic cylinder and circular-arc airfoil. *AIAA-83-1711*.

Jackson, D.J. (1952). Observations on the capacity for flight of water beetles. *Proc. R. Ent. Soc. Lond. (A)* **27**:57–70.

Jackson, D.J. (1956). Observations on flying and flightless water beetles. *J. Linn. Soc. Lond., Zool.* **43**:18–42.

Jacobs, M.E. (1985). Role of beta-alanine in cuticular tanning, sclerotization and temperature regulation in *Drosophila melanogaster. J. Insect Physiol.* **31**:509–515.

Jacot, A.P. (1935). The large winged mites of Florida. *Fla. Ent.* **19**:1–14.

Janet, C. (1899). Sur le mécanisme du vol chez les Insectes. *C.R. Acad. Sci. (Paris)* **128**:249–253.

Janssen, R. (1992). Thermal influences on nervous system function. *Neurosci. Biobehav. Rev.* **16**:399–413.

Janzen, D.H. (1971). Euglossine bees as long-distance pollinators of tropical plants. *Science* **171**:203–205.

Janzen, D.H. (1973). Sweep samples of tropical foliage insects: effects of seasons, vegetation types, elevation, time of day, and insularity. *Ecology* **54**:687–708.

Janzen, D.H. (1979). How to be a fig. *Ann. Rev. Ecol. Syst.* **10**:13–51.

Jaworowski, A. (1897). Zu meiner Extremitäten- und Kiementheorie bei den Arthropoden. *Zool. Anz.* **20**:177–184.

Jensen, M. (1956). Biology and physics of locust flight. III. The aerodynamics of locust flight. *Phil. Trans. R. Soc. Lond.* B **239**:511–552.

Jensen, M., and Weis-Fogh, T. (1962). Biology and physics of locust flight. V. Strength and elasticity of locust cuticle. *Phil. Trans. R. Soc. Lond.* B **245**:137–169.

Jepsen, G.L. (1970). Bat origins and evolution. In *Biology of Bats*, vol. 1 (ed. W.A. Wimsatt), pp. 1–64. London: Academic Press.

Jeram, A.J., Selden, P.A., and Edwards, D. (1990). Land animals in the Silurian: Arachnids and myriapods from Shropshire, England. *Science* **250**:658–661.

Jetten, T.H., Martens, W.J.M., and Takken, W. (1996). Model simulations to estimate malaria risk under climate change. *J. Med. Ent.* **33**:361–371.

Jewell, B.R., and Rüegg, J.C. (1966). Oscillatory contraction of insect fibrillar muscle after glycerol extraction. *Proc. R. Soc. Lond.* B **164**:427–459.

Johansson, A.S. (1957). The nervous system of the milkweed bug, *Oncopeltus fasciatus* (Heteroptera, Lygaeidae). *Trans. Ent. Soc. Amer.* **83**:119–1183.

Johnson, B. (1980). An electron microscopic study of flight muscle breakdown in an aphid *Megoura viciae*. *Tissue & Cell* **12**:529–539.

Johnson, C.G. (1957). The distribution of insects in the air and the empirical relation of density to height. *J. Anim. Ecol.* **26**:479–494.

Johnson, C.G. (1969). *Migration and Dispersal of Insects by Flight*. London: Methuen.

Johnson, C.G. (1976). Lability of the flight system: a context for functional adaptation. In *Insect Flight* (ed. R.C. Rainey), pp. 217–234. Oxford: Blackwell Scientific.

Johnston, I.A., and Altringham, J.D. (1988). Muscle function in locomotion. *Nature* **335**:767–768.

Jones, R.T. (1980). Wing flapping with minimal energy. *Aero. J.* **84**:214–217.

Jones, W.P. (1941). The virtual inertias of a tapered wing in still air. *Rep. Brit. Aero. Counc.*, no. 1946.

Joos, B., Young, P.A., and Casey, T.M. (1991). Wingstroke frequency of foraging and hovering bumblebees in relation to morphology and temperature. *Physiol. Ent.* **16**:191–200.

Joos, B., Lighton, J.R.B., Harrison, J.F., Suarez, R.K., and Roberts, S.P. (1997). Effects of ambient oxygen tension on flight performance, metabolism, and water loss of the honeybee. *Physiol. Zool.* **70**:167–174.

Jordanoglou, A.L. (1949a). The entomopter. *Amer. Helicopt.* **15(8)**:13–22.

Jordanoglou, A.L. (1949b). The entomopter-theory and experiments. Theory—Part II. *Amer. Helicopt.* **15(9)**:15–19.

Jordanoglou, A.L. (1949c). The entomopter-theory and experiments. Theory (cont'd)—Part III, Conclusion. *Amer. Helicopt.* **15(10)**:16–20.

Josephson, R.K. (1981). Temperature and the mechanical performance of insect muscle. In *Insect Thermoregulation* (ed. B. Heinrich), pp. 20–44. New York: John Wiley and Sons.

Josephson, R.K. (1984). Contraction dynamics of flight and stridulatory muscles of tettigoniid insects. *J. Exp. Biol.* **108**:77–96.

Josephson, R.K. (1985a). Mechanical power output from striated muscle during cyclic contraction. *J. Exp. Biol.* **114**:493–512.

Josephson, R.K. (1985b). The mechanical power output of a tettigoniid wing muscle during singing and flight. *J. Exp. Biol.* **117**:357–368.

Josephson, R.K. (1989). Power output from skeletal muscle during linear and sinusoidal shortening. *J. Exp. Biol.* **147**:533–537.

Josephson, R.K. (1993). Contraction dynamics and power output of skeletal muscle. *Ann. Rev. Physiol.* **55**:527–546.

Josephson, R.K. (1997a). Power output from a flight muscle of the bumblebee *Bombus terrestris*. II. Characterization of the parameters affecting power output. *J. Exp. Biol.* **200**:1227–1239.

Josephson, R.K. (1997b). Power output from a flight muscle of the bumblebee *Bombus terrestris*. III. Power during simulated flight. *J. Exp. Biol.* **200**:1241–1246.

Josephson, R.K., and Ellington, C.P. (1997). Power output from a flight muscle of the bumblebee *Bombus terrestris*. I. Some features of the dorso-ventral flight muscle. *J. Exp. Biol.* **200**:1215–1226.

Josephson, R.K., and Stevenson, R.D. (1991). The efficiency of a flight muscle from the locust *Schistocerca americana*. *J. Physiol.* **442**:413–429.

Josephson, R.K., and Young, D. (1981). Synchronous and asynchronous muscles in cicadas. *J. Exp. Biol.* **91**:219–237.

Josephson, R.K., and Young, D. (1985). A synchronous muscle with an operating frequency greater than 500 Hz. *J. Exp. Biol.* **118**:185–208.

Jutsum, A.R., Robinson, N.L., and Goldsworthy, G.J. (1982). Effects of flight training on flight speed and substrate utilization in locusts. *Physiol. Ent.* **7**:291–296.

Kahn, P. (1950). The Entomopter. *Aeroplane* **28** (April):496–497.

Kalmus, H. (1945). Correlations between flight and vision, and particularly between wings and ocelli, in insects. *Proc. R. Ent. Soc. Lond.* A **20**:84–96.

Kamada, T., and Kinosita, H. (1947). Regulation of flight in dragonflies. *Seiri Seitai* **1**:147–159. (In Japanese.)

Kammer, A.E. (1967). Muscle activity during flight in some large Lepidoptera. *J. Exp. Biol.* **47**:277–295.

Kammer, A.E. (1968). Motor patterns during flight and warm-up in Lepidoptera. *J. Exp. Biol.* **48**:89–109.

Kammer, A.E. (1971). The motor output during turning flight in a hawkmoth, *Manduca sexta*. *J. Insect Physiol.* **17**:1073–1086.

Kammer, A.E. (1985). Flying. In *Comprehensive Insect Physiology, Biochemistry, and Pharmacology,* vol. 5 (ed. G.A. Kerkut and L.I. Gilbert), pp. 491–552. Oxford: Pergamon Press.

Kammer, A.E., and Heinrich, B. (1978). Insect flight metabolism. *Adv. Insect Physiol.* **13**:133–228.

Kammer, A.E., and Nachtigall, W. (1973). Changing phase relationships among motor units during flight in a saturniid moth. *J. Comp. Physiol.* **83**:17–24.

Kammer, A.E., and Rheuben, M.B. (1976). Adult motor patterns produced by moth pupae during development. *J. Exp. Biol.* **65**:65–84.

Kammer, A.E., and Rheuben, M.B. (1981). Neuromuscular mechanisms of insect flight. In *Locomotion and Energetics in Arthropods* (ed. C.F. Herreid II and C.R. Fourtner), pp. 163–194. New York: Plenum Press.

Kane, S. (1982). Notes on the acoustic signals of a Neotropical satyrid butterfly. *J. Lep. Soc.* **36**:200–206.

Karpouzian, G., Spedding, G., and Cheng, H.K. (1990). Lunate-tail swimming propulsion. Part 2. Performance analysis. *J. Fluid Mech.* **210**:329–351.

Kastberger, G. (1990). The ocelli control the flight course in honeybees. *Physiol. Ent.* **5**:337–346.

Kastberger, G., and Schuhmann, K. (1993). Ocellar occlusion effect on the flight behavior of homing honeybees. *J. Insect Physiol.* **39**:589–600.

Katz, S.L., and Gosline, J.M. (1993). Ontogenetic scaling of jump performance in the African desert locust (*Schistocerca gregaria*). *J. Exp. Biol.* **177**:81–111.

Katzmayr, R. (1922). Über das Verhalten von Flügelfächen bei periodischen Änderungen der Geschwindigkeitsrichtung. *Z. f. Flugtechnik und Motorluftschiffahrt* **6**:80–82.

Kavanaugh, D.H. (1985). On wing atrophy in carabid beetles (Coleoptera:

Carabidae), with special reference to Nearctic *Nebria*. In *Taxonomy, Phylogeny and Zoogeography of Beetles and Ants* (ed. G.E. Ball), pp. 408–431. Dordrecht: Dr. W. Junk.

Kawachi, K. (1981). An extension of the local momentum theory to a distorted wake model of a hovering rotor. *NASA Technical Memorandum 81258*.

Kawano, K. (1997). Costs of evolving exaggerated mandibles in stag beetles (Coleoptera: Lucanidae). *Ann. Ent. Soc. Amer.* **90**:453–461.

Kawecki, Z. (1964). On the suitable term for the second pair of wings in male scale insects (Homoptera-Coccoidea). *Frustula Ent.* **7**:1–4.

Kelber, A., and Zeil, J. (1990). A robust procedure for visual stabilisation of hovering flight position in guard bees of *Trigona (Tetragonisca) angustula* (Apidae, Meliponinae). *J. Comp. Physiol.* A **167**:569–577.

Kelber, A., and Zeil, J. (1997). *Tetragonisca* guard bees interpret expanding and contracting patterns as unintended displacement in space. *J. Comp. Physiol.* A **181**:257–265.

Kendig, J.J. (1968). Motor neurone coupling in locust flight. *J. Exp. Biol.* **48**:389–404.

Kennedy, J.S. (1939). The visual responses of flying mosquitoes. *Proc. Zool. Soc. Lond.* **109**:221–242.

Kennedy, J.S. (1951). The migration of the desert locust (*Schistocerca gregaria* Forsk.). I. The behaviour of swarms. II. A theory of long-range migrations. *Phil. Trans. R. Soc. Lond.* B **235**:163–290.

Kennedy, J.S., and Booth, C.O. (1963). Free flight of aphids in the laboratory. *J. Exp. Biol.* **40**:67–85.

Kennedy, J.S., and Marsh, D. (1974). Pheromone-regulated anemotaxis in flying moths. *Science* **184**:999–1001.

Kennedy, J.S., and Thomas, A.A.G. (1974). Behaviour of some low-flying aphids in wind. *Ann. Appl. Biol.* **76**:143–159.

Kenrick, P., and Crane, P.R. (1997). The origin and early evolution of plants on land. *Nature* **389**:33–39.

Kerguelen, V., and Cardé, R.T. (1997). Manoeuvres of female *Bruchymeria intermedia* flying towards host-related odours in a wind tunnel. *Physiol. Ent.* **22**:344–356.

Kern, M.J. (1985). Metabolic rate of the insect brain in relation to body size and phylogeny. *Comp. Biochem. Physiol.* **81**A:501–506.

Kern, R. (1998). Visual position stabilization in the hummingbird hawk moth, *Macroglossum stellatarum*. II. Electrophysiological analysis of neurons sensitive to wide-field motion. *J. Comp. Physiol.* A **182**:239–249.

Kern, R., and Varjú, D. (1998). Visual position stabilization in the hummingbird hawk moth, *Macroglossum stellatarum*. I. Behavioural analysis. *J. Comp. Physiol.* A **182**:225–237.

Kesel, A.B. (1997). Einige Aspekte zur Statik der Insektenflügel. In *BIONA-report 11* (ed. A. Wisser, D. Bilo, A. Kesel, and B. Möhl), pp. 89–114. Stuttgart: Gustav Fischer.

Kestler, P. (1985). Respiration and respiratory water loss. In *Environmental Physiology and Biochemistry of Insects* (ed. K.H. Hoffman), pp. 137–183. Berlin: Springer-Verlag.

Keulegan, G.H., and Carpenter, L.H. (1958). Forces on cylinders and plates in an oscillating fluid. *J. Res. Natl. Bur. Stand.* **60**:423–440.

Kevan, P.G., and Baker, H.G. (1983). Insects as flower visitors and pollinators. *Ann. Rev. Ent.* **28**:407–453.

Kevan, P.G., Chaloner, W.G., and Savile, D.B.O. (1975). Interrelationships of early terrestrial arthropods and plants. *Paleontology* **18**:391–417.

Kien, J. (1977). Comparison of sensory input with motor output in the locust optomotor system. *J. Comp. Physiol.* A **113**:161–179.

Kien, J., and Altman, J.S. (1979). Connections of the locust wing tegulae with metathoracic flight motoneurons. *J. Comp. Physiol.* A **133**:299–310.

Kimmerle, B., Egelhaaf, M., and Srinivasan, M.V. (1996). Object detection by relative motion in freely flying flies. *Naturwissenschaften* **83**:380–381.

Kimmerle, B., Warzecha, A.-K., and Egelhaaf, M. (1997). Object detection in the fly during simulated translatory flight. *J. Comp. Physiol.* A **181**:247–255.

King, D.G. (1983). Evolutionary loss of a neural pathway from the nervous system of a fly (*Glossina morsitans*/Diptera). *J. Morphol.* **175**:27–32.

King, M.J., Buchmann, S.L., and Spangler, H. (1996). Activity of asynchronous flight muscle from two bee families during sonication (buzzing). *J. Exp. Biol.* **199**:2317–2321.

Kingsolver, J.G., and Koehl, M.A.R. (1985). Aerodynamics, thermoregulation, and the evolution of insect wings: differential scaling and evolutionary change. *Evolution* **39**:488–504.

Kingsolver, J.G., and Koehl, M.A.R. (1994). Selective factors in the evolution of insect wings. *Ann. Rev. Ent.* **39**:425–451.

Kinnamon, S.C., Klaassen, L.W., Kammer, A.E., and Claassen, D. (1984). Octopamine and chlordimeform enhance sensory responsiveness and production of the flight motor pattern in developing and adult moths. *J. Neurobiol.* **15**:283–293.

Kirchner, W.H., and Srinivasan, M.V. (1989). Freely flying honeybees use image motion to estimate object distance. *Naturwissenschaften* **76**:281–282.

Kirschfeld, K. (1989). Automatic gain control in movement detection of the fly. Implications for optomotor responses. *Naturwissenschaften* **76**:378–380.

Kirschfeld, K. (1994). Tracking of small objects in front of a textured background by insects and vertebrates: phenomena and neuronal basis. *Biol. Cybern.* **70**:407–415.

Kleinow, W. (1966). Untersuchungen zum Flügelmechanismus der Dermapteren. *Z. Morphol. Ökol. Tiere* **5**:363–416.

Kliss, M., Somps, C., and Luttges, M.W. (1989). Stable vortex structures: a flat plate model of dragonfly hovering. *J. theor. Biol.* **136**:209–228.

Kluge, N.Yu. (1989). A question of the homology of the tracheal gills and paranotal processes of mayfly larvae and insect wings with reference to the systematics and phylogeny of the mayfly order (Ephemeroptera). In *Reports of the Forty-first Annual Lecture Series in the Memory of N.A. Kholodkovsky*, pp. 48–77. Leningrad: Nauka. (In Russian.)

Kober, R., and Schnitzler, H.-U. (1990). Information in sonar echoes of fluttering insects available for echolocating bats. *J. Acoust. Soc. Amer.* **87**:882–896.

Koch, U. (1977). A miniature movement detector applied to recording of wingbeat in locusts. *Fortschr. Zool.* **24**:327–332.

Komárek, J. (1929). Die Reduktion des Dipterenkörpers aus flugtechnischen Ursachen und die morphologisch-systematischen Folgen. *Proc. Tenth Int. Congr. Zool., Budapest, 1927*, pp. 1060–1067.

Komai, Y. (1968). Augmented respiration in a flying insect. *J. Exp. Biol.* **201**:2359–2366.

Korschelt, E. (1932). Cuticularsehne und Bindegewebssehne. Eine vergleichende morphologische-histologische Betrachtung. *Z. wiss. Zool.* **150**:494–526.

Kozhov, M. (1963). *Lake Baikal and Its Life.* The Hague: Dr. W. Junk.

Kozlov, M.V., Ivanov, M.D., and Grodnitsky, D.L. (1986). Evolution of the wing apparatus and wing kinematics in Lepidoptera. *Uspekhi Sovr. Biolog.* **101**:291–305. (In Russian.)

Krämer, K., and Markl, H. (1978). Flight-inhibition on ground contact in the American cockroach, *Periplaneta americana*—I. Contact receptors and a model for their central connections. *J. Insect Physiol.* **24**:577–586.

Kramer, M. (1932). Die Zunahme des Maximalauftriebes von Tragflügeln bei plötzlicher Anstellwinkelvergrösserung (Böeneffekt). *Z. Flugtech. Motorluftschiff.* **23**:185–189.

Kramer, M.G., and Marden, J.H. (1997). Almost airborne. *Nature* **385**:403–404.

Kränzler, L., and Larink, O. (1980). Postembryonale Veränderungen und sensillen Muster der abdominalen Anhänge von *Thermobia domestica* (Packard) (Insecta: Zygentoma). *Braunschw. Naturk. Schr.* **1**:27–49.

Krapp, H.G., and Hengstenberg, R. (1996). Estimation of self-motion by optic flow processing in single visual neurons. *Nature* **384**:463–466.

Kraus, O. (1974). On the morphology of Paleozoic diplopods. *Symp. Zool. Soc. Lond.* **32**:13–22.

Krenn, H.W., and Pass, G. (1994). Morphological diversity and phylogenetic analysis of wing circulatory organs in insects, part I: Non-Holometabola. *Zoology* **98**:7–22.

Krenn, H.W., and Pass, G. (1994/95). Morphological diversity and phylogenetic analysis of wing circulatory organs in insects, part II: Holometabola. *Zoology* **98**:147–164.

Kristensen, N.P. (1974). On the evolution of wing transparency in Sesiidae (Insecta, Lepidoptera). *Vidensk. Medd. Dan. Naturhist. Foren.* **137**:125–134.

Kristensen, N.P. (1975). The phylogeny of hexapod "orders". A critical review of recent accounts. *Z. zool. Syst. Evolut.-forsch.* **13**:1–44.

Kristensen, N.P. (1981). Phylogeny of insect orders. *Ann. Rev. Ent.* **26**:135–157.

Kristensen, N.P. (1989). Insect phylogeny based on morphological evidence. In *The Hierarchy of Life* (ed. B. Fernholm, K. Bremer, and H. Jörnvall), pp. 295–306. Amsterdam: Elsevier (Biomedical Division).

Kristensen, N.P. (1991). Phylogeny of extant hexapods. In *The Insects of Australia*, 2d ed., vol. 1 (ed. CSIRO), pp. 125–140. Ithaca, N.Y.: Cornell University Press.

Kristensen, N.P. (1997). The groundplan and basal diversification of the hexapods. In *Arthropod Relationships* (ed. R.A. Fortey and R.H. Thomas), pp. 281–293. London: Chapman and Hall.

Krogh, A., and Weis-Fogh, W. (1951). The respiratory exchange of the desert locust (*Schistocerca gregaria*) before, during and after flight. *J. Exp. Biol.* **28**:344–357.

Krogh, A., and Zeuthen, E. (1941). The mechanism of flight preparation in some insects. *J. Exp. Biol.* **18**:1–10.

Krzelj, S., (1969). Structure anatomique comparée des élytres de Coléoptères. *Ann. Soc. R. Zool. Belg.* **99**:85–109.

Krzelj, S. and Jeuniaux, C. (1968). Propriétés physiques comparées des élytres de Coléoptères. *Ann. Soc. R. Zool. Belg.* **98**:87–99.

Küchemann, D., and von Holst, E. (1941). Zur Aerodynamik des Tierfluges. *Luftwissensch.* **8**:277–282.

Kuenen, L.P.S., and Cardé, R.T. (1993). Effects of moth size on velocity and steering during upwind flight toward a sex pheromone source by *Lymantria dispar* (Lepidoptera: Lymantriidae). *J. Insect Behav.* **6**:177–193.

Kuenen, L.P.S., and Cardé, R.T. (1994). Strategies for recontacting a lost pheromone plume: casting and upwind flight in the male gypsy moth. *Physiol. Ent.* **19**:15–29.

Kuethe, A.M. (1975). On the mechanics of flight of small insects. In *Swimming and Flying in Nature*, vol. 2 (ed. T.Y.-T. Wu, C.J. Brokaw, and C. Brennen), pp. 803–813. New York: Plenum Press.

Kukalová, J. (1968). Permian mayfly nymphs. *Psyche* **75**:310–327.

Kukalová-Peck, J. (1974). Wing-folding in the Paleozoic insect order Diaphanopterodea (Paleoptera), with a description of new representatives of the family Elmoidae. *Psyche* **81**:315–333.

Kukalová-Peck, J. (1978). Origin and evolution of insect wings and their relation to metamorphosis, as documented by the fossil record. *J. Morphol.* **156**:53–126.

Kukalová-Peck, J. (1983). Origin of the insect wing and wing articulation from the arthropodan leg. *Can. J. Zool.* **61**:1618–1669.

Kukalová-Peck, J. (1985). Ephemeroid wing venation based upon new gigantic Carboniferous mayflies and basic morphology, phylogeny, and metamorphosis of pterygote insects (Insecta, Ephemerida). *Can. J. Zool.* **63**:933–955.

Kukalová-Peck, J. (1987). New Carboniferous Diplura, Monura, and Thysanura, the hexapod ground plan, and the role of thoracic lobes in the origin of wings (Insecta). *Can. J. Zool.* **65**:2327–2345.

Kukalová-Peck, J. (1991). Fossil history and the evolution of hexapod structures. In *The Insects of Australia*, 2d ed., vol. 1 (ed. CSIRO), pp. 141–179. Ithaca, N.Y.: Cornell University Press.

Kukalová-Peck, J. (1992). The "Uniramia" do not exist: the ground plan of the Pterygota as revealed by Permian Diaphanopterodea from Russia (Insecta: Paleodictyopteroidea). *Can. J. Zool.* **70**:236–255.

Kukalová-Peck, J. (1997a). Arthropod phylogeny and 'basal' morphological structures. In *Arthropod Relationships* (ed. R.A. Fortey and R.H. Thomas), pp. 249–268. London: Chapman and Hall.

Kukalová-Peck, J. (1997b). Mazon Creek insect fossils: the origin of insect wings and clues about the origin of insect metamorphosis. In *Richardson's Guide to the Fossil Fauna of Mazon Creek* (ed. C.W. Shabica and A.A. Hay), pp. 194–207. Chicago: Northeastern Illinois University Press.

Kukalová-Peck, J., and Brauckmann, C. (1990). Wing folding in pterygote insects, and the oldest Diaphanopterodea from the early Late Carboniferous of West Germany. *Can. J. Zool.* **68**:1104–1111.

Kukalová-Peck, J., and Lawrence, J.F. (1993). Evolution of the hind wing in Coleoptera. *Can. Ent.* **125**:181–258.

Kunze, P. (1961). Untersuchung des Bewegungssehens fixiert fliegender Bienen. *Z. vergl. Physiol.* **44**:656–684.

Kuribayashi, S. (1981). *Insects in Flight.* Tokyo: Heybonsha. (In Japanese.)

Kutsch, W. (1971). The development of the flight pattern in the desert locust, *Schistocerca gregaria. Z. vergl. Physiol.* **74**:156–168.

Kutsch, W. (1973). The influence of age and culture temperature on the wingbeat frequency of the migratory locust, *Locusta migratoria. J. Insect Physiol.* **19**:763–772.

Kutsch, W. (1989). Development of the flight motor pattern. In *Insect Flight* (ed. G.J. Goldsworthy and C.H. Wheeler), pp. 51–73. Boca Raton, Fla.: CRC Press.

Kutsch, W., and Breidbach, O. (1994). Homologous structures in the nervous systems of Arthropoda. *Adv. Insect Physiol.* **24**:1–113.

Kutsch, W., and Gewecke, M. (1979). Development of flight behaviour in maturing adults of *Locusta migratoria*: II. Aerodynamic parameters. *J. Insect Physiol.* **25**:299–304.

Kutsch, W., and Hemmer, W. (1994a). Ontogenetic studies of flight initiation in *Locusta migratoria*: wind response of an identified interneurone (TCG). *J. Insect Physiol.* **40**:97–106.

Kutsch, W., and Hemmer, W. (1994b). Ontogenetic studies of flight initiation in *Locusta migratoria*: flight muscle activity. *J. Insect Physiol.* **40**:519–525.

Kutsch, W., and Hug, W. (1981). Dipteran flight motor pattern: invariabilities and changes during postlarval development. *J. Neurobiol.* **12**:1–14.

Kutsch, W., and Kittmann, R. (1991). Flight motor pattern in flying and nonflying Phasmida. *J. Comp. Physiol.* A **168**:483–490.

Kutsch, W., and Stevenson, P. (1981). Time-correlated flight of juvenile and mature locust: a comparison between free and tethered animals. *J. Insect Physiol.* **27**:455–459.

Kutsch, W., and Usherwood, P.N.R. (1970). Studies of the innervation and electrical activity of flight muscles in the locust, *Schistocerca gregaria. J. Exp. Biol.* **52**:299–312.

Kutsch, W., Camhi, J., and Sumbre, G. (1994). Close encounters among flying locusts produce wing-beat coupling. *J. Comp. Physiol.* A **174**:643–649.

Kutsch, W., Hanloser, H., and Reinecke, M. (1980). Light and electron-microscopic analysis of a complex sensory organ: the tegula of *Locusta migratoria. Cell Tiss. Res.* **210**:461–478.

Kutsch, W., Urbach, R., and Breidbach, O. (1993). Comparison of motor patterns in larval and adult stage of a beetle, *Zophobas morio. J. Exp. Zool.* **267**: 389–403.

Kutsch, W., Schwarz, G., Fischer, H., and Kautz, H. (1993). Wireless transmission of muscle potentials during free flight of a locust. *J. Exp. Biol.* **185**:367–373.

Labandeira, C.C. (1994). A compendium of fossil insect families. *Contrib. Biol. Geol. Milwaukee Publ. Mus.* **88**:1–71.

Labandeira, C.C. (1997). Insect mouthparts: ascertaining the paleobiology of insect feeding strategies. *Ann. Rev. Ecol. Syst.* **28**:153–193.

Labandeira, C.C., and Phillips, T.L. (1996a). A Carboniferous insect gall: Insight into early ecologic history of the Holometabola. *Proc. Natl. Acad. Sci. USA* **93**:8470–8474.

Labandeira, C.C., and Phillips, T.L. (1996b). Insect fluid-feeding on Upper Pennsylvanian tree ferns (Palaeodictyoptera, Marattiales) and the early history of the piercing-and-sucking functional feeding group. *Ann. Ent. Soc. Amer.* **89**:157–183.

Labandeira, C.C., and Sepkoski, J.J. (1993). Insect diversity in the fossil record. *Science* **261**:310–315.

Labandeira, C.C., Beall, B.S., and Hueber, F.M. (1988). Early insect diversification: evidence from a Lower Devonian bristletail from Québec. *Science* **242**:913–916.

Labeit, S., and Kolmerer, B. (1995). Titins: giant proteins in charge of muscle ultrastructure and elasticity. *Science* **270**:293–296.

La Greca, M. (1947). Morfologia funzionale dell'articolazione alare degli Ortotteri. *Arch. Zool. Ital.* **32**:271–327.

La Greca, M. (1954). Riduzione e scomparsa delle ali negli Insetti Pterigoti. *Arch. Zool. Ital.* **39**:361–440.

La Greca, M. (1980). Origin and evolution of wings and flight in insects. *Boll. Zool.* **47** (Suppl.):65–82.

La Greca, M. (1981). Origine ed evoluzione dell'articolazione alare degli insetti pterigoti. *Mem. Soc. ent. ital., Genova* **60**:221–226.

Lai-Fook, J. (1967). The structure of developing muscle insertions in insects. *J. Morphol.* **123**:503–528.

Lan, C.E. (1979). The unsteady quasi-vortex-lattice method with applications to animal propulsion. *J. Fluid Mech.* **93**:747–765.

Land, M.F. (1973). Head movement of flies during visually guided flight. *Nature* **243**:299–300.

Land, M.F. (1975). Head movements and fly vision. In *The Compound Eye and Vision of Insects* (ed. G.A. Horridge), pp. 469–489. Oxford: Clarendon Press.

Land, M.F. (1989). Variations in the structure and design of compound eyes. In *Facets of Vision* (ed. D.G. Stavenga and R.C. Hardie), pp. 90–111. Berlin: Springer-Verlag.

Land, M.F. (1992). Visual tracking and pursuit: humans and arthropods compared. *J. Insect Physiol.* **38**:939–951.

Land, M.F. (1993a). The visual control of courtship behavior in the fly *Poecilobothrus nobilitatus. J. Comp. Physiol.* A **173**:595–603.

Land, M.F. (1993b). Chasing and pursuit in the dolichopodid fly *Poecilobothrus nobilitatus. J. Comp. Physiol.* A **173**:605–613.

Land, M.F. (1997). Visual acuity in insects. *Ann. Rev. Ent.* **42**:147–177.

Land, M.F., and Collett, T.S. (1974). Chasing behaviour of houseflies (*Fannia canicularis*). A description and analysis. *J. Comp. Physiol.* **89**:331–357.

Lane, R.P. (1992). The 'new' taxonomy—Does it require new taxonomists or a new understanding? *Bull. Ent. Res.* **82**:437–440.

Langley, P.A. (1970). Post-teneral development of thoracic flight musculature in the tsetse flies *Glossina austeni* and *G. morsitans*. *Ent. Exp. Appl.* **13**: 133–140.

Langmuir, J. (1938). The speed of the deer fly. *Science* **87**:233–234.

Larimer, J.L., and Dudley, R. (1994). Centrifugal force and blood pressure elevation in the wings of flying hummingbirds. *J. theor. Biol.* **168**:233–236.

Larsen, O. (1966). On the morphology and function of the locomotor organs of the Gyrinidae and other Coleoptera. *Opusc. Ent. Suppl.* **30**:1–242.

Larsen, R.S. (1934). Der Flug der Insekten. Eine neue Methode zu dessen Erforschung. *Norsk Ent. Tidsskr.* **3**:306–315.

LaSalle, J., and Gauld, I.D. (1991). Parasitic Hymenoptera and the biodiversity crisis. *Redia* **74**:315–334.

Laurie-Ahlberg, C.C., Barnes, P.T., Curtsinger, J.W., Emigh, T.H., Karlin, B., Morris, R., Norman, R.A., and Wilton, A.N. (1985). Genetic variability of flight metabolism in *Drosophila melanogaster*. II. Relationship between power output and enzyme activity levels. *Genetics* **111**:845–868.

Lawson, F.A., and Chu, J. (1974). Wing coupling in a bark louse: a light and SEM study. *J. Kans. Ent. Soc.* **47**:136–140.

Lee, D.N. (1980). The optic flow field: the foundation of vision. *Phil. Trans. R. Soc. Lond.* B **290**:169–179.

Lee, D.N., Reddish, P.E., and Rand, D.T. (1991). Aerial docking by hummingbirds. *Naturwissenschaften* **78**:526–527.

Leech, R., and Cady, A. (1994). Function shift and the origin of insect flight. *Australian Biol.* **7**:160–168.

Lehane, M.J. (1991). *Biology of Blood-sucking Insects*. London: HarperCollins.

Lehmann, F.-O., and Götz, K.G. (1996). Activation phase ensures kinematic efficacy in flight-steering muscles of *Drosophila melanogaster*. *J. Comp. Physiol.* A **179**:311–322.

Lehmann, F.-O., and Dickinson, M.H. (1997). The changes in power requirements and muscle efficiency during elevated force production in the fruit fly *Drosophila melanogaster*. *J. Exp. Biol.* **200**:1133–1143.

Lehmann, F.-O., and Dickinson, M.H. (1998). The control of wing kinematics and flight forces in fruit flies (*Drosophila* spp.). *J. Exp. Biol.* **201**:385–401.

Lehrer, M. (1996). Small-scale navigation in the honeybee: active acquisition of visual information about the goal. *J. Exp. Biol.* **199**:253–261.

Lemche, H. (1940). The origin of winged insects. *Vidensk Medd. fra Dansk Naturh. Foren.* **104**:127–168.

Lévieux, J. (1982). A comparison of the ground dwelling ant populations between a Guinea savanna and an evergreen rain forest of the Ivory Coast. In *The Biology of Social Insects* (ed. M.D. Breed, C.D. Michener, and H.E. Evans), pp. 48–53. Boulder, Colo.: Westview Press.

Lewis, T., and Taylor, L.R. (1967). *Introduction to Experimental Ecology*. London: Academic Press.

L'Héritier, P., Neefs, Y., and Teissier, G. (1937). Aptérisme des Insectes et sélection naturelle. *C.R. Acad. Sci. (Paris)* **204**:907–909.

Libersat, F. (1992). Modulation of flight by the giant interneurons of the cockroach. *J. Comp. Physiol.* A **170**:379–392.

Libersat, F. (1994). The dorsal giant interneurons mediate evasive behavior in flying cockroaches. *J. Exp. Biol.* **197**:405–411.

Libersat, F., and Camhi, J. (1988). Control of sensory feedback by movement during flight in the cockroach. *J. Exp. Biol.* **136**:483–488.

Libersat, F., Levy, A., and Camhi, J.M. (1989). Multiple feedback loops in the flying cockroach: excitation of the dorsal and inhibition of the ventral giant interneurons. *J. Comp. Physiol.* A **165**:651–668.

Lighthill, J. (1977). Introduction to the scaling of aerial locomotion. In *Scale Effects in Animal Locomotion* (ed. T.J. Pedley), pp. 365–404. London: Academic Press.

Lighthill, J. (1979). A simple fluid-flow model of ground effect on hovering. *J. Fluid Mech.* **93**:781–797.

Lighthill, M.J. (1973). On the Weis-Fogh mechanism of lift generation. *J. Fluid Mech.* **60**:1–17.

Lighthill, M.J. (1978). A note on 'clap and fling' aerodynamics. *J. Exp. Biol.* **73**:279–280.

Lindquist, A.W., Yates, W.W., and Hoffman, R.A. (1951). Studies of the flight habits of three species of flies tagged with radioactive phosphorus. *J. Econ. Ent.* **44**:397–400.

Lindroth, C.H. (1946). Inheritance of wing dimorphism in *Pterostichus anthracinus* Ill. *Hereditas* **32**:37–40.

Linsley, E.G., Eisner, T., and Klots, A.B. (1961). Mimetic assemblages of sibling species of lycid beetles. *Evolution* **15**:15–29.

Liske, E. (1977). The influence of head position on the flight behaviour of the fly, *Calliphora erythrocephala*. *J. Insect Physiol.* **23**:375–379.

Lissaman, P.B.S. (1983). Low-Reynolds-number airfoils. *Ann. Rev. Fluid Mech.* **15**:223–239.

Little, C. (1983). *The Colonisation of Land: Origins and Adaptations of Terrestrial Animals.* Cambridge, U.K.: Cambridge University Press.

Little, C. (1990). *The Terrestrial Invasion: An Ecophysiological Approach to the Origins of Land Animals.* Cambridge, U.K.: Cambridge University Press.

Liu, H., and Kawachi, K. (1998). A numerical study of insect flight. *J. Comput. Phys.* **146**:124–156.

Liu, H., Ellington, C.P., Kawachi, K., Van den Berg, C., and Willmott, A. (1998). A computational fluid dynamic study of hawkmoth hovering. *J. Exp. Biol.* **201**:461–477.

Lombardi, E.C., and Kaplan, D.L. (1993). Preliminary characterization of resilin isolated from the cockroach, *Periplaneta americana*. *Mat. Res. Soc. Symp. Proc.* **292**:1–5.

Lorenz, E.N. (1993). *The Essence of Chaos.* Seattle: University of Washington Press.

Lorez, M. (1995). Neural control of hindleg steering in flight of the locust. *J. Exp. Biol.* **198**:869–875.

Loudon, C. (1988). Development of *Tenebrio molitor* in low oxygen levels. *J. Insect Physiol.* **34**:97–103.

Loudon, C. (1989). Tracheal hypertrophy in mealworms: design and plasticity in oxygen supply systems. *J. Exp. Biol.* **147**:217–235.

Loudon, C., Best, B.A., and Koehl, M.A.R. (1994). When does motion relative to neighboring surfaces alter the flow through arrays of hairs? *J. Exp. Biol.* **193**:233–254.

Louw, G.N., and Hadley, N.F. (1985). Water economy of the honeybee: a stoichiometric accounting. *J. Exp. Zool.* **235**:147–150.

Loxdale, H.D., Hardie, J., Halbert, S., Foottit, R., Kidd, N.A.C., and Carter, C.I. (1993). The relative importance of short- and long-range movement of flying aphids. *Biol. Rev.* **68**:291–311.

Lubbock, H. (1863). On two aquatic Hymenoptera, one of which uses its wings in swimming. *Trans. Linn. Soc. Lond.* **24**:135–142.

Luttges, M.W. (1989). Accomplished insect fliers. *Lect. Notes Eng.* **46**:429–456.

Lutz, F.E. (1927). Wind and the direction of insect flight. *Amer. Mus. Novit.*, no. 291.

Machin, K.E., and Pringle, J.W.S. (1959). The physiology of insect fibrillar flight muscle. II. Mechanical properties of a beetle flight muscle. *Proc. R. Soc. Lond.* B **151**:204–225.

Machin, K.E., and Pringle, J.W.S. (1960). The physiology of insect fibrillar muscle. III. The effect of sinusoidal changes of length on a beetle flight muscle. *Proc. R. Soc. Lond.* B **152**:311–330.

Machin, K.E., Pringle, J.W.S., and Tamasige, M. (1962). The physiology of insect fibrillar muscle. IV. The effect of temperature on a beetle flight muscle. *Proc. R. Soc. Lond.* B **155**:493–499.

Mackey, M.C., and Glass, L. (1977). Oscillation and chaos in physiological control systems. *Science* **197**:287–289.

MacLeod, J., and Donnelly, J. (1963). Dispersal and interspersal of blowfly populations. *J. Anim. Ecol.* **32**:1–32.

Maddison, W.P., and Maddison, D.R. (1992). *MacClade: Analysis of Phylogeny and Character Evolution.* Version 3. Sunderland, Mass: Sinauer Associates.

Madsen, B.M., and Miller, L.A. (1987). Auditory input to motor neurons of the dorsal longitudinal flight muscles in a noctuid moth (*Barathra brassicae* L.). *J. Comp. Physiol.* A **160**:23–31.

Mafra-Neto, A., and Cardé, R.T. (1994). Fine-scale structure of pheromone plumes modulates upwind orientation of flying moths. *Nature* **369**:142–144.

Magnan, A. (1934). *La Locomotion Chez les Animaux. I. Le Vol des Insectes.* Paris: Hermann et Cie.

Magnan, A., and Perrilliat-Botonet, C. (1932). Sur le poids relatif des muscles moteurs des ailes chez les Insectes. *C.R. Acad. Sci., Paris* **195**:559–561.

Magnan, A., and Planiol, A. (1933). Sur l'éxcedent de puissance des insectes. In *Exposés de Morphologie Dynamique et de Mécanique du Mouvement (Actualités Scientifiques et Industrielles #66)* (ed. M.A. Magnan), pp. 1–26. Paris: Hermann et Cie.

Magnan, A., and Sainte-Laguë, A. (1933). Le vol au point fixe. In *Exposés de morphologie Dynamique et de Mécanique du Mouvement (Actualités Scientifiques et Industrielles #60)* (ed. M. A. Magnan), pp. 1–31. Paris: Hermann et Cie.

Maki, T. (1938). Studies of the thoracic musculature of insects. *Mem. Fac. Sci. Agric. Taihoku Imp. Univ.* **24**:1–343.

Malamud, J.G., and Josephson, R.K. (1991). Force-velocity relationships of a locust flight muscle at different times during a twitch contraction. *J. Exp. Biol.* **159**:65–87.

Malamud, J.G., Mizisin, A.P., and Josephson, R.K. (1988). The effects of octopamine on contraction kinetics and power output of a locust flight muscle. *J. Comp. Physiol.* A **162**:827–835.

Mamayev, B.M. (1977). The gravitational hypothesis of the origin of insects. *Ent. Rev., Wash.* **54**:13–17.

Mani, M.S. (1962). *Introduction to High Altitude Entomology.* London: Methuen and Co.

Mani, M.S. (1968). *Ecology and Biogeography of High Altitude Insects.* The Hague: Dr. W. Junk.

Manton, S.M. (1966). The evolution of arthropodan locomotory mechanisms. Part 9. Functional requirements and body design in Symphyla and Pauropoda and the relationships between Myriapoda and pterygote Insects. *J. Linn. Soc. (Zool.)* **46**:103–141.

Manton, S.M. (1977). *The Arthropoda: Habits, Functional Morphology, and Evolution.* Oxford: Clarendon Press.

Marden, J.H. (1987). Maximum lift production during takeoff in flying animals. *J. Exp. Biol.* **130**:235–258.

Marden, J.H. (1989a). Bodybuilding dragonflies: costs and benefits of maximizing flight muscle. *Physiol. Zool.* **62**:505–521.

Marden, J.H. (1989b). Effects of load-lifting constraints on the mating system of a dance fly. *Ecology* **70**: 496–502.

Marden, J.H. (1990). Maximum load-lifting and induced power output of Harris' hawks are general functions of flight muscle mass. *J. Exp. Biol.* **149**:511–514.

Marden, J.H. (1994). From damselflies to pterosaurs: how burst and sustainable flight performance scale with size. *Amer. J. Physiol. (Regulatory Integrative Comp. Physiol.)* **266**:R1077-R1084.

Marden, J.H. (1995). Large-scale changes in thermal sensitivity of flight performance during adult maturation in a dragonfly. *J. Exp. Biol.* **198**:2095–2102.

Marden, J.H., and Chai, P. (1991). Aerial predation and butterfly design: how palatability, mimicry, and the need for evasive flight constrain mass allocation. *Amer. Nat.* **138**:15–36.

Marden, J.H., and Kramer, M.G. (1994). Surface-skimming stoneflies: a possible intermediate stage in insect flight evolution. *Science* **266**:427–430.

Marden, J.H., and Kramer, M.G. (1995). Locomotor performance of insects with rudimentary wings. *Nature* **377**:332–334.

Marden, J.H., and Waage, J.K. (1990). Escalated damselfly territorial contests are energetic wars of attrition. *Anim. Behav.* **39**:954–959.

Marden, J.H., Fitzhugh, G.H., and Wolf, M. (1998). From molecules to mating success: integrative biology of muscle maturation in a dragonfly. *Amer. Zool.* **38**:528–544.

Marden, J.H., Kramer, M.G., and Frisch, J. (1996). Age-related variation in body temperature, thermoregulation and activity in a thermally polymorphic dragonfly. *J. Exp. Biol.* **199**:529–535.

Marden, J.H., Wolf, M.R., and Weber, K.E. (1997). Aerial performance of *Drosophila melanogaster* from populations selected for upwind flight ability. *J. Exp. Biol.* **200**:2747–2755.

Maresca, C., Favier, D., and Rebont, J. (1979). Experiments on an aerofoil at high angle of incidence in longitudinal oscillations. *J. Fluid Mech.* **92**:671–690.

Marey, E.J. (1891). Le vol des insectes étudié par la Photochronographie. *C.R. Acad. Sci. (Paris)* **113**:15–18.

Margaria, R. (1968). Positive and negative work performances and their efficiencies in human locomotion. *Int. Z. Angew. Physiol. Einschl. Arbeitsphysiol.* **25**:339–351.

Marinelli, W. (1929). Über die Bedeutung des Flugvermögens der Tiere. *Biol. Gen.* **5**:110–156.

Markl, H. (1966). Peripheres Nervensystem und Muskulatur in Thorax der Arbeiterin von *Apis mellifica* L., *Formica polyctena* Foerster und *Vespa vulgaris* L. und der Grundplan der Innervierung des Insektenthorax. *Zool. Jb. Abt. Anat. Ontog. Tiere* **83**:107–184.

Marsh, D., Kennedy, J.S., and Ludlow, A.R. (1978). An analysis of anemotactic zigzagging flight in male moths stimulated by pheromone. *Physiol. Ent.* **3**:221–240.

Marshall, G.A.K., and Poulton, E.B. (1902). Five years' observations and experiments (1896–1901) on the bionomics of South African insects, chiefly directed to the investigation of mimicry and warning colours, with a discussion of the results and other subjects suggested by them. *Trans. Ent. Soc. Lond.* **1902**:287–584.

Martens, W.J.M., Niessen, L.W., Rotmans, J., Jetten, T.H., and McMichael, A.J. (1995). Potential impact of global climate change on malaria risk. *Environ. Health Perspect.* **103**:458–464.

Martin, L.J., and Carpenter, P.W. (1977). Flow-visualisation experiments on butterflies in simulated gliding flight. In *Physiology of Movement, Biomechanics* (ed. W. Nachtigall), pp. 307–316. Stuttgart: Gustav Fischer.

Martin, P.H., and Lefebvre, M.G. (1995). Malaria and climate: sensitivity of malaria potential transmission to climate. *Ambio* **24**:200–207.

Martynov, A. (1929). Ecological preconditions for the zoogeography of freshwater benthic animals. *Russk. Zool. Zh.* **9**:3–38. (In Russian.)

Martynov, A. (1935). The caddisflies of the Amur region. *Tr. Zool. Inst. Akad. Nauk. SSSR* **2**:305–395. (In Russian.)

Martynov, A.B. (1923). The interpretation of the wing venation and tracheation of the Odonata and Agnatha. *Rev. Russie Ent.* **18**:145–174. (In Russian.)

Martynov, A.B. (1925). Über zwei Grundtypen der Flügel bei den Insecten und ihre Evolution. *Z. Morphol. Ökol. Tiere* **4**:465–501.

Matheson, R., and Crosby, C.R. (1912). Aquatic Hymenoptera in America. *Ann. Ent. Soc. Amer.* **5**:65–71.

Mathieu-Costello, O., Suarez, R.K., and Hochachka, P.W. (1992). Capillary-to-fiber geometry and mitochondrial density in hummingbird flight muscle. *Respir. Physiol.* **89**:113–132.

Matsuda, R. (1963a). Evolution of thoracic musculature in insects. *Univ. Kans. Sci. Bull.* **44**:509–534.

Matsuda, R. (1963b). Some evolutionary aspects of the insect thorax. *Ann. Rev. Ent.* **8**:59–76.

Matsuda, R. (1970). Morphology and evolution of the insect thorax. *Mem. Ent. Soc. Can.* **76**:1–431.

Matsuda, R. (1979). Morphologie du thorax et des appendices thoraciques des insectes. In *Traité de zoologie. Vol. VIII, Fasc. II* (ed. P.-P. Grassé.), pp. 1–289. Paris: Masson et Cie.

Matsuda, R. (1981). The origin of insect wings (Arthropoda: Insecta). *Int. J. Insect Morphol. & Embryol.* **10**:387–398.

Maxworthy, T. (1979). Experiments on the Weis-Fogh mechanism of lift generation by insects in hovering flight. Part 1. Dynamics of the 'fling'. *J. Exp. Biol.* **93**:47–63.

Maxworthy, T. (1981). The fluid dynamics of insect flight. *Ann. Rev. Fluid Mech.* **13**:329–350.

May, M. (1991). Aerial defense tactics of flying insects. *Amer. Sci.* **79**:316–328.

May, M.L. (1976). Thermoregulation and adaptation to temperature in dragonflies (Odonata: Anisoptera). *Ecol. Monogr.* **46**:1–32.

May, M.L. (1981a). Allometric analysis of body and wing dimensions of male Anisoptera. *Odonatologica* **10**:279–291.

May, M.L. (1981b). Wingstroke frequency of dragonflies (Odonata: Anisoptera) in relation of temperature and body size. *J. Comp. Physiol.* B **144**:229–240.

May, M.L. (1982). Heat exchange and endothermy in Protodonata. *Evolution* **36**:1051–1058.

May, M.L. (1983). Thermoregulation. In *Comprehensive Insect Physiology, Biochemistry and Pharmacology*, vol. 4 (ed. G.A. Kerkut and L.I. Gilbert), pp. 507–552. Oxford: Pergamon Press.

May, M.L. (1991). Dragonfly flight: power requirements at high speed and acceleration. *J. Exp. Biol.* **158**:325–342.

May, M.L. (1995). Dependence of flight behavior and heat production on air temperature in the green darner dragonfly *Anax junius* (Odonata: Aeshnidae). *J. Exp. Biol.* **198**:2385–2392.

May, M.L., and Hoy, R.R. (1990). Leg-induced steering in flying crickets. *J. Exp. Biol.* **151**:485–488.

May, M.L., Brodfuehrer, P.D., and Hoy, R.R. (1988). Kinematic and aerodynamic aspects of ultrasound-induced negative phonotaxis in flying Australian field crickets (*Teleogryllus oceanicus*). *J. Comp. Physiol.* A **164**:243–249.

May, M.L., Wilkin, P.J., Heath, J.E., and Williams, B.A. (1980). Flight performance of the moth, *Manduca sexta*, at variable gravity. *J. Insect Physiol.* **26**:257–265.

May, R.M. (1978). The dynamics and diversity of insect faunas. In *Diversity of Insect Faunas* (ed. L.A. Mound and N. Waloff), pp. 188–204. Oxford: Blackwell Scientific.

May, R.M. (1988). How many species are there on Earth? *Science* **241**:1441–1449.

Maynard Smith, J. (1952). The importance of the nervous system in the evolution of animal flight. *Evolution* **6**:127–129.

Mayr, E. (1963). *Animal Species and Evolution.* Cambridge, Mass.: Harvard University Press.

McArthur, M., and Sohal, R.S. (1982). Relationship between metabolic rate, aging, lipid peroxidation, and age pigment in milkweed bug, *Oncopeltus fasciatus* (Hemiptera). *J. Gerontol.* **37**:268–274.

McCann, F.V., and Boettiger, E.G. (1961). Studies on the flight mechanism of insects. I. The electrophysiology of fibrillar flight muscle. *J. Gen. Physiol.* **45**:125–142.

McCann, G.D., and MacGintie, G.F. (1965). Optomotor response studies of insect vision. *Proc. R. Soc. Lond.* B **163**:369–401.

McCrae, A.W.R. (1975). Clicking in flight by an African fruit-piercing moth, *Achaea obvia* Hampson (Lep., Noctuidae). *Ent. Month. Mag.* **111**:161–164.

McLachlan, A.J. (1986a). Sexual dimorphism in midges: strategies for flight in the rain-pool dweller *Chironomus imicola* (Diptera: Chironomidae). *J. Anim. Ecol.* **55**:261–267.

McLachlan, A.J. (1986b). Survival of the smallest: advantages and costs of small size in flying animals. *Ecol. Ent.* **11**:237–240.

McLachlan, A.J., and Allen, D.F. (1987). Male mating success in Diptera: Advantages of small body size. *Oikos* **48**:11–14.

McLachlan, A.J., and Neems, R.M. (1995). Swarm-based mating systems. In *Insect Reproduction* (ed. S.R. Leather and J. Hardie), pp. 199–214. New York: CRC Press.

McLachlan, A.J., and Neems, R.M. (1996). Flight architecture determined by physical constraints or by natural selection?: the case of the midge *Chironomus plumosus*. *J. Zool., Lond.* **240**:301–308.

McNamara, K.J., and Trewin, N.H. (1993). A euthycarcinoid arthropod from the Silurian of Western Australia. *Paleontology* **36**:319–335.

Mees, A. (1986). Chaos in feedback systems. In *Chaos* (ed. A.V. Holden), pp. 99–110. Princeton, N.J.: Princeton University Press.

Mehta, U.B., and Lavan, Z. (1975). Starting vortex, separation bubbles and stall: a numerical study of laminar unsteady flow around an airfoil. *J. Fluid Mech.* **67**:227–256.

Melin, D. (1941). Contributions to the knowledge of the flight of insects. *Uppsala Univ. Årsskr.* **1**(4):1–247.

Messner, B. (1988). Sind die Insekten primäre oder sekundäre Wasserbewohner? *Dtsch. Ent. Z., N.F.* **35**:355–360.

Metcalfe, N.B., and Ure, S.E. (1995). Diurnal variation in flight performance and hence potential predation risk in small birds. *Proc. R. Soc. Lond.* B **261**:395–400.

Miall, R.C. (1978). The flicker fusion frequency of six laboratory insects, and the response of the compound eye to mains fluorescent 'ripple.' *Physiol. Ent.* **3**:99–106.

Mikkola, K. (1986). Direction of insect migrations in relation to the wind. In *Insect Flight: Dispersal and Migration* (ed. W. Danthanarayana), pp. 152–171. Berlin: Springer-Verlag.

Mill, P.J. (1985). Structure and physiology of the respiratory system. In *Comprehensive Insect Physiology, Biochemistry, and Pharmacology*, vol. 3 (ed. G.A. Kerkut and L.I. Gilbert), pp. 517–593. Oxford: Pergamon Press.

Miller, J.R., and Roelofs, W.L. (1978). Sustained-flight tunnel for measuring insect responses to wind-borne sex pheromones. *J. Chem. Ecol.* **4**:187–198.

Miller, L.A. (1983). How insects detect and avoid bats. In: *Neuroethology and Behavioral Physiology* (ed. F. Huber and H. Markl), pp. 251–266. Berlin: Springer-Verlag.

Miller, L.A., and Olsen, J. (1979). Avoidance behavior in green lacewings. I. Behavior of free flying lacewings to hunting bats and ultrasound. *J. Comp. Physiol.* A **131**:113–120.

Miller, P.L. (1960). Respiration in the desert locust. III. Ventilation and the spiracles during flight. *J. Exp. Biol.* **37**:264–278.

Miller, P.L. (1966). The supply of oxygen to the active flight muscles of some large beetles. *J. Exp. Biol.* **45**:285–304.

Miller, P.L. (1971). The possible stabilising function of the elytra of *Atractocerus brevicornis* (L.) (Lymexylidae: Coleoptera) in flight. *Entomologist* **104**:105–110.

Milton, J.G., Longtin, A., Beuter, A., Mackey, M.C., and Glass, L. (1989). Complex dynamics and bifurcations in neurology. *J. theor. Biol.* **138**:129–147.

Mimura, K. (1970). Integration and analysis of movement information by the visual system of flies. *Nature* **226**:964–966.

Mitchell, J.A., Heffron, J.J.A., and Hepburn, H.R. (1977). Comparison of the enzymatic activities of homologous muscles of flying and flightless beetles. *Comp. Biochem. Physiol.* **57**B:111–116.

Mittelstaedt, H. (1950). Physiologie des Gleichgewichtssinnes bei fliegenden Libellen. *Z. vergl. Physiol.* **32**:422–463.

Mitter, C., Farrell, B., and Wiegmann, B. (1988). The phylogenetic study of adaptive zones: has phytophagy promoted insect diversification? *Amer. Nat.* **132**:107–128.

Miyan, J.A., and Ewing, A.W. (1984). A wing synchronous receptor for the dipteran flight motor. *J. Insect Physiol.* **30**:567–574.

Miyan, J.A., and Ewing, A.W. (1985a). How Diptera move their wings: a reexamination of the wing base articulation and muscle systems concerned with flight. *Phil. Trans. R. Soc. Lond.* B **311**:271–302.

Miyan, J.A., and Ewing, A.W. (1985b). Is the 'click' mechanism of dipteran flight an artefact of CCl_4 anaesthesia? *J. Exp. Biol.* **116**:313–322.

Miyan, J.A., and Ewing, A.W. (1988). Further observations on dipteran flight: details of the mechanism. *J. Exp. Biol.* **136**:229–241.

Mizisin, A.P., and Josephson, R.K. (1987). Mechanical power output of locust flight muscle. *J. Comp. Physiol.* A **160**:413–419.

Mizunami, M. (1994). Information processing in the insect ocellar system: comparative approaches to the evolution of visual processing and neural circuits. *Adv. Insect Physiol.* **25**:151–265.

Möhl, B. (1985a). The role of proprioception in locust flight control. I. Asymmetry and coupling within the time pattern of motor units. *J. Comp. Physiol.* A **156**:93–101.

Möhl, B. (1985b). The role of proprioception in locust flight control. II. Information signalled by forewing stretch receptors during flight. *J. Comp. Physiol.* A **156**:103–116.

Möhl, B. (1985c). The role of proprioception in locust flight control. III. The influence of afferent stimulation of the stretch receptor nerve. *J. Comp. Physiol.* A **156**:281–291.

Möhl, B. (1988). Short-term learning during flight control in *Locusta migratoria*. *J. Comp. Physiol.* A **163**:803–812.

Möhl, B. (1989a). 'Biological noise' and plasticity of sensorimotor pathways in the locust flight system. *J. Comp. Physiol.* A **166**:75–82.

Möhl, B. (1989b). Sense organs and the control of flight. In *Insect Flight* (ed. G.J. Goldsworthy and C.H. Wheeler), pp. 75–97. Boca Raton, Fla.: CRC Press.

Möhl, B. (1991). Motor learning in the locust flight system requires proprioceptive feedback. *Naturwissenschaften* **78**:84–87.

Möhl, B. (1997). Verhaltungsphysiologie der Flugsteuerung bei Wanderheuschrecken. In *BIONA-report 11* (ed. A. Wisser, D. Bilo, A. Kesel, and B. Möhl), pp. 173–200. Stuttgart: Gustav Fischer.

Möhl, B., and Bacon, J. (1983). The tritocerebral commissure giant (TCG) windsensitive interneurone in the locust. II. Directional sensitivity and role in flight stabilisation. *J. Comp. Physiol.* A **150**:453–465.

Möhl, B., and Nachtigall, W. (1978). Proprioceptive input on the locust flight motor revealed by muscle stimulation. *J. Comp. Physiol.* A **128**:57–65.

Möhl, B., and Zarnack, W. (1977). Activity of the direct downstroke flight muscles of *Locusta migratoria* (L.) during steering behaviour in flight. II. Dynamics of the time shift and changes in the burst length. *J. Comp. Physiol.* A **118**:235–247.

Moiseff, A., Pollack, G.S., and Hoy, R.R. (1978). Steering responses of flying crickets to sound and ultrasound: Mate attraction and predator avoidance. *Proc. Natl. Acad. Sci. USA* **75**:4052–4056.

Möllenstädt, W. (1980). Einige Grundzüge der instationären Aerodynamik harmonisch schwingender Tragflügel in inkompressibler, reibungsfreier Strömung. In *Instationäre Effekte an schwingenden Tierflügeln* (ed. W. Nachtigall), pp. 9–34. Wiesbaden: Franz Steiner.

Møller, A.P. (1996). Sexual selection, viability selection, and developmental stability in the domestic fly *Musca domestica*. *Evolution* **50**:746–752.

Møller, A.P., and Swaddle, J.P. (1997). *Asymmetry, Developmental Stability, and Evolution*. Oxford: Oxford University Press.

Molloy, J.E., Kyrtatas, V., Sparrow, J.C., and White, D.C.S. (1987). Kinetics of flight muscles from insects with different wingbeat frequencies. *Nature* **328**:449–451.

Monge-Nájera, J. (1992). Clicking butterflies, *Hamadryas*, of Panama: Their biology and identification. In *Insects of Panama and Mesoamerica: Selected Studies* (ed. D. Quintero and A. Aiello), pp. 567–572. Oxford: Oxford University Press.

Monge-Nájera, J., and Hernández, F. (1991). A morphological search for the sound mechanism of *Hamadryas* butterflies (Lepidoptera: Nymphalidae). *J. Res. Lep.* **30**:196–208.

Moore, A.J. (1990). The evolution of sexual dimorphism by sexual selection: the separate effects of intrasexual selection and intersexual selection. *Evolution* **44**:315–331.

Mora, C.I., Driese, S.G., and Colarusso, L.A. (1996). Middle to late Paleozoic atmospheric CO_2 levels from soil carbonate and organic matter. *Science* **271**:1105–1107.

Moran, N.A. (1994). Adaptation and constraint in the complex life cycles of animals. *Ann. Rev. Ecol. Syst.* **25**:573–600.

Moran, V.C., and Ewer, D.W. (1966). Observations on certain characteristics of the flight motor of sphingid and saturniid moths. *J. Insect Physiol.* **12**:457–463.

Morse, D.H. (1975). Ecological aspects of adaptive radiation in birds. *Biol. Rev.* **50**:167–214.

Mpitsos, G.J., Burton, R.M., Creech, H.C., and Soinila, S.O. (1988). Evidence for chaos in spike trains of neurons that generate rhythmic motor patterns. *Brain Res. Bull.* **21**:529–538.

Müller, F. (1873). Beiträge zur Kenntniss der Termiten. III. Die "Nymphen mit kurzen Flügelscheiden" (Hagen), "nymphes de la deuxième forme" (Lespès). Ein Sultan in seinem Harem. *Jena. Zs. Med. Naturw.* **7**:451–463.

Müller, F. (1875). Beiträge zur Kenntniss der Termiten. IV. Die Larven von *Calotermes rugosus* Hag. *Jena. Zs. Med. Naturw.* **9**:241–264.

Mulloney, B. (1969). Interneurons in the central nervous system of flies and the start of flight. *Z. vergl. Physiol.* **64**:243–253.

Mulloney, B. (1970a). Organization of flight motoneurons of Diptera. *J. Neurophysiol.* **33**:86–95.

Mulloney, B. (1970b). Impulse patterns in the flight motor neurones of *Bombus californicus* and *Oncopeltus fasciatus*. *J. Exp. Biol.* **52**:59–77.

Nachtigall, W. (1964). Zur Aerodynamik des Coleopterenflugs: wirken die Elytren als Tragflügel? *Verh. Dtsch. Zool. Gesellsch. (Kiel)* **58**:319–326.

Nachtigall, W. (1965). Die aerodynamische Funktion der Schmetterlingsschuppen. *Naturwissenschaften* **52**:216–217.

Nachtigall, W. (1966). Die Kinematik der Schlagflügelbewegungen von Dipteren. Methodische und analytische Grundlagen zur Biophysik des Insektenflugs. *Z. vergl. Physiol.* **52**:155–211.

Nachtigall, W. (1967). Aerodynamische Messungen am Tragflügelsystem segelnder Schmetterlinge. *Z. vergl. Physiol.* **54**:210–231.

Nachtigall, W. (1968). Elektrophysiologische und kinematische Untersuchungen über Start und Stop des Flugmotors von Fliegen. *Z. vergl. Physiol.* **61**:1–20.

Nachtigall, W. (1973). *Insects in Flight: A Glimpse behind the Scene in Biophysical Research*. New York: McGraw-Hill.

Nachtigall, W. (1974a). *Biological Mechanisms of Attachment*. New York: Springer-Verlag.

Nachtigall, W. (1974b). Locomotion: mechanics and hydrodynamics of swimming in aquatic insects. In *The Physiology of Insects*, 2d ed., vol. 3 (ed. M. Rockstein), pp. 381–432. New York: Academic Press.

Nachtigall, W. (1977). Die aerodynamische Polare des Tipula-flügels und eine Einrichtung zur halbautomatischen Polarenaufnahme. In *The Physiology of Movement; Biomechanics* (ed. W. Nachtigall), pp. 347–352. Stuttgart: Gustav Fischer.

Nachtigall, W. (1978). Der Startsprung der Stubenfliege *Musca domestica* (Diptera: Muscidae). *Ent. Germ.* **4**:368–373.

Nachtigall, W. (1979a). Rasche Richtungsänderungen und Torsionen schwingender Fliegenflügel und Hypothesen über zugeordnete instationäre Strömungseffekte. *J. Comp. Physiol.* A **133**:351–355.

Nachtigall, W. (1979b). Schiebeflug bei der Schmeißfliege *Calliphora erythrocephala* (Diptera: Calliphoridae). *Ent. Gen.* **5**:255–265.

Nachtigall, W. (1981a). Der Vorderflügel großer Heuschrecken als Luftkrafterzeuger. I. Modellmessungen zur aerodynamischen Wirkung unterschiedlicher Flügelprofile. *J. Comp. Physiol.* A **142**:127–134.

Nachtigall, W. (1981b). Über den Einfluß von geometrischen Flügeländerungen auf die aerodynamische Funktion des Vorderflügels der Wüstenheuschrecke. Eine weiterführende Analyse der Jensenschen Untersuchungen. *Zool. Jb. Abt. Anat. Ontog. Tiere* **106**:1–11.

Nachtigall, W. (1991). Aerodynamic coefficients for hymenopteran bodies. *Naturwissenschaften* **78**:567–569.

Nachtigall, W. (1992). Behavioural and aerodynamic aspects in the searching flight of the scoliid wasp *Scolia flavifrons* (Hymenoptera: Scoliidae). *Ent. Gen.* **17**:1–8.

Nachtigall, W. (1997a). Dipterenflug. In *BIONA-report 11* (ed. A. Wisser, D. Bilo, A. Kesel, and B. Möhl), pp. 115–156. Stuttgart: Gustav Fischer.

Nachtigall, W. (1997b). Flug der Honigbiene. In *BIONA-report 11* (ed. A. Wisser, D. Bilo, A. Kesel, and B. Möhl), pp. 157–172. Stuttgart: Gustav Fischer.

Nachtigall, W., and Hanauer-Thieser, U. (1992). Flight of the honeybee. V. Drag and lift coefficients of the bee's body: implications for flight dynamics. *J. Comp. Physiol.* B **162**:267–277.

Nachtigall, W., and Roth, W. (1983). Correlations between stationary measurable parameters of wing movement and aerodynamic force production in the blowfly (*Calliphora vicina* R.-D.). *J. Comp. Physiol.* A **150**:251–260.

Nachtigall, W., and Wilson, D.M. (1967). Neuro-muscular control of dipteran flight. *J. Exp. Biol.* **47**:77–97.

Nachtigall, W., Hanauer-Thieser, U., and Mörz, M. (1995). Flight of the honey bee. VII: metabolic power versus flight speed relation. *J. Comp. Physiol.* B **165**:484–489.

Nachtigall, W., Widmann, R., and Renner, M. (1971). Über den "ortsfesten" freien Flug von Bienen in einem Saugkanal. *Apidologie* **2**:271–282.

Nachtigall, W., Wisser, A., and Eisinger, D. (1998). Flight of the honey bee. VIII. Functional elements and mechanics of the "flight motor" and the wing joint—one of the most complicated gear-mechanisms in the animal kingdom. *J. Comp. Physiol.* B **168**:323–344.

Nachtigall, W., Rothe, U., Feller, P., and Jungmann, R. (1989). Flight of the honey bee. III. Flight metabolic power calculated from gas analysis, thermoregulation and fuel consumption. *J. Comp. Physiol.* B **158**:729–737.

Nalbach, G. (1989). The gear change mechanism of the blowfly (*Calliphora erythrocephala*) in tethered flight. *J. Comp. Physiol.* A **165**:321–331.

Nalbach, G. (1993). The halteres of the blowfly *Calliphora*. I. Kinematics and dynamics. *J. Comp. Physiol.* A **173**:293–300.

Nalbach, G., and Hengstenberg, R. (1994). The halteres of the blowfly *Calliphora*. II. Three-dimensional organization of compensatory reactions to real and simulated rotations. *J. Comp. Physiol.* A **175**:695–708.

Nason, J.D., Herre, E.A., and Hamrick, J.L. (1998). The breeding structure of a tropical keystone plant resource. *Nature* **391**:685–687.

Neems, R.M., McLachlan, A.J., and Chambers, R. (1990). Body size and lifetime mating success of male midges (Diptera: Chironomidae). *Anim. Behav.* **40**: 648–652.

Neese, V. (1965). Zur Funktion der Augenborsten bei der Honigbiene. *Z. vergl. Physiol.* **49**:543–585.

Nelson, C.R., and Tidwell, W.D. (1987). *Brodioptera stricklani* n.sp. (Megasecoptera: Brodiopteridae), a new fossil insect from the Upper Manning Canyon Shale Formation, Utah (lowermost Namurian B). *Psyche* **94**:309–316.

Neuhaus, W., and Wohlgemuth, R. (1960). Über das Fächeln der Bienen und dessen Verhältnis zum Fliegen. *Z. vergl. Physiol.* **43**:615–641.

Neukirch, A. (1982). Dependence of life span of the honeybee (*Apis mellifera*) upon flight performance and energy consumption. *J. Comp. Physiol.* B **146**: 35–40.

Neumann, L., Möhl, B., and Nachtigall, W. (1982). Quick phase-specific influence of the tegula on the locust flight motor. *Naturwissenschaften* **69**:393–394.

Neville, A.C. (1963). Motor unit distribution of the dorsal longitudinal flight muscles in locusts. *J. Exp. Biol.* **40**:123–136.

Neville, A.C. (1984). Cuticle: organization. In *Biology of the Integument. I. Invertebrates* (ed. J. Bereiter-Hahn, A.G. Matoltsy, and K.S. Richards), pp. 611–625. Berlin: Springer-Verlag.

Neville, A.C. (1993). *Biology of Fibrous Composites: Development beyond the Cell Membrane.* Cambridge, U.K.: Cambridge University Press.

New, T.R. (1974). Structural variation in psocopteran wing-coupling mechanisms. *Int. J. Insect Morphol. & Embryol.* **3**:193–201.

Newman, B.G., Savage, S.B., and Schouella, D. (1977). Model tests on a wing section of an *Aeschna* dragonfly. In *Scale Effects in Animal Locomotion* (ed. T.J. Pedley), pp. 445–477. London: Academic Press.

Newman, D.J.S., and Wootton, R.J. (1986). An approach to the mechanics of pleating in dragonfly wings. *J. Exp. Biol.* **125**:361–372.

Newman, D.J.S., and Wootton, R.J. (1988). The role of the fulcroalar muscle in dragonfly flight. *Odonatologica* **17**:401–408.

Nicolson, S.W., and Louw, G.N. (1982). Simultaneous measurement of evaporative water loss, oxygen consumption, and thoracic temperature during flight in a carpenter bee. *J. Exp. Zool.* **222**:287–296.

Niehaus, M. (1981). Flight and flight control by the antennae in the Small Tortoiseshell (*Aglais urticae* L., Lepidoptera). II. Flight mill and free flight experiments. *J. Comp. Physiol.* A **145**:257–264.

Nikolaev, N.A. (1974). Speeds and types of flights of some cabbage butterflies/ Pieridae and Nymphalidae (Lepidoptera: Pieridae, Nymphalidae). *Vestn. Mosk. Univ.* **4**:17–20. (In Russian.)

Nilssen, A.C. and Anderson, J.R. (1995). Flight capacity of the reindeer warble fly, *Hypoderma tarandi* (L.), and the reindeer nose bot fly, *Cephenemyia trompe* (Modeer) (Diptera: Oestridae). *Can. J. Zool.* **73**:1228–1238.

Norberg, R.Å. (1972a). Evolution of flight in insects. *Zool. Scripta* **1**:247–250.

Norberg, R.Å. (1972b). Flight characteristics of two plume moths, *Alucita pentadactyla* L. and *Orneodes hexadactyla* L. (Microlepidoptera). *Zool. Scripta* **1**:241–246.

Norberg, R.Å. (1972c). The pterostigma of insect wings an inertial regulator of wing pitch. *J. Comp. Physiol.* **81**:9–22.

Norberg, R.Å. (1975). Hovering flight of the dragonfly *Aeschna juncea* L., kinematics and aerodynamics. In *Swimming and Flying in Nature*, vol. 2 (ed. T.Y.-T. Wu, C.J. Brokaw, and C. Brennen), pp. 763–781. New York: Plenum Press.

Norberg, R.Å. (1994). Swallow tail streamer is a mechanical device for self-deflection of tail leading edge, enhancing aerodynamic efficiency and flight manoeuvrability. *Proc. R. Soc. Lond.* B **257**:227–233.

Norberg, U.M. (1976). Aerodynamics, kinematics, and energetics of horizontal flapping flight in the long-eared bat *Plecotus auritus*. *J. Exp. Biol.* **65**:459–470.

Norberg, U.M. (1985). Evolution of vertebrate flight: an aerodynamic model for the transition from gliding to flapping flight. *Amer. Nat.* **126**:303–327.

Norberg, U.M. (1990). *Vertebrate Flight*. Berlin: Springer-Verlag.

Norberg, U.M. (1995). How a long tail and changes in mass and wing shape affect the cost for flight in animals. *Funct. Ecol.* **9**:48–54.

Norberg, U.M., and Rayner, J.M.V. (1987). Ecological morphology and flight in bats (Mammalia: Chiroptera): wing adaptations, flight performance, foraging strategy and echolocation. *Phil. Trans. R. Soc. Lond.* B **316**:335–427.

Novacek, M.J. (1985). Evidence for echolocation in the oldest known bats. *Nature* **315**:140–141.

Novicki, A. (1989a). Rapid postembryonic development of a cricket flight muscle. *J. Exp. Zool.* **250**:253–262.

Novicki, A. (1989b). Control of growth and ultrastructural maturation of a cricket flight muscle. *J. Exp. Zool.* **250**:263–272.

Noyes, J.S. (1978). On the numbers of genera and species of Chalcidoidea (Hymenoptera) in the world. *Ent. Gaz.* **29**:163–164.

Obara, Y. (1975). Mating behavior of the cabbage white butterfly, *Pieris rapae crucivora*. VI. Electrophysiological decision of muscle functions in wing and abdomen movements and muscle output patterns during flight. *J. Comp. Physiol.* A **102**:189–200.

O'Carroll, D.C., Bidwell, N.J., Laughlin, S.B., and Warrant, E.J. (1996). Insect motion detectors matched to visual ecology. *Nature* **382**:63–66.

O'Donnell, P.T., and Bernstein, S.I. (1988). Molecular and ultrastructural defects in a *Drosophila* myosin heavy chain mutant: differential effects on muscle function produced by similar thick filament abnormalities. *J. Cell. Biol.* **107**:2601–2612.

Oehme, H. (1965). Der Flug des Fahnendrongos (*Dicrurus macrocercus*). *J. Ornithol.* **106**:190–203.

Oehmichen, E. (1920). *Nos Maîtres les Oiseaux. Étude sur le Vol Animal et al Récupération de l'Énergie dans les Fluides.* Paris: Dunod.

Oertli, J.J. (1989). Relationship of wing beat frequency and temperature during take-off flight in temperate-zone beetles. *J. Exp. Biol.* **145**:321–328.

O'Gara, B.A., and Drewes, C.D. (1990). Modulation of tension production by octopamine in the metathoracic dorsal longitudinal muscle of the cricket *Teleogryllus oceanicus*. *J. Exp. Biol.* **149**:161–176.

Ohmi, K., Coutanceau, M., Daube, O., and Loc, T.P. (1991). Further experiments on vortex formation around an oscillating and translating airfoil at large incidences. *J. Fluid Mech.* **225**:607–630.

Ohmi, K., Coutanceau, M., Loc, T.P., and Dulieu, A. (1990). Vortex formation around an oscillating and translating airfoil at large incidences. *J. Fluid Mech.* **211**:37–60.

Ohsaki, N. (1995). Preferential predation of female butterflies and the evolution of batesian mimicry. *Nature* **378**:173–175.

Oka, H. (1930). Morphologie und Ökologie von *Clunio pacificus* Edwards. *Zool. Jb. Abt. Syst. Ökol. Geogr. Tiere* **59**:253–280.

Okamoto, M., Yasuda, K., and Azuma, A. (1996). Aerodynamic characteristics of the wings and body of a dragonfly. *J. Exp. Biol.* **199**:281–294.

Oken, L. (1809–1811). *Lehrbuch der Naturphilosophie.* 3 vols. Jena: Fromann.

Okubo, A., Bray, D.J., and Chiang, H.C. (1981). Use of shadows for studying the three-dimensional structure of insect swarms. *Ann. Ent. Soc. Amer.* **74**:48–50.

Okubo, A., Chiang, H.C., and Eggesmeyer, C.C. (1977). Acceleration field of individual midges, *Anarete pritchardi* (Diptera: Cecidomyiidae), within a swarm. *Can. Ent.* **109**:149–156.

Olberg, R.M., and Willis, M.A. (1990). Pheromone-modulated optomotor response in male gypsy moths, *Lymantria dispar* L.: Directionally sensitive visual interneurons in the ventral nerve cord. *J. Comp. Physiol.* A **167**:707–714.

Oliveira, E.G., Srygley, R.B., and Dudley, R. (1998). Do Neotropical migrant butterflies navigate by a solar compass? *J. Exp. Biol.* **201**:3317–3331.

Oliveira, C.R., and Malta, C.P. (1987). Bifurcations in a class of time delay equations. *Phys. Rev.* A **36**:3997–4001.

Olsen, J., and Miller, L.A. (1979). Avoidance behavior in green lacewings. II. Flight muscle activity. *J. Comp. Physiol.* A **131**:121–128.

Olsen, L.F., and Degn, H. (1985). Chaos in biological systems. *Quart. Rev. Biophys.* **18**:165–225.

Oosterveld, W.J., and Greven, A.J. (1975). Flight behaviour of pigeons during weightlessness. *Acta Otolaryngol.* **79**:233–241.

Opler, P.A. (1981). Polymorphic mimicry of polistine wasps by a Neotropical neuropteran. *Biotropica* **13**:165–176.

Orchard, I., Ramirez, J.-M., and Lange, A.B. (1993). A multifunctional role for octopamine in locust flight. *Ann. Rev. Ent.* **38**:227–249.

Orr, W.C., and Sohal, R.S. (1994). Extension of life-span by overexpression of superoxide dismutase and catalase in *Drosophila melanogaster*. *Science* **263**: 1128–1130.

Osborne, M.F.M. (1951). Aerodynamics of flapping flight with application to insects. *J. Exp. Biol.* **28**:221–245.

Ossiannilsson, R. (1950). On the wing-coupling apparatus of the Auchenorrhyncha (Hemiptera Homoptera). *Opus. Ent.* **15**:127–130.

Otero, L.D. (1990). The stridulatory organ in *Hamadryas* (Nymphalidae): Preliminary observations. *J. Lep. Soc.* **44**:285–288.

Otten, E. (1987). Optimal design of vertebrate and insect sarcomeres. *J. Morphol.* **191**:49–62.

Otzen, H., and Kuiper, J.W. (1983). The effect of air velocity on the wingstroke frequency of the blowfly *Calliphora erythrocephala*. *J. Comp. Physiol.* A **153**: 289–297.

Packard, A.S. (1898). *A Text-book of Entomology.* New York: Macmillan.

Paclt, J. (1956). *Biologie der Primär flügellosen Insekten.* Jena: Gustav Fischer.

Paganelli, C.V., Ar, A., Rahn, H., and Wangensteen, O.D. (1975). Diffusion in the gas phase: the effects of ambient pressure and gas composition. *Respir. Physiol.* **25**:247–258.

Panda, J., and Zaman, K.B.M.Q. (1992). Experimental investigation of the flowfield of an oscillating airfoil. *AIAA Paper No. 92–2622*.

Panda, J., and Zaman, K.B.M.Q. (1994). Experimental investigation of the flow field of an oscillating airfoil and estimation of lift from wake surveys. *J. Fluid Mech.* **265**:65–95.

Park, O.W. (1923). Flight studies of the honey bee. *Amer. Bee J.* **63**:71.

Parker, G.G. (1995). Structure and microclimate of forest canopies. In *Forest Canopies* (ed. M. D. Lowman and N.M. Nadkarni), pp. 73–106. San Diego: Academic Press.

Partridge, L., and Barton, N.H. (1993). Optimality, mutation and the evolution of ageing. *Nature* **362**:305–311.

Pashley, D.P., McPheron, B.A., and Zimmer, E.A. (1993). Systematics of holometabolous insect orders based on 18S ribosomal RNA. *Mol. Phylogenet. Evol.* **2**:132–142.

Paulus, H.F. (1975). The compound eyes of apterygote insects. In: *The Compound Eye and Vision of Insects* (ed. G.A. Horridge), pp. 3–19. Oxford: Clarendon Press.

Paulus, H.F. (1979). Eye structure and the monophyly of the Arthropoda. In *Arthropod Phylogeny* (ed. A.P. Gupta), pp. 299–383. New York: Van Nostrand Reinhold.

Pearson, D.L. (1989). What is the adaptive significance of multicomponent defensive repertoires? *Oikos* **54**:251–253.

Pearson, K.G., and Ramirez, J.M. (1990). Influence of input from the forewing stretch receptors on motoneurones in flying locusts. *J. Exp. Biol.* **151**: 317–340.

Pearson, K.G., and Wolf, H. (1987). Comparison of motor patterns in the intact and deafferented flight system of the locust. I. Electromyographic analysis. *J. Comp. Physiol.* A **160**:259–268.

Pearson, K.G., and Wolf, H. (1988). Connections of hindwing tegulae with flight neurones in the locust, *Locusta migratoria. J. Exp. Biol.* **135**:381–409.

Pearson, K.G., Reye, D.N., and Robertson, R.M. (1983). Phase-dependent influences of wing stretch receptors on flight rhythm in the locust. *J. Neurophysiol.* **49**:1168–1181.

Peckham, M., and White, D.C.S. (1991). Mechanical properties of demembranated flight muscle fibres from a dragonfly. *J. Exp. Biol.* **159**:135–147.

Peckham, M., Cripps, R., White, D., and Bullard, B. (1992). Mechanics and protein content of insect flight muscles. *J. Exp. Biol.* **168**:57–76.

Peckham, M., Molloy, J.E., Sparrow, J.C., and White, D.C.S. (1990). Physiological properties of the dorsal longitudinal flight muscle and the tergal depressor of the trochanter muscle of *Drosophila melanogaster. J. Muscle Res. Cell. Mot.* **11**:203–215.

Pedgley, D.S. (1982). *Windborne Pests and Diseases: Meteorology of Airborne Organisms.* Chichester, U.K.: Ellis Horwood Limited.

Pellmyr, O. (1992). Evolution of insect pollination and angiosperm diversification. *Trends Ecol. Evol.* **7**:46–49.

Pener, M.P. (1985). Hormonal effects on flight and migration. In *Comprehensive Insect Physiology, Biochemistry, and Pharmacology,* vol. 8 (ed. G.A. Kerkut and L.I. Gilbert), pp. 491–550. Oxford: Pergamon Press.

Pennycuick, C.J. (1968). Power requirements for horizontal flight in the pigeon *Columba livia. J. Exp. Biol.* **49**:527–555.

Pennycuick, C.J. (1969). The mechanics of bird migration. *Ibis* **111**:525–556.

Pennycuick, C.J. (1975). Mechanics of flight. In *Avian Biology,* vol. 5 (ed. D.S. Farner and J.R. King), pp. 1–75. London: Academic Press.

Pennycuick, C.J. (1978). Fifteen testable predictions about bird flight. *Oikos* **30**:165–176.

Pennycuick, C.J. (1982). The ornithodolite: an instrument for collecting large samples of bird speed measurements. *Phil. Trans. R. Soc. Lond.* B **300**:61–73.

Pennycuick, C.J. (1988). On the reconstruction of pterosaurs and their manner of flight, with notes on vortex wakes. *Biol. Rev.* **63**:299–331.

Pennycuick, C.J. (1989). *Bird Flight Performance.* New York: Oxford University Press.

Pennycuick, C.J., and Rezende, M.A. (1984). The specific power output of aerobic flight muscle, related to the power density of mitochondria. *J. Exp. Biol.* **108**:377–392.

Petersen, B., Lundgren, L., and Wilson, L. (1956). The development of flight capacity in a butterfly. *Behaviour* **10**:324–339.

Petersson, E. (1995). Male load-lifting capacity and mating success in the swarming caddis fly *Athripsodes cinereus. Physiol. Ent.* **20**:66–70.

Petrunkevitch, A. (1956). *Eohelea stridulans,* a striking example of paramorphism in a Baltic amber gnat. *Science* **123**:675.

Pettigrew, J.B. (1872). On the physiology of wings, being an analysis of the movements by which flight is produced in the insect, bat and bird. *Trans. R. Soc. Edinb.* **26**:321–448.

Pettigrew, J.B. (1873). *Animal Locomotion; or, Walking, Swimming, and Flying, with a Dissertation on Aeronautics.* London: H.S. King.

Pettigrew, J.D. (1994). Flying DNA. *Curr. Biol.* **4**:277–280.

Pfaff, M., and Varjú, D. (1991). Mechanisms of visual distance perception in the hawk moth *Macroglossum stellatarum. Zool. Jb. Abt. Allg. Zool. Physiol.* **95**:315–321.

Pfau, H.K. (1973). Fliegt unsere Schmeißfliege mit Gangschaltung? *Naturwissenschaften* **60**:160.

Pfau, H.K. (1977). Zur Morphologie und Funktion des Vorderflügels und Vorderflügelgelenks von *Locusta migratoria* L. *Fortschr. Zool.* **24**:342–345.

Pfau, H.K. (1978). Funktionsanatomische Aspekte des Insektenflugs. *Zool. Jb. Abt. Anat. Ontog. Tiere* **99**:99–108.

Pfau, H.K. (1986). Untersuchungen zur Konstruktion, Funktion und Evolution des Flugapparates der Libellen (Insecta, Odonata). *Tijdschr. Ent.* **129**:35–123.

Pfau, H.K. (1987). Critical comments on a 'novel mechanical model of dipteran flight' (Miyan & Ewing, 1985). *J. Exp. Biol.* **128**:463–468.

Pfau, H.K. (1991). Contributions of functional morphology to the phylogenetic systematics of Odonata. *Adv. Odonatol.* **5**:109–141.

Pfau, H.K., and Honomichl, K. (1979). Die campaniformen Sensillen des Flügels von *Cetonia aurata* L. und *Geotrupes silvaticus* Panz. (Insecta, Coleoptera) in ihrer Beziehung zur Flügelmechanik und Flugfunktion. *Zool. Jb. Abt. Anat. Ontog. Tiere* **102**:583–613.

Pfau, H.K., and Nachtigall, W. (1981). Der Vorderflügel großer Heuschrecken als Luftkrafterzeuger II. Zusammenspiel von Muskeln und Gelenkmechanik bei der Einstellung der Flügelgeometrie. *J. Comp. Physiol.* A **142**:135–140.

Pfau, H.K., Koch, U.T., and Möhl, B. (1989). Temperature dependence and response characteristics of the isolated wing hinge stretch receptor in the locust. *J. Comp. Physiol.* A **165**:247–252.

Pflüger, H.-J., and Elson, R. (1986). The central nervous organization of the motor neurones to a steering muscle in locusts. *J. Exp. Biol.* **120**:403–420.

Phlips, P.J., East, R.A., and Pratt, N.H. (1981). An unsteady lifting line theory of flapping wings with application to the forward flight of birds. *J. Fluid Mech.* **112**:97–125.

Pickard, W.F. (1974). Transition regime diffusion and the structure of the insect tracheolar system. *J. Insect Physiol.* **20**:947–956.

Picker, M.D. (1987). An unusual species of spoon-wing lacewing (Neuroptera: Nemopteridae) from South Africa, with notes on its biology. *Syst. Ent.* **12**:239–248.

Pierce, G. (1948). *The Songs of Insects.* Cambridge, Mass.: Harvard University Press.

Pix, W., Nalbach, G., and Zeil, J. (1993). Strepsipteran forewings are halterelike organs of equilibrium. *Naturwissenschaften* **80**:371–374.

Plateau, F. (1865). Sur la force musculaire des insectes. *Bull. Acad. R. Belg. Cl. Sci.* **34**:732–757.

Plateau, F. (1866). Sur la force musculaire des insectes (deuxième note). *Bull. Acad. R. Belg. Cl. Sci.* **35**:283–308.

Plateau, F. (1872). Recherches expérimentales sur la position du centre de gravité chez les insectes. *Arch. Sci. Soc. Phys. Hist. Nat. Geneve* **43**:5–57.

Poggio, T., and Reichardt, W. (1981). Visual fixation and tracking by flies: mathematical properties of simple control systems. *Biol. Cybern.* **40**:101–112.

Pomeroy, D. (1990). Why fly? The possible benefits for lower mortality. *Biol. J. Linn. Soc.* **40**:53–65.

Pond, C.M. (1972a). Neuromuscular activity and wing movements at the start of flight of *Periplaneta americana* and *Schistocerca gregaria. J. Comp. Physiol.* **78**:192–209.

Pond, C.M. (1972b). The initiation of flight in unrestrained locusts, *Schistocerca gregaria. J. Comp. Physiol.* **80**:163–178.

Pond, C. (1973). Initiation of flight and pre-flight behaviour of anisopterous dragonflies *Aeshna* spp. *J. Insect Physiol.* **19**:2225–2229.

Popov, A.V. (1981). Sound production and hearing in the cicada *Cicadetta sinuatipennis* Osh. (Homoptera, Cicadidae). *J. Comp. Physiol.* A **142**:271–280.

Popov, Yu.A., and Shcherbakov, D.E. (1996). Origin and evolution of the Coleorrhyncha as shown by the fossil record. In *Studies on Hemipteran Phylogeny* (ed. C.W. Schaefer), pp. 9–30. Lanham, Md.: Entomological Society of America.

Poujade, M.G.-A. (1884). Attitudes des Insectes pendant le vol. *Ann. Soc. Ent. Fr.* **4**:197–200.

Powell, J.R., and Dobzhansky, T. (1976). How far do flies fly? *Amer. Sci.* **64**:179–185.

Power, M.E. (1948). The thoraco-abdominal nervous system of *Drosophila melanogaster. J. Comp. Neurol.* **88**:347–410.

Preiss, R. (1987). Motion parallax and figural properties of depth control flight speed in an insect. *Biol. Cybern.* **57**:1–9.

Preiss, R. (1991). Separation of translation and rotation by means of eye-region specialization in flying gypsy moths (Lepidoptera: Lymantriidae). *J. Insect Behav.* **4**:209–219.

Preiss, R. (1992). Set point of retinal velocity of ground images in the control of swarming flight of desert locusts. *J. Comp. Physiol.* A **171**:251–256.

Preiss, R., and Futschek, L. (1985). Flight stabilization by pheromone-enhanced optomotor responses. *Naturwissenschaften* **72**:435–436.

Preiss, R., and Kramer, E. (1983). Stabilization of altitude and speed in tethered flying gypsy moth males: influence of (+) and (-)-disparlure. *Physiol. Ent.* **8**:55–68.

Preiss, R., and Spork, P. (1993). Flight-phase and visual-field related optomotor yaw responses in gregarious locusts during tethered flight. *J. Comp. Physiol.* A **172**:733–740.

Preiss, R., and Spork, P. (1994). Significance of reafferent information on yaw rotation in the visual control of translatory flight maneuvers in locusts. *Naturwissenschaften* **81**:38–40.

Preiss, R., and Spork, P. (1995). How locusts separate pattern flow into its rotatory and translatory components (Orthoptera: Acrididae). *J. Insect Behav.* **8**:763–779.

Pringle, J.W.S. (1948). The gyroscopic mechanism of the halteres of Diptera. *Phil. Trans. R. Soc. Lond.* B **233**:347–384.

Pringle, J.W.S. (1949). The excitation and contraction of the flight muscles of insects. *J. Physiol.* **108**:226–232.

Pringle, J.W.S. (1954). The mechanism of the myogenic rhythm of certain insect striated muscles. *J. Physiol.* **124**:269–291.

Pringle, J.W.S. (1957). *Insect Flight.* Cambridge, U.K.: Cambridge University Press.

Pringle, J.W.S. (1965). Locomotion: flight. In *The Physiology of Insecta*, vol. 2 (ed. M. Rockstein), pp. 283–329. New York: Academic Press.

Pringle, J.W.S. (1967). The contractile mechanism of insect fibrillar muscle. *Progr. Biophys. Mol. Biol.* **17**:1–60.

Pringle, J.W.S. (1968). Comparative physiology of the flight motor. *Adv. Insect Physiol.* **5**:163–227.

Pringle, J.W.S. (1978). Stretch activation of muscle: function and mechanism. *Proc. R. Soc. Lond.* B **201**:107–130.

Pringle, J.W.S. (1981). The evolution of fibrillar muscle in insects. *J. Exp. Biol.* **94**:1–14.

Pringle, J.W.S., and Tregear, R.T. (1969). Mechanical properties of insect fibrillar muscle at large amplitudes of oscillation. *Proc. R. Soc. Lond.* B **174**:33–50.

Pritchard, G., McKee, M.H., Pike, E.M., Scrimgeour, G.J., and Zloty, J. (1993). Did the first insects live in water or in air? *Biol. J. Linn. Soc.* **49**:31–44.

Proctor, M., Yeo, P., and Lack, A. (1996). *The Natural History of Pollination.* Portland, Ore.: Timber Press.

Provine, R.R. (1994). Pre- and postnatal development of wing-flapping and flight in birds: Embryological, comparative and evolutionary perspectives. In *Perception and Motor Control in Birds* (ed. M.N.O. Davies and P.R. Green), pp. 135–159. Berlin: Springer-Verlag.

Puchkova, L.V. (1971). The functions of the wings in the Hemiptera and trends in their specialization. *Ent. Rev., Wash.* **50**:303–309.

Purvis, A., and Rambaut, A. (1995). Comparative analysis by independent contrasts (CAIC): an Apple Macintosh application for analyzing comparative data. *Computer Appl. Biosciences* **11**:247–251.

Quartau, J.A. (1985). On some objections to the paranotal theory on the origin of insect wings. *Bol. Soc. Port. Ent. Supl.* **1**:359–371.

Quartau, J.A. (1986). An overview of the paranotal theory on the origin of the insect wings. *Publ. Inst. Zool. "Dr. Augusto Nobre", Fac. Cienc. Porto* **194**:1–42.

Rabaud, E. (1933). L'interdépendance des ailes des insectes et la capacité de vol. *Bull. Biol. Fr. Belg.* **77**:34–43.

Ragland, S.S., and Sohal, R.S. (1973). Mating behavior, physical activity and aging in the housefly, *Musca domestica. Exp. Geront.* **8**:135–145.

Rainey, R.C. (1974). Biometeorology and insect flight: some aspects of energy exchange. *Ann. Rev. Ent.* **19**:407–439.

Rainey, R.C. (1989). *Migration and Meteorology.* Oxford: Clarendon Press.

Rainey, R.C., Waloff, Z., and Burnett, G.F. (1957). The behaviour of the Red Locust (*Nomadacris septemfasciata* Serville) in relation to the topography, meteorology and vegetation of the Rukwa Rift Valley, Tanganyika. *Anti-Locust Bulletin* **26**:1–96.

Rall, J.A. (1982). Energetics of Ca^{2+} cycling during skeletal muscle contraction. *Fed. Proc.* **41**:155–160.

Ramirez, J.-M., and Orchard, I. (1990). Octopaminergic modulation of the forewing stretch receptor in the locust *Locusta migratoria*. *J. Exp. Biol.* **149**:255–279.

Ramirez, J.M., and Pearson, K.G. (1988). Generation of motor patterns for walking and flight in motoneurons supplying bifunctional muscles in the locust. *J. Neurobiol.* **19**:257–282.

Ramirez, J.M., and Pearson, K.G. (1989). Alteration of the respiratory system at the onset of locust flight. I. Abdominal pumping. *J. Exp. Biol.* **142**:401–424.

Rankin, M.A., and Burchsted, J.C.A. (1992). The cost of migration in insects. *Ann. Rev. Ent.* **37**:533–559.

Rankin, M.A., McAnelly, M.L., and Bodenhamer, J.E. (1986). The oogenesis-flight syndrome revisited. In *Insect Flight: Dispersal and Migration* (ed. W. Danthanarayana), pp. 27–48. Berlin: Springer-Verlag.

Rashevsky, N. (1944). Studies in the physicomathematical theory of organic form. *Bull. Math. Biophys.* **6**:1–59.

Rasnitsyn, A.P. (1981). A modified paranotal theory of insect wing origin. *J. Morphol.* **168**:331–338.

Rasnitsyn, A.P., and Krassilov, V.A. (1996). First find of pollen grains in the gut of Permian insects. *Paleont. J.* **30**:484–490.

Raven, J.A. (1986). Evolution of plant life forms. In *On the Economy of Plant Form and Function* (ed. T.J. Givnish), pp. 421–492. Cambridge, U.K.: Cambridge University Press.

Raw, F. (1956). Origin of winged insects. *Ann. Mag. Nat. Hist.*, ser. 12 **9**:673–685.

Rayner, J.M.V. (1979a). A new approach to animal flight mechanics. *J. Exp. Biol.* **80**:17–54.

Rayner, J.M.V. (1979b). A vortex theory of animal flight. II. The forward flight of birds. *J. Fluid Mech.* **91**:731–763.

Rayner, J.M.V. (1987). Form and function in avian flight. In *Current Ornithology*, vol. 5 (ed. R.F. Johnston), pp. 1–66. New York: Plenum Press.

Rayner, J.M.V. (1990). The mechanics of flight and bird migration performance. In *Bird Migrations* (ed. E. Gwinner), pp. 284–299. Berlin: Springer-Verlag.

Rayner, J.M.V. (1991). On the aerodynamics of animal flight in ground effect. *Phil. Trans. R. Soc. Lond.* B **334**:119–128.

Rayner, J.M.V. (1994). Aerodynamic corrections for the flight of birds and bats in wind tunnels. *J. Zool., Lond.* **234**:537–563.

Rayner, J.M.V. (1995). Flight mechanics and constraints of flight performance. *Isr. J. Zool.* **41**:321–342.

Rayner, J.M.V. (1996). Biomechanical constraints on size in flying vertebrates. *Symp. Zool. Soc. Lond.* **69**:83–109.

Rayner, J.M.V., and Aldridge, H.D.J.N. (1985). Three-dimensional reconstruction of animal flight paths and the turning flight of microchiropteran bats. *J. Exp. Biol.* **118**:247–265.

Rayner, J.M.V., and Thomas, A.L.R. (1991). On the vortex wake of an animal flying in a confined volume. *Phil. Trans. R. Soc. Lond.* B **334**:107–117.

Ready, N.E., and Josephson, R.K. (1982). Flight muscle development in a hemimetabolous insect. *J. Exp. Zool.* **220**:49–56.

Ready, N.E., and Najm, R.E. (1985). Structural and functional development of cricket wing muscles. *J. Exp. Zool.* **233**:35–50.

Reavis, M.A., and Luttges, M.W. (1988). Aerodynamic forces produced by a dragonfly. *AIAA Paper 88-0330*.

Reed, S.C., Williams, C.M., and Chadwick, L.E. (1942). Frequency of wing-beat as a character for separating species, races, and geographic varieties of *Drosophila. Genetics* **27**:349–361.

Reedy, M.K. (1967). Cross-bridges and periods in insect flight muscle. *Amer. Zool.* **7**:465–481.

Rees, C.J.C. (1975a). Aerodynamic properties of an insect wing section and a smooth airfoil compared. *Nature* **258**:141–142.

Rees, C.J.C. (1975b). Form and function in corrugated insect wings. *Nature* **256**:200–203.

Rees, C.J.C. (1983). Microclimate and the flying Hemiptera fauna of a primary lowland rain forest in Sulawesi. In *Tropical Rain Forest: Ecology and Management* (ed. S.L. Sutton, T.C. Whitmore, and A.C. Chadwick), pp. 121–136. Oxford: Blackwell Scientific.

Regal, P.J. (1977). Ecology and evolution of flowering plant dominance. *Science* **196**:622–629.

Reichardt, W. (1973). Musterinduzierte Flugorientierung. Verhaltens-Versuche an der Fliege *Musca domestica. Naturwissenschaften* **60**:122–138.

Reichardt, W., and Poggio, T. (1979). Figure-ground discrimination by relative movement in the visual system of the fly. Part I. Experimental results. *Biol. Cybern.* **35**:81–100.

Reichardt, W., Poggio, T., and Hausen, K. (1983). Figure-ground discrimination by relative movement in the visual system of the fly. Part II. Towards the neural circuitry. *Biol. Cybern.* **46** (suppl.):1–30.

Reichert, H. (1993). Sensory inputs and flight orientation in locusts. *Comp. Biochem. Physiol.* **104**A:647–657.

Reichert, H., and Rowell, C.H.F. (1985). Integration of nonphaselocked exteroceptive information in the control of rhythmic flight in the locust. *J. Neurophysiol.* **53**:1201–1218.

Reichert, H., and Rowell, C.H.F. (1989). Invariance of oscillator interneurone activity during variable motor output by locusts. *J. Exp. Biol.* **141**:231–239.

Reichert, H., Rowell, C.H.F., and Griss, C. (1985). Course correction circuitry translates feature detection into behavioural action in locusts. *Nature* **315**:147–149.

Reid, R.C., Prausnitz, J.M., and Poling, B.E. (1987). *The Properties of Gases and Liquids*. 4th ed. New York: McGraw-Hill.

Remsen, J.V., and Robinson, S.K. (1990). A classification scheme for foraging behavior of birds in terrestrial habitats. In *Avian Foraging: Theory, Methodology, and Applications* (ed. M.L. Morrison, C.J. Ralph, J. Verner, and J.R. Jehl), pp. 144–160. Los Angeles: Cooper Ornithological Society.

Ren, D. (1998). Flower-associated Brachycera flies as fossil evidence for Jurassic angiosperm origins. *Science* **280**:85–88.

Rensch, B. (1948). Histological changes correlated with evolutionary changes of body size. *Evolution* **2**:218–230.

Resh, V.H., and Solem, J.O. (1984). Phylogenetic relationships and evolutionary adaptations of aquatic insects. In *An Introduction to the Aquatic Insects of North America*, 2d ed. (ed. R.W. Merritt and K.W. Cummins), pp. 66–75. Dubuque, Iowa: Kendall/Hunt.

Rheuben, M.B., and Kammer, A.E. (1987). Structure and innervation of the third axillary muscle of *Manduca* relative to its role in turning flight. *J. Exp. Biol.* **131**:373–402.

Richards, O.W., and Davies, R.G. (1977). *Imm's General Textbook of Entomology, Tenth Edition.* London: Chapman and Hall.

Ricklefs, R.E., and Renner, S.S. (1994). Species richness within families of flowering plants. *Evolution* **48**:1619–1636.

Ridgeley, R.S., and Gwynne, J.A. (1989). *A Guide to the Birds of Panama: with Costa Rica, Nicaragua, and Honduras.* Princeton, N.J.: Princeton University Press.

Riek, E.F. (1977). Four-winged Diptera from the Upper Permian of Australia. *Proc. Linn. Soc. NSW* **101**:250–255.

Riley, J.R. (1993). Flying insects in the field. In *Video Techniques in Animal Ecology and Behaviour* (ed. S.D. Wratten), pp. 1–15. London: Chapman and Hall.

Riley, J.R., and Reynolds, D.R. (1986). Orientation at night by high-flying insects. In *Insect Flight: Dispersal and Migration* (ed. W. Danthanarayana), pp. 71–87. Berlin: Springer-Verlag.

Riley, J.R., Downham, M.C.A., and Cooter, R.J. (1997). Comparison of the performance of *Cicadulina* leafhoppers on flight mills with that to be expected in free flight. *Ent. Exp. Appl.* **83**:317–322.

Riley, J.R., Smith, A.D., and Bettany, B.W. (1990). The use of video equipment to record in three dimensions the flight trajectories of *Heliothis armigera* and other moths at night. *Physiol. Ent.* **15**:73–88.

Riley, J.R., Krueger, U., Addison, C.M., and Gewecke, M. (1988). Visual detection of wind-drift by high-flying insects at night: a laboratory study. *J. Comp. Physiol.* A **162**:793–798.

Riley, J.R., Smith, A.D., Reynolds, D.R., Edwards, A.S., Osborne, J.L., Williams, I.H., Carreck, N.L., and Poppy, G.M. (1996). Tracking bees with harmonic radar. *Nature* **379**:29–30.

Riley, J.R., Valeur, P., Smith, A.D., Reynolds, D.R., Poppy, G.M., and Löfstedt, C. (1998). Harmonic radar as a means of tracking the pheromone-finding and pheromone-following flight of male moths. *J. Insect Behav.* **11**:287–296.

Rind, F.C. (1983). The organization of flight motoneurones in the moth, *Manduca sexta*. *J. Exp. Biol.* **102**:239–251.

Ripper, W. (1931). Versuch einer Kritik der Homologiefrage der Arthropodentracheen. *Z. wiss. Zool.* **138**:303–369.

Ritter, W. (1911). The flying apparatus of the blow-fly. *Smithsonian Misc. Collec.* **56**:1–79.

Ritzmann, R.E. (1984). The cockroach escape response. In *Neural Mechanisms of Startle Behavior* (ed. R.C. Eaton), pp. 93–131. New York: Plenum Press.

Ritzmann, R.E., and Fourtner, C.R. (1980). Flight activity initiated via giant interneurons of the cockroach: evidence for bifunctional trigger interneurons. *Science* **210**:443–445.

Ritzmann, R.E., Fourtner, C.R., and Pollack, A.J. (1983). Morphological and physiological identification of motor neurons innervating flight musculature in the cockroach *Periplaneta americana*. *J. Exp. Zool.* **225**:347–356.

Ritzmann, R.E., Pollack, A.J., and Tobias, M.L. (1982). Flight activity mediated by intracellular stimulation of dorsal giant interneurons of the cockroach *Periplaneta americana*. *J. Comp. Physiol.* A **147**:313–322.

Robbins, R.K. (1981). The "false head" hypothesis: predation and wing pattern variation of lycaenid butterflies. *Amer. Nat.* **34**:194–208.

Robert, D. (1988). Visual steering under closed-loop conditions by flying locusts: flexibility of optomotor response and mechanisms of correctional steering. *J. Comp. Physiol.* A **164**:15–24.

Robert, D. (1989). The auditory behaviour of flying locusts. *J. Exp. Biol.* **147**:279–301.

Robert, D., and Rowell, C.H.F. (1992a). Locust flight steering. I. Head movements and the organization of correctional manoeuvres. *J. Comp. Physiol.* A **171**:41–51.

Robert, D., and Rowell, C.H.F. (1992b). Locust flight steering. II. Acoustic avoidance manoeuvres and associated head movements, compared with correctional steering. *J. Comp. Physiol.* A **171**:53–62.

Roberts, S.P., and Harrison, J.F. (1998). Mechanisms of thermoregulation in flying bees. *Amer. Zool.* **38**:492–502.

Roberts, S.P., Harrison, J.F., and Hadley, N.F. (1998). Mechanisms of thermal balance in flying *Centris pallida* (Hymenoptera: Anthophoridae). *J. Exp. Biol.* **201**:2321–2331.

Robertson, M. (1989). Idiosyncratic computational units generating innate motor patterns: neurones and circuits in the locust flight system. In *The Computing Neuron* (ed. R. Durban and C. Miall), pp. 262–277. Reading, Mass.: Addison-Wesley.

Robertson, M., and Olberg, R.M. (1988). A comparison of the activity of flight interneurones in locusts, crickets, dragonflies and mayflies. *Experientia* **44**:735–738.

Robertson, R.M., and Johnson, A.G. (1993a). Collision avoidance of flying locusts: steering torques and behaviour. *J. Exp. Biol.* **183**:35–60.

Robertson, R.M., and Johnson, A.G. (1993b). Retinal image size triggers obstacle avoidance in flying locusts. *Naturwissenschaften* **80**:176–178.

Robertson, R.M., and Pearson, K.G. (1982). A preparation for intracellular analysis of neuronal activity during flight in the locust. *J. Comp. Physiol.* A **146**:311–320.

Robertson, R.M., and Pearson, K.G. (1983). Interneurons in flight system of the locust: distribution, connections and resetting properties. *J. Comp. Neurol.* **215**:33–50.

Robertson, R.M., and Pearson, K.G. (1984). Interneuronal organization in the flight system of the locust. *J. Insect Physiol.* **30**:95–101.

Robertson, R.M., and Pearson, K.G. (1985). Neural circuits in the flight system of the locust. *J. Neurophysiol.* **53**:110–128.

Robertson, R.M., and Reye, D.N. (1992). Wing movements associated with collision-avoidance manoeuvres during flight in the locust *Locusta migratoria. J. Exp. Biol.* **163**:231–258.

Robertson, R.M., Kuhnert, C.T., and Dawson, J.W. (1996). Thermal avoidance during flight in the locust *Locusta migratoria. J. Exp. Biol.* **199**:1383–1393.

Robertson, R.M., Pearson, K.G., and Reichert, H. (1982). Flight interneurons in the locust and the origin of insect wings. *Science* **217**:177–179.

Robinson, M.C., and Luttges, M.W. (1983). Unsteady flow separation and attachment induced by pitching airfoils. *AIAA Paper 83-0131.*

Roeder, K.D. (1951). Movements of the thorax and potential changes in the thoracic muscles of insects during flight. *Biol. Bull.* **100**:95–106.

Roeder, K.D. (1962). The behaviour of free flying moths in the presence of artificial ultrasonic pulses. *Anim. Behav.* **10**:300–304.

Roeder, K.D. (1967). Turning tendency of moths exposed to ultrasound while in stationary flight. *J. Insect Physiol.* **13**:873–888.

Roff, D.A. (1977). Dispersal in dipterans: its costs and benefits. *J. Anim. Ecol.* **46**:443–456.

Roff, D.A. (1986). The evolution of wing dimorphism in insects. *Evolution* **40**:1009–1020.

Roff, D.A. (1989). Exaptation and the evolution of dealation in insects. *J. Evol. Biol.* **2**:109–123.

Roff, D.A. (1990). The evolution of flightlessness in insects. *Ecol. Monogr.* **60**:389–421.

Roff, D.A. (1994a). Habitat persistence and the evolution of wing dimorphism in insects. *Amer. Nat.* **144**:772–798.

Roff, D.A. (1994b). The evolution of flightlessness: is history important? *Evol. Ecol.* **8**:639–657.

Roff, D.A., and Fairbairn, D.J. (1991). Wing dimorphisms and the evolution of migratory polymorphisms among the Insecta. *Amer. Zool.* **31**:243–251.

Rohdendorf, B.B. (1943). On the evolution of flight in insects. *C.R. Acad. Sci. URSS* **40**:170–172.

Rohdendorf, B.B. (1949). Evolution and classification of the flight apparatus of insects. *Trud. Paleontol. Inst. Akad. Nauk SSSR* **16**:1–176. (In Russian.)

Rohdendorf, B.B. (1958/59a). Die Bewegungsorgane der Zweiflügler-Insekten und ihre Entwicklung (I). *Wiss. Z. Humboldt-Univ. Berlin, Math.-Nat. R.* **8**:73–119.

Rohdendorf, B.B. (1958/59b). Die Bewegungsorgane der Zweiflügler-Insekten und ihre Entwicklung (II). *Wiss. Z. Humboldt-Univ. Berlin, Math.-Nat. R.* **8**:269–308.

Rohdendorf, B.B. (1958/59c). Die Bewegungsorgane der Zweiflügler-Insekten und ihre Entwicklung (III). *Wiss. Z. Humboldt-Univ. Berlin, Math.-Nat. R.* **8**:435–454.

Rohdendorf, B.B. (1961). Description of the first winged insect from the Devonian beds of the Timan (Insecta, Pterygota). *Ent. Obozr.* **40**:485–489.

Rohdendorf, B.B. (1970a). A second find of remains of winged Devonian insects. *Ent. Rev., Wash.* **49**:508–509.

Rohdendorf, B.B. (1970b). The importance of insects in the evolution of land vertebrates. *Paleont. J.* **1970**:5–11.

Rohdendorf, B.B. (1972). Devonian Eopterids were not insects but eumalacostracans (Crustaceae). *Ent. Rev.* **51**:58–59.

Rohdendorf, B.B., and Rasnitsyn, A.P. (1980). Historical development of the Class Insecta. *Trud. Paleontol. Inst. Akad. Nauk SSSR* **175**:1–269. (In Russian.)

Roland, J., McKinnon, G., Backhouse, C., and Taylor, P.D. (1996). Even smaller radar tags on insects. *Nature* **381**:120.

Rolfe, W.D. (1980). Early invertebrate terrestrial fossils. In *The Terrestrial Environment and the Origin of Land Vertebrates* (ed. A.L. Panchen), pp. 117–157. Systematics Association Special Volume No. 15. London: Academic Press.

Rolfe, W.D. (1985). Early terrestrial arthropods: a fragmentary record. *Phil. Trans. R. Soc. Lond.* B **309**:207–218.

Ronacher, B., Wolf, H., and Reichert, H. (1988). Locust flight behavior after hemisection of individual thoracic ganglia: evidence for hemiganglionic premotor centers. *J. Comp. Physiol.* A **163**:749–759.

Rose, M.R. (1991). *Evolutionary Biology of Aging.* New York: Oxford University Press.

Rosenberg, N.J., Blad, B.L., and Verma, S.B. (1983). *Microclimate: The Biological Environment.* New York: John Wiley and Sons.

Ross, A.J., and Jarzembowski, E.A. (1993). Arthropoda (Hexapoda; Insecta). In *The Fossil Record 2* (ed. M.J. Benton), pp. 363–426. London: Chapman and Hall.

Ross, M.H. (1964). Pronotal wings in *Blattella germanica* (L.) and their possible evolutionary significance. *Amer. Midl. Nat.* **71**:161–180.

Rothe, U., and Nachtigall, W. (1989). Flight of the honey bee. IV. Respiratory quotients and metabolic rates during sitting, walking and flying. *J. Comp. Physiol.* B **158**:739–749.

Roubik, D.W. (1989). *Ecology and Natural History of Tropical Bees.* Cambridge, U.K.: Cambridge University Press.

Rowan, A.N., and Newsholme, E.A. (1979). Changes in the contents of adenine nucleotides and intermediates of glycolysis and the citric acid cycle in flight muscle of the locust upon flight and their relationship to the control of the cycle. *Biochem. J.* **178**:209–216.

Rowell, C.H.F. (1988). Mechanisms of flight steering in locusts. *Experientia* **44**:389–395.

Rowell, C.H.F. (1989). Descending interneurones of the locust reporting deviation from flight course: what is their role in steering? *J. Exp. Biol.* **146**:177–194.

Rowell, C.H.F., and Pearson, K.G. (1983). Ocellar input to the flight motor system of the locust: structure and function. *J. Exp. Biol.* **103**:265–288.

Rowell, C.H.F., and Reichert, H. (1986). Three descending interneurons reporting deviation from course in the locust. *J. Comp. Physiol.* A **158**:775–794.

Ruck, P. (1958). A comparison of the electrical responses of compound eyes and dorsal ocelli in four insect species. *J. Insect Physiol.* **2**:261–274.

Ruck, P. (1961). Photoreceptor cell response and flicker fusion frequency in the compound eye of the fly, *Lucilia sericata* (Meigen). *Biol. Bull.* **120**:373–383.

Rudolph, R. (1978). Aerodynamic properties of *Libellula quadrimaculata* L. (Anisoptera: Libellulidae), and the flow around smooth and corrugated wing section models during gliding flight. *Odonatologica* **7**:49–58.

Ruffieux, L., Elouard, J.-M., and Sartori, M. (1998). Flightlessness in mayflies and its relevance to hypotheses on the origin of insect flight. *Proc. R. Soc. Lond.* B **265**:2135–2140.

Rumney, G.R. (1968). *Climatology and the World's Climates.* New York: Macmillan.

Rüppell, G. (1985). Kinematic and behavioural aspects of flight of the male banded agrion, *Calopteryx* (*Agrion*) *splendens* L. In *Insect Locomotion* (ed. M. Gewecke and G. Wendler), pp. 195–204. Berlin: Verlag Paul Parey.

Rüppell, G. (1989). Kinematic analysis of symmetrical flight manoeuvres of Odonata. *J. Exp. Biol.* **144**:13–42.

Rüppell, G., and Fincke, O. (1989). *Megaloprepus coerulatus* (Pseudostigmatidae)—Flug- und Fortpflanzungsverhalten. *Publ. Wiss. Fil., Sekt. Biol., IWF Göttingen*, ser. 20, no. 10/E 2976, pp. 1–20.

Rüschkamp, P.F. (1927). Der Flugapparat der Käfer. Vorbedingungen, Ursache und Verlauf seiner Rückbildung. *Zoologica, Stuttg.* **75**:1–88.

Russenberger, H., and Russenberger, M. (1959/1960). Bau und Wirkungsweise des Flugapparates von Libellen mit besonderer Berücksichtigung von *Aeschna cyanea*. *Mitt. Naturforsch. Gesellsch. Schaffhausen* **26**:1–88.

Rutten, M.G. (1966). Geologic data on atmospheric history. *Palaeogeogr. Palaeoclimatol. Palaeoecol.* **2**:47–57.

Ryazanova, G.I. (1966). Comparative aspects of dragonfly flight. *Zh. obsch. biol.* **27**:349–359. (In Russian.)

Sabelis, M.W., and Schippers, P. (1984). Variable wind directions and anemotactic strategies of searching for an odour plume. *Oecologia* **63**:225–228.

Sacktor, B. (1965). Energetics and respiratory metabolism of muscular contraction. In *The Physiology of Insecta*, vol. 2 (ed. M. Rockstein), pp. 483–580. New York: Academic Press.

Saharon, D., and Luttges, M. (1987). Three-dimensional flow produced by a pitching-plunging model dragonfly wing. *AIAA Paper 87-0121.*

Saharon, D., and Luttges, M.W. (1988). Visualization of unsteady separated flow produced by mechanically driven dragonfly wing kinematics model. *AIAA Paper 88-0569.*

Sakami, Y., and Akimoto, S.-I. (1997). Wing shape of gelechiid moths and its functions: analysis by elliptic Fourier transformation. *Ann. Ent. Soc. Amer.* **90**:447–452.

Samways, M.J. (1994). 'Sailing' on the water surface by adult male *Enallagma nigridorsum* Selys (Zygoptera: Coenagrionidae). *Odonatologica* **23**:175–178.

Samways, M.J. (1996). Skimming and insect evolution. *Trends Ecol. Evol.* **11**:471.

Sandeman, D.C. (1980). Angular acceleration, compensatory head movements and the halteres of flies (*Lucilia serricata*). *J. Comp. Physiol.* A **136**:361–367.

Sander, K. (1957). Bau und Funktion des Sprungapparates von *Pyrilla perpusilla* Walker (Homoptera-Fulgoroidae). *Zool. Jb. Abt. Anat. Ontog. Tiere* **25**:383–388.

Sargent, W.D. (1937). The internal thoracic skeleton of the dragonflies (Odonata; Suborder Anisoptera). *Ann. Ent. Soc. Amer.* **30**:81–95.

Sarpkaya, T. (1996). Unsteady flows. In *Handbook of Fluid Dynamics and Fluid Machinery. Volume I: Fundamentals of Fluid Dynamics* (ed. J.A. Schetz and A.E. Fuhs), pp. 697–732. New York: John Wiley and Sons.

Sato, M., and Azuma, A. (1997). The flight performance of a damselfly *Ceriagrion melanurum* Selys. *J. Exp. Biol.* **200**:1765–1779.

Sattler, K. (1991). A review of wing reduction in Lepidoptera. *Bull. Brit. Mus. Nat. Hist. (Ent.)* **60**:243–288.

Savage, S.B., Newman, B.G., and Wong, D.T.-M. (1979). The role of vortices and unsteady effects during the hovering flight of dragonflies. *J. Exp. Biol.* **83**:59–77.

Sayter, H.J. (1965). The determination of flight performance of insects and birds and the associated wind structure of the atmosphere. *Anim. Behav.* **13**:337–341.

Schaefer, C.W. (ed.) (1996). *Studies on Hemipteran Phylogeny.* Lanham, Md.: Entomological Society of America.

Schaefer, G.W. (1976). Radar observations of insect flight. In *Insect Flight* (ed. R.C. Rainey), pp. 157–197. Oxford: Blackwell Scientific.

Schelver, F.J. (1802). Entomologische Beobachtungen, Versuche und Muthmaßungen über den Flug und das Gesumme einiger zweiflüglichter Insecten und insbesondere über die Schwingkölbchen und Schüppchen unter den Flügeln derselben. *Archiv für Zoologie und Zootomie* **2**:210–218.

Scherenstein, J. (1968). Zum Kurvenflug des Tapfauenauges. *Naturwissenschaften* **55**:43–44.

Schidlowski, M. (1971). Probleme der atmosphärischen Evolution im Präkambrium. *Geol. Rundsch.* **60**:1351–1384.

Schmidt, E. (1938). Bewegung und Wirkungsweise von Insektenflügeln. *Ent. Rundsch.* **55**:605–606.

Schmidt, E. (1939). Bewegung und Wirkungsweise von Insektenflügeln. (Schluß). *Ent. Rundsch.* **56**:61–64.

Schmidt, J., and Zarnack, W. (1987). The motor pattern of locusts during visually induced rolling in long-term flight. *Biol. Cybern.* **56**:397–410.

Schmidt-Nielsen, K. (1972). Locomotion: energy cost of swimming, flying, and running. *Science* **177**:222–228.

Schmidt-Nielsen, K. (1984). *Scaling: Why Is Animal Size So Important?* Cambridge, U.K.: Cambridge University Press.

Schmitt, J.B. (1962). The comparative anatomy of the insect nervous system. *Ann. Rev. Ent.* **7**:137–156.

Schneider, G. (1953). Die Halteren der Schmeißfliege (*Calliphora*) als Sinnesorgane und als mechanische Flugstabilisatoren. *Z. vergl. Physiol.* **35**:416–458.

Schneider, G. (1965). Vergleichende Untersuchungen zur Steuerung der Fluggeschwindigkeit bei *Calliphora vicina* Rob.-Desvoidy (Diptera). *Z. wiss. Zool.* **173**:114–173.

Schneider, P. (1975a). Die Flugtypen der Käfer (Coleoptera). *Ent. Germ.* **1**:222–231.

Schneider, P. (1975b). Zum Faltungsmechanismus der Hinterflügel einiger Blatthornkäfer. *Ent. Germ.* **1**:232–248.

Schneider, P. (1978a). Die Flug- und Faltungstypen der Käfer (Coleoptera). *Zool. Jb. Abt. Anat. Ontog. Tiere* **99**:174–210.

Schneider, P. (1978b). Eine geniale Erfindung der Nature: Flugmotor der Käfer. *Umschau* **22**:683–688.

Schneider, P. (1980a). Beiträge zur Flugbiologie der Käfer. 4. Körpertemperatur, Flugverhalten und Flügelschlagfrequenz. *Zool. Anz., Jena* **205**:1–19.

Schneider, P. (1980b). Beiträge zur Flugbiologie der Käfer. 5. Kinematik der Alae und vertikale Richtungsänderungen. *Zool. Anz., Jena* **205**:188–198.

Schneider, P. (1981a). Beiträge zur Flugbiologie der Käfer. 1. Einstellung der Flügelschlagfrequenz im fixierten und freien Flug. *Zool. Jb. Abt. Allg. Zool. Physiol.* **85**:126–145.

Schneider, P. (1981b). Beiträge zur Flugbiologie der Käfer. 3. Start- und Flugmanöver des Maikäfers *Melolontha melolontha* L. *Zool. Jb. Abt. Allg. Zool. Physiol.* **85**:146–163.

Schneider, P. (1981c). Flugmanöver der Käfer. *Mitt. Dtsch. Gesellsch. Allg. Angew. Ent.* **3**:259–263.

Schneider, P. (1982). Beiträge zur Flugbiologie der Käfer. 2. Steuerung der Flügelschlagamplitude und des Kurvenfluges. *Zool. Jb. Abt. Allg. Zool. Physiol.* **86**:371–399.

Schneider, P. (1987). Mechanik des Auf- und Abschlages der Hinterflügel bei Käfern (*Coleoptera*). *Zool. Anz., Jena* **218**:25–32.

Schneider, P., and Hermes, M. (1976). Die Bedeutung der Elytren bei Vertretern des *Melolontha*- Flugtyps (Coleoptera). *J. Comp. Physiol.* A **106**:39–49.

Schneider, P., and Hermes, M. (1977). Aerodynamische Eigenschaften der Käferelytren. *Zool. Jb. Abt. Allg. Zool. Physiol.* **81**:273–280.

Schneider, P., and Krämer, B. (1974). Die Steuerung des Fluges beim Sandlaufkäfer (*Cicindela*) und beim Maikäfer (*Melolontha*). *J. Comp. Physiol.* **91**:377–386.

Schneider, P., and Meurer, I. (1975). Die mittelbar-indirekte Bewegungen der Elytren beim Nashornkäfer *Oryctes boas* Fabr. (Coleoptera). *Zool. Jb. Abt. Allg. Zool. Physiol.* **79**:297–310.

Schneider, P., and Schill, R. (1978). Der Gleitdoppelmechanismus bei vierflügeligen Insekten mit asynchronem Flugmotor. *Zool. Jb. Abt. Allg. Zool. Physiol.* **82**:365–382.

Schnitzler, H.-U., Menne, D., Kober, R., and Heblich, K. (1983). The acoustical image of fluttering insects in echolocating bats. In *Neuroethology and Behavioral Physiology* (ed. F. Huber and H. Markl), pp. 235–250. Berlin: Springer-Verlag.

Schoof, H.F., and Silverly, R.E. (1954). Multiple release studies on the dispersion of *Musca domestica* at Phoenix, Arizona. *J. Econ. Ent.* **47**:830–838.

Schouest, L.P., Anderson, M., and Miller, T.A. (1986). The ultrastructure and physiology of the tergotrochanteral depressor muscle of the housefly, *Musca domestica. J. Exp. Zool.* **239**:147–158.

Schram, F.R., and Emerson, M.J. (1991). Arthropod pattern theory: a new approach to arthropod phylogeny. *Mem. Queensl. Mus.* **31**:1–18.

Schreck, S.J., and Luttges, M.W. (1988). Unsteady separated flow structure: extended K range and oscillations through zero pitch angle. *AIAA Paper 88-0325.*

Schreiner, H.R. (1966). Interaction of inert gases with molecular cell processes. *Biometeorology* **2**:339–349.

Schreiner, H.R. (1968). General biological effects of the helium-xenon series of elements. *Fed. Proc.* **27**:872–878.

Schuchmann, K.-L. (1979). Metabolism of flying hummingbirds. *Ibis* **121**:85–86.

Schuppe, H., and Hengstenberg, R. (1993). Optical properties of the ocelli of *Calliphora erythrocephala* and their role in the dorsal light response. *J. Comp. Physiol* A **173**:143–149.

Schwanwitsch, B.N. (1943). Subdivision of Insecta Pterygota into subordinate groups. *Nature* **1943**:727–728.

Schwanwitsch, B.N. (1946). On the interrelations of the orders of Insecta as dependent on the origin of flight. *Zool. Zh.* **25**:529–542. (In Russian.)

Schwanwitsch, B.N. (1958). Alary musculature as a basis of the system of pterygote insects. *Proc. 10th Int. Congr. Ent.* **1**:605–610.

Schwenne, T., and Zarnack, W. (1987). Movements of the hindwings of *Locusta migratoria*, measured with miniature coils. *J. Comp. Physiol.* A **160**:657–666.

Scott, A.C., and Taylor, T.N. (1983). Plant/animal interactions during the Upper Carboniferous. *Bot. Rev.* **49**:259–307.

Scott, A.C., Chaloner, W.G., and Paterson, S. (1985). Evidence of pteridophyte-arthropod interactions in the fossil record. *Proc. R. Soc. Edinb.* B **86**:133–140.

Scott, A.C., Stephenson, J., and Chaloner, W.G. (1992). Interaction and coevolution of plants and arthropods during the Paleozoic and Mesozoic. *Phil. Trans. R. Soc. Lond.* B **335**:129–165.

Scruton, C. (1941). Some experimental determinations of the apparent additional mass effect for an aerofoil and for flat plates. *Rep. Brit. Aero. Counc.,* no. 1931, pp. 1–9.

Scudder, S.H. (1876). The mode in which cockroaches and earwigs fold their wings. *Amer. Nat.* **10**:521–529.

Séguy, A. (1973). L'aile des insectes. In *Traité de Zoologie, Tome VIII, Fascicule I* (ed. P.P. Grassé), pp. 595–766. Paris: Masson et Cie.

Séguy, E. (1950). *La Biologie des Diptères.* Paris: P. Lechevalier.

Séguy, E. (1959). Introduction a l'étude morphologique de l'aile des insectes. *Mem. Mus. Natl. Hist. Nat. Ser. A (Paris)* **21**:1–248.

Send, W. (1992). The mean power of forces and moments in unsteady aerodynamics. *Z. angew. Math. Mech.* **72**:113–132.

Sepkoski, J.J., and Hulver, M.L. (1985). An atlas of Phanerozoic clade diversity diagrams. In *Phanerozoic Diversity Patterns* (ed. J.W. Valentine), pp. 11–39. Princeton, N.J.: Princeton University Press.

Sharov, A.G. (1966). *Basic Arthropodan Stock.* Oxford: Pergamon Press.

Shear, W. (1991). The early development of terrestrial ecosystems. *Science* **351**:283–289.

Shear, W.A., and Kukalová-Peck, J. (1990). The ecology of Paleozoic terrestrial arthropods: the fossil evidence. *Can. J. Zool.* **68**:1807–1834.

Shear, W.A., Gensel, P.G., and Jeram, A.J. (1996). Fossils of large terrestrial arthropods from the Lower Devonian of Canada. *Nature* **384**:555–557.

Shear, W.A., Grierson, J.D., Rolfe, W.D.I., Smith, E.L., and Norton, R.A. (1984). Early land animals in North America: evidence from Devonian age arthropods from Gilboa, New York. *Science* **224**:492–494.

Shelly, T.E. (1982). Comparative foraging behavior of light- versus shade-seeking adult damselflies in a lowland Neotropical forest. *Physiol. Zool.* **55**:335–343.

Shiga, S., Kogawauchi, S., Yasuyama, K., and Yamaguchi, T. (1991). Flight behaviour and selective degeneration of flight muscles in the adult cricket (*Gryllus bimaculatus*). *J. Exp. Biol.* **155**:661–667.

Shigenaga, M.K., Hagen, T.M., and Ames, B.G. (1994). Oxidative damage and mitochondrial decay in aging. *Proc. Natl. Acad. Sci. USA* **91**:10771–10778.

Shubin, N., Tabin, C., and Carroll, S. (1997). Fossils, genes and the evolution of animal limbs. *Nature* **388**:639–648.

Sicilia, S., and Smith, D.A. (1991). Theory of asynchronous oscillations in loaded insect flight muscle. *Math. Biosci.* **106**:159–201.

Siedlecki, M. (1917). Spadochronowy lot niektórych owadów.—Der Fallschirmflug einiger Insekten. *Bull. Int. Acad. Pol. Sci. Lett.* B **1917**:230–256.

Siemann, E., Tilman, D., and Haarstad, J. (1996). Insect species diversity, abundance and body size relationships. *Nature* **380**:704–706.

Silberglied, R.E. (1984). Visual communication and sexual selection among butterflies. In *The Biology of Butterflies* (ed. R.I. Vane-Wright and P.R. Ackery), pp. 207–223. London: Academic Press.

Silberglied, R.E., and Eisner, T.E. (1969). Mimicry of Hymenoptera by beetles with unconventional flight. *Science* **163**:486–488.

Simmons, P. (1977a). The neuronal control of locust flight. I. Anatomy. *J. Exp. Biol.* **71**:123–140.

Simmons, P. (1977b). The neuronal control of dragonfly flight. II. Physiology. *J. Exp. Biol.* **71**:141–155.

Sinervo, B., Doughty, P., Huey, R.B., and Zamudio, K. (1992). Allometric engineering: a causal analysis of natural selection on offspring size. *Science* **258**:1927–1930.

Sirovich, L., and Karlsson, S. (1997). Turbulent drag reduction by passive mechanisms. *Nature* **388**:753–755.

Smart, J. (1971). Palaeoecological factors affecting the origin of winged insects. In *Proceedings of the XIIIth International Congress of Entomology, vol.1 (Moscow, 1968)*, pp. 304–306. Leningrad: Nauka.

Smart, J., and Hughes, N.F. (1972). The insect and the plant: progressive palaeoecological integration. In *Insect/Plant Relationships* (ed. H.F. van Emden), pp. 143–155. New York: John Wiley and Sons.

Smith, A.H. (1992). Centrifuges: their development and use in gravitational biology. *Amer. Soc. Grav. Space Biol. Bull.* **5**:33–41.

Smith, D.S. (1961a). The organization of the flight muscle in a dragonfly, *Aeshna* sp. (Odonata). *J. Biophysic. Biochem. Cytol.* **11**:119–144.

Smith, D.S. (1961b). The structure of insect fibrillar flight muscle. A study made with special reference to the membrane systems of the fiber. *J. Biophysic. Biochem. Cytol.* **10(4)** (suppl.), pp. 123–158.

Smith, D.S. (1962). Cytological studies on some insect muscles (with special reference to the sarcoplasmic reticulum.). *Rev. Can. Biol.* **21**:279–301.

Smith, D.S. (1964). The structure and development of flightless Coleoptera: a light and electron microscopic study of the wings, thoracic exoskeleton and rudimentary flight musculature. *J. Morphol.* **114**:107–184.

Smith, D.S. (1965). The organization of flight muscle in an aphid, *Megoura viciae* (Homoptera) with a discussion on the structure of synchronous and asynchronous striated muscle fibers. *J. Cell. Biol.* **27**:379–393.

Smith, D.S. (1966a). The organization of flight muscle fibers in the Odonata. *J. Cell Biol.* **28**:109–126.

Smith, D.S. (1966b). The organisation and function of the sarcoplasmic reticulum and T-system of muscle cells. *Progr. Biophys. Molec. Biol.* **16**:107–143.

Smith, D.S. (1983). 100 Hz remains upper limit of synchronous muscle contraction—an anomaly resolved. *Nature* **303**:539–540.

Smith, D.S. (1984). The structure of insect muscles. In *Insect Ultrastructure*, vol. 2 (ed. R.C. King and H. Akai), pp. 111–150. New York: Academic Press.

Smith, D.S., and Kathirithamby, J. (1984). Atypical 'fibrillar' flight muscle in Strepsiptera. *Tissue & Cell* **16**:929–940.

Smith, M.J.C. (1996). Simulating moth wing aerodynamics: towards the development of flapping-wing technology. *AIAA J.* **34**:1348–1355.

Smith, M.J.C., Wilkin, P.J., and Williams, M.H. (1996). The advantages of an unsteady panel method in modelling the aerodynamic forces on rigid flapping wings. *J. Exp. Biol.* **199**:1073–1083.

Smola, U. (1970). Untersuchung zur Topographie, Mechanik und Strömungsmechanik der Sinneshaare auf dem Kopf der Wanderheuschrecke *Locusta migratoria. Z. vergl. Physiol.* **67**:382–402.

Snodgrass, R.E. (1909). The thorax of insects and the articulation of wings. *Proc. U.S. Natl. Mus.* **36**:511–595.

Snodgrass, R.E. (1910). The thorax of the Hymenoptera. *Proc. U.S. Natl. Mus.* **39**:37–91.

Snodgrass, R.E. (1927). Morphology and mechanism of the insect thorax. *Smithsonian Misc. Collec.* **80**:1–108.

Snodgrass, R.E. (1930). How insects fly. *Ann. Rep. Smithsonian Inst. 1929*, pp. 383–420.

Snodgrass, R.E. (1935). *Principles of Insect Morphology.* New York: McGraw-Hill.

Snodgrass, R.E. (1958). Evolution of arthropod mechanisms. *Smithsonian Misc. Collec.* **138**:1–77.

Sohal, R.S. (1986). Aging in insects. In *Comprehensive Insect Physiology, Biochemistry and Pharmacology*, vol. 10 (ed. G.A. Kerkut and L.I. Gilbert), pp. 595–631. Oxford: Pergamon Press.

Sohal, R.S., and Weindruch, R. (1996). Oxidative stress, caloric restriction, and aging. *Science* **273**:59–63.

Sombati, S., and Hoyle, G. (1984). Generation of specific behaviour in a locust by local release into neuropil of the natural neuromodulator octopamine. *J. Neurobiol.* **15**:481–506.

Somps, C., and Luttges, M.W. (1985). Dragonfly flight: novel uses of unsteady separated flows. *Science* **228**:1326–1329.

Sorensen, J.T., Campbell, B.C., Gill, R.J., and Steffen-Campbell, J.D. (1995). Non-monophyly of Auchenorrhyncha ("Homoptera") based on 18S rDNA phylogeny: eco-evolutionary and cladistic implications within pre-Heteropterodea Hemiptera (S.L.) and a proposal for new monophyletic suborders. *Pan-Pacif. Ent.* **71**:31–60.

Sörensson, M. (1997). Morphological and taxonomical novelties in the world's smallest beetles, and the first Old World record of Nanosellini (Coleoptera: Ptiliidae). *Syst. Ent.* **22**:257–283.

Sotavalta, O. (1941). Some studies on the flying tones of insects and the determination of the frequency of the wing strokes. *Ann. Ent. Fenn.* **7**:32–52.

Sotavalta, O. (1947). The flight-tone (wing-stroke frequency) of insects. *Acta Ent. Fenn.* **4**:1–117.

Sotavalta, O. (1952a). Flight-tone and wing-stroke frequency of insects and the dynamics of insect flight. *Nature* **170**:1057–1060.

Sotavalta, O. (1952b). The essential factor regulating the wing-stroke frequency of insects in wing mutilation and loading experiments and in experiments at subatmospheric pressure. *Ann. Zool. Soc. Vanamo* **15**:1–67.

Sotavalta, O. (1953). Recordings of the high wing-stroke and thoracic vibration frequency in some midges. *Biol. Bull.* **104**:439–444.

Sotavalta, O. (1954). The effect of wing inertia on the wing-stroke frequency of moths, dragonflies and cockroach. *Ann. Ent. Fenn.* **20**:93–101.

Sotavalta, O., and Laulajainen, E. (1961). On the sugar consumption of the drone fly (*Eristalis tenax* L.) in flight experiments. *Ann. Acad. Sci. Fenn. A IV* **53**:1–25.

Spangler, H. (1988). Hearing in tiger beetles (Cicindelidae). *Physiol. Ent.* **13**: 447–452.

Sparrow, C.T. (1980). Bifurcation and chaotic behaviour in simple feedback systems. *J. theor. Biol.* **83**:93–105.

Sparrow, J., Drummond, D., Peckham, M., Hennessey, E., and White, D. (1991). Protein engineering and the study of muscle contraction in *Drosophila* flight muscles. *J. Cell Sci.* (suppl.) **14**:73–78.

Spedding, G.R. (1992). The aerodynamics of flight. In *Mechanics of Animal Locomotion* (ed. R.M. Alexander), pp. 51–111. Berlin: Springer-Verlag.

Spedding, G.R. (1993). On the significance of unsteady effects in the aerodynamic performance of flying animals. *Contemp. Math.* **141**:401–419.

Spedding, G.R., and DeLaurier, J.D. (1996). Animal and ornithopter flight. In *Handbook of Fluid Dynamics and Fluid Machinery. Volume III: Applications of Fluid Dynamics* (ed. J.A. Schetz and A.E. Fuhs), pp. 1951–1967. New York: John Wiley and Sons.

Spedding, G.R., and Maxworthy, T. (1986). The generation of circulation and lift in a rigid two-dimensional fling. *J. Fluid Mech.* **165**:247–272.

Spork, P., and Preiss, R. (1993). Control of flight by means of lateral visual stimuli in gregarious desert locusts, *Schistocerca gregaria. Physiol. Ent.* **18**:195–203.

Spork, P., and Preiss, R. (1994). Adjustment of flight speed of gregarious desert locusts (Orthoptera: Acrididae) flying side by side. *J. Insect Behav.* **7**:217–232.

Squire, J.M. (1992). Muscle filament lattices and stretch-activation: the match-mismatch model reassessed. *J. Muscle Res. Cell Mot.* **13**:183–189.

Srere, P.A. (1985). Organisation of proteins within the mitochondrion. In: *Organised Multienzyme Systems* (ed. G.R. Welch), pp. 1–66. New York: Academic Press.

Srinivasan, M.V., and Bernard, G.D. (1977). The pursuit response of the housefly and its interaction with the optomotor response. *J. Comp. Physiol.* A **115**:101–117.

Srinivasan, M.V., and Davey, M. (1995). Strategies for active camouflage of motion. *Proc. R. Soc. Lond.* B **259**:19–25.

Srinivasan, M.V., and Lehrer, M. (1984). Temporal acuity of honeybee vision: behavioural studies using moving stimuli. *J. Comp. Physiol.* A **155**:297–312.

Srinivasan, M.V., Zhang, S.W., and Bidwell, N.J. (1997). Visually mediated odometry in honeybees. *J. Exp. Biol.* **200**:2513–2522.

Srinivasan, M.V., Lehrer, M., Kirchner, W.H., and Zhang, S.W. (1991). Range perception through apparent image speed in freely flying honeybees. *Vis. Neurosci.* **6**:519–535.

Srinivasan, M.V., Zhang, S.W., Lehrer, M., and Collett, T.S. (1996). Honeybee navigation *en route* to the goal: visual flight control and odometry. *J. Exp. Biol.* **199**:237–244.

Srygley, R.B. (1994a). Locomotor mimicry in butterflies? The associations of positions of centres of mass among groups of mimetic, unprofitable prey. *Phil. Trans. R. Soc. Lond.* B **343**:145–155.

Srygley, R.B. (1994b). Shivering and its cost during reproductive behaviour in Neotropical owl butterflies, *Caligo* and *Opsiphanes* (Nymphalidae: Brassolinae). *Anim. Behav.* **47**:23–32.

Srygley, R.B. (1999). Locomotor mimicry in *Heliconius* butterflies: Contrast analyses of flight morphology and kinematics. *Phil. Trans. R. Soc. Lond.* B **354**:203–214.

Srygley, R.B., and Chai, P. (1990a). Flight morphology of Neotropical butterflies: palatability and distribution of mass to the thorax and abdomen. *Oecologia* **84**:491–499.

Srygley, R.B., and Chai, P. (1990b). Predation and the elevation of thoracic temperature in brightly colored Neotropical butterflies. *Amer. Nat.* **135**:766–787.

Srygley, R.B., and Dudley, R. (1993). Correlations of the position of center of body mass with butterfly escape tactics. *J. Exp. Biol.* **174**:155–166.

Srygley, R.B., and Kingsolver, J.G. (1998). Red-wing blackbird reproductive behaviour and the palatability, flight performance, and morphology of temperate pierid butterflies (*Colias, Pieris*, and *Pontia*). *Biol. J. Linn. Soc.* **64**:41–55.

Srygley, R.B., Oliveira, E.G., and Dudley, R. (1996). Wind drift compensation, flyways, and conservation of diurnal, migrant Neotropical Lepidoptera. *Proc. R. Soc. Lond.* B **263**:1351–1357.

444 REFERENCES

Stalker, H.D. (1980). Chromosome studies in wild populations of *Drosophila melanogaster*. II. Relationships of inversion frequencies to latitude, season, wing-loading and flight activity. *Genetics* **95**:211–223.

Stalker, H.D., and Carson, H.L. (1948). An altitudinal transect of *Drosophila robusta* Sturtevant. *Evolution* **2**:295–305.

Stamhuis, E.J., and Videler, J.J. (1995). Quantitative flow analysis around aquatic animals using laser sheet particle image velocimetry. *J. Exp. Biol.* **198**:283–294.

Stange, G. (1981). The ocellar component of flight equilibrium control in dragonflies. *J. Comp. Physiol.* A **141**:335–347.

Stange, G., and Howard, J. (1979). An ocellar dorsal light response in a dragonfly. *J. Exp. Biol.* **83**:351–355.

Staples, J.F., and Suarez, R.K. (1997). Honeybee flight muscle phosphoglucose isomerase: matching enzyme capacities to flux requirements at a near-equilibrium reaction. *J. Exp. Biol.* **200**:1247–1254.

Starmer, W.T., and Wolf, L.L. (1989). Causes of variation in wing loading among *Drosophila* species. *Biol. J. Linn. Soc.* **37**:247–261.

Stavenga, D.G., Schwering, P.B.W., and Tinbergen, J. (1993). A three-compartment model describing temperature changes in tethered flying blowflies. *J. Exp. Biol.* **185**:325–333.

Steele, J.E. (1981). The role of carbohydrate metabolism in physiological function. In *Energy Metabolism in Insects* (ed. R.G.H. Downer), pp. 101–133. New York: Plenum Press.

Steiger, G.J., and Rüegg, J.C. (1969). Energetics and "efficiency" in the isolated contractile machinery of an insect fibrillar muscle at various frequencies of operation. *Pflügers Arch.* **307**:1–21.

Steiner, G. (1953). Zur Duftorientierung fliegender Insekten. *Naturwissenschaften* **40**:514–515.

Stellwaag, F. (1914a). Der Flugapparat der Lamellicornier. *Z. wiss. Zool.* **108**: 359–429.

Stellwaag, F. (1914b). Die Alula der Käfer. *Dtsch. Ent. Z.* **1914**:419–434.

Stellwaag, F. (1916). Wie steuern die Insekten während des Fluges? *Biol. Zentralbl.* **36**:30–43.

Stern, D.L., and Dudley, R. (1991). Wing buzzing by male orchid bees, *Eulaema meriana* (Hymenoptera: Apidae). *J. Kans. Ent. Soc.* **64**:88–94.

Stevens, B.L., and Lewis, F.L. (1992). *Aircraft Control and Simulation*. New York: John Wiley and Sons.

Stevenson, E. (1968). The carnitine-independent oxidation of palmitate plus malate by moth flight-muscle mitochondria. *J. Biochem.* **110**:105–110.

Stevenson, P.A. (1997). Reflex activation of locust flight motoneurones by proprioceptors responsive to muscle contractions. *J. Comp. Physiol.* A **180**:91–98.

Stevenson, P.A., and Kutsch, W. (1987). A reconsideration of the central pattern generator concept for locust flight. *J. Comp. Physiol.* A **161**:115–129.

Stevenson, P.A., and Kutsch, W. (1988). Demonstration of functional connectivity of the flight motor system in all stages of the locust. *J. Comp. Physiol.* A **162**:247–259.

Stevenson, P.A., and Meuser, S. (1997). Octopaminergic innervation and modulation of a locust flight steering muscle. *J. Exp. Biol.* **200**:633–642.

Stevenson, R.D., and Josephson, R.K. (1990). Effects of operating frequency and temperature on mechanical power output from moth flight muscle. *J. Exp. Biol.* **149**:61–78.

Stevenson, R.D., Corbo, K., Baca, L.B., and Le, Q.D. (1995). Cage size and flight speed of the tobacco hawkmoth *Manduca sexta*. *J. Exp. Biol.* **198**:1665–1672.

Stewart, W.N. (1983). *Paleobotany and the Evolution of Plants.* Cambridge, U.K.: Cambridge University Press.

Stiles, F.G. (1995). Behavioral, ecological and morphological correlates of foraging for arthropods by the hummingbirds of a tropical wet forest. *Condor* **97**:853–878.

Stokes, D.R. (1987). Insect muscles innervated by single motoneurons: structural and biochemical features. *Amer. Zool.* **27**:1001–1010.

Stone, G.N., and Willmer, P.G. (1989). Warm-up rates and body temperatures in bees: the importance of body size, thermal regime and phylogeny. *J. Exp. Biol.* **147**:303–328.

Stone, J.V., Mordue, W., Batley, K.E., and Morris, H.R. (1976). Structure of locust adipokinetic hormone, a neurohormone that regulates lipid utilization during flight. *Nature* **263**:207–211.

Stork, N.E. (1988). Insect diversity: facts, fiction and speculation. *Biol. J. Linn. Soc.* **35**:321–337.

Størmer, L., Petrunkevitch, A., and Hedgpeth, J. (eds.) (1955). *Treatise on Invertebrate Paleontology. Part P, Arthropoda 2.* Lawrence: University of Kansas Press.

Strausfeld, N.J., and Gronenberg, W. (1990). Descending neurons supplying the neck and flight motor of Diptera: organization and neuroanatomical relationships with visual pathways. *J. Comp. Neurol.* **302**:954–972.

Strauss, R., and Heisenberg, M. (1993). A higher control center of locomotor behavior in the *Drosophila* brain. *J. Neurosci.* **13**:1852–1861.

Stride, G.O. (1958). The application of a Bernoulli equation to problems of insect respiration. *Proc. Xth Int. Congr. Ent.* **2**:335–336.

Strong, D.R., Lawton, J.H., and Southwood, R. (1984). *Insects on Plants.* Oxford: Blackwell Scientific.

Suarez, R.K. (1992). Hummingbird flight: Sustaining the highest mass-specific rates among vertebrates. *Experientia* **48**:565–570.

Suarez, R.K. (1998). Oxygen and the upper limits to animal design and performance. *J. Exp. Biol.* **201**:1065–1072.

Suarez, R.K., and Moyes, C.D. (1992). Mitochondrial respiration in locust flight muscles. *J. Exp. Zool.* **263**:351–355.

Suarez, R.K., Lighton, J.R.B., Brown, G.S., and Mathieu-Costello, O. (1991). Mitochondrial respiration in hummingbird flight muscles. *Proc. Natl. Acad. Sci. USA* **88**:4870–4873.

Suarez, R.K., Lighton, J.R.B., Joos, B., Roberts, S.P., and Harrison, J.F. (1996). Energy metabolism, enzymatic flux capacities, and metabolic flux rates in flying honeybees. *Proc. Natl. Acad. Sci. USA* **93**:12616–12620.

Suarez, R.K., Staples, J.F., Lighton, J.R.B., and West, T.G. (1997). Relationships between enzymatic flux capacities and metabolic flux rates: Nonequilibrium reactions in muscle glycolysis. *Proc. Natl. Acad. Sci. USA* **94**:7065–7069.

Šulc, K. (1927). Das Tracheensystem von *Lepisma* (Thysanura) und Phylogenie der Pterygogenea. *Acta Soc. Sci. Nat. Morav.* **4**:1–108.

Sullivan, R.T. (1981). Insect swarming and mating. *Fla. Ent.* **64**:44–65.

Sunada, S., Kawachi, K., Watanabe, I., and Azuma, A. (1993a). Fundamental analysis of three-dimensional 'near fling.' *J. Exp. Biol.* **183**:217–248.

Sunada, S., Kawachi, K., Watanabe, I., and Azuma, A. (1993b). Performance of a butterfly in take-off flight. *J. Exp. Biol.* **183**:249–277.

Sutton, S.L. (1983). The spatial distribution of flying insects in tropical rain forests. In *Tropical Rain Forest: Ecology and Management* (ed. S.L. Sutton, T.C. Whitmore, and A.C. Chadwick), pp. 77–91. Oxford: Blackwell Scientific.

Sutton, S.L., and Hudson, P.J. (1980). The vertical distribution of small flying insects in the lowland rain forest of Zaire. *Zool. J. Linn. Soc.* **68**:111–123.

Suzuki, K., Shimoyama, I., and Miura, H. (1994). Insect-model based micro-robot with elastic hinges. *J. Microelectromech. Syst.* **3**:4–9.

Swihart, S.L., and Baust, J.G. (1965). A technique for the recording of bioelectric potentials from free-flying insects (Lepidoptera: *Heliconius erato*). *Zoologica* **50**:255–258.

Tanaka, S. (1991). De-alation and its influences on egg production and flight muscle histolysis in a cricket (*Velarifictorus parvus*) that undergoes inter-reproductive migration. *J. Insect Physiol.* **37**:517–523.

Tanaka, S., and Suzuki, Y. (1998). Physiological trade-offs between reproduction, flight capability and longevity in a wing-dimorphic cricket, *Modicogryllus confirmatus*. *J. Insect Physiol.* **44**:121–129.

Tannert, W. (1958). Die Flügelgelenkung bei Odonaten. *Dtsch. Ent. Z., N.F.* **5**:394–495.

Tanouye, M.A., and Wyman, R.J. (1980). Motor outputs of giant nerve fiber in *Drosophila*. *J. Neurophysiol.* **44**:405–421.

Tanouye, M.A., and Wyman, R.J. (1981). Inhibition between flight motoneurons in *Drosophila*. *J. Comp. Physiol.* A **144**:345–356.

Tappan, H. (1974). Molecular oxygen and evolution. In *Molecular Oxygen in Biology* (ed. O. Hayaishi), pp. 81–135. Amsterdam: North-Holland.

Tauber, M.J., Tauber, C.A., and Masaki, S. (1986). *Seasonal Adaptations of Insects*. New York: Oxford University Press.

Taylor, C.P. (1981a). Contribution of compound eyes and ocelli to steering of locusts in flight. I. Behavioural analysis. *J. Exp. Biol.* **93**:1–18.

Taylor, C.P. (1981b). Contribution of compound eyes and ocelli to steering of locusts in flight. II. Timing changes in flight motor units. *J. Exp. Biol.* **93**:19–31.

Taylor, L.R. (1958). Aphid dispersal and diurnal periodicity. *Proc. Linn. Soc. Lond.* **169**:67–73.

Taylor, L.R. (1974). Insect migration, flight periodicity, and the boundary layer. *J. Anim. Ecol.* **43**:225–238.

Taylor, L.R., Brown, E.S., and Littlewood, S.C. (1979). The effect of size on the height of flight of migrant moths. *Bull. Ent. Res.* **69**:605–609.

Taylor, P. (1978). Radioisotopes as metabolic labels for *Glossina* (Diptera, Glossinidae). II. The excretion of ^{137}Cs under field conditions as a means of estimating energy utilisation, activity and temperature regulation. *Bull. Ent. Res.* **68**:331–340.

Taylor, P.D., and Merriam, G. (1995). Wing morphology of a forest damselfly is related to landscape structure. *Oikos* **73**:43–48.

Taylor, T.N., and Millay, M.A. (1979). Pollination biology and reproduction in early seed plants. *Rev. Palaeobot. Palynol.* **27**:329–355.

Taylor, T.N., and Scott, A.C. (1983). Interactions of plants and animals during the Carboniferous. *BioScience* **33**:488–493.

Termier, M. (1970a). Consequence des capacités d'accéleration en vol des Dipteres syrphides et bombylides. *C.R. Acad. Sci. Paris (D)* **271**:2361–2363.

Termier, M. (1970b). Essai d'interpretation du vol des Coléoptères. *C.R. Acad. Sci. Paris (D)* **270**:1157–1160.

Termier, M., Devillers, J.-F., and Diep, G.B. (1971). Étude de la force instantanée de sustentation des Coléoptères. *C.R. Acad. Sci. Paris (A)* **273**:1327–1330.

Thiele, H.-U. (1977). *Carabid Beetles in Their Environment: A Study on Habitat Selection by Adaptations in Physiology and Behaviour.* Berlin: Springer-Verlag.

Thom, A., and Swart, P. (1940). The forces on an aerofoil at very low speeds. *J. R. Aero. Soc.* **44**:761–770.

Thomas, A.A.G., Ludlow, A.R., and Kennedy, J.S. (1977). Sinking speeds of falling and flying *Aphis fabae* Scopoli. *Ecol. Ent.* **2**:315–326.

Thomas, A.L.R., and Norberg, R.Å. (1996). Skimming the surface—the origin of flight in insects? *Trends Ecol. Evol.* **11**:187–188.

Thomas, J.G. (1952). A comparison of the pterothoracic skeleton and flight muscles of male and female *Lamarckiana* species (Orthoptera, Acrididae). *Proc. R. Ent. Soc. Lond. (A)* **27**:1–12.

Thomas, J.G. (1953). A comparison of the flight muscles of Acrididae with different wing development. *Proc. R. Ent. Soc. Lond. (A)* **28**:47–56.

Thomas, N., and Thornhill, R.A. (1995a). A theory of tension fluctuations due to muscle cross-bridges. *Proc. R. Soc. Lond.* B **259**:235–242.

Thomas, N., and Thornhill, R.A. (1995b). Negative viscosity and nonlinear elasticity of muscle cross-bridges. *Chaos, Solitons & Fractals.* **5**:393–406.

Thompson, D.A.W. (1917). *On Growth and Form.* Cambridge, U.K.: University Press.

Thompson, O.E., and Pinker, R.T. (1975). Wind and temperature profile characteristics in a tropical evergreen forest in Thailand. *Tellus* **27**:562–573.

Thornhill, R., and Alcock, J. (1983). *The Evolution of Insect Mating Systems.* Cambridge, Mass.: Harvard University Press.

Thorson, J., and White, D.C.S. (1969). Distributed representations for actin-myosin interaction in the oscillatory contraction of muscle. *Biophys. J.* **9**:360–390.

Thorson, J., and White, D.C.S. (1983). Role of cross-bridge distortion in the small-signal mechanical dynamics of insect and rabbit striated muscle. *J. Physiol.* **343**:59–84.

Thüring, D.A. (1986). Variability of motor output during flight steering in locusts. *J. Comp. Physiol.* A **158**:653–664.

Tiegs, O.W. (1949). The problem of the origin of insects. *Aust. N.Z. Assoc. Adv. Sci. Congr. Pap.* **1949**D:47–56.

Tiegs, O.W. (1955). The flight muscles of insects—Their anatomy and histology; with some observations on the structure of striated muscle in general. *Phil. Trans. R. Soc. Lond.* B **238**:221–348.

Tietze, F. (1963a). Untersuchungen über die Beziehungen zwischen Flügelreduktion und Ausbildung des Metathorax bei Carabiden unter besonderer Berücksichtigung der Flugmuskulatur. *Beitr. Ent.* **13**:88–167.

Tietze, F. (1963b). Zur Flügelausbildung und Flugfähigkeit von *Clivina fossor* L. (Carabidae). *Dtsch. Ent. Z.* **10**:175–179.

Tillyard, R.J. (1918). The panorpoid complex. Part i. The wing-coupling apparatus, with special reference to the Lepidoptera. *Proc. Linn. Soc. NSW* **43**:286–319.

Tohtong, R., Yamashita, H., Graham, M., Haeberle, J., Simcox, A., and Maughan, D. (1995). Impairment of muscle function caused by mutations of phosphorylation sites in myosin regulatory light chain. *Nature* **374**:650–653.

Tokunaga, M. (1932). Morphological and biological studies on a new chironomid fly, *Pontomyia pacifica*, from Japan. Part I. Morphology and taxonomy. *Mem. Coll. Agric. Kyoto Imp. Univ.* **19**:1–56.

Toms, R.B. (1984). Were the first insects terrestrial or aquatic? *S. Afr. J. Sci.* **80**:319–323.

Toms, R.B. (1986). Evolution of insect wings: Ensiferan (Orthoptera) wings used only for communication. *S. Afr. J. Sci.* **82**:477–479.

Townsend, C. (1927). On the *Cephenemyia* flight mechanism and the daylight-day circuit of the Earth by flight. *J. N.Y. Ent. Soc.* **35**:245–252.

Townsend, C. (1939). Speed of *Cephenemyia*. *J. N.Y. Ent. Soc.* **47**:43–46.

Townsend, C. (1942). *Manual of Myiology. Part XII. General Consideration of the Oestromuscaria.* São Paulo: Charles Townsend and Filhos.

Treat, A.E. (1955). The response to sound in certain Lepidoptera. *Ann. Ent. Soc. Amer.* **48**:272–284.

Tregear, R.T. (1983). Physiology of insect flight muscle. In *Handbook of Physiology. Section 10: Skeletal Muscle* (ed. L.D. Peachey, R.H. Adrian, and S.R. Geiger), pp. 487–506. Bethesda, Md.: American Physiological Society.

Tregear, R.T. (ed.) (1977). *Insect Flight Muscle.* Amsterdam: North-Holland.

Tregear, R.T., Townes, E., Gabriel, J., and Ellington, C. (1993). Inferences concerning crossbridges from work on insect muscle. In *Mechanism of Myofilament Sliding in Muscle Contraction* (ed. H. Sugi and G.H. Pollack), pp. 557–564. New York: Plenum Press.

Triantafyllou, M.S., and Triantafyllou, G.S. (1991). Wake mechanics for thrust generation in oscillating foils. *Phys. Fluids* A **3**:2835–2837.

Triantafyllou, G.S., Triantafyllou, M.S., and Grosenbaugh, M.A. (1993). Optimal thrust development in oscillating foils with application to fish propulsion. *J. Fluids Struct.* **7**:205–224.

Trimarchi, J.R., and Schneiderman, A.M. (1993). Giant fiber activation of an intrinsic muscle in the mesothoracic leg of *Drosophila melanogaster*. *J. Exp. Biol.* **177**:149–167.

Trimarchi, J.R., and Schneiderman, A.M. (1995a). Different neural pathways coordinate *Drosophila* flight initiations evoked by visual and olfactory stimuli. *J. Exp. Biol.* **198**:1099–1104.

Trimarchi, J.R., and Schneiderman, A.M. (1995b). Flight initiations in *Drosophila melanogaster* are mediated by several distinct motor patterns. *J. Comp. Physiol.* A **176**:355–364.

Trimarchi, J.R., and Schneiderman, A.M. (1995c). Initiation of flight in the unrestrained fly, *Drosophila melanogaster*. *J. Zool., Lond.* **235**:211–222.

Trinick, J. (1991). Elastic filaments and giant proteins in muscle. *Curr. Opin. Cell Biol.* **3**:112–118.

Trombitas, K., and Tigyi-Sebes, A. (1979). The continuity of thick filaments between sarcomeres in honey bee flight muscle. *Nature* **281**:319–320.

Trueman, J.W.H. (1990). Comment—Evolution of insect wings: a limb exite plus endite model. *Can. J. Zool.* **68**:1333–1335.

Tsuji, J.S., Kingsolver, J.G., and Watt, W.B. (1986). Thermal physiological ecology of *Colias* butterflies in flight. *Oecologia* **69**:161–170.

Tu, M.S., and Dickinson, M.H. (1994). Modulation of negative work output from a steering muscle of the blowfly *Calliphora vicina*. *J. Exp. Biol.* **192**:207–224.

Tu, M.S., and Dickinson, M.H. (1996). The control of wing kinematics by two steering muscles of the blowfly (*Calliphora vicina*). *J. Comp. Physiol.* A **178**:813–830.

Tucker, V.A. (1975). The energetic cost of moving about. *Amer. Sci.* **63**:413–419.

Tucker, V.A. (1991). Stereoscopic views of three-dimensional rectangular flight paths in descending African White-backed Vultures (*Gyps africanus*). *Auk* **108**:1–6.

Turillazzi, S. (1983). Patrolling behavior in males of *Parischnogaster nigricans serrei* (Du Buysson) and *P. mellyi* (Saussure) (Hymenoptera: Stenogastrinae). *Atti Accad. Naz. Lincei. Cl. Sci. Fis., Mat. Nat., Rend.* **72**:153–157.

Tuttle, M.D., and Stevenson, D. (1982). Growth and survival of bats. In *Ecology of Bats* (ed. T. H. Kunz), pp. 105–150. New York: Plenum Press.

Tyrer, N.M., and Altman, J.S. (1974). Motor and sensory flight neurones in a locust demonstrated using cobalt chloride. *J. Comp. Neurol.* **157**:117–138.

Ulrich, W. (1930). Die Strepsipteren-Männchen als Insekten mit Halteren an Stelle der Vorderflügel. *Z. Morphol. Ökol. Tiere* **17**:552–624.

Unwin, D.M., and Corbet, S.A. (1984). Wingbeat frequency, temperature and body size in bees and flies. *Physiol. Ent.* **9**:115–121.

Unwin, D.M., and Ellington, C.P. (1979). An optical tachometer for measurement of the wing-beat frequency of free-flying insects. *J. Exp. Biol.* **82**:377–378.

Usherwood, P.N.R. (ed.) (1975). *Insect Muscle*. London: Academic Press.

Van den Berg, C., and Ellington, C.P. (1997a). The three-dimensional leading-edge vortex of a 'hovering' model hawkmoth. *Phil. Trans. R. Soc. Lond.* B **352**:329–340.

Van den Berg, C., and Ellington, C.P. (1997b). The vortex wake of a 'hovering' model hawkmoth. *Phil. Trans. R. Soc. Lond.* B **352**:317–328.

Vanderplank, F.L. (1950). Air-speed/wing-tip speed ratios of insect flight. *Nature* **165**:806–807.

Van Dyck, H., Matthysen, E., and Dhondt, A.A. (1997). Mate-locating strategies are related to relative body length and wing colour in the speckled wood butterfly *Pararge aegeria*. *Ecol. Ent.* **22**:116–120.

Van Gheluwe, B. (1978). Computerized three-dimensional cinematography for any arbitrary camera setup. In *Biomechanics VI*, pp. 343–348. Baltimore: University Park Press.

Van Handel, E., and Nayar, J.K. (1972). Turnover of diglycerides during flight and rest in the moth *Spodoptera frugiperda*. *Insect Biochem.* **2**:8–12.

Van Leeuwen, J.L. (1991). Optimum power output and structural design of sarcomeres. *J. theor. Biol.* **149**:229–256.

Vater, G. (1962). Vergleichende Untersuchungen über die Morphologie des Nervensystems der Dipteren. *Z. wiss. Zool.* **167**:137–196.

Vermeij, G.J. (1982). Unsuccessful predation and evolution. *Amer. Nat.* **120**:701–720.

Vermeij, G.J. (1987). *Evolution and Escalation*. Princeton, N.J.: Princeton University Press.

Vermeij, G.J. (1995). Economics, volcanoes, and Phanerozoic revolutions. *Paleobiology* **21**:125–152.

Vest, M.S., and Katz, J. (1996). Unsteady aerodynamic model of flapping wings. *AIAA J.* **34**:1435–1440.

Vickers, N.J., and Baker, T.C. (1994). Visual feedback in the control of pheromone-mediated flight of *Heliothis virescens* males (Lepidoptera: Noctuidae). *J. Insect Behav.* **7**:605–631.

Vigoreaux, J.O., Hernandez, C., Moore, J., Ayer, G., and Maughan, D. (1998). A genetic deficiency that spans the flightin gene of *Drosophila melanogaster* affects the ultrastructure and function of the flight muscles. *J. Exp. Biol.* **201**:2033–2044.

Villa, J., McDiarmid, R.W., and Gallardo, J.M. (1983). Arthropod predators of leptodactylid frog foam nests. *Brenesia* **19/20**:577–589.

Virsik, R.P., and Reichardt, W. (1974). Tracking of moving objects by the fly *Musca domestica*. *Naturwissenschaften* **61**:132–133.

Virsik, R.P., and Reichardt, W. (1976). Detection and tracking of moving objects by the fly *Musca domestica*. *Biol. Cybern.* **23**:83–98.

Vogel, S. (1962). A possible role of the boundary layer in insect flight. *Nature* **193**:1201–1202.

Vogel, S. (1966). Flight in *Drosophila*. I. Flight performance of tethered flies. *J. Exp. Biol.* **44**:567–578.

Vogel, S. (1967a). Flight in *Drosophila*. II. Variations in stroke parameters and wing contour. *J. Exp. Biol.* **46**:383–392.

Vogel, S. (1967b). Flight in *Drosophila*. III. Aerodynamic characteristics of fly wings and wing models. *J. Exp. Biol.* **46**:431–443.

Vogel, S. (1981). *Life in Moving Fluids*. Boston: Willard Grant Press.

Vogel, S. (1983). How much air flows through a silkmoth's antenna? *J. Insect Physiol.* **29**:597–602.

Vogel, S. (1988). *Life's Devices: The Physical World of Animals and Plants*. Princeton, N.J.: Princeton University Press.

Vogel, S. (1994). *Life in Moving Fluids: The Physical Biology of Flow*. Princeton, N.J.: Princeton University Press.

von Dohlen, C.D., and Moran, N.A. (1995). Molecular phylogeny of the Homoptera: a paraphyletic taxon. *J. Molec. Evol.* **40**:211–223.

von Frisch, K., and Lindauer, M. (1955). Über die Fluggeschwindigkeit der Bienen und über ihre Richtungsweisung bei Seitenwind. *Naturwissenschaften* **42**:377–385.

von Helverson, O. (1993). Adaptations of flowers to the pollination by glossophagine bats. In *Animal-Plant Interactions in Tropical Environments* (ed. W. Barthlott), pp. 41–59. Bonn: Museum Koenig.

von Holst, E. (1943). Untersuchungen über Flugbiophysik. I. Messungen zur Aerodynamik kleiner schwingender Flügel. *Biol. Zentralbl.* **63**:289–326.

von Holst, E., and Küchemann, D. (1941). Biologische und aerodynamische Probleme des Tierfluges. *Naturwissenschaften* **29**:348–362.

von Holst, E., and Küchemann, D. (1942). Biological and aerodynamical problems in animal flight. *J. R. Aero. Soc.* **46**:39–56.

Voss, F. (1913). Vergleichende Untersuchungen über die Flugwerkzeuge der Insekten. Einleitendes. *Verh. Dtsch. Zool. Gesellsch.* **23**:118–142.

Voss, F. (1914). Vergleichende Untersuchungen über die Flugwerkzeuge der Insekten. 2. Abhandlungen. *Verh. Dtsch. Zool. Gesellsch.* **24**:59–90.

Voss, R., and Zeil, J. (1995). Automatic tracking of complex objects under natural conditions. *Biol. Cybern.* **73**:415–423.

Voss, R., and Zeil, J. (1998). Active vision in insects: an analysis of object-directed zig-zag flights in wasps (*Odynerus spinipes*, Eumenidae). *J. Comp. Physiol.* A **182**:377–387.

Waage, J.K. (1979). The evolution of insect/vertebrate associations. *Biol. J. Linn. Soc.* **12**:187–224.

Wagner, D.L., and Liebherr, J.K. (1992). Flightlessness in insects. *Trends Ecol. Evol.* **7**:216–220.

Wagner, H. (1925). Über die Entstehung des dynamischen Auftriebes von Tragflügeln. *Z. Angew. Math. Mech.* **5**:17–35.

Wagner, H. (1982). Flow-field variables trigger landing in flies. *Nature* **297**:147–148.

Wagner, H. (1986a). Flight performance and visual control of flight of the free-flying housefly (*Musca domestica* L.) I. Organization of the flight motor. *Phil. Trans. R. Soc. Lond.* B **312**:527–551.

Wagner, H. (1986b). Flight performance and visual control of flight of the free-flying housefly (*Musca domestica* L.) II. Pursuit of targets. *Phil. Trans. R. Soc. Lond.* B **312**:553–579.

Wagner, H. (1986c). Flight performance and visual control of flight of the free-flying housefly (*Musca domestica* L.) III. Interactions between angular movement induced by wide- and smallfield stimuli. *Phil. Trans. R. Soc. Lond.* B **312**:581–595.

Wagner, P.D. (1996). Determinants of maximal oxygen transport and utilization. *Ann. Rev. Physiol.* **58**:21–50.

Wainwright, S.A., Biggs, W.D., Currey, J.D., and Gosline, J.M. (1976). *Mechanical Design in Organisms.* Princeton, N.J.: Princeton University Press.

Wakeling, J.M. (1993). Dragonfly aerodynamics and unsteady mechanisms: a review. *Odonatologica* **22**:319–334.

Wakeling, J.M. (1997). Odonatan wing and body morphologies. *Odonatologica* **26**:35–52.

Wakeling, J.M., and Ellington, C.P. (1997a). Dragonfly flight. I. Gliding flight and steady-state aerodynamic forces. *J. Exp. Biol.* **200**:543–556.

Wakeling, J.M., and Ellington, C.P. (1997b). Dragonfly flight. II. Velocities, accelerations and kinematics of flapping flight. *J. Exp. Biol.* **200**:557–582.

Wakeling, J.M., and Ellington, C.P. (1997c). Dragonfly flight. III. Lift and power requirements. *J. Exp. Biol.* **200**:583–600.

Waldmann, B., and Zarnack, W. (1988). Forewing movements and motor activity during roll manoeuvers in flying desert locusts. *Biol. Cybern.* **59**:325–335.

Walker, G.T. (1925). The flapping flight of birds. I. *J. R. Aero. Soc.* **29**:590–594.

Walker, G.T. (1927). The flapping flight of birds. II. *J. R. Aero. Soc.* **31**:337–342.

Walker, J.A. (1998). Estimating velocities and accelerations of animal locomotion: a simulation experiment comparing numerical differentiation algorithms. *J. Exp. Biol.* **201**:981–995.

Walker, P.B. (1931). A new instrument for the measurement of fluid motion; with an application to the development of the flow around a wing started impulsively from rest. *Rep. Memo. Aero. Res. Counc.*, no. 1402.

Walker, T.J. (1985). Butterfly migration in the boundary layer. In *Migration: Mechanisms and Adaptive Significance. Contrib. Mar. Sci. Suppl.*, vol. 27 (ed. M.A. Rankin), pp. 704–723.

Walker, T.J., and Riordan, A.J. (1981). Butterfly migration: are synoptic-scale wind systems important? *Ecol. Ent.* **6**:433–440.

Wallace, A.R. (1865). 1. On the phenomena of variation and geographical distribution as illustrated by the Papilionidae of the Malayan region. *Trans. Linn. Soc. Lond.* **xxv**:1–71.

Wallace, A.R. (1867). Mimicry, and other protective resemblances among animals. *Westminster Rev.* (n.s.) **32**:1–43.

Waloff, N. (1983). Absence of wing polymorphism in the arboreal, phytophagous species of some taxa of temperate Hemiptera; an hypothesis. *Ecol. Ent.* **8**:229–232.

Waloff, Z. (1972a). Observations on the airspeeds of freely flying locusts. *Anim. Behav.* **20**:367–372.

Waloff, Z. (1972b). Orientation of flying locusts, *Schistocerca gregaria* (Forsk.), in migrating swarms. *Bull. Ent. Res.* **62**:1–72.

Walsberg, G.E. (1990). Problems inhibiting energetic analyses of migration. In *Bird Migration* (ed. E. Gwinner), pp. 413–421. Berlin: Springer-Verlag.

Walsh, M.J. (1980). Drag characteristics of V-groove and transverse curvature riblets. In *Progress in Aeronautics and Astronautics*, vol. 72 (ed. G.R. Hough), pp. 168–184. New York: American Institute of Aeronautics and Astronautics.

Wang, K. (1985). Sarcomere-associated cytoskeletal lattices in striated muscle: Review and hypothesis. In *Cell and Muscle Motility*, vol. 6 (ed. J.W. Shay), pp. 315–369. New York: Plenum Publishing Corporation.

Wang, K., McCarter, R., Wright, J., Beverly, J., and Ramirez-Mitchell, R. (1991). Regulation of skeletal muscle stiffness and elasticity by titin isoforms: A test of the segmental extension model of resting tension. *Proc. Natl. Acad. Sci. USA* **88**:7101–7105.

Wang, K., McCarter, R., Wright, J., Beverly, J., and Ramirez-Mitchell, R. (1993). Viscoelasticity of the sarcomere matrix of skeletal muscle. The titin-myosin composite filament is a dual-stage molecular spring. *Biophys. J.* **64**:1161–1177.

Ward, J.P., and Baker, P.S. (1982). The tethered flight performance of a laboratory colony of *Triatoma infestans* (Klug.) (Hemiptera: Reduviidae). *Bull. Ent. Res.* **75**:17–28.

Warmke, J., Yamakawa, M., Molloy, J., Falkenthal, S., and Maughan, D. (1992). Myosin light chain-2 mutation affects flight, wing beat frequency, and indirect flight muscle contraction kinetics in *Drosophila*. *J. Cell Biol.* **119**:1523–1539.

Warneck, P. (1988). *Chemistry of the Natural Atmosphere*. San Diego: Academic Press.

Washburn, J.O., and Washburn, L. (1984). Active aerial dispersal of minute wingless arthropods: exploitation of boundary-layer velocity gradients. *Science* **223**:1088–1089.

Wasserthal, L.T. (1974). Funktion und Entwicklung der Flügel der Federmotten (Lepidoptera, Pterophoridae). *Z. Morphol. Tiere* **77**:127–155.

Wasserthal, L.T. (1982). Antagonism between haemolymph transport and tracheal ventilation in an insect wing (*Attacus atlas* L.). *J. Comp. Physiol. B* **147**: 27–40.

Wasserthal, L.T. (1996). Interaction of circulation and tracheal ventilation in holometabolous insects. *Adv. Insect Physiol.* **26**:297–351.

Watt, W.B. (1992). Eggs, enzymes, and evolution: Natural genetic variants change insect fecundity. *Proc. Natl. Acad. Sci. USA* **89**:10608–10612.

Watt, W.B., Cassin, R.C., and Swan, M.S. (1983). Adaptation at specific loci. III. Field behavior and survivorship differences among *Colias* PGI genotypes are predictable from *in vitro* biochemistry. *Genetics* **103**:725–739.

Webb, J.C., Sharp, J.L., Chambers, D.L., and Benner, J.C. (1976). Acoustical properties of the flight activities of the Caribbean fruit fly. *J. Exp. Biol.* **64**:761–772.

Weber, H. (1924a). Das Grundschema des Pterygotenthorax. Erste Mitteilung. *Zool. Anz.* **60**:17–37.

Weber, H. (1924b). Das Grundschema des Pterygotenthorax. Zweite Mitteilung. *Zool. Anz.* **60**:57–83.

Weber, H. (1929). Kopf und Thorax von *Psylla mali* Schmidh. (Hemiptera-Homoptera). *Z. Morphol. Oekol. Tiere* **14**:59–165.

Weber, K.E. (1988). An apparatus for selection on flying speed. *Drosophila Inf. Serv.* **67**:92–93.

Weber, K.E. (1996). Large genetic change at small fitness cost in large populations of *Drosophila melanogaster* selected for wind tunnel flight: rethinking fitness surfaces. *Genetics* **144**:205–213.

Wegener, G. (1990). Elite invertebrate athletes: flight in insects, its metabolic requirements and regulation and its effects on life span. In *International Perspectives in Exercise Physiology* (ed. K. Nazar, R.L. Terjung, H. Kaciuba-Uscilko, and L. Budohoski), pp. 83–87. Champaign, Ill.: Human Kinetics.

Wegener, G. (1996). Flying insects: model systems in exercise physiology. *Experientia* **52**:404–412.

Wegener, G., Bolas, N.M., and Thomas, A.A.G. (1991). Locust flight metabolism studied by ^{31}P NMR spectroscopy. *J. Comp. Physiol.* B **161**:247–256.

Wehner, R. (1981). Spatial vision in insects. In *Handbook of Sensory Physiology*, vol. VII/6C (ed. H. Autrum), pp. 287–616. Berlin: Springer-Verlag.

Wehrhahn, C. (1978). The angular orientation of the movement detectors acting on the flight lift response in flies. *Biol. Cybern.* **31**:169–173.

Wehrhahn, C. (1979). Sex-specific differences in the chasing behaviour of houseflies (*Musca*). *Biol. Cybern.* **32**:239–241.

Wehrhahn, C. (1981). Fast and slow torque responses in flies and their possible role in visual orientation behavior. *Biol. Cybern.* **40**:213–221.

Wehrhahn, C. (1986). Motion sensitive yaw torque responses of the housefly *Musca*: a quantitative study. *Biol. Cybern.* **55**:275–280.

Wehrhahn, C., and Hausen, K. (1980). How is tracking and fixation accomplished in the nervous system of the fly? A behavioural analysis based on short time stimulation. *Biol. Cybern.* **38**:179–186.

Wehrhahn, C., Hausen, K., and Zanker, J. (1981). Is the landing response of the housefly (*Musca*) driven by motion of a flow field? *Biol. Cybern.* **41**:91–99.

Wehrhahn, C., Poggio, T., and Bülthoff, H. (1982). Tracking and chasing in houseflies (*Musca*). An analysis of 3-D flight trajectories. *Biol. Cybern.* **45**:123–130.

Weibel, E.R., Taylor, C.R., and Bolis, L. (eds.) (1998). *Principles of Animal Design: The Optimization and Symmorphosis Debate.* Cambridge, U.K.: Cambridge University Press.

Weis-Fogh, T. (1949). An aerodynamic sense organ stimulating and regulating flight in locusts. *Nature* **164**:873–874.

Weis-Fogh, T. (1952a). Fat combustion and metabolic rate of flying locusts (*Schistocerca gregaria* Forskål). *Phil. Trans. R. Soc. Lond.* B **237**:1–36.

Weis-Fogh, T. (1952b). Weight economy of flying insects. *Trans. 9th Int. Congr. Ent.* **1**:341–347.

Weis-Fogh, T. (1956a). Biology and physics of locust flight. II. Flight performance of the desert locust (*Schistocerca gregaria*). *Phil. Trans. R. Soc. Lond.* B **239**:459–510.

Weis-Fogh, T. (1956b). Biology and physics of locust flight. IV. Notes on sensory mechanisms in locust flight. *Phil. Trans. R. Soc. Lond.* B **239**:553–584.

Weis-Fogh, T. (1956c). Tetanic force and shortening in locust flight muscle. *J. Exp. Biol.* **33**:668–684.

Weis-Fogh, T. (1956d). The ventilatory mechanism during flight of insects in relation to the call for oxygen. In *Proceedings of the 14th International Congress of Zoology*, pp. 283–285. Copenhagen: Danish Science Press.

Weis-Fogh, T. (1959). Elasticity in arthropod locomotion: a neglected subject, illustrated by the wing system of insects. *XVth Int. Congr. Zool.*, vol. 4, pp. 393–395.

Weis-Fogh, T. (1960). A rubber-like protein in insect cuticle. *J. Exp. Biol.* **37**:889–907.

Weis-Fogh, T. (1961a). Thermodynamic properties of resilin, a rubber-like protein. *J. Molec. Biol.* **3**:520–531.

Weis-Fogh, T. (1961b). Molecular interpretation of the elasticity of resilin, a rubber-like protein. *J. Molec. Biol.* **3**:648–667.

Weis-Fogh, T. (1961c). Power in flapping flight. In *The Cell and Organism* (ed. J.A. Ramsay and V.B. Wigglesworth), pp. 283–300. Cambridge, U.K.: Cambridge University Press.

Weis-Fogh, T. (1964a). Biology and physics of locust flight. VIII. Lift and metabolic rate of flying insects. *J. Exp. Biol.* **41**:257–271.

Weis-Fogh, T. (1964b). Diffusion in insect wing muscle, the most active tissue known. *J. Exp. Biol.* **41**:229–256.

Weis-Fogh, T. (1964c). Functional design of the tracheal system of flying insects as compared with the avian lung. *J. Exp. Biol.* **41**:207–227.

Weis-Fogh, T. (1965). Elasticity and wing movements in insects. *Proc. XIIth Int. Congr. Ent.* (ed. P. Freeman), pp. 186–188. London: Regal Entomological Society.

Weis-Fogh, T. (1967a). Metabolism and weight economy in migrating animals, particularly birds and insects. In *Insects and Physiology* (ed. J.W.L. Beament and J.E. Treherne), pp. 143–159. Edinburgh: Oliver and Boyd.

Weis-Fogh, T. (1967b). Respiration and tracheal ventilation in locusts and other flying insects. *J. Exp. Biol.* **47**:561–587.

Weis-Fogh, T. (1972). Energetics of hovering flight in hummingbirds and in *Drosophila*. *J. Exp. Biol.* **56**:79–104.

Weis-Fogh, T. (1973). Quick estimates of flight fitness in hovering animals, including novel mechanisms for lift production. *J. Exp. Biol.* **59**:169–230.

Weis-Fogh, T. (1975). Flapping flight and power in birds and insects, conventional and novel mechanisms. In *Swimming and Flying in Nature*, vol. 2 (ed. T.Y.-T. Wu, C.J. Brokaw, and C. Brennen), pp. 729–762. New York: Plenum Press.

Weis-Fogh, T. (1976). Energetics and aerodynamics of flapping flight: a synthesis. In *Insect Flight* (ed. R.C. Rainey), pp. 48–72. Oxford: Blackwell Scientific.

Weis-Fogh, T. (1977). Dimensional analysis of hovering flight. In *Scale Effects in Animal Locomotion* (ed. T.J. Pedley), pp. 405–420. London: Academic Press.

Weis-Fogh, T., and Alexander, R.M. (1977). The sustained power output from striated muscle. In *Scale Effects in Animal Locomotion* (ed. T.J. Pedley), pp. 511–525. London: Academic Press.

Weis-Fogh, T., and Jensen, M. (1956). Biology and physics of locust flight. I. Basic principles in insect flight. A critical review. *Phil. Trans. R. Soc. Lond.* B **239**:415–458.

Weisel-Eichler, A., and Libersat, F. (1996). Neuromodulation of flight initiation by octopamine in the cockroach *Periplaneta americana*. *J. Comp. Physiol.* A **179**:103–112.

Weissflog, A., Maschwitz, U., Disney, R.H.L., and Rościszewski, K. (1995). A fly's ultimate con. *Nature* **378**:137.

Wells, D.J., and Ellington, C.P. (1994). Beyond the vertebrates: achieving maximum power during flight in insects and hummingbirds. *Adv. Vet. Sci. Comp. Med.* **38B**:219–232.

Wendler, G. (1974). The influence of proprioceptive feedback on locust flight co-ordination. *J. Comp. Physiol.* **88**:173–200.

Wendler, G. (1978). The possible role of fast wing reflexes in locust flight. *Naturwissenschaften* **65**:65–66.

Wendler, G. (1983). The locust flight system: functional aspects of sensory input and methods of investigation. In *BIONA-report 2* (ed. W. Nachtigall), pp. 113–125. Stuttgart: Gustav Fischer.

Wendler, G., Müller, M., and Dombrowski, U. (1993). The activity of pleurodorsal muscles during flight and at rest in the moth *Manduca sexta* (L.). *J. Comp. Physiol.* A **173**:65–75.

Wenner, A.M. (1963). The flight speed of honeybees: a quantitative approach. *J. Apic. Res.* **2**:25–32.

Wensler, R.J. (1977). The ultrastructure of the indirect flight muscles of the monarch butterfly, *Danaus plexippus* (L.) with implications for fuel utilization. *Acta Zool.* **58**:157–167.

West-Eberhard, M.J. (1983). Sexual selection, social competition, and speciation. *Quart. Rev. Biol.* **58**:155–183.

West-Eberhard, M.J. (1984). Sexual selection, competitive communication and species-specific signals in insects. In *Insect Communication* (ed. T. Lewis), pp. 283–324. New York: Academic Press.

Whalley, P., and Jarzembowski, E.A. (1981). A new assessment of *Rhyniella*, the earliest known insect, from the Devonian of Rhynie, Scotland. *Nature* **291**:317.

Whalley, P.E.S. (1979). New species of Protorthoptera and Protodonata (Insecta) from the Upper Carboniferous of Britain, with a comment on the origin of wings. *Bull. Brit. Mus. Nat. Hist. (Geol.)* **32**:85–90.

Wheeler, C.H. (1989). Mobilization and transport of fuels to the flight muscles. In *Insect Flight* (ed. G.J. Goldsworthy and C.H. Wheeler), pp. 273–303. Boca Raton, Fla.: CRC Press.

Wheeler, W.C. (1997). Sampling, groundplans, total evidence and the systematics of arthropods. In *Arthropod Relationships* (ed. R.A. Fortey and R.H. Thomas), pp. 87–96. London: Chapman and Hall.

Wheeler, W.C., Cartwright, P., and Hayashi, C.Y. (1993). Arthropod phylogeny: a combined approach. *Cladistics* **9**:1–39.

Wieschaus, E., and Gehring, W.J. (1976). Clonal analysis of primordial disc cells in the developing embryo of *Drosophila melanogaster. Dev. Biol.* **50**:249–263.

White, D.C.S. (1983). The elasticity of relaxed insect fibrillar flight muscle. *J. Physiol.* **343**:31–57.

White, D.C.S., Lund, J., and Webb, M.R. (1988). Cross-bridge kinetics in asynchronous insect flight muscle. In *Molecular Mechanisms of Muscle Contraction* (ed. H. Sugi and G.H. Pollack), pp. 169–178. New York: Plenum Press.

White, J.A., Ganter, P., McFarland, R., Stanton, N., and Lloyd, M. (1983). Spontaneous, field tested and tethered flight in healthy and infected *Magicicada septendecim* L. *Oecologia* **57**:281–286.

Whiting, M.F., and Wheeler, W.C. (1994). Insect homeotic transformation. *Nature* **368**:696.

Whiting, M.F., Carpenter, J.C., Wheeler, Q.D., and Wheeler, W.C. (1997). The Strepsiptera problem: Phylogeny of the holometabolous insect orders inferred from 18S and 28S ribosomal DNA sequences and morphology. *Syst. Biol.* **46**:1–68.

Wickman, P.-O. (1992). Sexual selection and butterfly design—a comparative study. *Evolution* **46**:1525–1536.

Wiebes, J.T. (1979). Co-evolution of figs and their insect pollinators. *Ann. Rev. Ecol. Syst.* **10**:1–12.

Wigglesworth, V.B. (1946). Organs of equilibrium in flying insects. *Nature* **157**:655.

Wigglesworth, V.B. (1963). Origin of wings in insects. *Nature* **197**:97–98.

Wigglesworth, V.B. (1973). Evolution of insect wings and flight. *Nature* **246**:127–129.

Wigglesworth, V.B. (1976). The evolution of insect flight. In *Insect Flight* (ed. R.C. Rainey), pp. 255–269. Oxford: Blackwell Scientific.

Wigglesworth, V.B. (1983). The physiology of insect tracheoles. *Adv. Insect Physiol.* **17**:85–149.

Wigglesworth, V.B. (and contributors) (1963). The origin of flight in insects. *Proc. R. Ent. Soc. Lond.* **28**:23–32.

Wigglesworth, V.B., and Lee, W.M. (1982). The supply of oxygen to the flight muscles of insects: a theory of tracheole physiology. *Tissue & Cell* **14**:501–518.

Wilkerson, R.C., and Butler, J.F. (1984). The Immelmann turn, a pursuit maneuver used by hovering male *Hybomitra hinei wrighti* (Diptera: Tabanidae). *Ann. Ent. Soc. Amer.* **77**:293–295.

Wilkin, P.J. (1990). The instantaneous force on a desert locust, *Schistocerca gregaria* (Orthoptera: Acrididae), flying in a wind tunnel. *J. Kans. Ent. Soc.* **63**: 316–328.

Wilkin, P.J. (1991). Instantaneous aerodynamic forces developed by an Indian moon moth, *Actias selene*, in near-hovering flight. *Physiol. Zool.* **64**:193–211.

Wilkin, P.J. (1993). Comparison of the aerodynamic forces on a flying sphingid moth with those predicted by quasi-steady theory. *Physiol. Zool.* **66**:1015–1044.

Will, K.W. (1995). Plecopteran surface-skimming and insect flight evolution. *Science* **270**:1684–1685.

Wille, A. (1960). The phylogeny and relationships between the insect orders. *Rev. Biol. Trop.* **8**:93–123.

Williams, B. (1994). Models of trap seeking by tsetse flies: anemotaxis, klinokinesis and edge detection. *J. theor. Biol.* **168**:105–115.

Williams, C.B. (1930). *The Migration of Butterflies*. London: Oliver and Tweed.

Williams, C.B. (1958). *Insect Migration*. London: Collins.

Williams, C.B. (1960). The range and pattern of insect abundance. *Amer. Nat.* **94**:137–151.

Williams, C.M., and Galambos, R. (1950). Oscilloscopic and stroboscopic analysis of the flight sounds of *Drosophila*. *Biol. Bull.* **99**:300–307.

Williams, J.A., and Carroll, S.B. (1993). The origin, patterning and evolution of insect appendages. *BioEssays* **15**:567–577.

Williams, J.A., Paddock, S.W., Vorwerk, K., and Carroll, S.B. (1994). Organization of wing formation and induction of a wing-patterning gene at the dorsal/ventral compartment boundary. *Nature* **368**:299–305.

Williams, P.H. (1985). A preliminary cladistic investigation of relationships among the bumble bees (Hymenoptera, Apidae). *Syst. Ent.* **10**:239–225.

Williams, P.H. (1991). The bumble bees of the Kashmir Himalaya (Hymenoptera: Apidae, Bombini). *Bull. Brit. Mus. Nat. Hist. (Ent.)* **60**:1–204.

Williams, P.H. (1994). Phylogenetic relationships among bumble bees (*Bombus* Latr.): a reappraisal of morphological evidence. *Syst. Ent.* **19**:327–344.

Williamson, M.H., and Lawton, J.H. (1991). Fractal geometry of ecological habitats. In *Habitat Structure: The Physical Arrangement of Objects in Space* (ed. S.S. Bell, E.D. McCoy, and H.R. Mushinsky), pp. 69–86. London: Chapman and Hall.

Willis, M.A., and Arbas, E.A. (1991). Odor-modulated upwind flight of the sphinx moth, *Manduca sexta* L. *J. Comp. Physiol.* A **169**:427–440.

Willis, M.A., and Arbas, E.A. (1998). Variability in odor-modulated flight by moths. *J. Comp. Physiol.* A **182**:191–202.

Willis, M.A., and Baker, T.C. (1994). Behaviour of flying oriental fruit moth males during approach to sex pheromone sources. *Physiol. Ent.* **19**:61–69.

Willis, M.A., and Cardé, R.T. (1990). Pheromone-modulated optomotor response in male gypsy moths, *Lymantria dispar* L.: Upwind flight in a pheromone plume in different wind velocities. *J. Comp. Physiol.* A **167**:699–706.

Willmer, P.G. (1982). Microclimate and the environmental physiology of insects. *Adv. Insect Physiol.* **16**:1–57.

Willmer, P.G. (1983). Thermal constraints on activity patterns in nectar-feeding insects. *Ecol. Ent.* **8**:455–469.

Willmer, P.G. (1986a). Foraging patterns and water balance: problems of optimization for a xerophilic bee, *Chalicodoma sicula*. *J. Anim. Ecol.* **55**:941–962.

Willmer, P.G. (1986b). Microclimatic effects on insects at the plant surface. In *Insects and Plant Surfaces* (ed. B.E. Juniper and T.R.E. Southwood), pp. 65–80. London: Edward Arnold.

Willmer, P.G. (1991). Thermal biology and mate acquisition in ectotherms. *Trends Ecol. Evol.* **6**:396–399.

Willmer, P.G., and Unwin, D.M. (1981). Field analyses of insect heat budgets: reflectance, size and heating rates. *Oecologia* **50**:250–255.

Willmott, A.P., and Ellington, C.P. (1997a). Measuring the angle of attack of beating insect wings: Robust 3-dimensional reconstruction from 2-dimensional images. *J. Exp. Biol.* **200**:2693–2704.

Willmott, A.P., and Ellington, C.P. (1997b). The mechanics of flight in the hawkmoth *Manduca sexta*. I. Kinematics of hovering and forward flight. *J. Exp. Biol.* **200**:2705–2722.

Willmott, A.P. and Ellington, C.P. (1997c). The mechanics of flight in the hawkmoth *Manduca sexta*. II. Aerodynamic consequences of kinematic and morphological variation. *J. Exp. Biol.* **200**:2723–2745.

Willmott, A.P., Ellington, C.P., and Thomas, A.L.R. (1997). Flow visualization and unsteady aerodynamics in the flight of the hawkmoth *Manduca sexta*. *Phil. Trans. R. Soc. Lond.* B **352**:303–316.

Wilson, D.M. (1961). The central nervous control of flight in a locust. *J. Exp. Biol.* **38**:471–490.

Wilson, D.M. (1962). Bifunctional muscles in the thorax of grasshoppers. *J. Exp. Biol.* **39**:669–677.

Wilson, D.M. (1968). Inherent asymmetry and reflex modulation of the locust flight motor pattern. *J. Exp. Biol.* **48**:631–641.

Wilson, D.M., and Gettrup, E. (1963). A stretch reflex controlling wingbeat frequency in grasshoppers. *J. Exp. Biol.* **40**:171–185.

Wilson, D.M., and Weis-Fogh, T. (1962). Patterned activity of co-ordinated motor units, studied in flying locusts. *J. Exp. Biol.* **39**:643–667.

Wilson, D.M., and Wyman, R.J. (1965). Motor output patterns during random and rhythmic stimulation of locust thoracic ganglia. *Biophys. J.* **5**:121–143.

Wilson, E.O. (1971). *The Insect Societies.* Cambridge, Mass: Harvard University Press.

Wilson, M. (1978). The functional organisation of locust ocelli. *J. Comp. Physiol.* A **124**:317–331.

Windsor, D.M. (1990). Climate and moisture variability in a tropical forest: long-term records from Barro Colorado Island, Panamá. *Smithsonian. Contrib. Earth Sci.* **29**:1–145.

Winter, Y., Voigt, C., and von Helverson, O. (1998). Gas exchange during hovering flight in a nectar-feeding bat *Glossophaga soricina*. *J. Exp. Biol.* **201**:237–244.

Wise, D.H. (1993). *Spiders in Ecological Webs.* Cambridge, U.K.: Cambridge University Press.

Wisser, A. (1987). Mechanisms of wing rotating regulation in *Calliphora erythrocephala* (Insecta, Diptera). *Zoomorphology* **106**:261–268.

Wisser, A. (1988). Wing beat of *Calliphora erythrocephala*: Turning axis and gearbox of the wing base (Insecta, Diptera). *Zoomorphology* **107**:359–369.

Wisser, A. (1997). Funktionsmorphologie der Flügel und Flügelgelenke. In *BIONA-report 11* (ed. A. Wisser, D. Bilo, A. Kesel, and B. Möhl), pp. 57–88. Stuttgart: Gustav Fischer.

Wisser, A., and Nachtigall, W. (1984). Functional-morphological investigations on the flight muscles and their insertion points in the blowfly *Calliphora erythrocephala* (Insecta, Diptera). *Zoomorphology* **104**:188–195.

Withers, P.C. (1981). The effects of ambient air pressure on oxygen consumption of resting and hovering honeybees. *J. Comp. Physiol.* B **141**:433–437.

Wittekind, W.C. (1988). The landing response of tethered flying *Drosophila* is induced at a critical object angle. *J. Exp. Biol.* **135**:491–493.

Witter, M.S., and Cuthill, I.C. (1993). The ecological costs of avian fat storage. *Phil. Trans. R. Soc. Lond.* B **340**:73–92.

Wittman, D. (1985). Aerial defense of the nest by workers of the stingless bee *Trigona (Tetragonisca) angustula* (Latreille) (Hymenoptera: Apidae). *Behav. Ecol. Sociobiol.* **16**:111–114.

Wohlgemuth, R. (1962). Die Schlagform des Bienenflügels beim Sterzeln im Vergleich zur Bewegungsweise beim Fliegen und Fächeln. *Z. vergl. Physiol.* **45**:581–589.

Woledge, R.C. (1989). Energy transformation in living muscle. In *Energy Transformation in Cells and Organisms* (ed. W. Wieser and E. Gnaiger), pp. 36–45. Stuttgart: Georg Thieme.

Wolf, H. (1990). On the function of a locust flight steering muscle and its inhibitory innervation. *J. Exp. Biol.* **150**:55–80.

Wolf, H. (1993). The locust tegula: significance for flight rhythm generation, wing movement control and aerodynamic force production. *J. Exp. Biol.* **182**:229–253.

Wolf, R., and Heisenberg, M. (1986). Visual orientation in motion-blind flies is an operant behaviour. *Nature* **323**:154–156.

Wolf, R., and Heisenberg, M. (1990). Visual control of straight flight in *Drosophila melanogaster*. *J. Comp. Physiol.* A **167**:269–283.

Wolf, H., and Pearson, K.G. (1987). Flight motor patterns recorded in surgically isolated sections of the ventral nerve cord of *Locusta migratoria*. *J. Comp. Physiol.* A **161**:103–114.

Wolf, H., and Pearson, K.G. (1989). Comparison of motor patterns in the intact and deafferented flight system of the locust. III. Patterns of interneuronal activity. *J. Comp. Physiol.* A **165**:61–74.

Wolf, T.J., and Schmid-Hempel, P. (1989). Extra loads and foraging life span in honeybee workers. *J. Anim. Ecol.* **58**:943–954.

Wolf, T.J., Ellington, C.P., Davis, S., and Feltham, M.J. (1996). Validation of the doubly labelled water technique for bumblebees *Bombus terrestris* (L.). *J. Exp. Biol.* **199**:959–972.

Wolf, T.J., Schmid-Hempel, P., Ellington, C.P., and Stevenson, R.D. (1989). Physiological correlates of foraging efforts in honey-bees: oxygen consumption and nectar load. *Funct. Ecol.* **3**:417–424.

Wood, J. (1970). A study of the instantaneous air velocities in a plane behind the wings of certain Diptera flying in a wind tunnel. *J. Exp. Biol.* **52**:17–25.

Wood, J. (1972). An experimental determination of the relationship between lift and aerodynamic power in *Calliphora erythrocephala* and *Phormia regina*. *J. Exp. Biol.* **56**:31–36.

Wood, S.P., Panchen, A.L., and Smithson, T.R. (1985). A terrestrial fauna from the Scottish Lower Carboniferous. *Nature* **314**:355–356.

Woodring, J.P. (1962). Oribatid (Acari) pteromorphs, pterogasterine phylogeny, and evolution of wings. *Ann. Ent. Soc. Amer.* **55**:394–403.

Woodworth, C.W. (1906). The wing veins of insects. *Univ. Calif. Bull. Ent.* **1**:1–152.

Wootton, R.J. (1979). Function, homology and terminology in insect wings. *Syst. Ent.* **4**:81–93.

Wootton, R.J. (1981a). Palaeozoic insects. *Ann. Rev. Ent.* **26**:319–344.

Wootton, R.J. (1981b). Support and deformability in insect wings. *J. Zool., Lond.* **193**:447–468.

Wootton, R.J. (1986). The origin of insect flight: where are we now? *Antenna* **10**:82–86.

Wootton, R.J. (1988). The historical ecology of aquatic insects: an overview. *Palaeogeogr. Palaeoclimatol. Palaeoecol.* **62**:477–492.

Wootton, R.J. (1990). Major insect radiations. In *Major Evolutionary Radiations* (ed. P.D. Taylor and G.P. Larwood), pp. 187–208. Systematics Association Special Volume, no. 42. Oxford: Clarendon Press.

Wootton, R.J. (1991). The functional morphology of the wings of Odonata. *Adv. Odonatol.* **5**:153–169.

Wootton, R.J. (1992). Functional morphology of insect wings. *Ann. Rev. Ent.* **37**:113–140.

Wootton, R.J. (1993). Leading edge section and asymmetric twisting in the wings of flying butterflies. *J. Exp. Biol.* **180**:105–117.

Wootton, R.J. (1995). Geometry and mechanics of insect hindwing fans: a modelling approach. *Proc. R. Soc. Lond.* B **262**:181–187.

Wootton, R.J. (1996). Functional wing morphology in Hemiptera systematics. In *Studies on Hemipteran Phylogeny* (ed. C.W. Schaefer), pp. 179–198. Lanham, Md.: Entomological Society of America.

Wootton, R.J., and Betts, C.R. (1986). Homology and function in the wings of Heteroptera. *Syst. Ent.* **11**:389–400.

Wootton, R.J., and Ellington, C.P. (1991). Biomechanics and the origin of insect flight. In *Biomechanics in Evolution* (ed. J.M.V. Rayner and R.J. Wootton), pp. 99–112. Cambridge, U.K.: Cambridge University Press.

Wootton, R.J., and Ennos, A.R. (1989). The implications of function on the origin and homologies of the dipterous wing. *Syst. Ent.* **14**:507–520.

Wootton, R.J., and Newman, D.J.S. (1979). Whitefly have the highest contraction frequencies yet recorded in non-fibrillar flight muscles. *Nature* **280**:402–403.

Wootton, R.J., Kukalová-Peck, J., Newman, D.J.S., and Muzón, J. (1998). Smart engineering in the mid-Carboniferous: How well could Palaeozoic dragonflies fly? *Science* **282**:749–751.

Wortmann, M., and Zarnack, W. (1993). Wing movements and lift regulation in the flight of desert locusts. *J. Exp. Biol.* **182**:57–69.

Wray, J.S. (1979). Filament geometry and the activation of insect flight muscles. *Nature* **280**:325–326.

Wu, J.Z., Vakili, A.D., and Wu, J.M. (1991). Review of the physics of enhancing vortex lift by unsteady excitation. *Progr. Aerospace Sci.* **28**:73–131.

Wyatt, G., and Kalf, G. (1957). The chemistry of insect hemolymph. II. Trehalose and other carbohydrates. *J. Gen. Physiol.* **40**:833–847.

Wygodzinsky, P. (1987a). Order Microcoryphia. In *Immature Insects*, vol. 1 (ed. F.W. Stehr), pp. 68–70. Dubuque, Iowa: Kendall/Hunt.

Wygodzinsky, P. (1987b). Order Thysanura. In *Immature Insects*, vol. 1 (ed. F.W. Stehr), pp. 71–74. Dubuque, Iowa: Kendall/Hunt.

Wyman, R.J., and Tanouye, M.A. (1982). *Drosophila* flight motor pattern: the evidence from interspike intervals. *J. Exp. Biol.* **96**:413–416.

Wyman, R.J., Thomas, J.B., Salkoff, L., and King, D.G. (1984). The *Drosophila* giant fiber system. In *Neural Mechanisms of Startle Behavior* (ed. R.C. Eaton), pp. 133–161. New York: Plenum Press.

Xu, H., and Robertson, R.M. (1994). Effects of temperature on properties of flight neurons in the locust. *J. Comp. Physiol.* A **175**:193–202.

Yager, D.D., and May, M.L. (1990). Ultrasound-triggered, flight-gated evasive maneuvers in the praying mantis *Parasphendale agrionina*. II. Tethered flight. *J. Exp. Biol.* **152**:41–58.

Yager, D.D., and Spangler, H.G. (1997). Behavioral response to ultrasound by the tiger beetle *Cicindela marutha* Dow combines aerodynamic changes and sound production. *J. Exp. Biol.* **200**:649–659.

Yager, D.D., May, M.L., and Fenton, B. (1990). Ultrasound-triggered, flight-gated evasive maneuvers in the praying mantis *Parasphendale agrionina*. I. Free flight. *J. Exp. Biol.* **152**:17–39.

Yates, G.T. (1986). Optimum pitching axes in flapping wing propulsion. *J. theor. Biol.* **120**:255–276.

Young, A.M. (1971). Wing coloration and reflectance in *Morpho* butterflies as related to reproductive behavior and escape from avian predators. *Oecologia* **7**:209–222.

Young, B.P. (1921). Attachment of the abdomen to the thorax in Diptera. *Mem. Cornell Agric. Exp. Stat.* **44**:255–282.

Young, D., and Josephson, R.K. (1985). 100 Hz is not the upper limit of synchronous muscle contraction. *Nature* **309**:286–287.

Young, S., Hardie, J., and Gibson, G. (1993). Flying insects in the laboratory. In *Video Techniques in Animal Ecology and Behaviour* (ed. S.D. Wratten), pp. 17–32. London: Chapman and Hall.

Yuval, B., Holliday-Hanson, M.L., and Washino, R. (1994). Energy budget of swarming mosquitoes. *Ecol. Ent.* **19**:74–78.

Zalessky, Y.M. (1949). The origin of wings and flight in insects, with reference to environmental conditions. *Usp. Sovr. Biolog.* **28**:400–414. (In Russian.)

Zalessky, Y.M. (1953). The role of wind in the origin of insect flight. *Priroda* **11**:85–90. (In Russian.)

Zalucki, M.P., Kitching, R.L., Abel, D., and Pearson, J. (1980). A novel device for tracking butterflies in the field. *Ann. Ent. Soc. Amer.* **73**:262–265.

Zanen, P.O., Sabelis, M.W., Buonaccorsi, J.P., and Cardé, R.T. (1994). Search strategies of fruit flies in steady and shifting winds in the absence of food odours. *Physiol. Ent.* **19**:335–341.

Zanker, J.M. (1988a). How does lateral abdomen deflection contribute to flight control of *Drosophila melanogaster*? *J. Comp. Physiol.* A **162**:581–588.

Zanker, J.M. (1988b). On the mechanism of speed and altitude control in *Drosophila melanogaster*. *Physiol. Ent.* **13**:351–361.

Zanker, J.M. (1990a). The wing beat of *Drosophila melanogaster*. I. Kinematics. *Phil. Trans. R. Soc. Lond.* B **327**:1–18.

Zanker, J.M. (1990b). The wing beat of *Drosophila melanogaster*. III. Control. *Phil. Trans. R. Soc. Lond.* B **327**:45–64.

Zanker, J.M., and Götz, K.G. (1990). The wing beat of *Drosophila melanogaster*. II. Dynamics. *Phil. Trans. R. Soc. Lond.* B **327**:19–44.

Zanker, J.M., Egelhaaf, M., and Warzecha, A.-K. (1991). On the coordination of motor output during visual flight control of flies. *J. Comp. Physiol.* A **169**:127–134.

Zarnack, W. (1972). Flugbiophysik der Wanderheuschrecke (*Locusta migratoria* L.). I. Die Bewegungen der Vorderflügel. *J. Comp. Physiol.* **78**:356–395.

Zarnack, W. (1975). Aerodynamic forces and their calculation in insect flight. In *Swimming and Flying in Nature*, vol. 2 (ed. T.Y.-T. Wu, C.J. Brokaw, and C. Brennen), pp. 797–801. New York: Plenum Press.

Zarnack, W. (1978a). A transducer recording continuously 3-dimensional rotations of biological objects. *J. Comp. Physiol.* A **126**:161–168.

Zarnack, W. (1978b). Locust flight control. On-line measurements of phase shifting in fore-wing movements. *Naturwissenschaften* **65**:64–65.

Zarnack, W. (1983). Untersuchungen zum Flug von Wanderheuschrecken. Die Bewegungen räumlichen Lagebeziehungen sowie Formen und Profile von Vorder- und Hinterflügeln. In *BIONA-report 1* (ed. W. Nachtigall), pp. 79–102. Stuttgart: Gustav Fischer.

Zarnack, W. (1988). The effect of forewing depressor activity on wing movement during locust flight. *Biol. Cybern.* **59**:55–70.

Zarnack, W. (1997). Kinematik und Aerodynamik des Heuschreckenflugs. In *BIONA-report 11* (ed. A. Wisser, D. Bilo, A. Kesel, and B. Möhl), pp. 173–200. Stuttgart: Gustav Fischer.

Zarnack, W., and Möhl, B. (1977). Activity of the direct downstroke flight muscles of *Locusta migratoria* (L.) during steering behaviour in flight. I. Patterns of time shift. *J. Comp. Physiol.* A **118**:215–233.

Zarnack, W., and Wortmann, M. (1989). On the so-called constant-lift reaction of migratory locusts. *J. Exp. Biol.* **147**:111–124.

Zebe, E. (1954). Über den Stoffwechsel der Lepidopteren. *Z. vergl. Physiol.* **36**:290–317.

Zebe, E., and Gäde, G. (1993). Flight metabolism in the African fruit beetle, *Pachnoda sinuata*. *J. Comp. Physiol.* B **163**:107–112.

Zeil, J. (1986). The territorial flight of male houseflies (*Fannia canicularis* L.). *Behav. Ecol. Sociobiol.* **19**:213–219.

Zeil, J. (1993a). Orientation flights of solitary wasps (*Cerceris*; Sphecidae; Hymenoptera). I. Description of flight. *J. Comp. Physiol.* A **172**:189–205.

Zeil, J. (1993b). Orientation flights of solitary wasps (*Cerceris*; Sphecidae; Hymenoptera). II. Similarities between orientation and return flights and the use of motion parallax. *J. Comp. Physiol.* A **172**:207–222.

Zeil, J. (1993c). Sexual dimorphism in the visual system of flies: the compound eyes and neural superposition in Bibionidae (Diptera). *J. Comp. Physiol.* A **150**:379–393.

Zeil, J. (1993d). Sexual dimorphism in the visual system of flies: the free flight behaviour of male Bibionidae (Diptera). *J. Comp. Physiol.* A **150**:395–412.

Zeil, J. (1997). The control of optic flow during learning flights. *J. Comp. Physiol.* A **180**:25–37.

Zeil, J., and Wittmann, D. (1989). Visually controlled station-keeping by hovering guard bees of *Trigona* (*Tetragonisca*) *angustula* (Apidae, Meliponinae). *J. Comp. Physiol.* A **165**:711–718.

Zeil, J., Kelber, A., and Voss, R. (1996). Structure and function of learning flights in bees and wasps. *J. Exp. Biol.* **199**:245–252.

Zeil, J., Nalbach, G., and Nalbach, H.-O. (1989). Spatial vision in a flat world: optical and neural adaptations in arthropods. In *Neurobiology of Sensory Systems* (ed. R.N. Singh and N.J. Strausfeld), pp. 123–147. New York: Plenum Press.

Zeng, L., Liu, H., and Kawachi, K. (1996). Measurement and flow visualization of a beating bumblebee wing. *J. Flow Vis. Image Process.* **3**:319–327.

Zeng, L., Matsumoto, H., and Kawachi, K. (1996a). A fringe shadow method for measuring flapping angle and torsional angle of a dragonfly wing. *Meas. Sci. Tech.* **7**:776–781.

Zeng, L., Matsumoto, H., and Kawachi, K. (1996b). Angle-compensation sensor for measuring the shape of a dragonfly wing. *Sens. and Actuators* A **55**:87–90.

Zeng, L., Matsumoto, H., and Kawachi, K. (1996c). Divergent-ray projection method for measuring the flapping angle, lag angle, and torsional angle of a bumblebee wing. *Opt. Eng.* **35**:3135–3139.

Zeng, L., Matsumoto, H., and Kawachi, K. (1996d). Scanning beam collimation method for measuring dynamic angle variations using an acousto-optic deflector. *Opt. Eng.* **35**:1662–1667.

Zeng, L., Matsumoto, H., and Kawachi, K. (1996e). Simultaneous measurement of the shape and thickness of a dragonfly wing. *Meas. Sci. Tech.* **7**:1728–1732.

Zeng, L., Matsumoto, H., and Kawachi, K. (1996f). Two-colour compensation method for measuring unsteady vertical force of an insect in a wind tunnel. *Meas. Sci. Tech.* **7**:515–519.

Zeng, L., Matsumoto, H., Sunada, S., and Kawachi, K. (1996). High-resolution method for measuring the torsional deformation of a dragonfly wing by combining a displacement probe with an acousto-optic deflector. *Opt. Eng.* **35**:507–513.

Zera, A.J., and Denno, R.F. (1997). Physiology and ecology of dispersal polymorphism in insects. *Ann. Rev. Ent.* **42**:207–231.

Zera, A.J., and Tobe, S.S. (1990). Juvenile hormone-III biosynthesis in presumptive long-winged and short-winged *Gryllus rubens*: implications for the endocrine regulation of wing dimorphism. *J. Insect Physiol.* **36**:271–280.

Zera, A.J., Mole, S., and Rokke, K. (1994). Lipid, carbohydrate nitrogen content of long- and short-winged *Gryllus firmus*: implications for the physiological cost of flight capability. *J. Insect Physiol.* **40**:1037–1044.

Zera, A.J., Sall, J., and Grudzinski, K. (1997). Flight-muscle polymorphism in the cricket *Gryllus firmus*: muscle characteristics and their influence on the evolution of flightlessness. *Physiol. Zool.* **70**:519–529.

Zeuner, F.E. (1940). Biology and evolution of fossil insects. *Proc. Geol. Assoc.* **51**:44–48.

Ziegler, C. (1994). Titin-related proteins in invertebrate muscles. *Comp. Biochem. Physiol.* **109**A:823–833.

Ziegler, R., and Schulz, M. (1986a). Regulation of carbohydrate metabolism during flight in *Manduca sexta*. *J. Insect Physiol.* **32**:997–1001.

Ziegler, R., and Schulz, M. (1986b). Regulation of lipid metabolism during flight in *Manduca sexta*. *J. Insect Physiol.* **32**:903–908.

Ziegler, R., Ryan, R.O., Arbas, E.D., and Law, J.H. (1988). Adipokinetic response of a flightless grasshopper (*Barytettix psolus*): functional components, defective response. *Arch. Insect Biochem. Physiol.* **9**:255–268.

Zimmerman, E.C. (1948). *Insects of Hawaii*, vol. III. *Heteroptera*. Honolulu: University of Hawaii Press.

INDEX

abdomen, 8, 73, 104, 111, 174, 232–233
abundance, of individual insects, 13, 303
acceleration, 82–83, 135, 147, 242, 252, 319; allometry of, 243–246; of body, 82–83, 147; cost of, 156–157; and flight muscle mass, 247–248; maximum value of, 332. *See also* virtual mass
acceleration reaction, 19–20
accessory pulsatile organs, 53
acoustic power, 87
Acrididae (Orthoptera), 72, 332. *See also* locusts
actin, 150, 177, 180, 250
actuator disk, 121, 147, 251. *See also* induced power; induced velocity
added mass. *See* virtual mass
adipokinetic hormone, 170
advance ratio, 94; and shape of power curve, 154–155
aeolian dispersal, 312, 322, 323–327
aerial plankton, 323
aerobic metabolism, 159–168. *See also* metabolic rate
aerodynamic power, 149. *See also* induced power; profile power
Agaristidae (Lepidoptera), 72
aggregations, 104–105. *See also* swarming
aging. *See* senescence
Agrion (Odonata), 76
air sac, 160, 162
airspeed, 75, 77, 82, 319, 332; allometry of, 80, 81, 325; direct measurement of, 79; and effects of chamber dimensions, 78; regulation of, 62–63
alary buds, 52
alary mode, 46. *See also* aptery; diptery; tetraptery
Aleyrodoidea (Homoptera), 185, 186
aliasing, 87
allometric engineering, 253
alpine habitat. *See* altitudinal gradients
altitudinal gradients, 342–343; and air density, 15; and flight, 311; and hummingbird diversity, 336; and nectar rewards, 312
alula, of beetles, 216
amino acids, 169

Amphibia, as insectivores, 285, 316
amplexiform wings, 69
anaerobic metabolism , 163
anal vein, 53, 54
anemotaxis, 238–240
angiosperms, 12, 309–311
angle of attack, 21, 27, 112–113, 230; measurement of, 92–93, 348
angle of incidence. *See* angle of attack
Anisoptera (Odonata), 41, 64, 69, 83, 99
antagonistic muscles, 43–44, 174, 179
Antarctica, 3, 299
antennae, 104, 107, 213–214
anterior notal process, 39, 40
anteromotorism, 45, 48, 66, 289, 290; and wing coupling, 69. *See also* bimotorism, posteromotorism
ants. *See* Formicidae
Aphididae (Homoptera), 76, 78, 248, 298, 326. *See also* Aphidoidea
Aphidoidea (Homoptera), and asynchronous muscle, 185–186
Apis (Hymenoptera): airspeed of, 79; and body lift, 107; dispersal of, 327; and distance assessment, 211; metabolism of, 164, 165, 171, 195; muscle efficiency of, 193; sound production by, 71; streamlining of, 107; thermoregulation in, 198–199; tracking by, 235; and vertical ascent, 82; and vertical force production, 134; water balance of, 170
apodeme, 49, 271
Apodidae (Aves), 314
aposematic coloration, 319
aptery, 46, 48, 295. *See also* flightlessness
Apterygota, 6–7, 9, 10, 263, 271, 273; jumping of, 284–285. *See also* Archaeognatha, Thysanura
apterygote insects. *See* Apterygota
aquatic insects, 1, 3, 269, 270. *See also* swimming
Arachnida: as insectivores, 284, 285, 316–317; as prey of insects, 248, 313
Archaeognatha, 7, 213, 271, 272, 276; jumping of, 224, 285
Archaeorrhyncha (Homoptera), 185–186
Arctiidae (Lepidoptera), 316

QL 496.7 .D83 2000
Dudley, Robert, 1961-
The biomechanics of insect
 flight